7th AHS Technical Meeting on VTOL Unmanned Rotorcraft Systems 2017

Mesa, Arizona, USA
24 - 26 January 2017

ISBN: 978-1-5108-4675-3

Printed from e-media with permission by:

Curran Associates, Inc.
57 Morehouse Lane
Red Hook, NY 12571

Some format issues inherent in the e-media version may also appear in this print version.

Printed by Curran Associates, Inc. (2017)

For permission requests, please contact American Helicopter Society International
at the address below.

American Helicopter Society International
2701 Prosperity Ave, Suite 210
Fairfax, VA 22031

Phone (703) 684-6777
Fax: (703) 739-9279

staff@vtol.org

Additional copies of this publication are available from:

Curran Associates, Inc.
57 Morehouse Lane
Red Hook, NY 12571 USA
Phone: 845-758-0400
Fax: 845-758-2633
Email: curran@proceedings.com
Web: www.proceedings.com

TABLE OF CONTENTS

OPENING GENERAL SESSION

UAV AERODYNAMICS SESSION

MICRO AIR VEHICLES SESSION

VTOL UAV FLIGHT CONTROL & NAVIGATION SESSION

Regulations that Allow Growth & Safe Integration of UAS: How do we get there?

Presented to: AHS VTOL Unmanned Aircraft Systems Technical Conference

Presented by: Jim Viola, Manager, FAA Flight Standards Service – General Aviation and Commercial Division

Date: January 24, 2017

Overview

- **FAA Authority**
- **Hobby/Recreational Operations**
- **Small UAS Rule (Part 107)**
  - Becoming a Pilot
  - Operating Rules
- **Next Steps in Integration**
  - Focus Area Pathfinders & Extended Operations
  - UAS Over People Rulemaking
- **UAS Registration**
- **Outreach Efforts**

What is the FAA's Authority?

- **U.S. airspace is public space**
  - 49 U.S.C. §40103(a)(1)
- **UAS are aircraft subject to regulation**
  - 49 U.S.C. §40102(a)(6); 14 CFR 1.1; PL 112-95 §331, §336
  - An aircraft is any device used, or intended to be used, for flight
- **UAS must comply with FAA regulations**

Types of UAS Operations

	Recreational Only Operations	Commercial and Other Operations
Pilot Requirements	• No FAA pilot requirements	• Must have Remote Pilot Airman Certification • Must be 16 years or older • Must pass TSA vetting
Aircraft Requirements	• Must be registered if over 0.55 pounds	• Must be less than 55 pounds • Must be registered if over 0.55 pounds • Must undergo pre-flight checklist
Location Requirements	• Must notify all airports and air traffic control (if applicable) within five miles of proposed area of operations	• Class G airspace without ATC permission • Class B, C, D, and E require ATC permission
Operating Rules	• Must ALWAYS yield right of way to manned aircraft • Must keep aircraft in visual line-of-sight • Must follow community-based safety guidelines	• Must keep aircraft in visual line-of-sight* • Must fly under 400 feet* • Must fly only during daylight hours* • Must fly at or below 100 mph* • Must yield right of way to manned aircraft* • Must NOT fly over people* • Must NOT fly from a moving vehicle*
Definitions	• Education or recreational flying only	• Flying for commercial use • Flying incidental to a business • Flying public aircraft operations

*These requirements are subject to waiver.

Hobby/Recreational Aircraft

- **Generally, hobby/recreational operators do not need FAA authorization to fly, but they must fly safely at all times:**
  - Avoid manned aircraft
  - Maintain visual line-of-sight
  - Fly only for hobby/recreation
- **They must register and mark their UAS before flying outdoors**
  - UAS between 0.55 pounds and 55 pounds may register online

Interpretive Rule

- **FAA published guidance in June 2014 for hobby or recreational use of UAS**
- **This guidance clarifies that:**
  1. Model aircraft must satisfy the criteria in the Act to qualify as model aircraft and to be exempt from future FAA rulemaking action
  2. Consistent with the Act, if a model aircraft operator endangers the safety of the NAS, the FAA has the authority to take enforcement action against those operators for safety violations
- **Status: FAA evaluating comments to determine where clarification is needed**

https://www.federalregister.gov/articles/2014/06/25/2014-14948/interpretation-of-the-special-rule-for-model-aircraft

Authorizing Small UAS Operations (Part 107)

- UAS must weigh less than 55 lbs. and be registered
- UAS operators must obtain a Remote Pilot Certificate
- Visual line-of-sight, daylight or civil twilight operations
- Must yield right-of-way to manned aircraft
- External load operations permitted if the load does not affect flight operations or control
- One UAS per Remote Pilot
- No operations over people
- 400 feet or below

Becoming a Pilot under Part 107

- Must be 16 years old or older
- Must read, write, speak English
- Must pass an aeronautical knowledge exam at an FAA-approved Knowledge Testing Center
  - Part 61 certificate holders can take online training at faasafety.gov instead of the knowledge exam
- Must undergo TSA background security screening

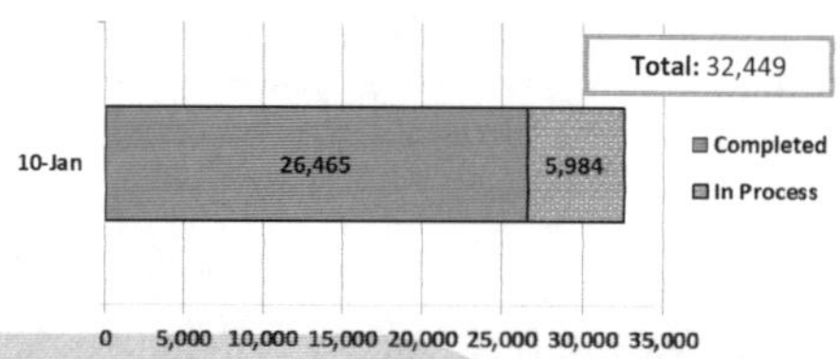

Waivable Provisions of Part 107

- **Operation from a moving vehicle or aircraft** (§ 107.25)
- **Daylight operation** (§ 107.29)
- **Visual line of sight aircraft operation** (§ 107.31)
- **Visual observer** (§ 107.33)
- **Operation of multiple small UAS** (§ 107.35)
- **Yielding the right of way** (§ 107.37(a))
- **Operation over people** (§ 107.39)
- **Operation in certain airspace** (§ 107.41)
- **Operating limitations for small UAS** (§ 107.51)

Online portal available at www.faa.gov/uas/request_waiver/

Waiver Criteria

Two key documents for submitting Part 107 waiver requests:

1. Performance-Based Standards – explains what information is necessary for the FAA to evaluate waiver requests
2. Waiver Form Instructions – how to fill out the form

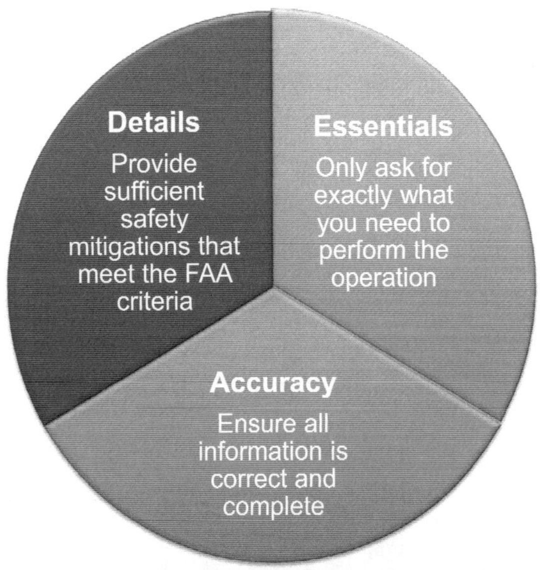

Part 107 Airspace Requirements

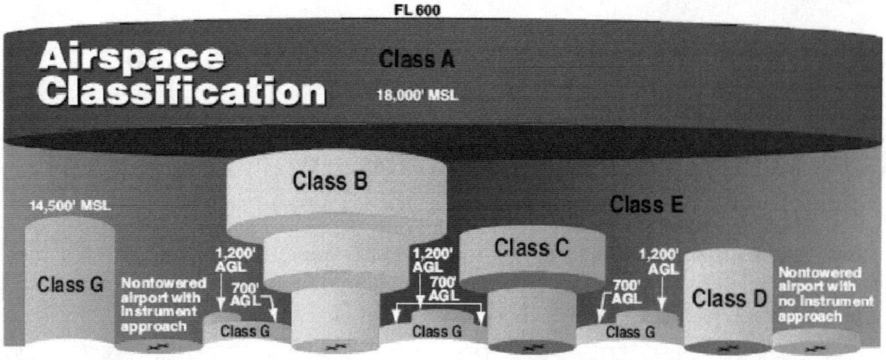

- Operations in Class G without ATC authorization
- Operations in Class B, C, D & Class E surface areas require ATC authorization
- Phased approach to airspace authorizations
- Online portal available at www.faa.gov/uas/request_waiver/

Strategies for Airspace Authorization Requests

Only Ask for What You Need

- Airspace waivers generally take longer than authorizations
- More complex requests = longer processing time

Stay Low

- Lower altitudes enable operations closer to airports
- Consider additional risk mitigations for requests in high traffic locations

Location Location Location

- Be precise with your latitude and longitude
- Be discrete and specific – smaller operating areas are easier to approve

Focus Area Pathfinders – Expanding Operations

- **3 Focus Area Pathfinder Partners:**

1. CNN
   - Exploring visual line-of-sight operations over people
2. Precision Hawk
   - Exploring extended visual line-of-sight operations in rural areas
3. BNSF Railways
   - Exploring beyond visual line-of-sight operations in rural areas

UAS Over People Rulemaking

- **Stakeholder committee recommended regulatory framework for UAS operations over people to FAA on April 1, 2016**
- **Report outlines four categories of small UAS operations**
  - Defined primarily by level of risk of injury posed
- **The FAA is considering these recommendations and developing a performance-based rule that would allow operations over people under part 107**
- **Proposed rule expected in 2017**

Online UAS Registration

- **Applies to small UAS 0.55-55 lbs. flown outside**
- **Owner must provide name, address, email**
  - Non-recreational owners must provide make, model, and serial number (if available) of each sUAS

UAS Outreach and Education

KNOW BEFORE YOU FLY

FAA B4U FLY

Questions?

www.faa.gov/uas

Backup Slides

Aeronautical Knowledge Exam Topics

- Applicable regulations relating to small unmanned aircraft system rating privileges, limitations, and flight operation
- Airspace classification and operating requirements, and flight restrictions affecting small unmanned aircraft operation
- Aviation weather sources and effects of weather on small unmanned aircraft performance
- Small unmanned aircraft loading and performance
- Emergency procedures
- Crew resource management
- Radio communication procedures
- Determining the performance of small unmanned aircraft
- Physiological effects of drugs and alcohol
- Aeronautical decision-making and judgment
- Airport operations
- Maintenance and preflight inspection procedures

AHS International Specialists' Meeting
Unmanned Rotorcraft Systems
24 January 2017

Challenges in Unmanned (Helicopter) Aviation
- Lessons learned in the Development of the MQ-8C RDC

Robert Ernst
NAVAIR (PMA-266)
Chief Engineer
301-757-0442
robert.ernst@navy.mil

Outline

- **MQ-8C Background**
- **RDC lessons Learned**
  - Sense and Avoid
  - (Timely) Ice detection and Avoidance
  - Redundant and Robust Landing Systems
  - Agile + Software Environment
  - Enhanced Reliability and Maintainability
  - UAS appropriate Diagnostics/Prognostics
- **Closing Comments**

Unclassified

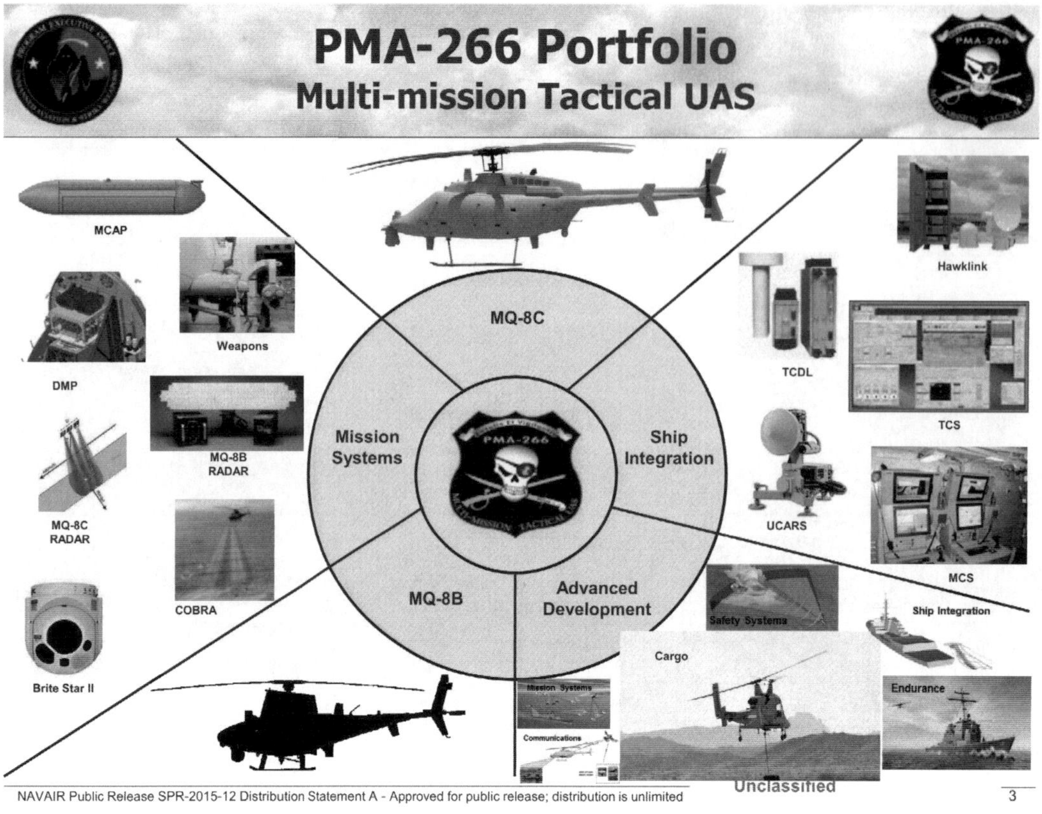
PMA-266 Portfolio
Multi-mission Tactical UAS
MCAP
Weapons
DMP
MQ-8B
RADAR
MQ-8C
RADAR
COBRA
Brite Star II
MQ-8C
Mission
Systems
Ship
Integration
MQ-8B
Advanced
Development
Hawklink
TCDL
TCS
UCARS
MCS
Safety Systems
Ship Integration
Cargo
Endurance
Mission Systems
Communications
NAVAIR Public Release SPR-2015-12 Distribution Statement A - Approved for public release; distribution is unlimited
Unclassified
3

MQ-8 System Overview
GIG
CSG
APKWS II
SUW
TAC-ISR
To RVT
COBRA
MCM
EO/IR
Radar
AIS
Communications
Relay
LOS
DMP
MCM
AIS
Voice
Data
GCCS-M
NAVAIR
6

Fire Scout
System Description

Air Vehicle

- Autonomous
- Fully Digital, Dual Redundant Control System and C2 links
- Open System Architecture facilities integration and testing

Mission Control System

- Interoperable Control Station
- Tactical Control System software integrated
- Open Architecture
- GCCS-M, JDISS, AFATDS, CCTV & JSIPS-N
- NATO STANAG 4586 Compliant
- Multi-vehicle control

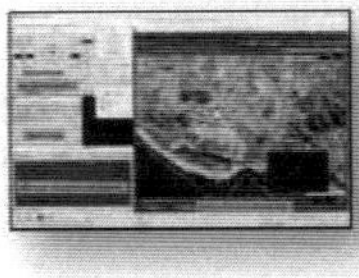

Launch & Recovery

Tactical Control Data Link (TCDL)

UCARS-V2 for Ship Launch & Recovery

Payloads

EO/IR

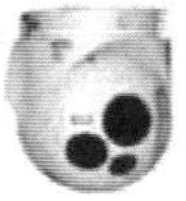

AN/ZPY-8 RADAR

(MQ-8C)

AN/ZPY-4 RADAR

(MQ-8B)

AIS

TAC-ISR

(Omni)

Future Payloads

COBRA

Data Mission

NAVAIR

NAVAIR Public Release SPR-2014-491: Distribution Statement A - Approved for public release; distribution is unlimited

MQ-8 UAV
MQ-8B/MQ-8C Comparison

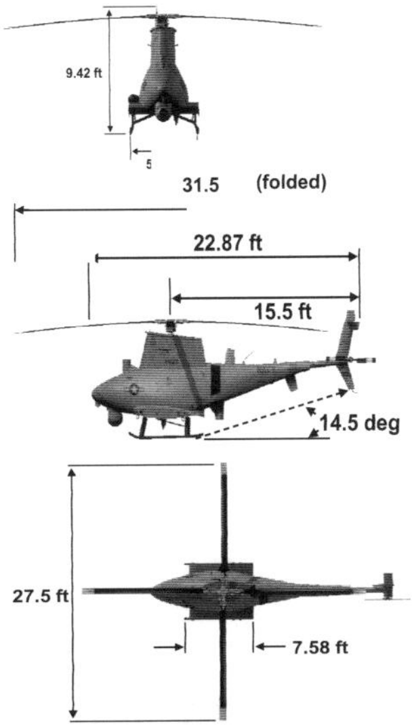

MQ-8B	Parameter	MQ-8C
85 kts	Maximum Speed	135 kts
80 kts	Cruise Speed	115 kts
12,500 ft	Service Ceiling	16,000 ft
5.5 hrs	Std Day Maximum Endurance (with 300lb payload)	12 hrs
4.5 hrs	Hot Day Maximum Endurance (with 300lb payload)	10 hrs
2,000 lbs	Empty Weight	3,200 lbs
3,150 lbs	Std Day Fuel & Payload	6,000 lbs
31.5 ft	Length (folded)	34.7 ft

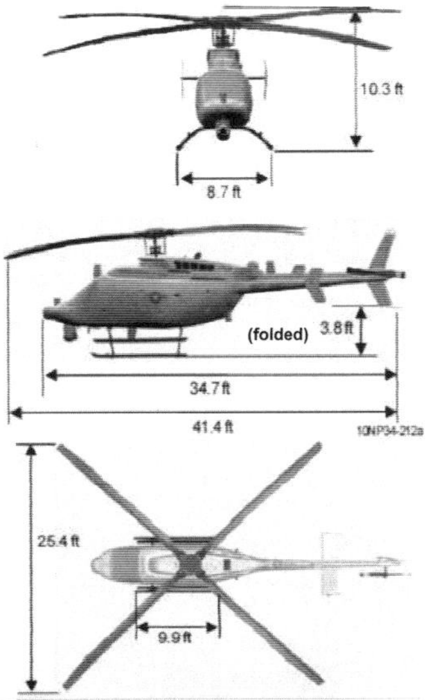

MQ-8C: 3 ft Longer (folded), 1 ft Taller, 2.5 ft Wider & 1200# Heavier than MQ-8B

Unclassified

NAVAIR Public Release SPR-2015-12 Distribution Statement A - Approved for public release; distribution is unlimited

6

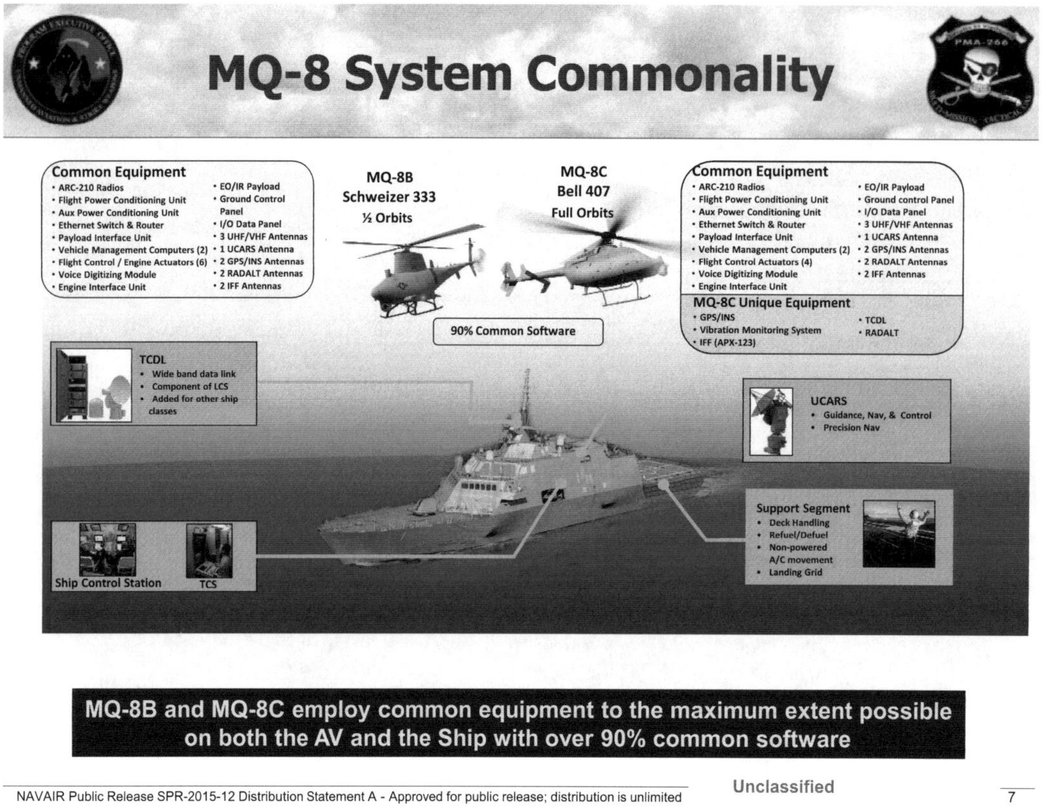
MQ-8 System Commonality
Common Equipment
• ARC-210 Radios
• Flight Power Conditioning Unit
• Aux Power Conditioning Unit
• Ethernet Switch & Router
• Payload Interface Unit
• Vehicle Management Computers (2)
• Flight Control / Engine Actuators (6)
• Voice Digitizing Module
• Engine Interface Unit
• EO/IR Payload
• Ground Control Panel
• I/O Data Panel
• 3 UHF/VHF Antennas
• 1 UCARS Antenna
• 2 GPS/INS Antennas
• 2 RADALT Antennas
• 2 IFF Antennas
MQ-8B
Schweizer 333
½ Orbits
MQ-8C
Bell 407
Full Orbits
Common Equipment
• ARC-210 Radios
• Flight Power Conditioning Unit
• Aux Power Conditioning Unit
• Ethernet Switch & Router
• Payload Interface Unit
• Vehicle Management Computers (2)
• Flight Control Actuators (4)
• Voice Digitizing Module
• Engine Interface Unit
• EO/IR Payload
• Ground control Panel
• I/O Data Panel
• 3 UHF/VHF Antennas
• 1 UCARS Antenna
• 2 GPS/INS Antennas
• 2 RADALT Antennas
• 2 IFF Antennas
MQ-8C Unique Equipment
• GPS/INS
• Vibration Monitoring System
• IFF (APX-123)
• TCDL
• RADALT
90% Common Software
TCDL
• Wide band data link
• Component of LCS
• Added for other ship classes
UCARS
• Guidance, Nav, & Control
• Precision Nav
Support Segment
• Deck Handling
• Refuel/Defuel
• Non-powered A/C movement
• Landing Grid
Ship Control Station
TCS
MQ-8B and MQ-8C employ common equipment to the maximum extent possible on both the AV and the Ship with over 90% common software
NAVAIR Public Release SPR-2015-12 Distribution Statement A - Approved for public release; distribution is unlimited
Unclassified
7

Fire Scout MQ-8B Accomplishments
USN MQ-8B Deployments
✓ Land Based Operations...7,425 Flt Hours
• Test, Training and in Afghanistan
✓ FFG Non-Integrated Operations...8,185 Flt Hours
• 10 deployments on six different ships
✓ LCS Integrated Operations...143 Flt Hours
• USS Fort Worth (LCS 3) was the 1st deployment with an MQ-8B and MH-60R
• USS Coronado (LCS 4) is the first deployment of a MQ-8B equipped radar
ISR Task Force Support in Afghanistan
✓ 2 A/C, 2 GCS, 300 hrs/mo FMV
✓ First flight 2 May 2011
✓ Last flight 1 August 2013
✓ 1,438 flights for 5,084.3 flight hours completed
✓ Mission completed August 2013
Support LCS Mission Packages in conjunction with the MH-60
✓ LCS-1 Dynamic Interface (DI) testing (Nov 2010);
✓ LCS-3 DI expansion (Nov 2013)
✓ COBRA MCM Capability land testing (Apr 2013)
✓ SUW Mission Package Assessment (Apr 2014)
✓ LCS-4 DI testing (Sep 2014)
2006
2021
Maritime ISR Support to SOF
✓ Emergent Requirement approved in January 2012
✓ Deployments continue aboard FFGs
✓ DDG installation complete
✓ MQ-8B 10th FFG Deployment aboard USS Simpson complete
Radar integration
✓ Provides wide-area maritime search capability
✓ Quick Reaction Assessment
✓ LCS 4 deployment
as of 1 Nov 2016
MQ-8B has flown over 16,060 flight hours since 2006, with continuous growth in Flight Hours, Reliability and Operational Availability
NAVAIR Public Release SPR-2015-124 Distribution Statement A - Approved for public release; distribution is unlimited

MQ-8 Top Challenges

- Sense and Avoid
- Operations in/around icing environment
- Redundant and Robust Landing Systems
- Agile + Software Environment
- Enhanced Reliability and Maintainability
- UAS appropriate Diagnostics/Prognostics

MQ-8 Sense and Avoid System

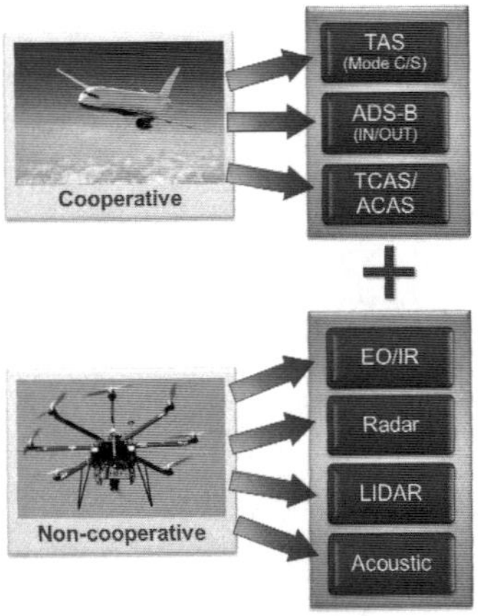

- **UAS integration into NAS**
  - SAA = Sensing + Detecting (**Tracking + Link Transmission + AVO Cueing/Alerting**) + Collision Avoidance
- **Key Attributes**
  - Adverse Weather Performance
  - Detection Accuracy (Pd; false alarm rate)
  - Range (Detection range, azimuth, elevation)
  - Detection Type (cooperative / non-cooperative)
  - Tech Readiness Level
  - Minimal Shipboard Mods & Aircraft SWAP
- **UAS Helicopters in that "sweet spot"**
  - ~ 50 lbs sensor package
  - ~ 5 nmiles detection range
  - Need it now

MQ-8 Ice Detection Capability

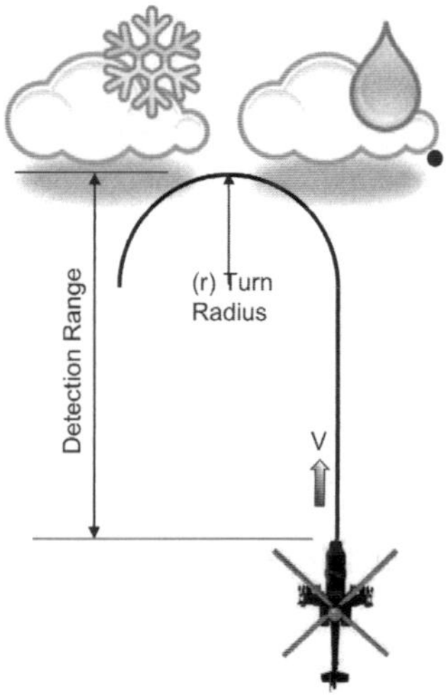

- **Early Ice Detection**
  - Operations in Icing = Potential Loss of AV
  - Early Detection provides Loss Risk Mitigation
  - **Aircrew Cueing**
- **Key Attributes**
  - Adverse Weather Performance
  - Accuracy (Prob Of detection; false alarm rate)
  - Range (Detection range, azimuth, elevation)
  - Time to notify AVO
  - Tech Readiness Level
  - Minimal Ship Mods/SWAP

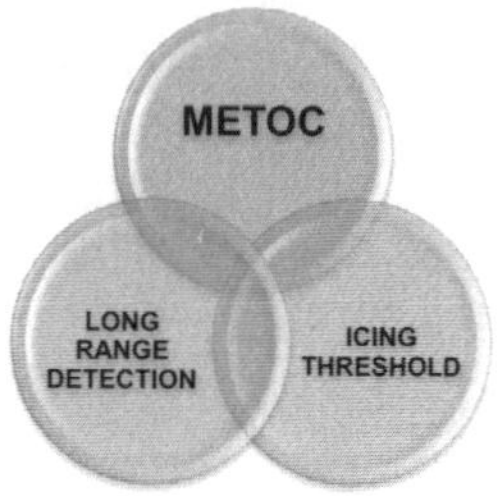

MQ-8 Ice Detection Trade Study

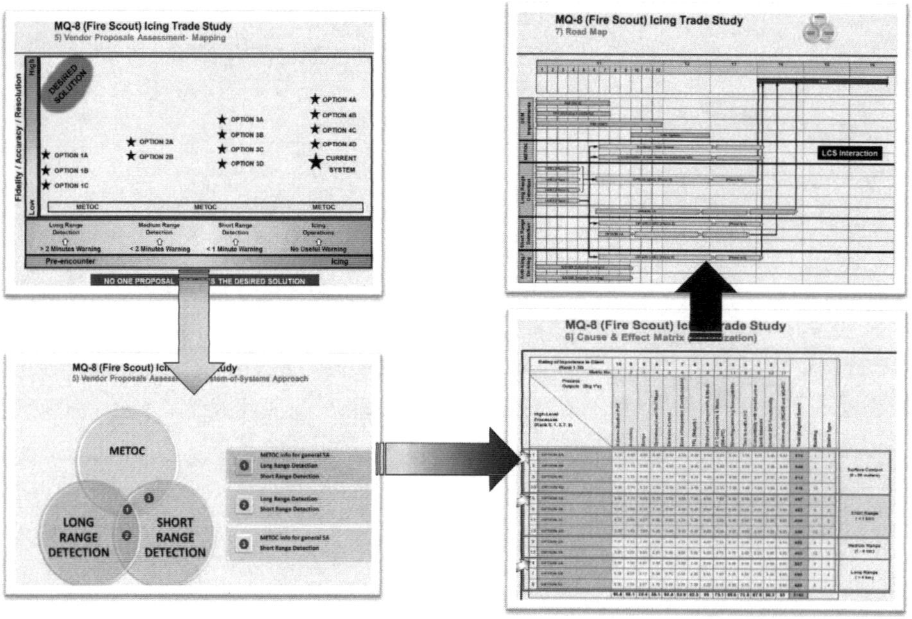

MQ-8 Redundant Landing System

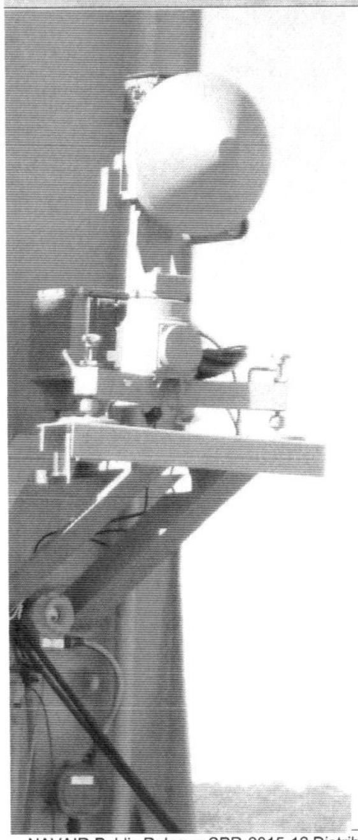

- **Blue Water Reliability**
  - Loss of Capability = Loss of Air Vehicle
  - Redundancy Provides Loss Risk Mitigation
- **Key Attributes**
  - Redundancy/Reliability
  - High TRL
  - Anti-Jam/Anti-Spoof
  - Minimal Shipboard Mods/SWAP
  - Adverse Weather Performance
- **Updating proven UCARS system**
  - Add Dual Aircraft Components
  - Upgrade TS systems for better reliability
  - Incorporate Dual Shipboard Systems

MQ-8 Redundant Landing System

INAV-SL uses passive EO/IR sensors to detect and analyze existing ship deck markings to provide Relative Navigation (RelNav) measurements

INAV-SL (optical landing system) has been demonstrated to meet MQ-8 requirements for landing system accuracy.

Expanding Agile Software Development

- **UxS systems Integration Requires More Agile Processes**
  - Cyber, TTP Integration, Emerging CONOPS
- **Agile Software Processes Require**
  - "Bucketing of Work" Into backlog
  - Establishment of Rough Priorities
  - Delegation of ownership to OEM
  - Development of automatic testing
- **MQ-8 Software Development has seen a 300% increase in Capability**

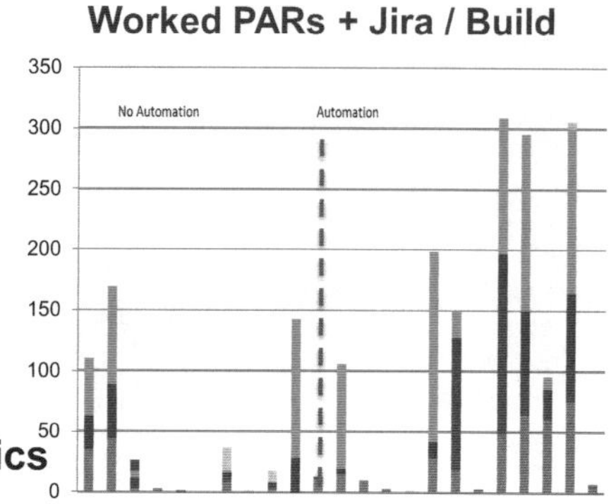

Challenges

- **SRR Like Event to understand Verification (FT, Lab...)**
- **Evaluation of Logistics Products** *upfront*
- **Concurrent Review of Test and logistics**

1/20/2017 11:47 AM

Unclassified

15

Expanding Agile Software Development Including Logistics and Test

1. Establish Sprint PARS from "backlog of PARS
   - All "must" PARS
   - 50-60% of "should"
   - Others when CSCI is open
2. Weekly TIMs between OEM/TAE to review "how" portion of PAR
3. Documented in living TIM document
4. Unit testing as software requires

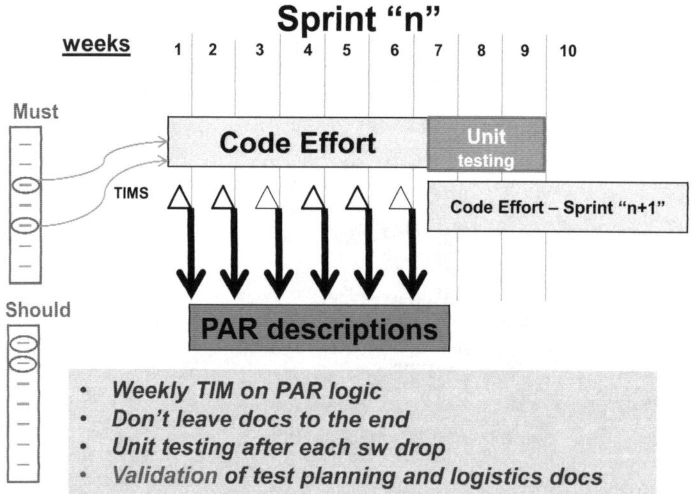

- *Weekly TIM on PAR logic*
- *Don't leave docs to the end*
- *Unit testing after each sw drop*
- *Validation of test planning and logistics docs*

- *Requires Trust between OEM and Govt teams*
- *Constantly challenge Testing and documentation*
- *Move to rule based Testing with automation*

Unclassified

MQ-8 Enhanced R & M

- **Key attribute of an unmanned system is ENDURANCE**
  - i.e. 12 hour vs 5 hour mission (factor of 2.4)
- **To achieve equivalent number of missions, reliability and maintainability must be enhanced**
  - MFHBOMF (measure of reliability) must also be factored by 2.4, typically increased to 50-60 hours/FH
  - MMH/FH (measure of maintainability) must also be factored by 2.4, typically reduced to 2-3 hours/FH
  - Typical LCS aviation maintenance crew = 15 (appr)

R&M must be enhanced to achieve equivalent number of missions using smaller crew sizes on ships due to limited manning capability

MQ-8 Enhanced R & M

- **Enhance R & M is achieved by using**
  - Condition-Based Maintenance (CBM)
  - Active RCM program
  - Aggressive Corrosion Control Program
  - Transitioning to Augmented Reality/JPA
    - Ease of troubleshooting
    - 21st century construct

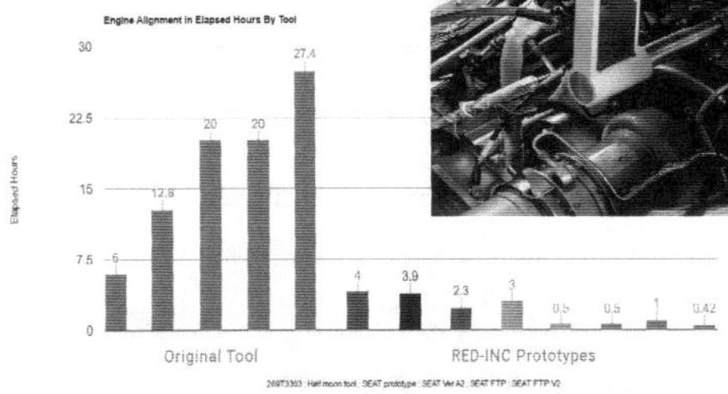

R&M must be enhanced to achieve equivalent number of missions using smaller crew sizes on ships due to limited manning capability

MQ-8 Diagnostics/Prognostics

Adapting Health and Usage Monitoring System (HUMS) to unmanned system- Because we have:

- Repeatable landing data
- Better-defined mission profile/spectrum

We can derive

- Increased fatigue life (typically 25-30% higher)
- Less "off condition" inspections

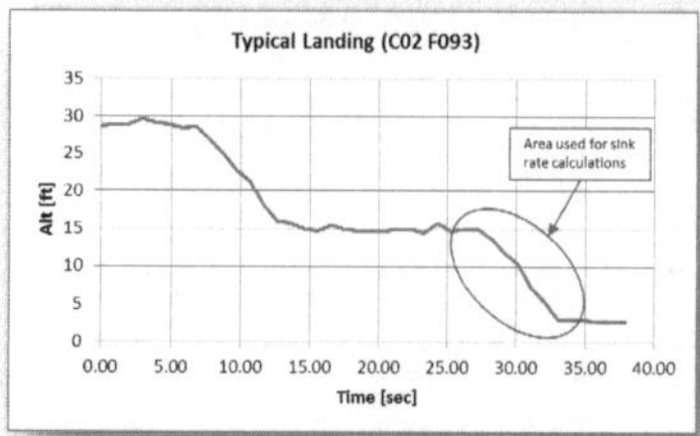

45 Landings	
Mean (μ)	2.20
Std. Dev. (σ)	0.141
Max / Min	2.62 / 1.93

Land-based slope landings (includes slopes up to 6.9°)

27 Landings	
Mean (μ)	2.21
Std. Dev. (σ)	0.060
Max / Min	2.32 / 2.11

Shipboard Landings (DI test- DDG)

Need better sensors and industry standards on methods

Actual visuals on TCS

- **Achieved thru better real-time measurements**
  - Aircraft
  - Ship

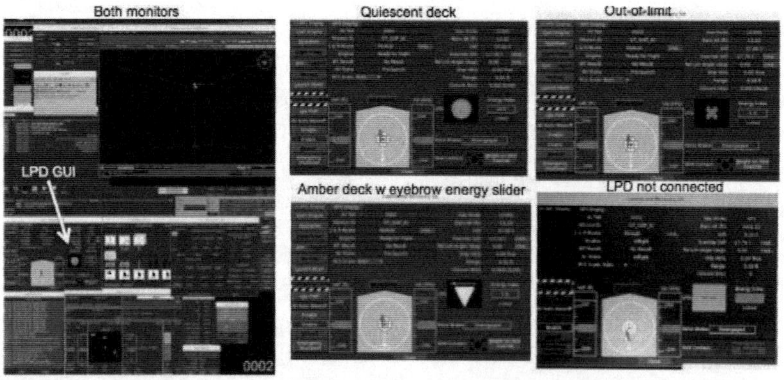

Launch and Recovery SA Page

Emphasis must be placed on prognostics due to limited resources on ships deployed in "remote" locations

Final Comments

- **Unmanned Systems Require Additional Thoughts and Processes**
  - New Considerations (Communicate, Navigate, Aviate)
  - "Opportunity" to Define Processes
  - Requires Transition to Agile Processes to support Asymmetric Threats and Cyber
- **More Endurance/ Less Maintenance**
  - Think "Millennial" in your support concepts
- **UxS offers potential to fully exploit prognostics**

21st Century Challenges for a Talented Workforce

Questions

Fire Scout MQ-8C Accomplishments

MQ-8C aboard DDG-109 during Dynamic Interface Testing

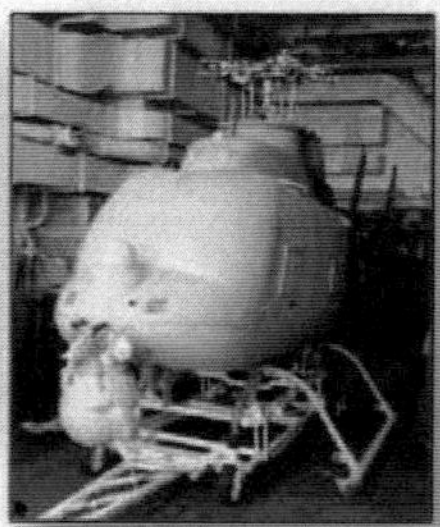

MQ-8C Fire Scout demonstrates a long range, long endurance flight part of a capability based test at Naval Base Ventura County, Pt Mugu

Apr 2012	First Contract Awarded 2 Test and 17 Production Air Vehicles
Oct 2012	Preliminary Design Review successfully conducted
Jan 2013	Critical Design Review successfully conducted
Oct 2013	First MQ-8C Flight
Jul 2014	Electromagnetic (E3) testing complete
Dec 2014	DDG Dynamic Interface completed aboard DDG-109
Apr 2015	Developmental Test Flight Test completed (345 Sorties; >525 Flight Hours)
Aug 2015	Long Range/Long Endurance Flight (150NM/11 hrs.)
Sep 2015	MQ-8C Capabilities Based Testing completed (4 Sep)
Sep 2015	U.S. Navy accepted the first production MQ-8C at Point Mugu, CA
May 2016	MQ-8C RADAR awarded to Selex ES Ltd. (26 May)
Jun 2016	Capabilities Production Document (CPD) approved by CNO; stable requirements
Sep 2016	MQ-8C Production Contract and Long Term Supply Support in place
Sep 2016	MQ-8B / MQ-8C Test Team consolidation aboard Webster Field (Pax River area)

1st Contract Award to 1st Flight – 18 months!

NAVAIR Public Release SPR-2015-124 Distribution Statement A - Approved for public release; distribution is unlimited

Radar Integration (MQ-8B)

Rapid Deployment Capability Goals

- Modification of commercial Telephonics RDR-1700B+ radar to AN/ZPY-4(V)1 configuration
- Multi-mode maritime surface search radar
  - Long and short range detection
  - Spot SAR
  - ISAR
  - AIS track processing
- Radar track cue to EO/IR sensor

MQ-8B Radar Performance

- EO/IR Cueing
- Detection Accuracies:
  - Small boat: 25 nmi
  - Large boat: 70 nmi
- Tracks: up to 150 targets
- ISAR – Maritime Classification/ID aid
- Spot SAR - weather independent high resolution imagery
- IOC: FY15

NAVAIR Public Release SPR-2015-12 Distribution Statement A - Approved for public release; distribution is unlimited

Unclassified

24

MQ-8B Weapons RDC

- **Employs Advanced Precision Kill Weapon System II (APKWS II)**
  - Mk 66 Mod 4, 2.75 inch rocket system
- **LAU-153/A Launcher**
  - Each 92 pound launcher carries three rockets
  - MQ-8B unique
- **BRU-59/A rack**
  - Common to other Navy/Marine Corps aircraft
- **Live Fire Integration and QRA complete Jun 13**
  - 12 total shots/11 hits
- **Currently no decision to deploy weaponized MQ-8B**

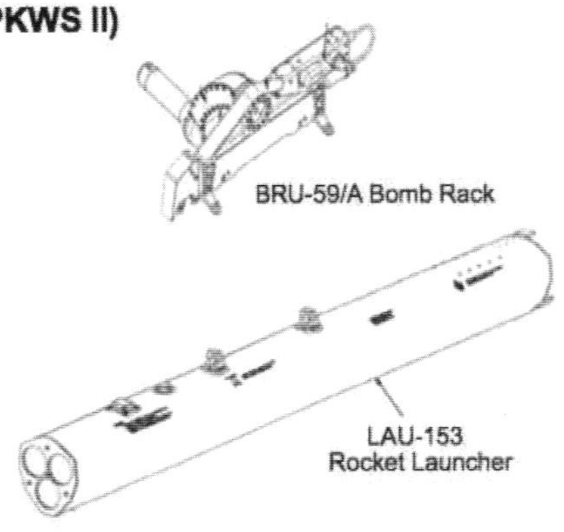

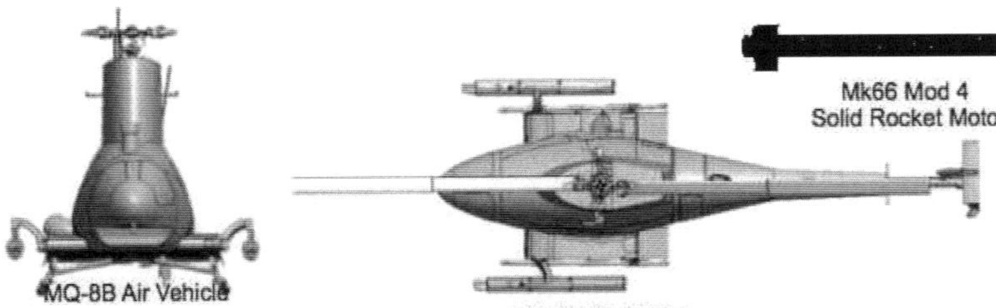

NAVAIR Public Release SPR-2015-12 Distribution Statement A - Approved for public release; distribution is unlimited

NAVAIR 13

U.S. ARMY RDECOM

Army S&T Activities Leading to Future UAS Capabilities

AMRDEC

This information product has been reviewed and approved for public release, distribution A (Unlimited). Review completed by the AMRDEC Public Affairs office PR2252 Sept. 13, 2016.

TECHNOLOGY DRIVEN. WARFIGHTER FOCUSED.

24 January 2017

Presented by:
Layne Merritt
Chief Engineer
Aviation Development Directorate
U.S. Army Aviation and Missile Research, Development, and Engineering Center

U.S. ARMY RDECOM

Overview

AMRDEC

- Current fielded Unmanned Aerial systems have served the Army well in OEF/OIF

- Future fights require operations in complex and contested airspace in a Multi Domain Battle concept

- Future UAS require a holistic design approach to be effective and survivable
- The air vehicle is not simply another federated technology independent from the mission equipment

TECHNOLOGY DRIVEN. WARFIGHTER FOCUSED.

FileName.pptx

2

U.S. ARMY RDECOM — AMRDEC

Possibilities

- Effective and efficient teaming
  - Intelligent, intuitive interface
  - Natural interaction for mission- & objective-level commands
  - UAS is part of the formation
- Shared situational awareness
  - Exchange information, not just data
  - Distributed secure processing
- Execute tactical missions in hybrid operating environments
  - GPS denied / Data link free operations
  - Systems work with good, bad or no networked comms
- Rapid autonomy enhancements and upgrades
  - Open System Architectures, data management
- Complex, intelligent systems certified quickly
  - Established, benchmarked procedures

TECHNOLOGY DRIVEN. WARFIGHTER FOCUSED.

3

U.S. ARMY RDECOM — AMRDEC

Vision for Autonomy

Manned Unmanned Teaming

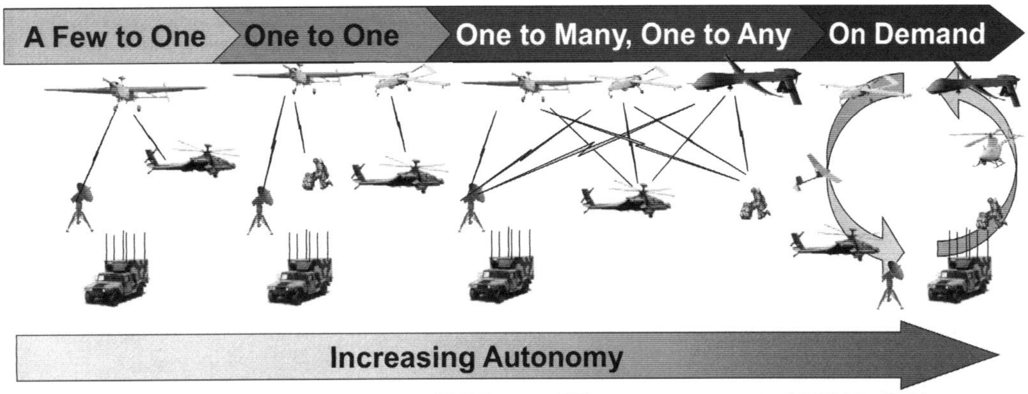

UAV as a Pointman	*UAV as a Wingman*	*UAV Task Force*
• Sensor Guided Flight	• Close Proximity Flight	• UAV Platoons
• Reliable, Autonomous Flight	• Tactical Perception	• Distributed Control
• Shared Airspace Flight	• Intuitive Interface	• Self Organizing/Tasking
• Adaptable, Tactical Behaviors	• Team Survivability & Lethality	• Autonomous Over-watch
	• Weaponized	• Swarming

TECHNOLOGY DRIVEN. WARFIGHTER FOCUSED.

4

Future Tactical UAS Technology Demonstration

U.S. ARMY RDECOM — AMRDEC

Purpose:

- Develop and demonstrate game-changing performance enabling technologies to meet the next generation Tactical UAS range, efficiency, reliability, and survivability requirements and operational capabilities

Products:

- Innovative low drag, long endurance concepts
- Runway independent, confined area solutions
- Platform integration of non-traditional engine designs (RASPS)
- System level SWAP allocation
- Mission flexibility and efficiency demonstrations
  - Adaptive and reconfigurable subsystems
- Autonomous navigation control algorithms
- Wingman concepts for FVL manned systems
- Enhanced survivability sufficient for emerging A2AD environments
- Draft Model Performance Specifications (MPS)
  - Used as basis for solicitation of aircraft from industry; provides quantifiable metrics for technical evaluation of proposals

Schedule

Elements	FY18	FY19	FY20	FY21	FY22	FY23
Studies and Analysis, simulation		4				
Configuration Trades & Analysis						
Next Gen UAS Demo						6

Payoffs:

- Expanding scope of portfolio to include emerging ASSP Classes 2 and 3 platforms
- A refined set of technologically feasible and affordable capabilities that enable next generation UAS requirements in POR EMD phase
- Operational parity with manned fleet for advanced teaming
- Data readily available to support future acquisitions

TECHNOLOGY DRIVEN. WARFIGHTER FOCUSED.

FileName.pptx

5

Reliable Advanced Small Power Systems (RASPS)

U.S. ARMY RDECOM — AMRDEC

Engine Attributes

40-400 Horsepower
Reliable / Durable
Low Noise
Reduced Fuel Consumption
Improved Hp/Wt
Heavy Fuel Capable

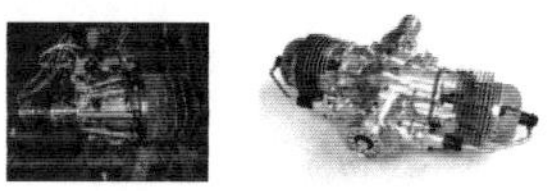

Purpose:

- Demonstrate high reliability (MTBO), efficient, low-noise small engines / power systems to provide improved performance, reliability, readiness, and affordability for increased operational capability of future Army UAS.

Products:

- Full system demonstration of JP8 fueled, reliable, fuel efficient, high power to weight engine concepts for group 3 and 4 UAS.

Potential Concepts/Technologies

Innovative Piston/Diesel Engine Concepts	Recuperated Turbine Engine
Innovative Rotary Engine Concepts	Hybrid Electric with Energy Storage

Schedule

Elements	FY17	FY18	FY19	FY20	FY21	FY22
Design		Potential for multiple awards				
Fabrication						
Test						
Redesign/Fab						
Final Test						6

Payoff:

- Provides improved mission capability for unmanned Army aircraft
- Increased readiness/availability, reduced logistic footprint and reduced O&S Cost
- Potential applications include future unmanned air systems or ground systems

TECHNOLOGY DRIVEN. WARFIGHTER FOCUSED.

FileName.pptx

6

Synergistic Unmanned Manned Intelligent Teaming (SUMIT)

U.S. ARMY RDECOM | AMRDEC

MILESTONE	FY14	FY15	FY16	FY17	FY18	FY19	FY20	FY21
SUMIT Part 1								
Initial HMI Definition & Req.								
Simulation Development								
Automation & HMI Development and Assessment								
SUMIT PART 2								
Develop Advanced Automation Technology								
Integrate into SUMIT Simulation Environment								
User Assessments								
SUMIT PART 3								
Test Planning and Preparation								
Integrate into Cockpit								
Ground and flight tests								

Purpose:

Mature and demonstrate advanced Manned/Unmanned Teaming (MUMT) capabilities through the integration and assessment of new technologies in the areas of the Human Machine Interface (HMI), decision aiding, and autonomy, and the implementation and assessment of new employment concepts. Focus is on an Air Mission Commander using one crew station managing up to eight UAS to execute Scout/Attack and Air Assault Missions.

Product:

PART 1 – *MUMT Development Environment Build*
- Government based MUMT development environment to assess MUMT, HMI, Decision Aiding, and Autonomy
- Engineering evaluations of decision aiding, supervisory control of multiple UAVs, and HMI to support the Air Mission Commander's role
- Produce experimentation plan

PART 2 – *Advanced MUMT Experimentation*
- Decision aiding, autonomous behaviors, and advanced HMI integrated into the SUMIT simulation environment
- MUMT experimentation and user evaluations.
- Findings and Recommendations report

PART 3 – *Flight Demonstration*
- Demonstrate key technologies assessed by SUMIT

Payoffs:
- Increased combat team effectiveness
  - Supports advanced decision aiding for MUM teams
  - High levels of autonomy to increase operator throughput, enabling tactical MUM teaming
  - One aviator effectively utilizing multiple unmanned aircraft to complete mission
- MUMT technologies aligned with FACE and JCA standards to facilitate transition
- Simulation environment development for future mission systems S&T programs

TECHNOLOGY DRIVEN. WARFIGHTER FOCUSED.

7

Holistic Situational Awareness and Decision Making

U.S. ARMY RDECOM | AMRDEC

Schedule

Elements	FY19	FY20	FY21	FY22	FY23	FY24
HSA & DM Open Arc	3		4			
Next Gen Crew Station						
• Advanced SA and DM Algorithms				4		
• Survivability Planner				5		
• Engagement Planner	3		4	5		
• SIL Int. and Demos			4		6	
• Aircraft Int. and Flt Test.						6

Purpose:
Holistic Situational Awareness and Decision Making (HSA-DM) focuses on the integration of currently federated mission threads into a pilotage and decision making system that allows for freedom of maneuver in highly contested environments.

Products:
Demonstration of Next Generation Crew Station concepts to include:
- **Comprehensive human machine interface for all SA domains (terrain & obstacles, threat, engagement, weather & environment)**
- **Decision making technologies (Survivability and Engagement planner) to reduce cognitive loading of air crews during operations in complex and hostile environments**
- **Multi spectral data correlation and fusion algorithms for target, threat and hazard identification**
- **Integrated program with RDECOM partners (AMRDEC Missile, CERDEC, ARL)**

Payoff:
- **Enable freedom of maneuver in complex and hostile environments**
- **Reduced risks to pilots and aircrews through managed cognitive work loads**
- **Decide faster**
- **Decision making algorithms for future autonomous systems**
- **Increased mission effectiveness**
- **Increased aircraft survivability**

TECHNOLOGY DRIVEN. WARFIGHTER FOCUSED.

8

U.S. ARMY RDECOM

Advanced Teaming for Tactical Aviation Operations

AMRDEC

Schedule

Elements	FY19	FY20	FY21	FY22	FY23	FY24	FY25
Develop Algorithms & identify mission sets	2 ———	———	———	3 ———	———	———	——— 4
Architecture Development	2 ———	———	———	3 ———	———	———	——— 4
Configuration Trades and Analysis				3 ——— 4			
System Design & Build / Simulation Tests				4 ———	———	——— 5	
Aircraft Integration / Flight Tests							——— 6

Purpose:
- Develop and demonstrate teaming behaviors and autonomous decision making for mixed platform formations in combined arms operations

Products:
- Resilient autonomy algorithms
- Self-organizing, self-healing unmanned formations
- Distributed command and control algorithms to enable operations in complex environments,
- Navigation systems to enable GPS denied operations
- Framework for airworthiness certification of autonomy algorithms and mixed operations
- Develop combined arms world model

Payoff:
- Effective operations and enhanced survivability of aviation operations in contested environments
- Enhanced reliability of autonomy that adapts to failures
- Enables true autonomous mission capabilities beyond MUM-T tele-control
- Reduced risk for critical technologies in the mid to far term

TECHNOLOGY DRIVEN. WARFIGHTER FOCUSED.

FileName.pptx

9

U.S. ARMY RDECOM

Questions?

AMRDEC

TECHNOLOGY DRIVEN. WARFIGHTER FOCUSED.

FileName.pptx

10

Presented to:

7th AHS Technical Meeting on VTOL UAS Mesa, AZ, 24-26 January 2017

U.S. ARMY RDECOM

NATO STANAG Airworthiness Code Development for small VTOL Unmanned Aircraft Systems

AMRDEC

DISTRIBUTION A. Approved for public release: distribution unlimited

TECHNOLOGY DRIVEN. WARFIGHTER FOCUSED.

Presented by:

Leo Flynn

Unmanned Aircraft Systems Division

Aviation Engineering Directorate

U.S. Army Aviation and Missile Research, Development, and Engineering Center

24-26 January 2017

U.S. ARMY RDECOM

Unmanned Aircraft Systems (UAS) Airspace Integration

AMRDEC

How do we open up all Airspace to UAS?

Currently majority of operations are Risk based and over limited areas

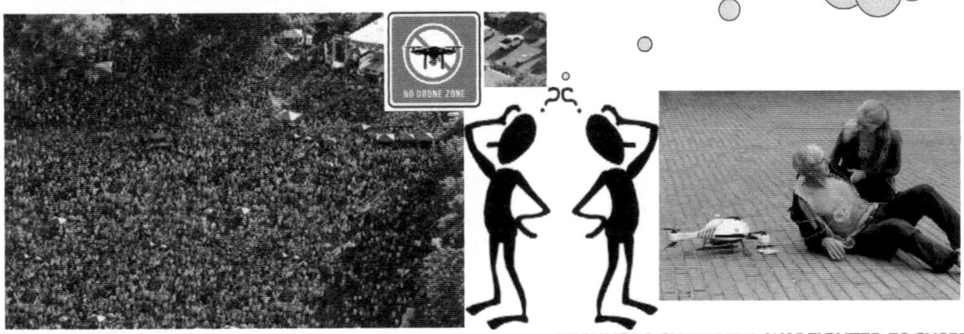

TECHNOLOGY DRIVEN. WARFIGHTER FOCUSED.

FileName.pptx

2

DOD Airspace Integration Approach

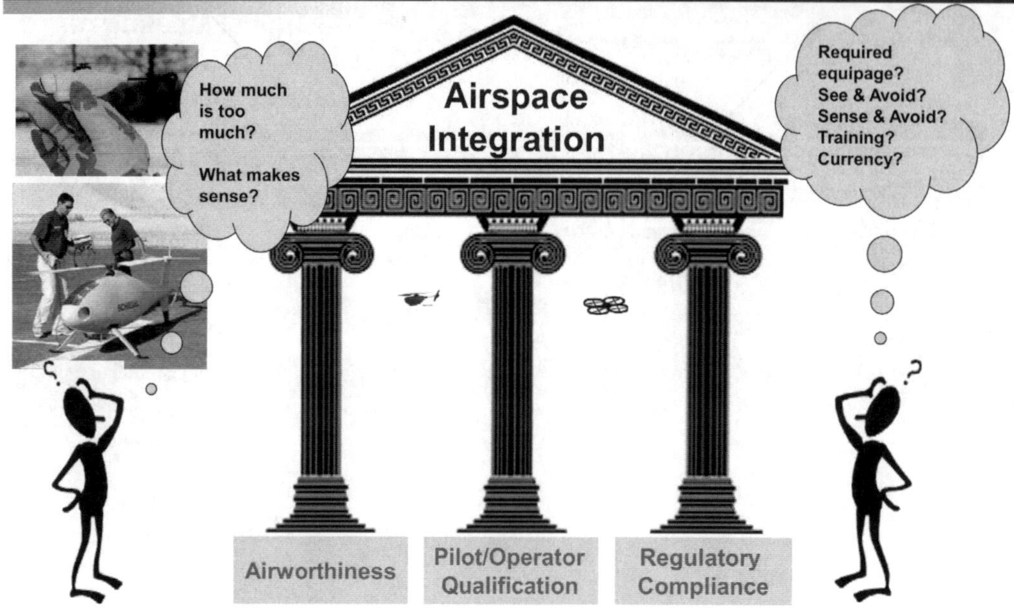

Department of Defense: Unmanned Aircraft System Airspace Integration Plan
Version 2.0, March 2011
Prepared by: UAS Task Force Airspace Integration Integrated Product Team

TECHNOLOGY DRIVEN. **WARFIGHTER FOCUSED.**

3

Airworthiness Cert Operations

- **Aircraft that have an Airworthiness certification will operate with less or no restrictions (equipage and regulations dependent)**
  - **Equivalent Level of Safety (ELOS) presented to other aircraft**
  - **Operations in all classes of airspace benefits Commercial and Public Use UAS operators**
  - **Operates over all population densities to include open air assemblies by presenting ELOS 3rd parties on the ground**

TECHNOLOGY DRIVEN. **WARFIGHTER FOCUSED.**

4

No Airworthiness Cert Operations

- Aircraft that do not have an Airworthiness Certification can be safely operated on a limited basis: (Risk based approach)
  - Must not present a navigational hazard to other aircraft
    - Airborne navigational hazards (risks) mitigated by:
      - ✓ Segregated airspace operations
      - ✓ UAS operating crew must provide See and Avoid
  - Must not present a hazard to 3rd parties
    - Hazards to people on the ground mitigated by:
      - ✓ Operations over sparsely populated areas or;
      - ✓ Ground space controlled via exclusion of people

TECHNOLOGY DRIVEN. WARFIGHTER FOCUSED.

5

NATO FINAS STANAG APPROACH

- NATO FINAS = Flight In Non-Segregated Air Space

Mission Statement: "To recommend and document NATO-wide guidelines to allow the cross-border operation of unmanned aerial vehicles (UAVs) in non-segregated airspace"

- Determine the *minimum* set of Airworthiness requirements (use EASA CS as a basis for the code (bilateral equivalent to the FARs))
- UAS require no less nor a greater level of safety than what is required for manned aviation
- Military Qualification/Certification typically requires different levels of Certification rigor than Civil aircraft based on operational needs
- Civil aircraft Certification does not however consider Military Utility
- The STANAGs do not consider Military Utility
- Military Utility may be considered by use of Special Conditions that the Certifying Authority approves

TECHNOLOGY DRIVEN. WARFIGHTER FOCUSED.

6

RATIFIED NATO STANAGS

- STANAG 4671 UAV Systems Airworthiness Requirements (USAR)
  - Large Fixed wing UAS 150kg to 20,000kg based on CS/FAR-23 and CS/FAR-25
  - In use by AU, DE, ES, FR, GB, IL, IT, and US due to maturity of the code
- STANAG 4702 USAR-RW
  - Large Rotor Wing UAS 150kg to 3,175kg based on CS/FAR-27 and CS/FAR-29
- STANAG 4703 USAR-LFW
  - Light Fixed Wing UAS < 150kg to 66Joules of energy not closely CS/FAR based
  - Architecturally and technology simple, broad configuration variety of possible in this category of UAS

TECHNOLOGY DRIVEN. WARFIGHTER FOCUSED.

7

DRAFT Allied Engineering Publication -89 (AEP-89)

- Unmanned Aerial Vehicle (UAV) Systems Airworthiness Requirements (USAR) for Light Vertical Take Off and Landing (VTOL) Aircraft
- Not closely CS/FAR based
- UAVs < 150kg to 66Joules of energy
- Large variety of possible configurations and technology in this category of UAS
- Most of these systems will be architecturally simple, broad configuration variety of possible in this category of UAS

TECHNOLOGY DRIVEN. WARFIGHTER FOCUSED.

8

U.S. ARMY RDECOM

TERMS OF REFERENCE (TORs)

USAR Vertical Take Off and Landing Light Specialists Team (UVLST)

AMRDEC

- DRAFT AEP-89 is based on STANAG 4703
- Maximum harmonization with STANAG 4703 required
- Light as possible in the amount of certification evidence required
- Rigorous and complete in design attributes which may endanger safety
- Flexible and non prescriptive, not limit design solutions
- Code must address all design attributes which may endanger safety
- Based on the Essential requirements for Airworthiness as required by Regulation (EC) No 216/2008 of the European Parliament and of the Council of 20 February 2008 on common rules in the field of civil aviation and establishing a European Aviation Safety Agency Article 5 Annex I

TECHNOLOGY DRIVEN. **WARFIGHTER FOCUSED.**

FileName.pptx

9

U.S. ARMY RDECOM

Snapshot of the Code

AMRDEC

NOTE

DETAILED ARGUMENTS: Compliance with the Essential Requirements may be shown by the Applicant through these detailed arguments or by any other argument which meets the intent behind them with a comparable level of safety, to be as agreed by the Certifying Authority, wherever a "should" statement appears.

AIRWORTHINESS ESSENTIAL REQUIREMENTS	*DETAILED ARGUMENTS*	*MEANS OF EVIDENCE*
ER.1 System integrity System integrity must be assured for all anticipated flight conditions and ground operations for the operational life of the UAS. Compliance with all requirements must be shown by assessment or analysis, supported, where necessary, by tests.	*UL.0 The Applicant must identify the design usage spectrum as the set of all the foreseen operational conditions of the UAS:* *- typical design missions;* *- in-flight operation conditions;* *- on-ground operation conditions;* *- operational modes (automatic, speed-hold, altitude hold, direct manual, etc.);* *- take-off / launch / ramp conditions;* *- landing / recovery conditions;* *- locations and platforms (e.g. land vehicle, water vessel, aircraft, building, etc.) from which launch, command and control, and recovery operations will be performed (e.g., land, littoral/maritime, air);* *- number of air vehicles to be operated simultaneously;* *- transport conditions (define the transportation and storage environment of the UAS e.g. bag, package, truck or whatever is required);* *- operating environmental conditions:* *- natural climate (altitude, temperature, pressure, humidity, wind, rainfall rate, lightning, ice, salt fog, fungus, hail, bird strike, sand and dust, etc.);* *- electromagnetic environmental effects (electromagnetic environment among all sub-systems and equipment, electromagnetic effects caused by external environment, electromagnetic interference among more than one UAS operated in proximity);* *- lighting conditions (e.g., day, night, dawn, dusk, mixed, etc.);* *- identify all the possible mass configurations (minimum and maximum flying weight, empty CG, most forward CG, most rearward CG must be identified).* *In all the identified conditions the Applicant must verify to the satisfaction of the Certifying Authority the requirements of the following paragraphs.*	*ME0 Description of the design usage spectrum*

DRAFT

Essential Requirements

The Detailed Arguments address how the code breaks down the ER's and what's required to fulfill it

The Means of Evidence presents an acceptable set of evidence that may be provided to the Certifying Authority in order to demonstrate compliance with the detailed arguments

TECHNOLOGY DRIVEN. **WARFIGHTER FOCUSED.**

FileName.pptx

10

U.S. ARMY RDECOM

Snapshot of Code ANNEX

AMRDEC

ANNEX B TO
AEP-83

LANDING GEAR

UL.GL.1 Shock absorption tests

The landing inertia load factor and the reserve energy absorption capacity of the landing gear must be substantiated by the tests prescribed in UL.GL.2 and UL.GL.3, respectively or by analysis. These tests must be conducted on the complete rotorcraft or on undercarriage units in their proper relation.

UL.GL.2 Limit drop test

The limit drop test must be conducted as follows:

(a) The drop height must be 0.20 m (8 inches) from the lowest point of the landing gear to the ground; or
(b) If considered, the rotor lifting specified in UL.6.2 (a) must be introduced into the drop test by appropriate energy absorbing devices or by the use of an effective mass.
(c) Each landing gear unit must be tested in the attitude simulating the landing condition that is most critical from the standpoint of the energy to be absorbed by it.
(d) When an effective mass is used in showing compliance with sub-paragraph (b) the following formula may be used instead of more rational computations:

where:

$$W_e = W\frac{h + (1 - L)d}{h + d}$$

$$n = n_j\frac{W_e}{W} + L$$

DRAFT

The Annexes are designed to present an acceptable means of compliance but not the only means of compliance.

11

TECHNOLOGY DRIVEN. **WARFIGHTER FOCUSED.**

FileName.pptx

U.S. ARMY RDECOM

Flexibility is the Key!

AMRDEC

DRAFT AEP-89

Requirements (page 1-3)

"The Applicant should follow the Detailed Arguments. Nevertheless, if it is difficult for a particular application to comply with the detailed request, the Applicant may propose to the Certifying Authority *compelling alternative arguments with a rational demonstration that a comparable level of safety is assured."*

12

TECHNOLOGY DRIVEN. **WARFIGHTER FOCUSED.**

FileName.pptx

REQUEST FOR INPUT

- During the development of all of the UAS STANAGs Industry has been invited to review them and submit comments or questions to the Code development Specialists Teams
- This typically takes place before the code is finalized and presented to NATO for Ratification
- The Specialists Team reviews the comments/questions
- Comments/questions are categorized, adjudicated and the Team authors a written response
- The Team issues the response to Industries that provided input and anyone else requesting it
- Within the next year DRAFT AEP-89 will be at a state of maturity to request review by Industry
- In the United States a NOTICE is posted in the Federal Register or sent directly to Industry if they request it
- In the past comment from US Industry has been light compared to comment from European Industry

TECHNOLOGY DRIVEN. WARFIGHTER FOCUSED.

13

Questions?

QUESTIONS?

TECHNOLOGY DRIVEN. WARFIGHTER FOCUSED.

14

U.S. ARMY RDECOM

AMRDEC

AMRDEC Web Site
www.amrdec.army.mil

Facebook
www.facebook.com/rdecom.amrdec

YouTube
www.youtube.com/user/AMRDEC

Twitter
@usarmyamrdec

Public Affairs
AMRDEC-PAO@amrdec.army.mil

15

TECHNOLOGY DRIVEN. ***WARFIGHTER FOCUSED.***

FileName.pptx

VTUAV Autonomous Operations and Collaboration in the Maritime Domain

THE VALUE OF PERFORMANCE.

NORTHROP GRUMMAN

AHS VTOL UAS 24 Jan 2017

Dan Redman

Manager, Advanced Autonomous Systems, NGAS

Fire Scout

NORTHROP GRUMMAN

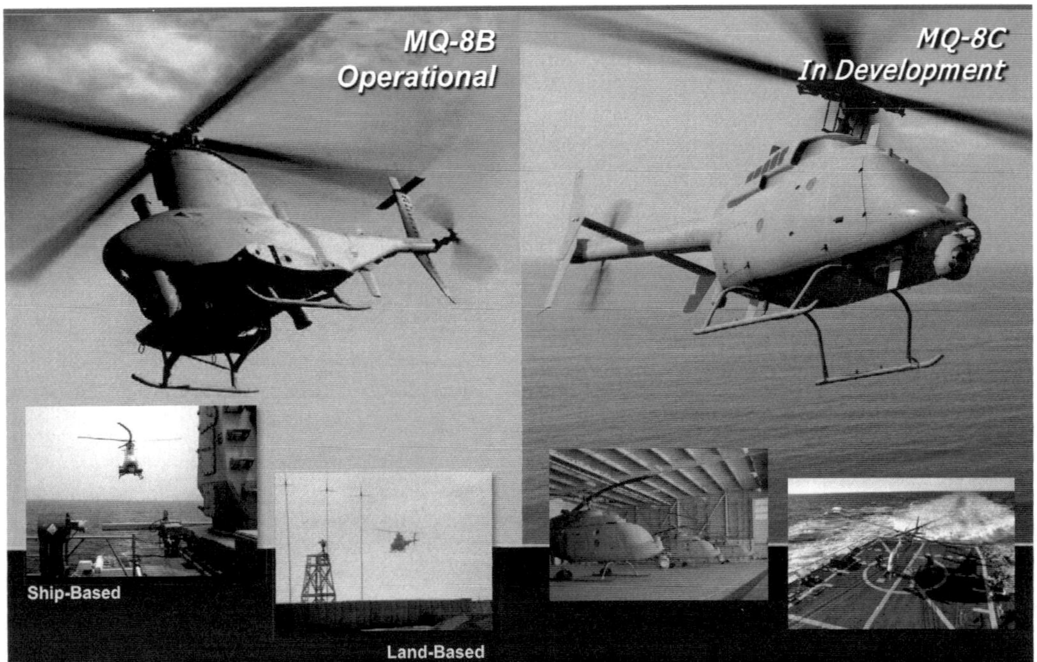

2

Approved for Public Release, Distribution Unlimited: NAVAIR 2016-92; Northrop Grumman 16-0100

Fire Scout Description

NORTHROP GRUMMAN

Mission: To provide ship-based and/or land-based intelligence, surveillance, target acquisition, reconnaissance (ISTAR) to tactical users

Providing:

- Unmatched Tactical ISR Persistence
- Complement to manned operations
- Increased mission effectiveness
- Reduced overall cost per flight hour

3

MQ-8C System Evolution

NORTHROP GRUMMAN

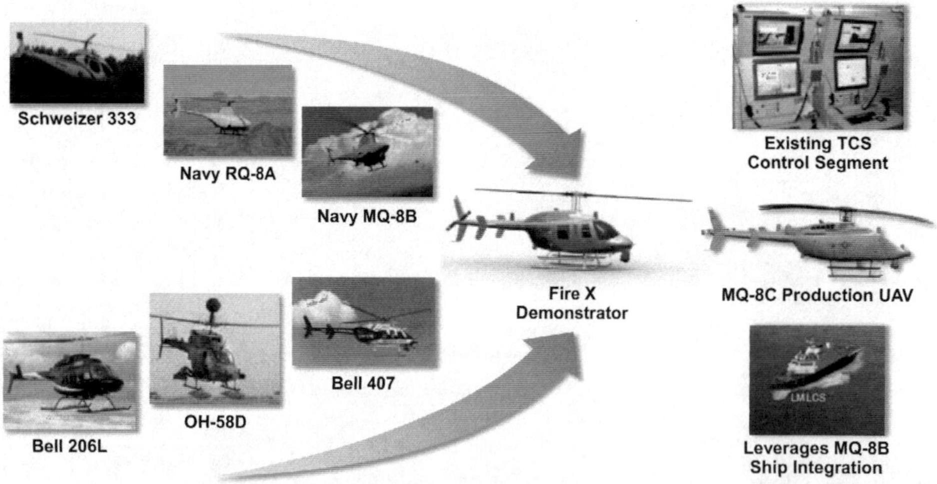

Leverages current Navy program of record and over 10 years of development

4

MQ-8 UAV
MQ-8B/MQ-8C Comparison

NORTHROP GRUMMAN

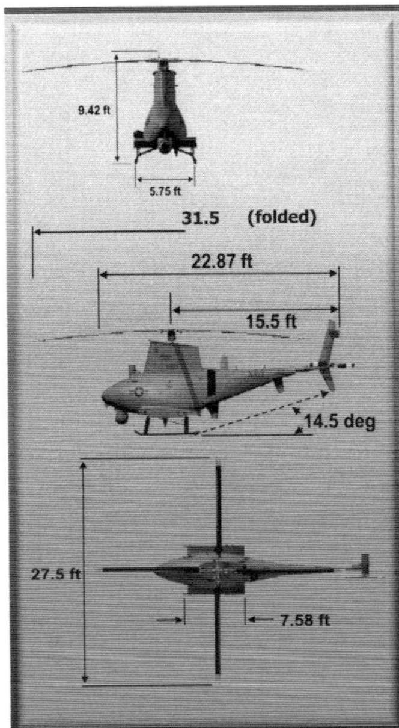

MQ-8B	Parameter	MQ-8C
85 kts	Maximum Speed	135 kts
80kts	Cruise Speed	115kts
12,500 ft	Service Ceiling	16,000ft
5.5 hrs	Std Day Maximum Endurance (with 300lb payload)	12 hrs
4.5 hrs	Hot Day Maximum Endurance (with 300lb payload)	10 hrs
2,000 lbs	Empty Weight	3,200 lbs
3,150 lbs	Std Day Fuel & Payload	6,000 lbs
31.5 ft	Length (folded)	34.7 ft

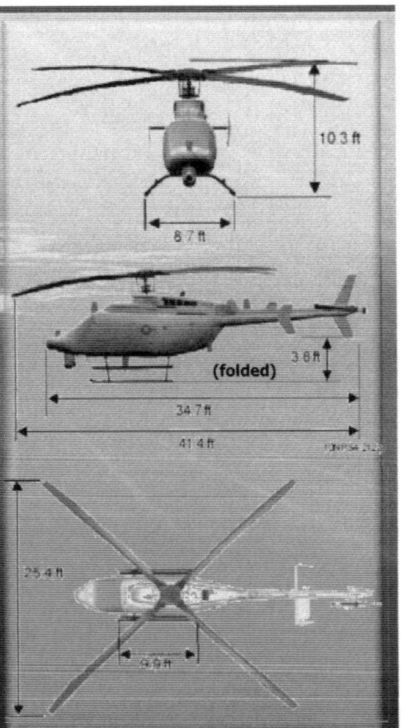

MQ-8C: 3 ft Longer (folded), 1 ft Taller, 2.5 ft Wider & 1200# Heavier than MQ-8B

5

Distribution Statement A – Approved for public release; distribution is unlimited, as submitted under NAVAIR Public Release Authorization 2013-687

MQ-8C Fire Scout Testing at Point Mugu, CA

NORTHROP GRUMMAN

MQ-8C Fire Scout First Flight
Point Mugu, CA – 31 October 2013

NORTHROP GRUMMAN

MQ-8C Fire Scout Slopes Testing
Point Mugu, CA

NORTHROP GRUMMAN

MQ-8C Fire Scout on USS Dunham (DDG-109) December 2014

NORTHROP GRUMMAN

MQ-8C Fire Scout Accomplishment Summary

NORTHROP GRUMMAN

MQ-8C aboard DDG-109 during Dynamic Interface Testing

MQ-8C Fire Scout demonstrates a long range, long endurance flight part of a capability based test at Naval Base Ventura County, Pt Mugu

Apr 2012	First Contract Awarded 2 Test and 17 Production Air Vehicles
Oct 2012	Preliminary Design Review successfully conducted
Jan 2013	Critical Design Review successfully conducted
Oct 2013	First MQ-8C Flight
Jul 2014	Electromagnetic (E3) testing complete
Dec 2014	DDG Dynamic Interface completed aboard DDG-109
Apr 2015	Developmental Test Flight Test completed (345 Sorties; >525 Flight Hours)
Aug 2015	Long Range/Long Endurance Flight (150NM/11 hrs.)
Sep 2015	MQ-8C Capabilities Based Testing completed (4 Sep)
Sep 2015	U.S. Navy accepted the first production MQ-8C at Point Mugu, CA
Dec 2015	18 MQ-8Cs accepted by the U.S. Navy

1st Contract Award to 1st Flight – 18 Months!

10

Fire Scout Missions

NORTHROP GRUMMAN

- Maritime ISR
- Surface Warfare
- Mine Counter Measures
- Anti-Submarine Warfare
- Anti-Piracy
- Counter Drug Ops
- Logistics
- Communications Relay
- Search & Rescue
- Littoral ISR
- Special Operations support
- Humanitarian Assistance / Disaster Relief

11

Summary

NORTHROP GRUMMAN

- Persistent Intelligence, Surveillance, Reconnaissance (ISR) and targeting
- Mature, reliable and maintainable systems and airframe
- On-station now (MQ-8B)
  - 16,477 Hours; 6,210 Sorties
- Increased endurance (MQ-8C)
  - 946.6 Hours; 558 Sorties
- Will complement and reduce the burden on manned systems
- Designed to operate from all air capable ships and shore facilities
- Low risk

MQ-8 Fire Scout program supports the Littoral Combat Ship (LCS) and other suitably equipped air capable ships (SEACS)

12

THE VALUE OF PERFORMANCE.

NORTHROP GRUMMAN

ANTX 2016 Demonstration Autonomous Collaboration

13

NORTHROP GRUMMAN

ANTX Demonstration Overview

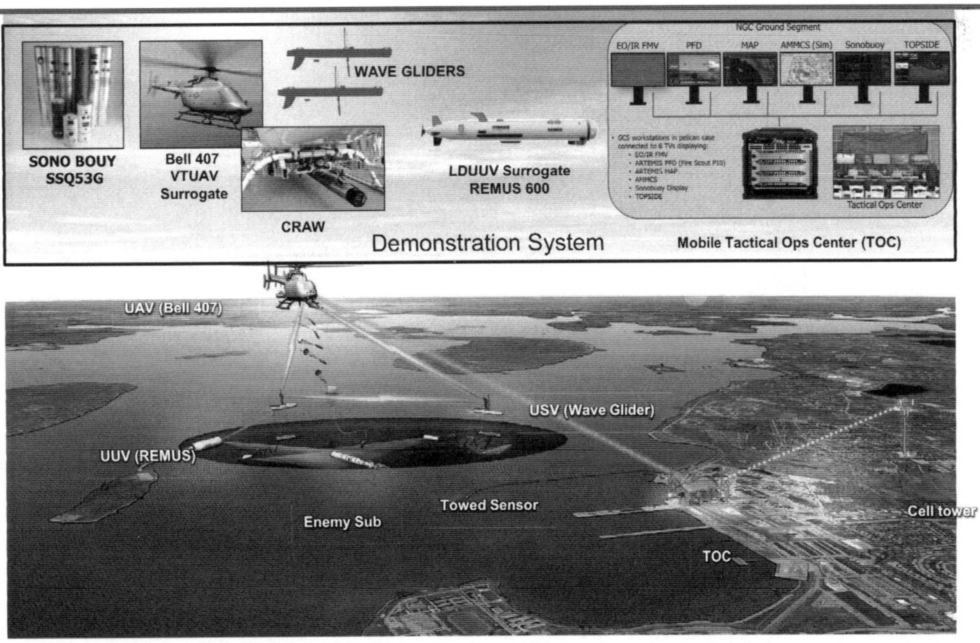

14

Heterogeneous Vehicle Interconnectivity

NORTHROP GRUMMAN

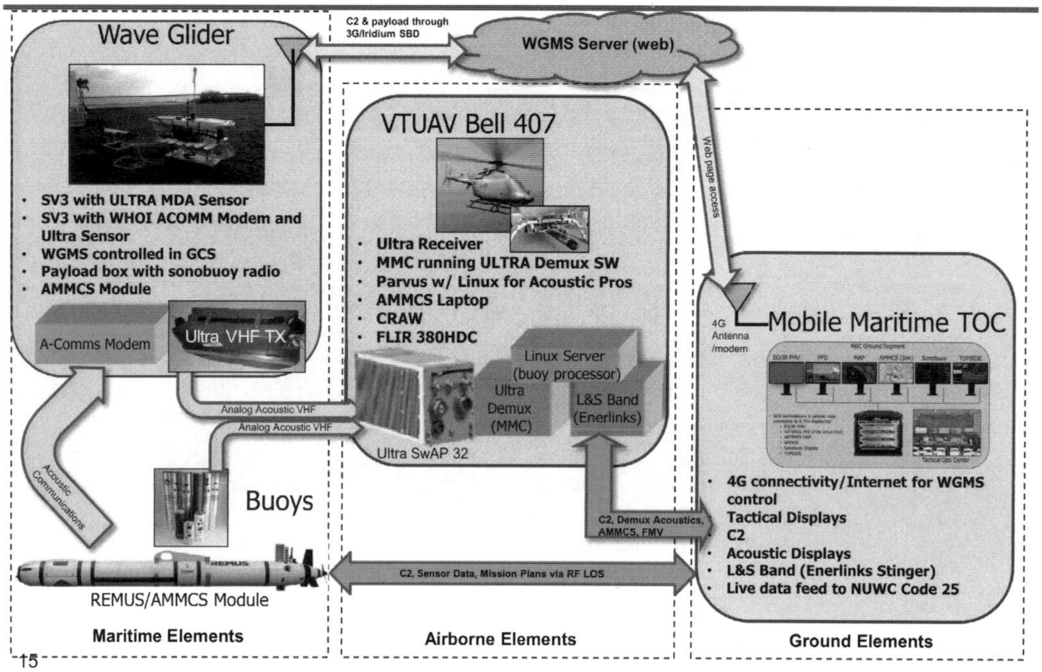

15

NORTHROP GRUMMAN

Multi-Vehicle Control via a Single Software Architecture

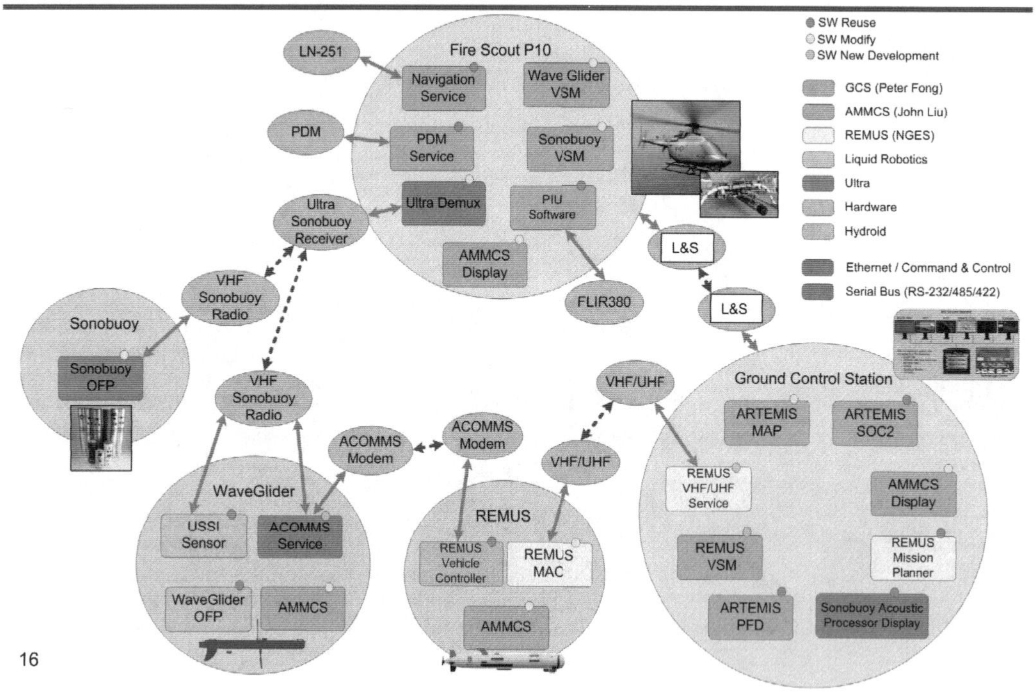

16

Choke Point ASW Barrier Scenario

NORTHROP GRUMMAN

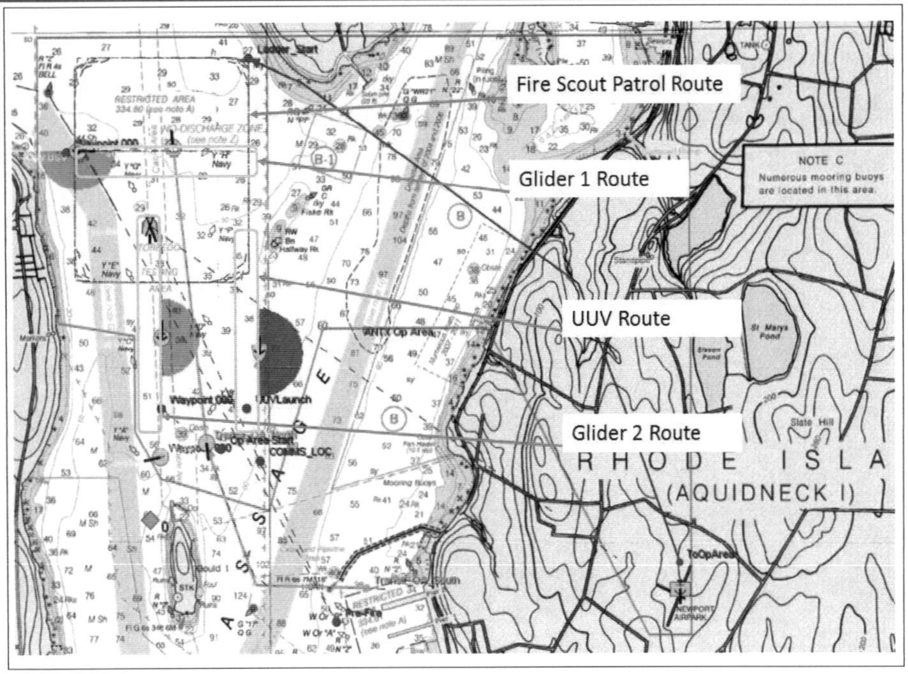

17

VTUAV Surrogate (Bell 407 armed with CRAW)

NORTHROP GRUMMAN

- Demonstrated Capabilities
  - Gross Weight 6,000 lbs
  - Range 330 Nautical Miles
  - Cruise Speed 135 MPH
  - Altitude 16 K Feet
  - CRAW
    - Electro-Mechanical release
    - Max drop altitude: 500ft

18

LDUUV Surrogate (Hydroid REMUS 600)

NORTHROP GRUMMAN

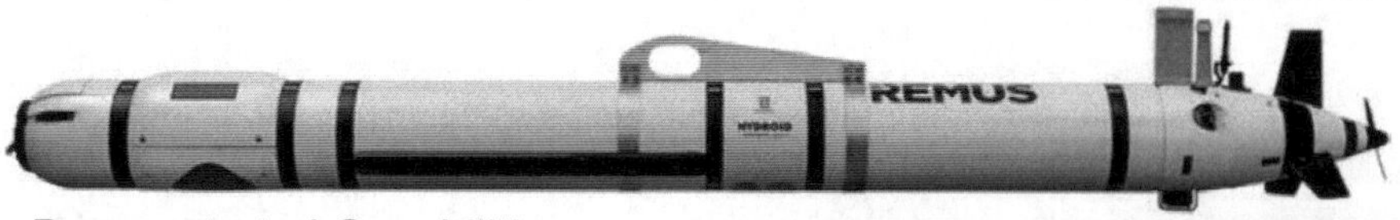

- Demonstrated Capabilities
  - Autonomous Controller
  - Vehicle Controller
  - Payload Controller
  - AMMCS Controller
  - Acoustic communications
  - RF Line of site Communications
  - NGSC 3 D Bathymetric Sensor
  - ASW Sensor
  - BWISR Sensor

19

USVs – Liquid Robotics Wavegliders

NORTHROP GRUMMAN

- Demonstrated Capabilities
  - Water Speed: 1kt to 3kts
  - Operating Water Depth: > 15m
  - Payload: BWISR, RF/ACOMMS
  - Tow Capability: Up to 500kg mass
  - Average Continuous Power: 5W – 20W
  - Max Solar Collection: 156W
  - Battery Storage: 0.9kWh – 4.5kWh
  - Communications: Satellite, Cell, Wi-Fi

20

Mobile Tactical Operations Center

NORTHROP GRUMMAN

- Demonstrated Capabilities
  - Tactical operation Center
  - Mission Planning
  - Mission Validation
  - Track Fusion
  - TOPSIDE
  - Command and Control
  - Situational Awareness
  - Live data feed to NUWC code 25

21

TOC Multi-Source Data Fusion for Targeting

NORTHROP GRUMMAN

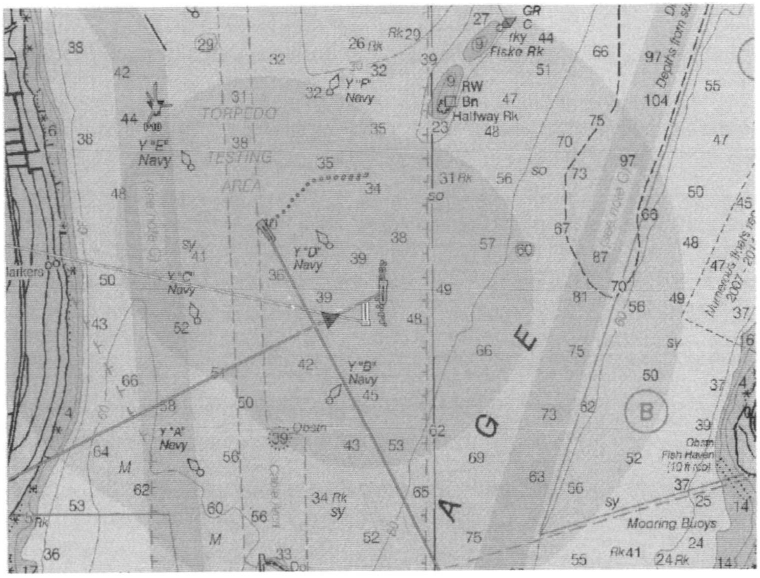

22

Heterogeneous Prosecution of an ASW Target

NORTHROP GRUMMAN

23

ANTX Takeaways

NORTHROP GRUMMAN

- "Tipping Point" for autonomous systems contribution to mission accomplishment
  - Open architecture command and control enables collaboration of heterogeneous vehicles with complementary capabilities – "best-of-breed" approach can combine features developed independently for air and maritime systems
  - Trained personnel "on-the-loop" can manage multiple vehicles operating across air, surface and subsurface domains to provide mission tasking and make key decisions
- Modeling, simulation, integration and testing are key building blocks of successful experimentation
  - ANTX plan included dedicated periods for each, including dry run events in Annapolis and Newport
- Distributed autonomy reduces reliance on robust communications and networking, but sufficient connectivity is needed to enable human oversight/decision-making
  - Significant integration time spent ensuring communications systems were robust enough to support necessary information exchange between platforms and tactical operations center
  - Balance required between trust in autonomy and need for operator oversight

24

THE VALUE OF PERFORMANCE.

NORTHROP GRUMMAN

25

Autonomy *IS* Pervasive

The Washington Post

Drone sales soaring this Christmas, capping a record year for the industry

MilitaryTimes

Pentagon successfully tests world's largest micro-drone swarm

Los Angeles Times

Why the driverless car industry is happy (so far) with Trump's pick for Transportation secretary

AINonline

'Angel One' Predator C Avenger Made Available for Aid Missions

Autonomy at Sikorsky

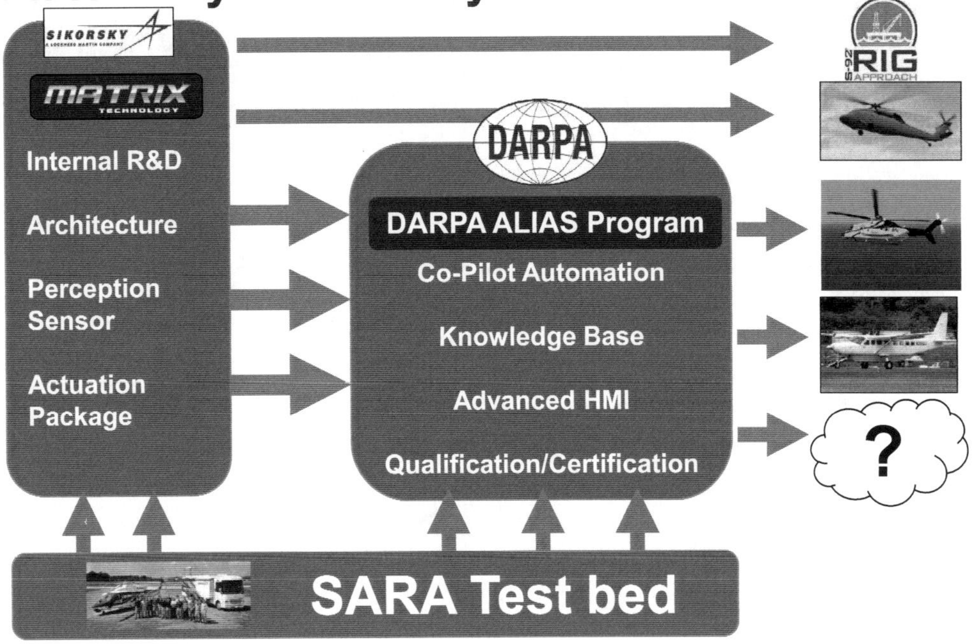

DARPA ALIAS Introduction

DARPA Aircrew Labor In-Cockpit Automation System (ALIAS)

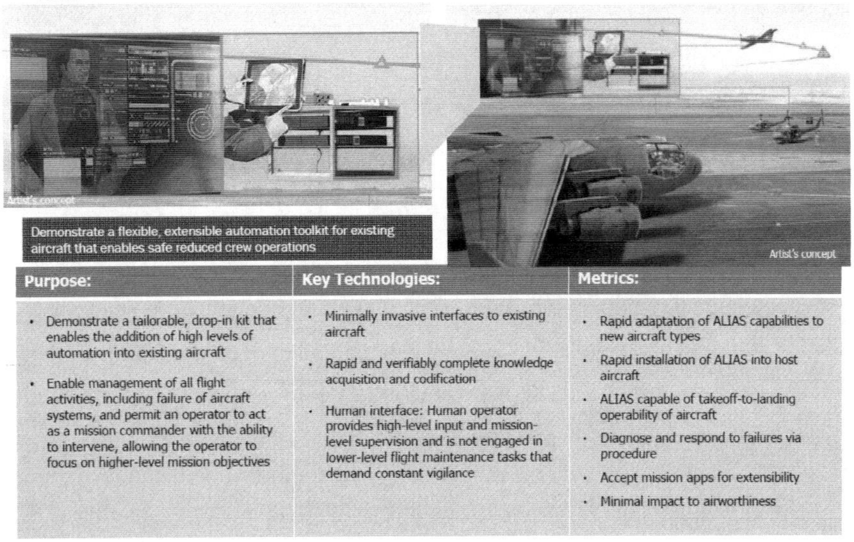

Purpose:	Key Technologies:	Metrics:
• Demonstrate a tailorable, drop-in kit that enables the addition of high levels of automation into existing aircraft • Enable management of all flight activities, including failure of aircraft systems, and permit an operator to act as a mission commander with the ability to intervene, allowing the operator to focus on higher-level mission objectives	• Minimally invasive interfaces to existing aircraft • Rapid and verifiably complete knowledge acquisition and codification • Human interface: Human operator provides high-level input and mission-level supervision and is not engaged in lower-level flight maintenance tasks that demand constant vigilance	• Rapid adaptation of ALIAS capabilities to new aircraft types • Rapid installation of ALIAS into host aircraft • ALIAS capable of takeoff-to-landing operability of aircraft • Diagnose and respond to failures via procedure • Accept mission apps for extensibility • Minimal impact to airworthiness

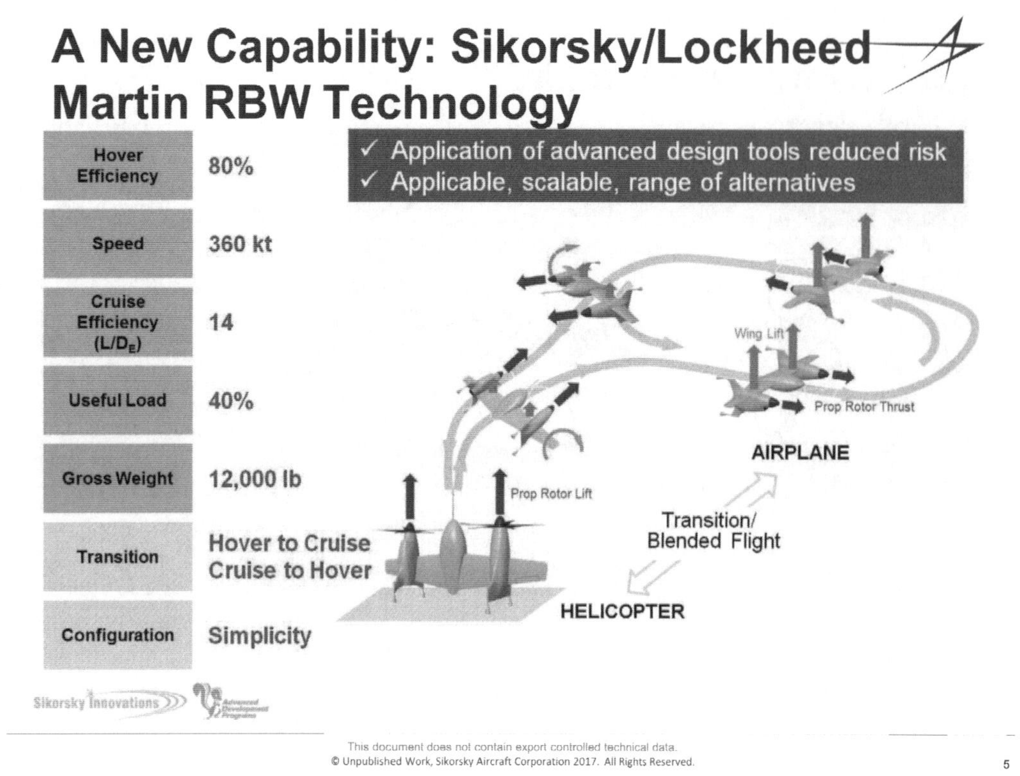
A New Capability: Sikorsky/Lockheed Martin RBW Technology
Hover Efficiency
80%
Speed
360 kt
Cruise Efficiency (L/D$_E$)
14
Useful Load
40%
Gross Weight
12,000 lb
Transition
Hover to Cruise
Cruise to Hover
Configuration
Simplicity
✓ Application of advanced design tools reduced risk
✓ Applicable, scalable, range of alternatives
Wing Lift
Prop Rotor Thrust
AIRPLANE
Prop Rotor Lift
Transition/
Blended Flight
HELICOPTER
Sikorsky Innovations
This document does not contain export controlled technical data.
© Unpublished Work, Sikorsky Aircraft Corporation 2017. All Rights Reserved.
5

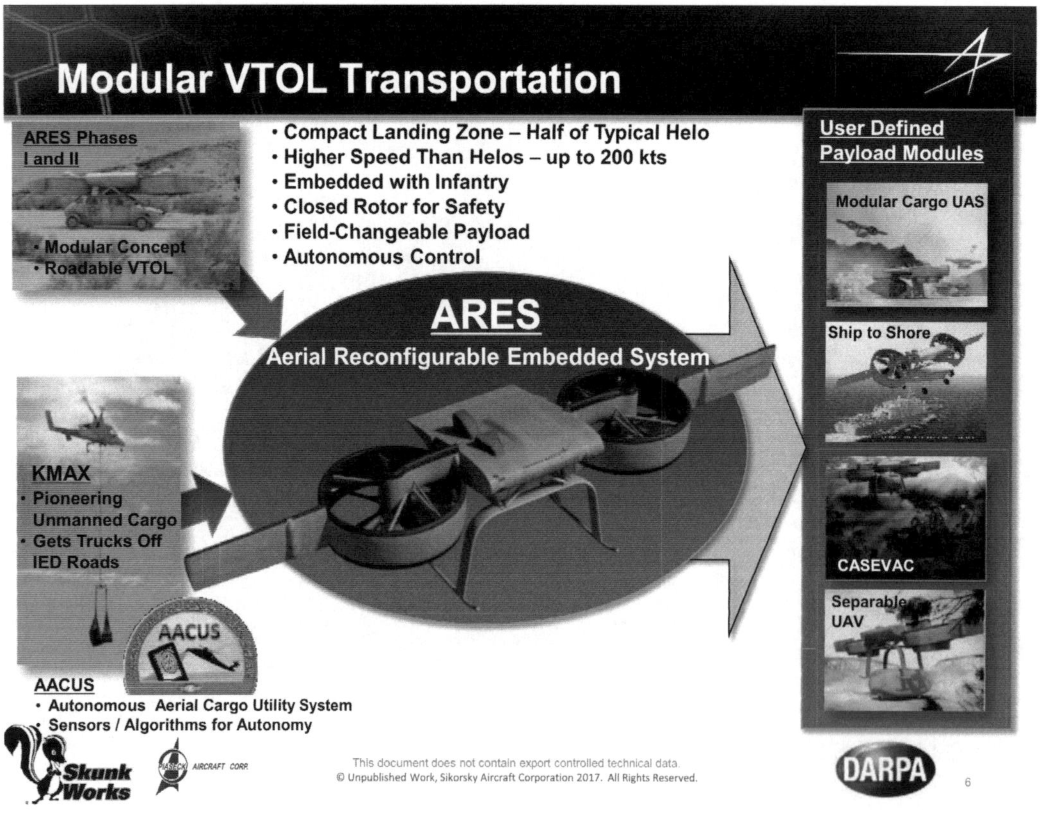
Modular VTOL Transportation
ARES Phases I and II
• Modular Concept
• Roadable VTOL
• Compact Landing Zone – Half of Typical Helo
• Higher Speed Than Helos – up to 200 kts
• Embedded with Infantry
• Closed Rotor for Safety
• Field-Changeable Payload
• Autonomous Control
ARES
Aerial Reconfigurable Embedded System
KMAX
• Pioneering Unmanned Cargo
• Gets Trucks Off IED Roads
AACUS
AACUS
• Autonomous Aerial Cargo Utility System
• Sensors / Algorithms for Autonomy
User Defined Payload Modules
Modular Cargo UAS
Ship to Shore
CASEVAC
Separable UAV
Skunk Works
AIRCRAFT CORP.
This document does not contain export controlled technical data.
© Unpublished Work, Sikorsky Aircraft Corporation 2017. All Rights Reserved.
DARPA
6

Future Autonomy

UNCLASSIFIED

US ARMY RDECOM

JMR Mission Systems Vision

AMRDEC

Crew Station
- Fully Integrated Next-Gen cockpit
- HMI Designed-In
- Advanced Controls and Displays
  - Multi-Function
  - Helmet Mounted
  - Heads up
  - Effective cueing
- EDM (Electronic Data Manager)

Flexibility
- Mission reconfigurable
- Upgradable
- Land based and maritime operations
- Tactical and peace-time operations

Navigation/Pilotage
- Integrated GPS
- Transponders
- Integrated IFF
- Decision Aiding/OPV
- Integrated DVE/CA/OA

Engagement & Effects
- Scalable effect warheads & weapons
- Directed Energy
- Counter-UAS and Air-to-Air
- Hostile Fire Detection
- Next Gen Integrated Targeting, SA, DVE sensor suite
- Decision Aiding for optimal, synchronized use of on- and off-board effects
- Maritime search, track, and identification of surface and subsurface targets
- IED Detection
- CBR Detection

Avionics Architecture
- Advanced Mission Processing
- High Speed Backplanes
- Data Concentrators
- Solid State Recording Devices
- High Speed Interconnects
- Open Systems Standards (JCA / FACE)
- Information Assurance
- Multi-level Security

JMR Aircraft TD
- Position with aircraft primes

JMR Mission System TD Architecture
- Conceptual Design of Full Architectures
- Model based approach, Modeling & Analysis, AADL Model Validations, System Architecture Virtual Integration
- Portability, interoperability, software reuse, product line approach

Communication
- LOS Voice and Data
- BLOS Voice and Data
- Integrated Tactical Networks (SRW, WNW, JTRS, etc)
- Advanced Antennas (Conformal, Multi-band, etc.)
- Interoperability with DHS
- Backwards compatibility

Situation Awareness
- Real-time information (threat, weather, a/c state, BC, etc.)
- 360 Spherical Sensing
- Information Mgmt
- Data Fusion, Decision Aiding
- COP (GIG, BFT, etc.)

MUM-T
- LOI 4 / LOI 2
- Decision Aiding
- Wingman

Survivability
- Signature Management / Suppression
- Survivability SA and Planning
- Advanced Countermeasures
- Tunable Pyrotechnics
- Hostile Fire Indication
- IASE
- Advanced Warning Receivers
- EMP Hardening

We must identify the Key concepts and technologies to pursue in the Tech Demo effort

TECHNOLOGY DRIVEN. WARFIGHTER FOCUSED.

On the Horizon: USMC MUX Program

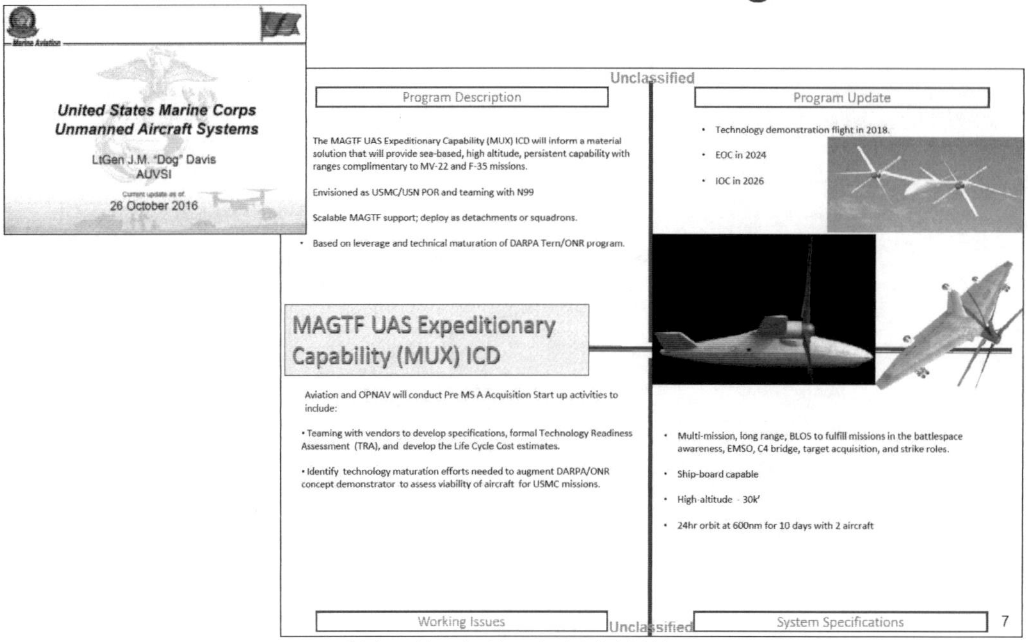

9

Army's Segmented UAS Needs

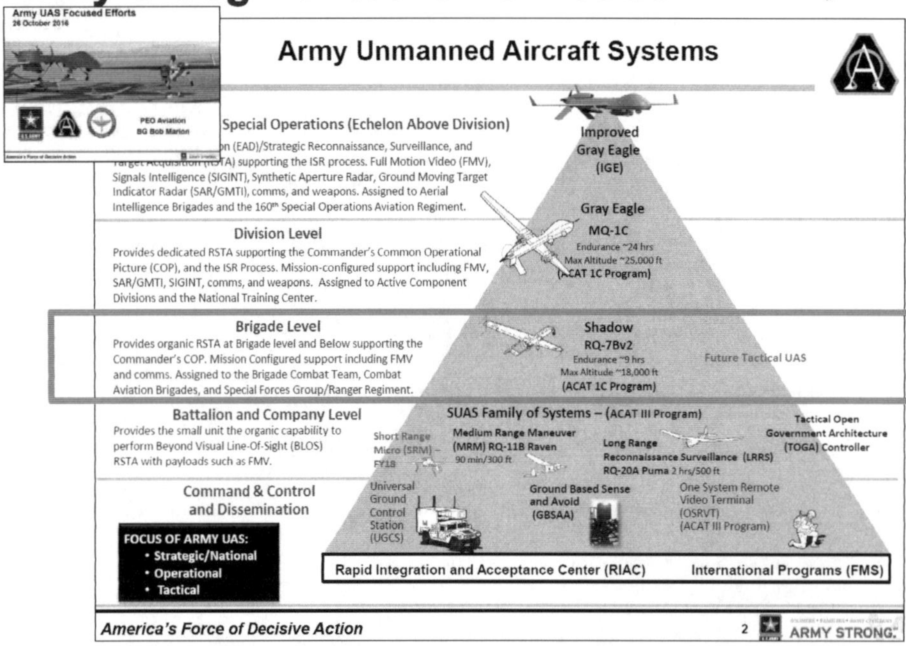

10

Future Autonomy – Beyond Part 107

Part 107 – Rule that governs the use of small UAS in National Airspace in effect **August 20th, 2016**

Drone Advisory Committee (DAC) announced May 4th, 2016

- Provides an open venue for FAA/key decision-makers for safe introduction of UAS into the NAS
- 35 members from stakeholders for two year term

Current Status: Considering forming three task groups

1. Roles and responsibilities of federal, state, and local governments
2. Technological and regulatory mechanisms to expand airspace access
3. Funding the expanded provision of services for UAS integration

BOEING

Engineering, Operations & Technology

Boeing Test & Evaluation

OPA: An Alternative Development Method for UAV Hardware and Software

"Standard" Development Process

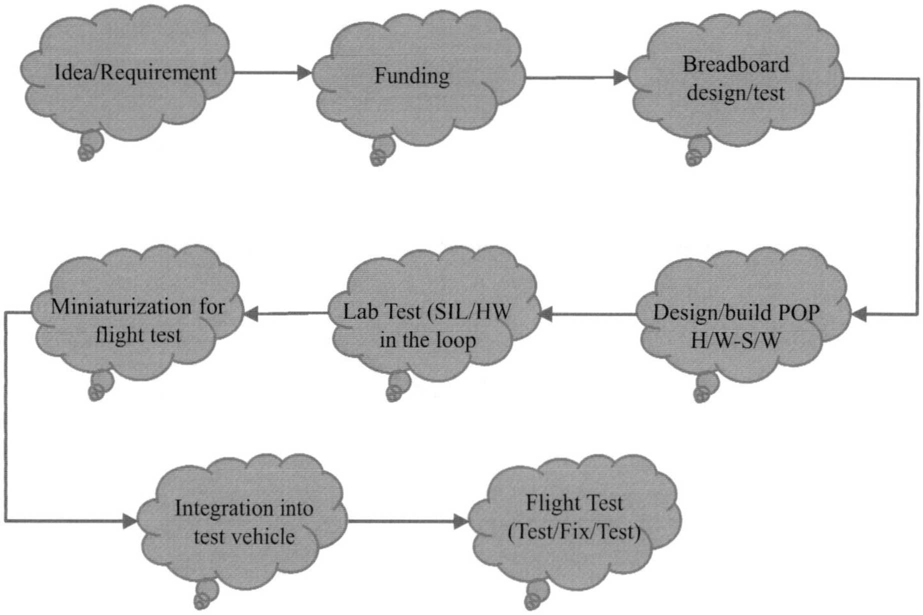

Engineering, Operations & Technology | Boeing Test & Evaluation |

OPA Development Option

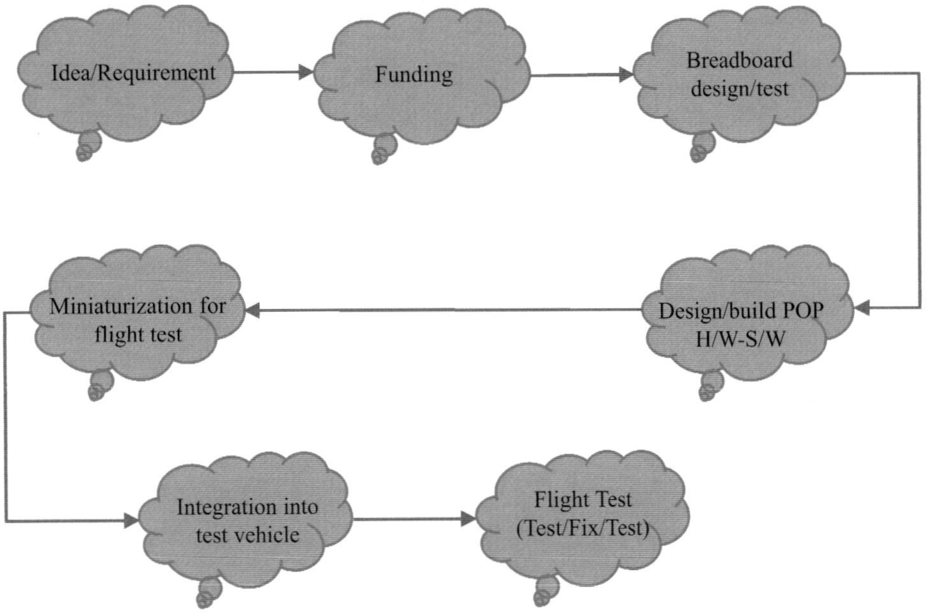

2/6/2017 | 3

Engineering, Operations & Technology | Boeing Test & Evaluation |

OPA Development Option

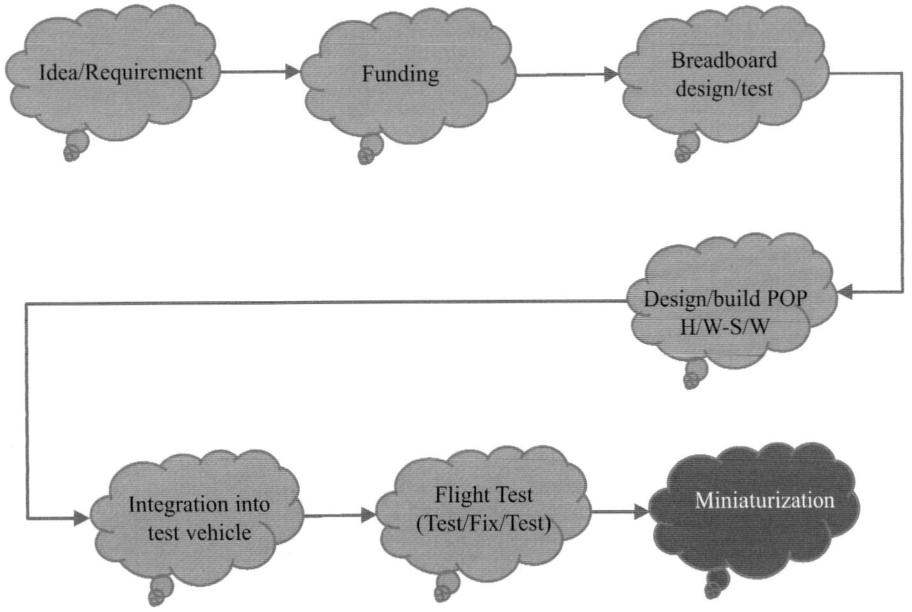

2/6/2017 | 4

Boeing's OPA Aircraft Description

What is it?

A test platform for rapid prototyping, developing and deploying technologies and capabilities for VTOL UAS.

Has an optionally piloted capability that speeds development while reducing operational risk and cost: Protects valuable test vehicle investments.

Engineering, Operations & Technology
Boeing Test & Evaluation |

History (Partial Sampling)

H-6 OPA

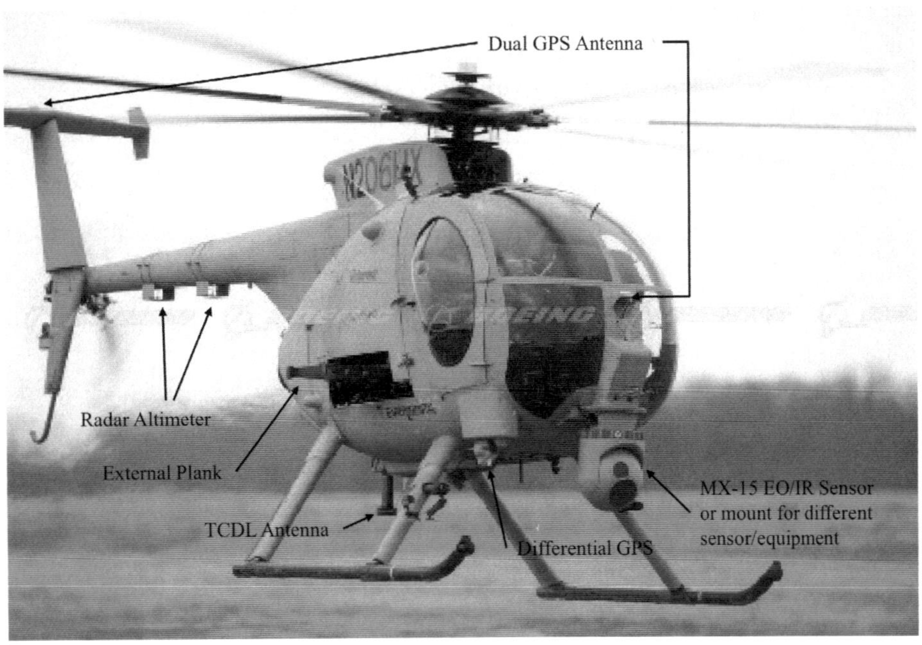

H-6 OPA

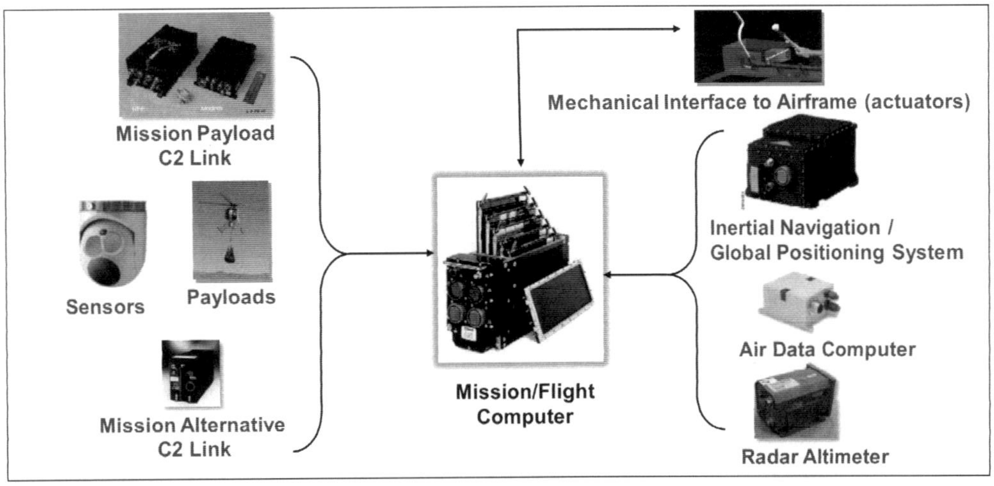

Optionally Piloted Capability Can Be a Feature for Any Aircraft

H-6 OPA

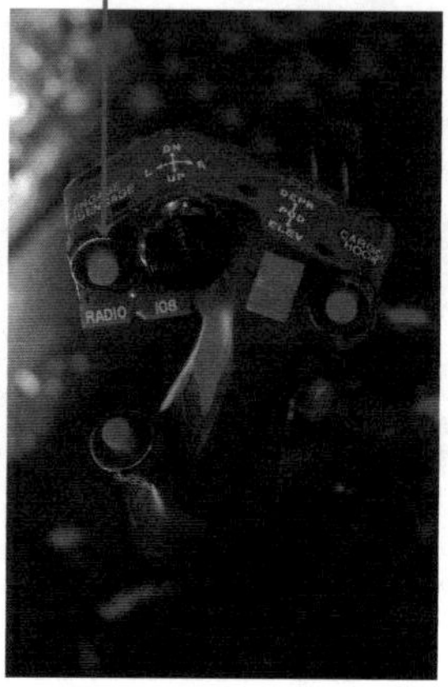

Risk

Risk = Hazard Level + Probability

UAV:

- Risk analysis is performed and risk is mitigated but risk is not confined to the test asset if testing over unpopulated area, limited to loss of asset. This could be $ or $$$$$$$.

OPA:

- Risk is analyzed in a manner similar to UAV but risk is not limited to loss of $.
- With pilot in the aircraft, risk can be minimized or eliminated by pilot actions.

H-6 OPA Risk Mitigation

Sensor/Avionics Switches

All important LRB

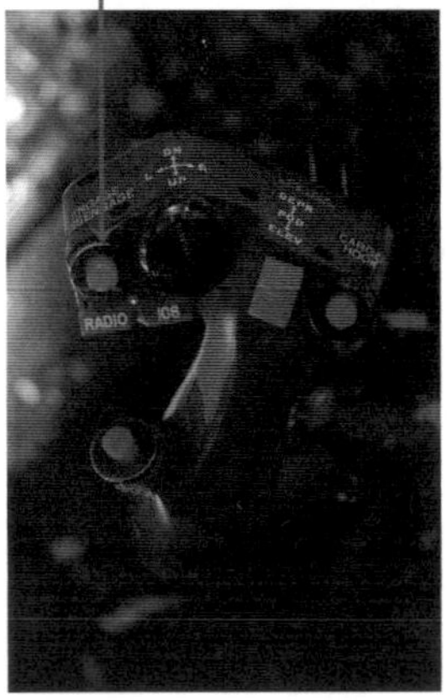

Surrogate System

OPA can be used to represent another UAV

- POP program for UAV integration into NAS
- H-6 control laws modified to represent Scan Eagle constraints (attitudes, rates, accels)
- Scan Eagle integrated so it controlled H-6 navigation via input from Scan Eagle ground station operator.

System Development

- Aircraft used in Test-Fix-Test environment for development of hardware and software.
- Often integrate more than one system for development and test.

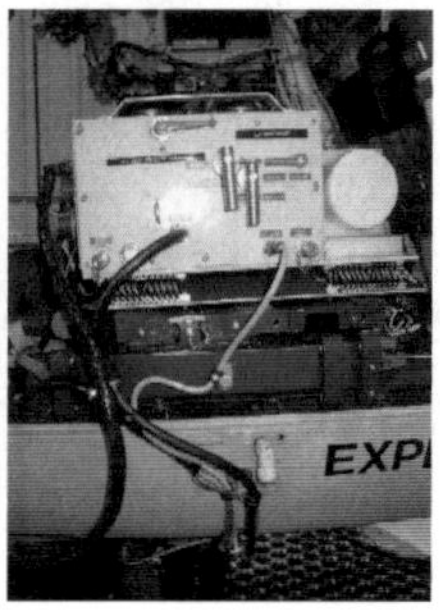

Surrogate System

Another surrogate system that included integrated LIDAR sensor, navigation, LZ evaluation, local obstacle avoidance and local short range control.

UAV Registration and Airspace Restrictions

Title 14 CFR Part 107 (29 Aug 16)

- Weight less than 55 Lbs. (22kg)
- VLOS required (camera not does not satisfy this requirement
- Max speed = 87 kts
- Max altitude = 400' AGL
- Ops in Class B, C, D, and E airspace only with ATC permission
- Establishes pilot requirements and ratings.

FAA Order 8130.34C (2 Aug 13)

- Establishes airworthiness certification requirements for three classes of "UAV's
  - UAV
  - OPA (206HX)
  - UAV/OPA
- UAV and UAV/OPA have very restrictive and difficult to achieve requirements.

Example of OPA Approved Test Area

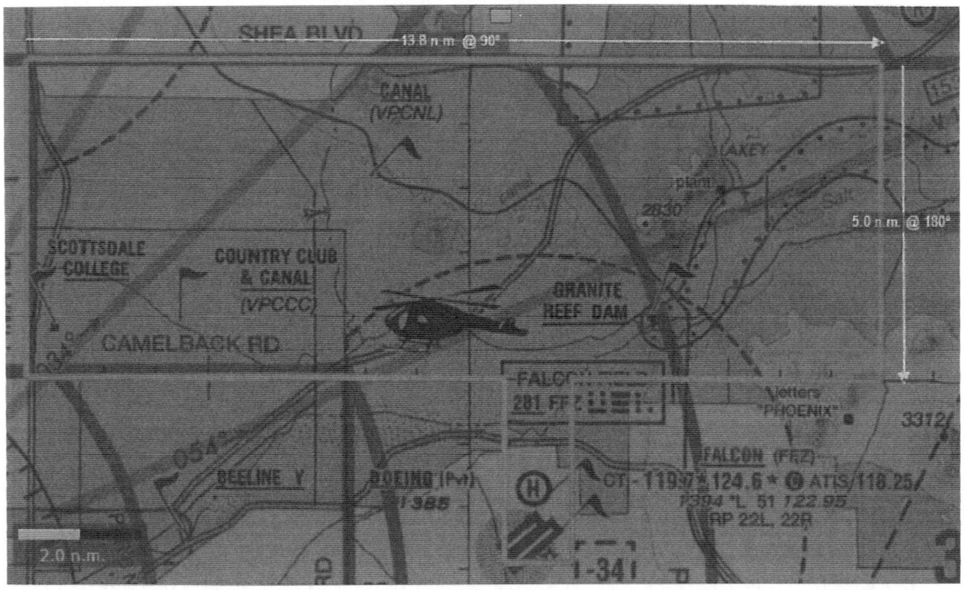

OPA flight test area just down the street!

Example of OPA Approved Test Area

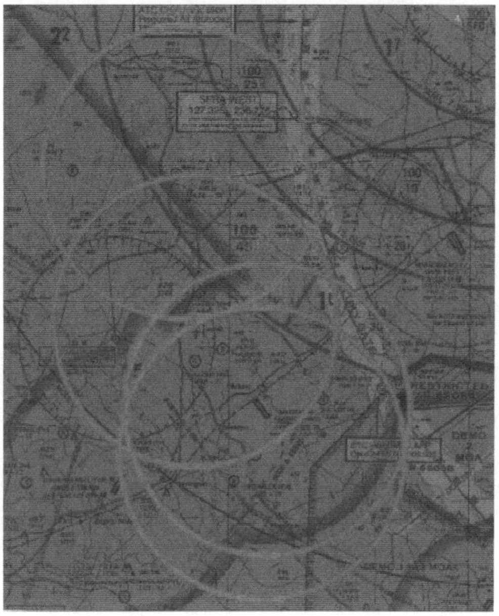

OPA flight test area just outside DC SFRA

Video: Trailer Autonomous Landing

Videos: Allure Autonomous Landing

Sampling Based Collision Avoidance for UAVs

Srikanth Saripalli
Unmanned Systems Lab
Texas A&M University

2/6/2017 1

outline

- Background
- Problem formulation
- Algorithm description
- Experiment results
- Conclusion and future work

An UAV encounters a commercial aircraft

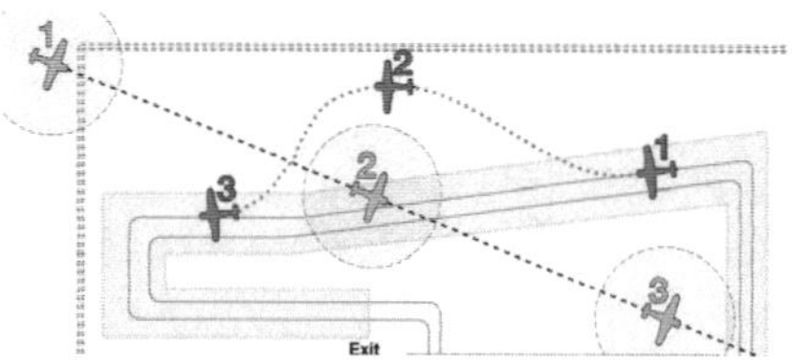

Concept Demo of Path Planning for UAV Collision Avoidance

2/6/2017 2

outline

- Background
- Problem formulation
- Algorithm description
- Experiment results
- Conclusion and future work

An UAV encounters a commercial aircraft

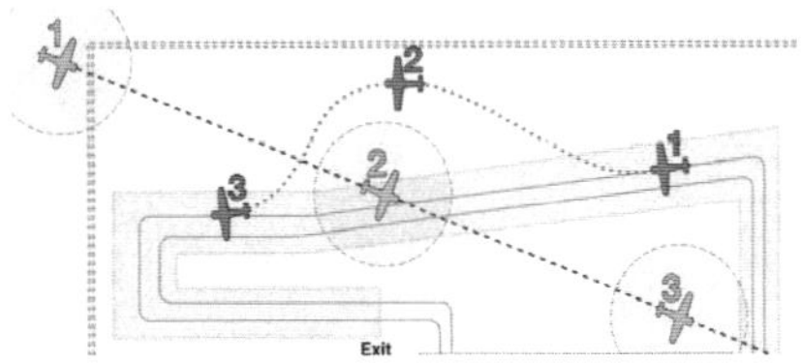

Concept Demo of Path Planning for UAV Collision Avoidance

Why is collision avoidance important for UAVs

Expanded applications of UAVs in civilian domain

Delivery

Photography

Agriculture

Infrastructure Monitor

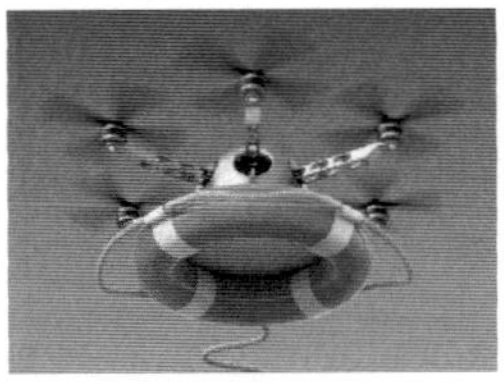

Search and Rescue

Surveying

Why is collision avoidance important for UAVs

Integrate UAVs into the national airspace:
To avoid collisions with other aircrafts is necessary!

Congested airspace

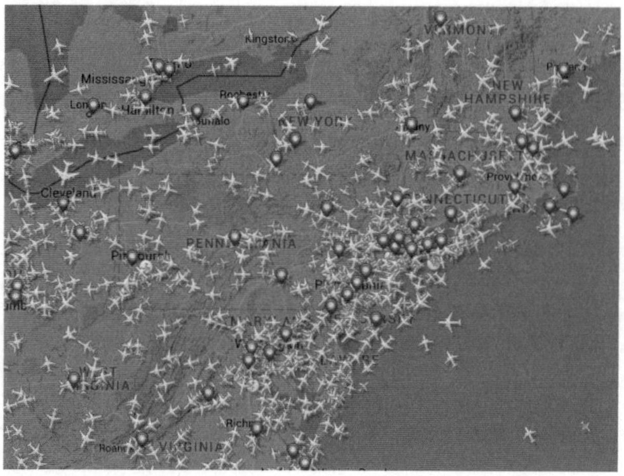

2:08pm. Jun 25. 2016. from Flightradar24.com

outline

- Background
- Problem formulation
- Algorithm description
- Experiment results
- Conclusion and future work

An UAV encounters a commercial aircraft

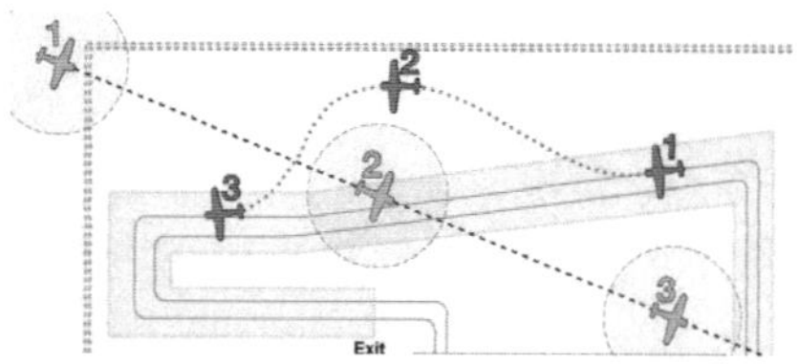

Concept Demo of Path Planning for UAV Collision Avoidance

The Sampling-Based Path Planner
for UAV Moving Obstacle Avoidance

Problem Formulation

Target: a path $f(x, y, z) = 0$

Satisfy: 1.given the start and goal

2. separation from the obstacle $\|x_{obs} - x_{uav}\| > R$

3. achieve some optimal target function $f^* = \arg\min_f \int_f q(f)ds$

4. vehicle's model $\dot{\mathbf{y}}(t) = \mathbf{f}(\mathbf{y}(t), \mathbf{u}(t)),\ \mathbf{u}(t) \in U$

5. other physical constraints: $S(\mathbf{x}(t))<0$

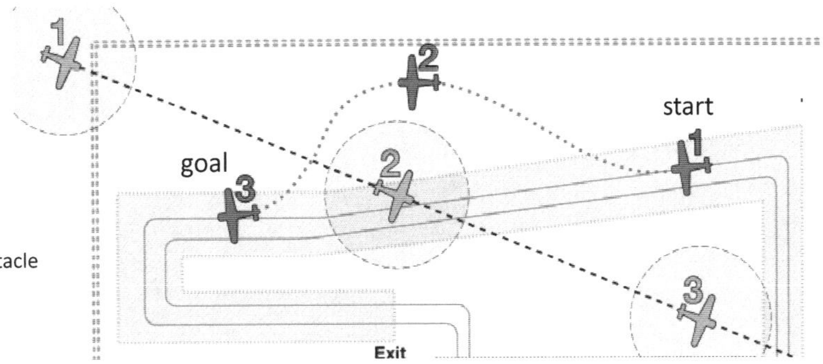

Red: UAV
Blue: obstacle

2/6/2017 7

outline

- Background
- Problem formulation
- Algorithm description
- Experiment results
- Conclusion and future work

An UAV encounters a commercial aircraft

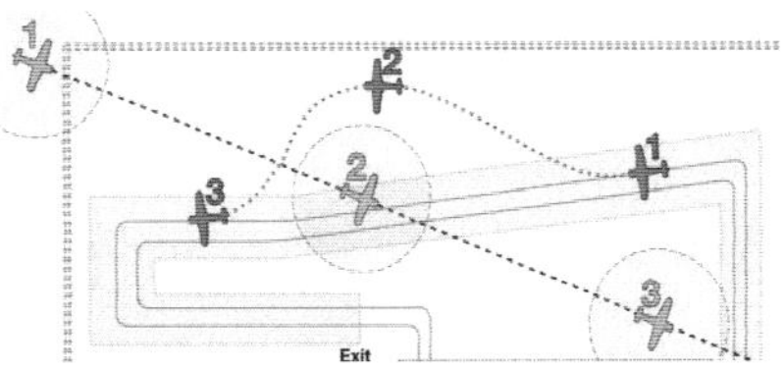

Concept Demo of Path Planning for UAV Collision Avoidance

2/6/2017 8

A Sampling Based Path Planning Algorithm: illustration

Adding a Candidate Trajectory:

$$\mathbf{x}_{O_i}(t) = \mathbf{x}_{O_i}(t_0) + \mathbf{v}(t_0) * (t - t_0)$$

Connect the sample to the start

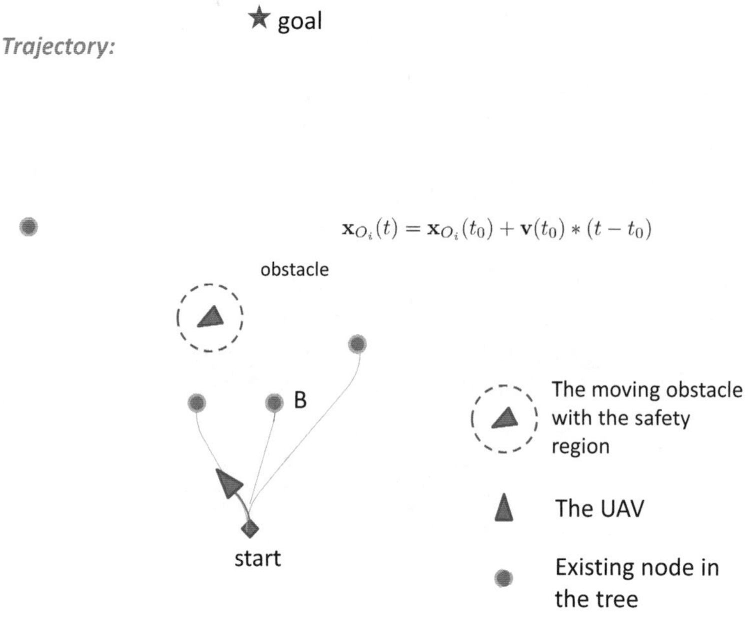

A Sampling Based Path Planning Algorithm: illustration

Adding a Candidate Trajectory:

$$\mathbf{x}_{O_i}(t) = \mathbf{x}_{O_i}(t_0) + \mathbf{v}(t_0) * (t - t_0)$$

Connect the sample to the start

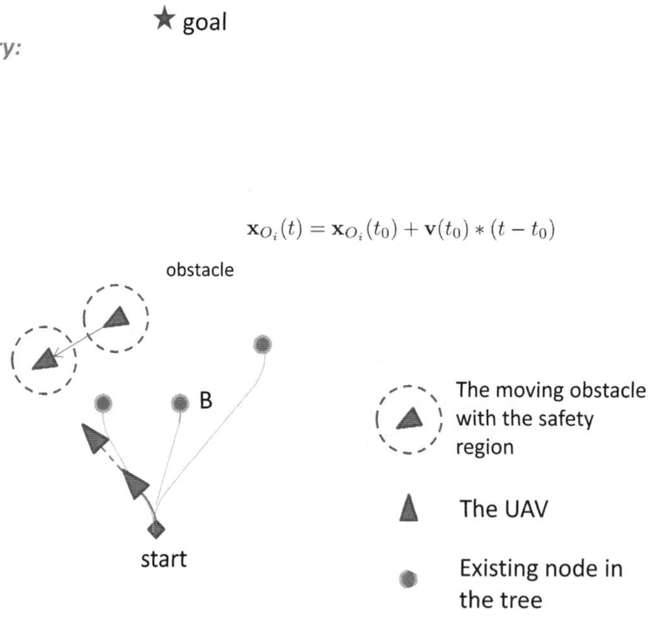

A Sampling Based Path Planning Algorithm: illustration

Adding a Candidate Trajectory:

Connect the sample to the start

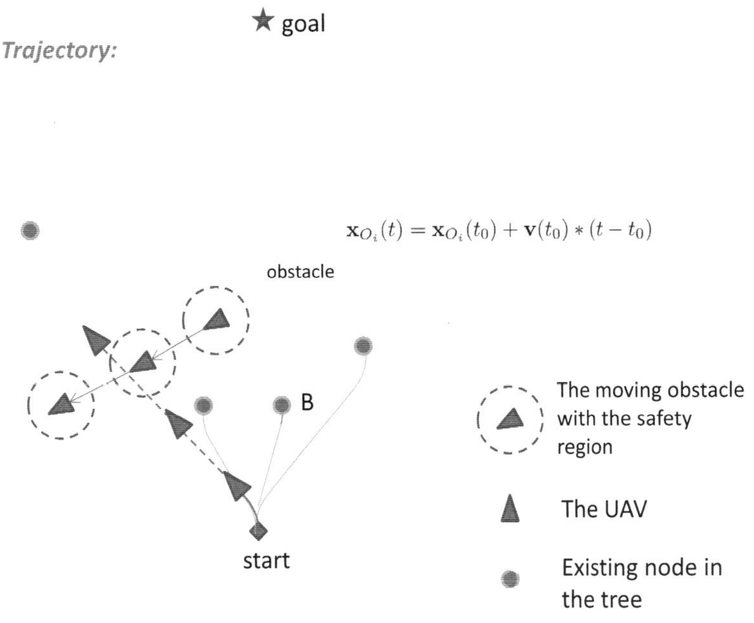

A Sampling Based Path Planning Algorithm: illustration

Adding a Candidate Trajectory:

Connect the sample to the start

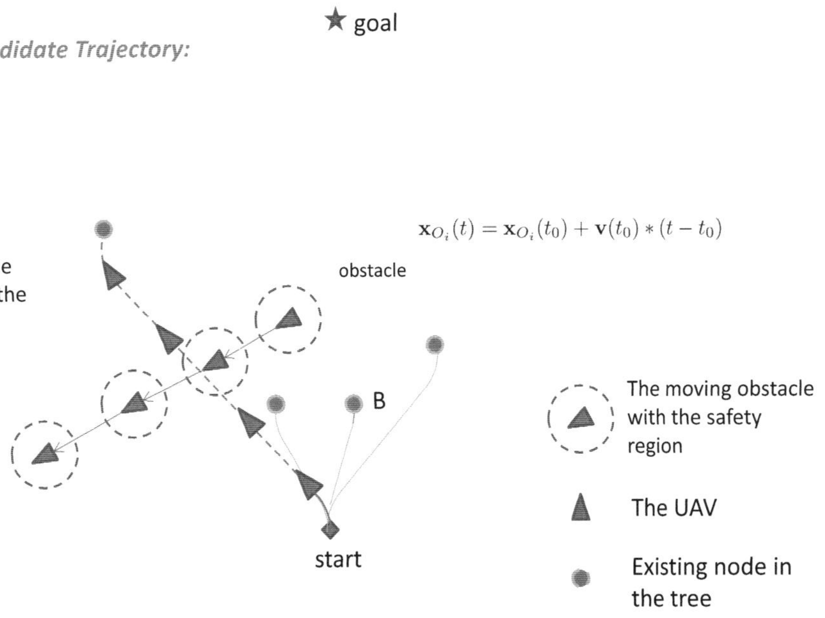

A Sampling Based Path Planning Algorithm: illustration

Adding a Candidate Trajectory:

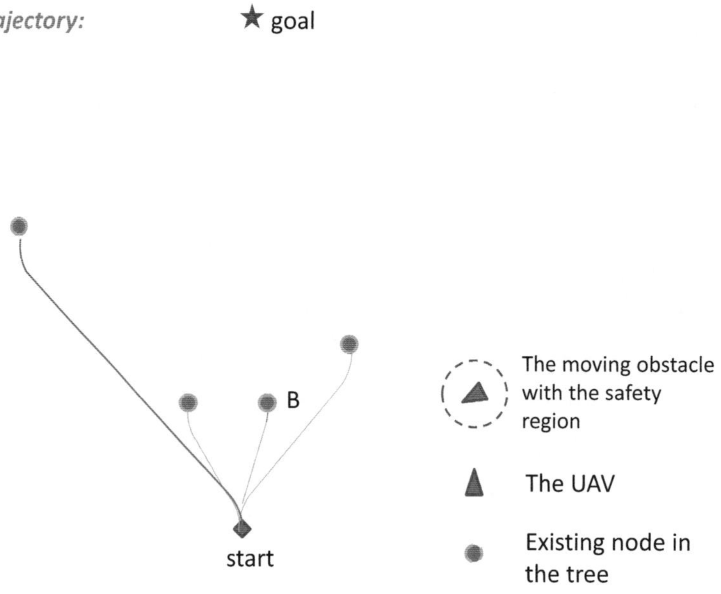

A Sampling Based Path Planning Algorithm: illustration

Adding a Candidate Trajectory:

Connect the sample to the goal

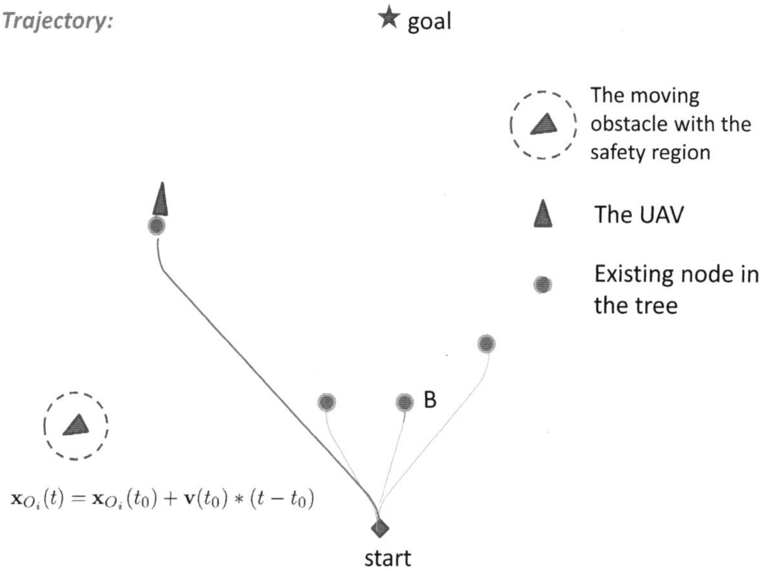

A Sampling Based Path Planning Algorithm: illustration

Adding a Candidate Trajectory:

The moving obstacle with the safety region

The UAV

Existing node in the tree

Connect the sample to the goal

$$\mathbf{x}_{O_i}(t) = \mathbf{x}_{O_i}(t_0) + \mathbf{v}(t_0) * (t - t_0)$$

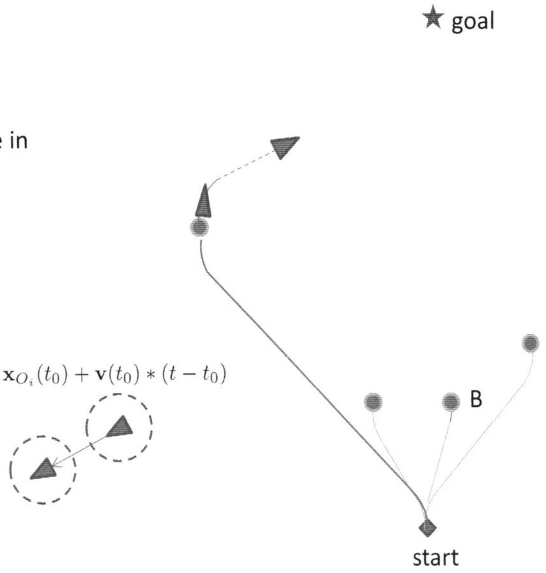

2/6/2017 15

A Sampling Based Path Planning Algorithm: illustration

Adding a Candidate Trajectory:

The moving obstacle with the safety region

The UAV

Existing node in the tree

Connect the sample to the goal

$$\mathbf{x}_{O_i}(t) = \mathbf{x}_{O_i}(t_0) + \mathbf{v}(t_0) * (t - t_0)$$

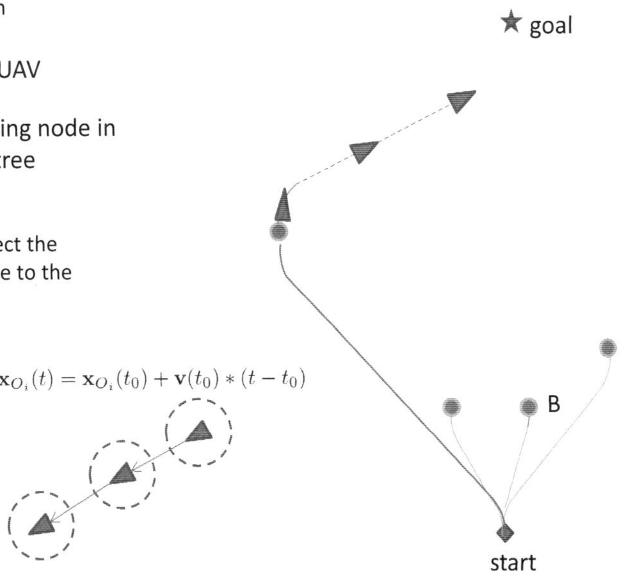

2/6/2017 16

A Sampling Based Path Planning Algorithm: illustration

Adding a Candidate Trajectory:

The moving obstacle with the safety region

The UAV

Existing node in the tree

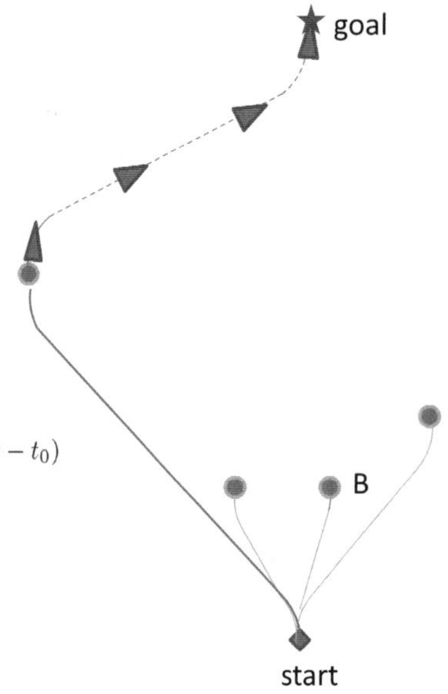

Connect the sample to the goal

$$\mathbf{x}_{O_i}(t) = \mathbf{x}_{O_i}(t_0) + \mathbf{v}(t_0) * (t - t_0)$$

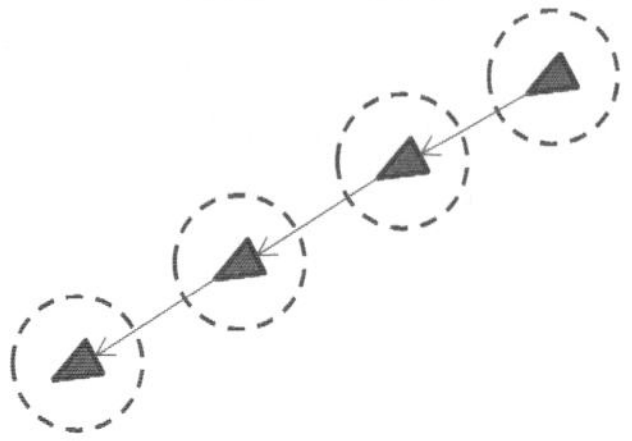

2/6/2017 17

A Sampling Based Path Planning Algorithm: illustration

Adding a Candidate Trajectory:

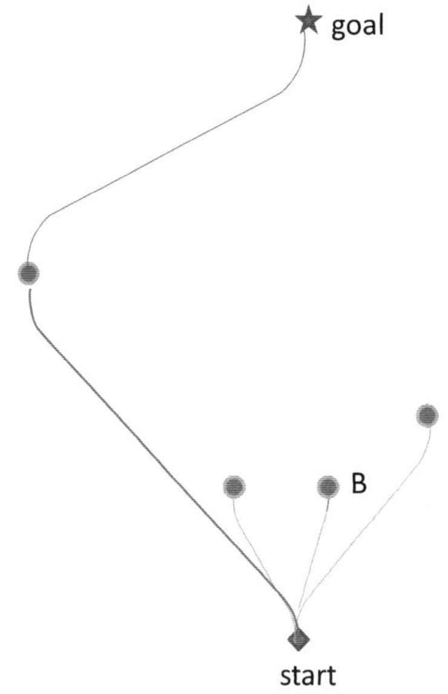

A full candidate trajectory

2/6/2017 18

A Sampling Based Path Planning Algorithm: illustration

Created more candidate trajectories using the same method via intermediate points

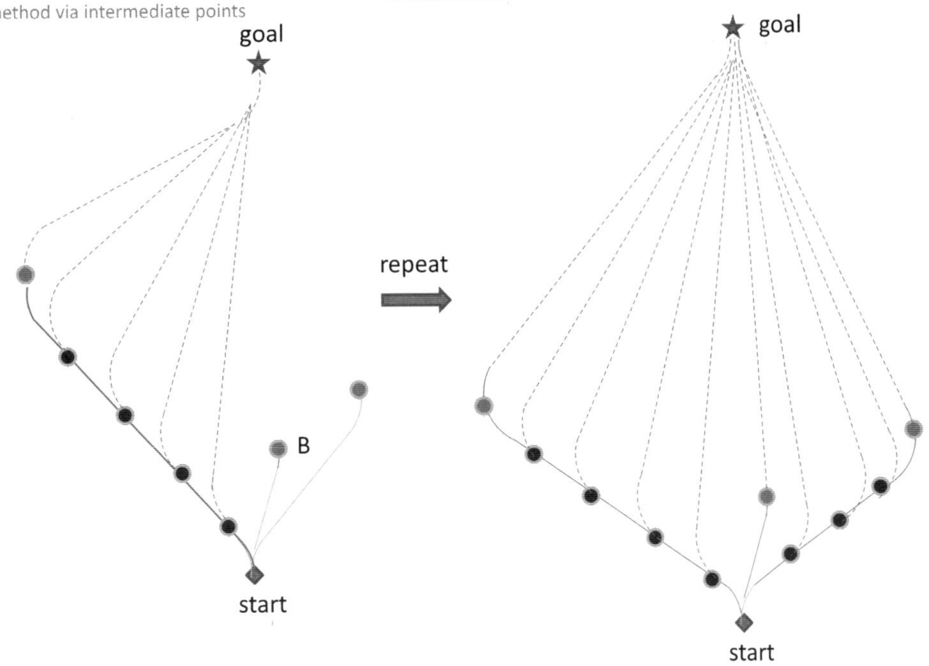

2/6/2017 19

A Sampling Based Path Planning Algorithm: illustration

Select the shortest path from candidates

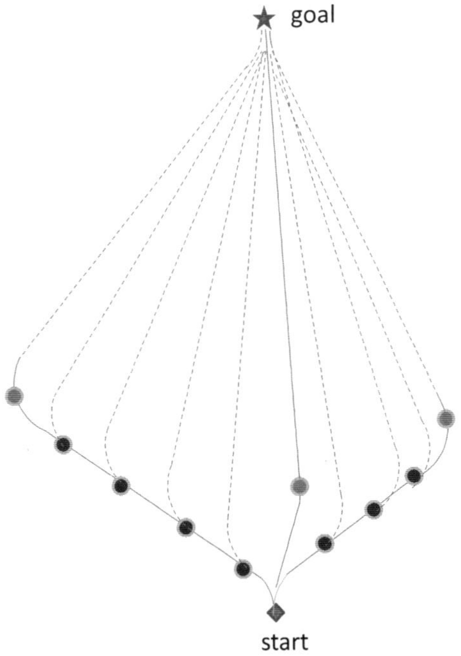

2/6/2017

Presentation for the 2016 American Control Conference.

20

A Sampling Based Path Planning Algorithm: illustration

The execution loop

The actual UAV starts to follow the path

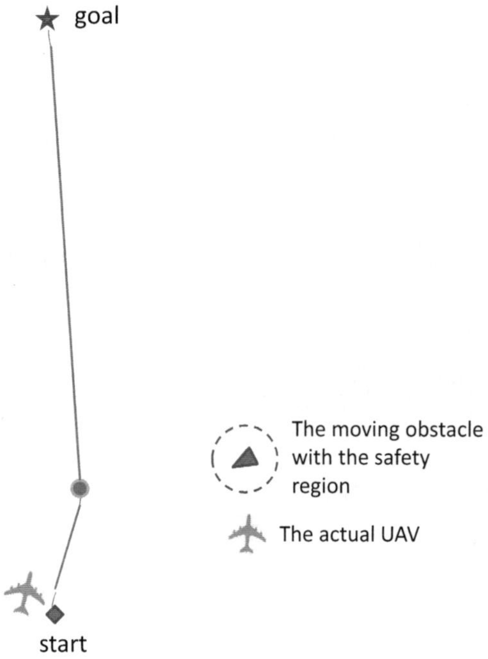

2/6/2017 21

A Sampling Based Path Planning Algorithm: illustration

The execution loop

Predict the collision in advance while following

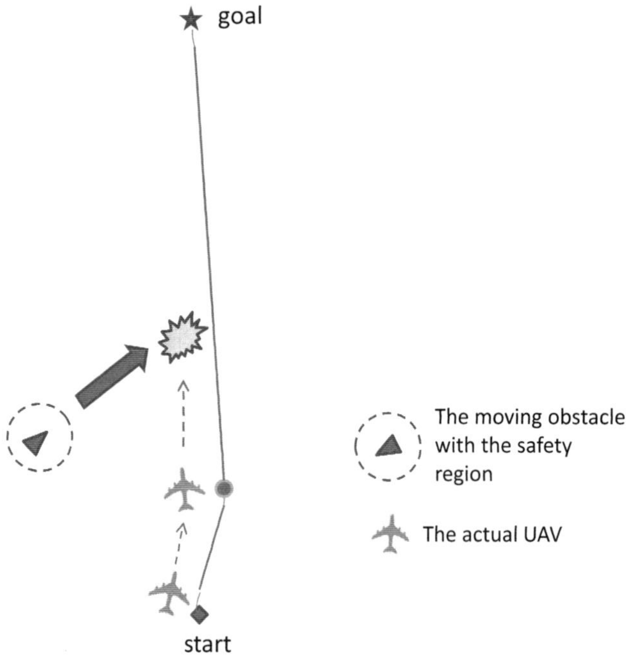

2/6/2017 22

A Sampling Based Path Planning Algorithm: illustration

The execution loop

Re-planning
given
the predicted
collision

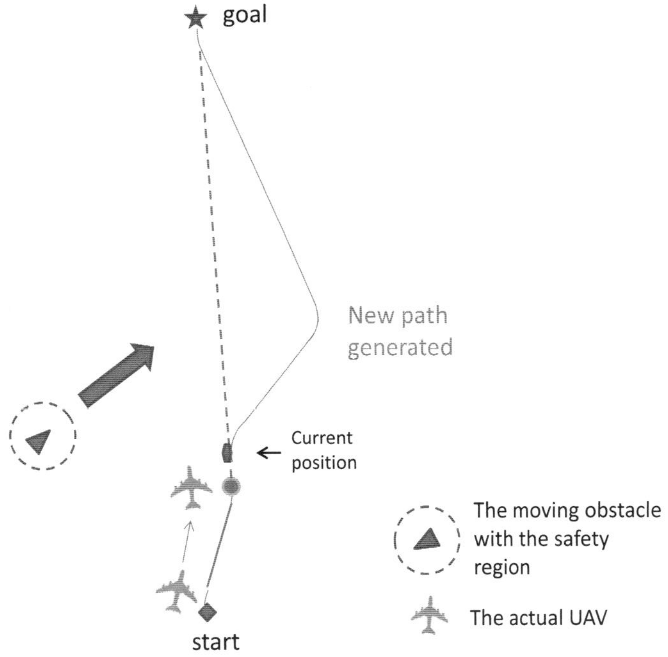

outline

- Background
- Problem formulation
- Algorithm description
- **Experiment results**
- Conclusion and future work

An UAV encounters a
commercial aircraft

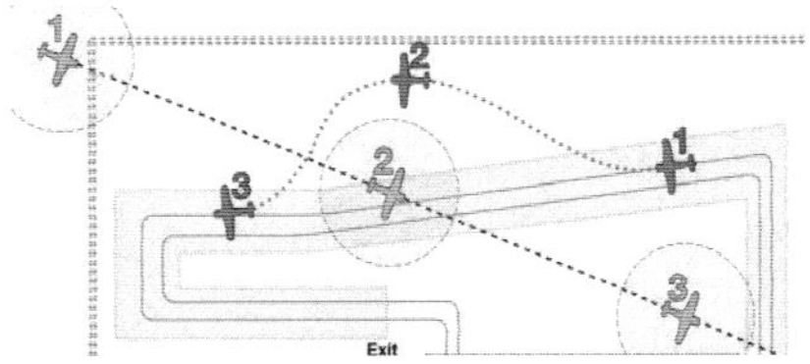

Concept Demo of Path Planning for
UAV Collision Avoidance

Experimental Validation

Hardware-in-the-Loop Simulation Setup

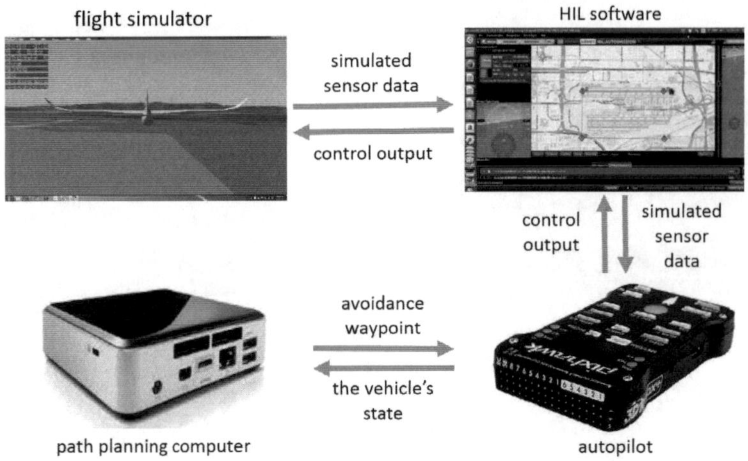

Pixhawk firmware is used

Experimental Validation

Hardware-in-the-Loop Simulation

Path generation to
avoid double
obstacle aircraft

	1
2	3

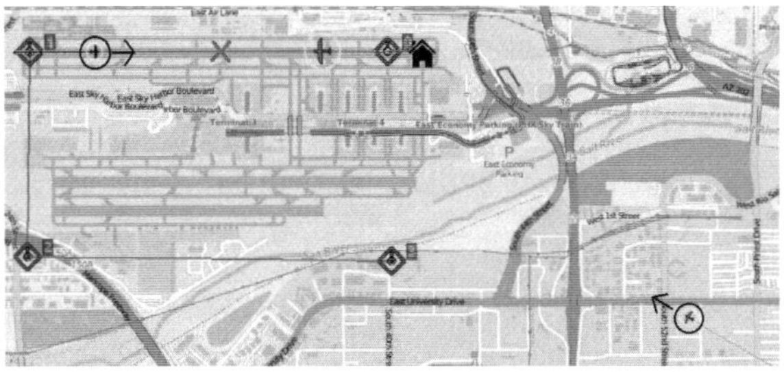

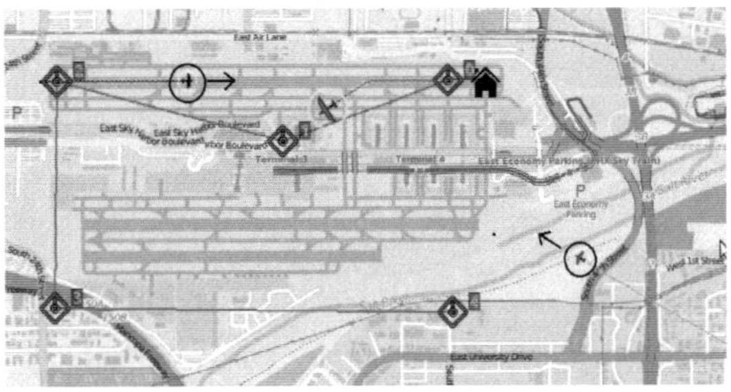

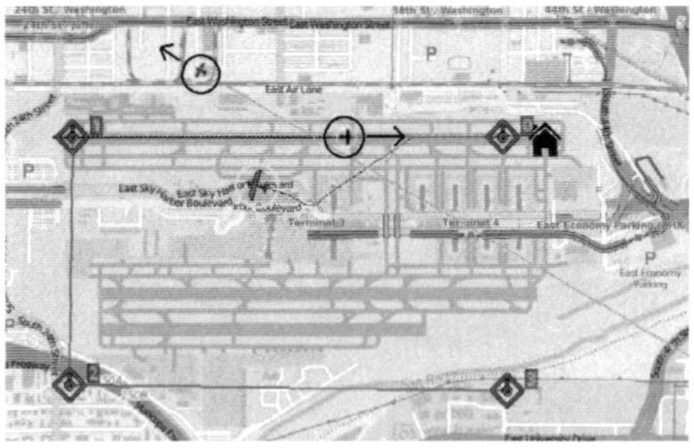

Experimental Validation

Hardware-in-the-Loop Simulation Result

1. 20 tests single obstacles, 20 tests double, 20 tests three.
2. Random selection and combinations of obstacles and offsets.
3. Safety zone: vertical 50m, horizontal 300m.

The horizontal and vertical distance between the UAV and each moving obstacle when they are closest to each other.

95.83% successful rate

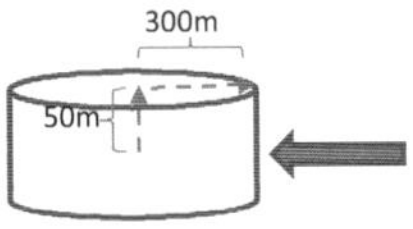

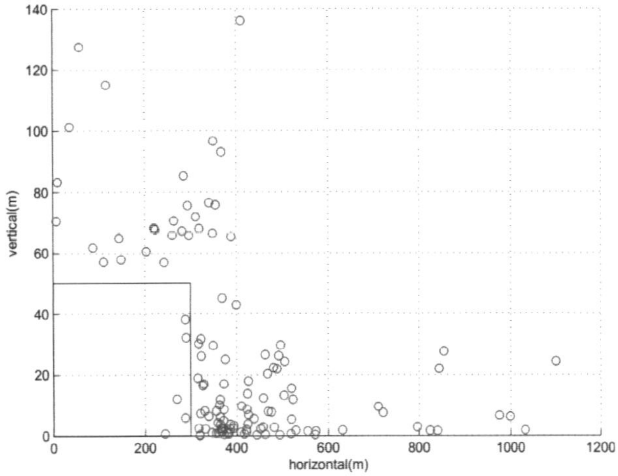

Experimental Validation

Outdoor Flight Experiment Setup

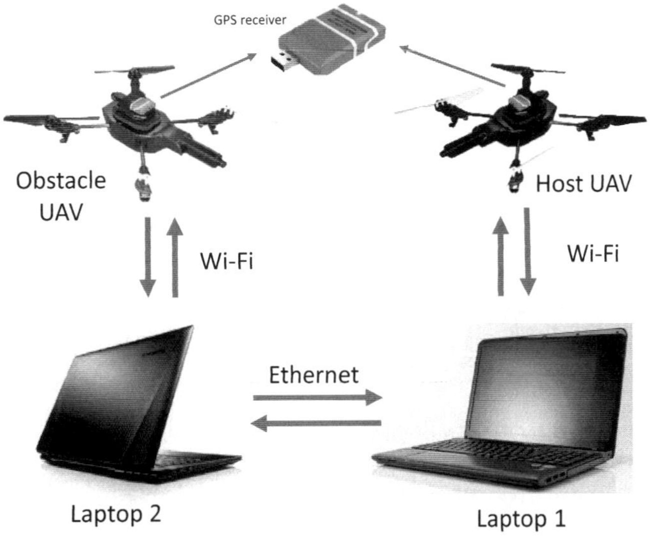

Experimental Validation

2/6/2017 29

Experimental Validation

Results of Outdoor Flight Experiments

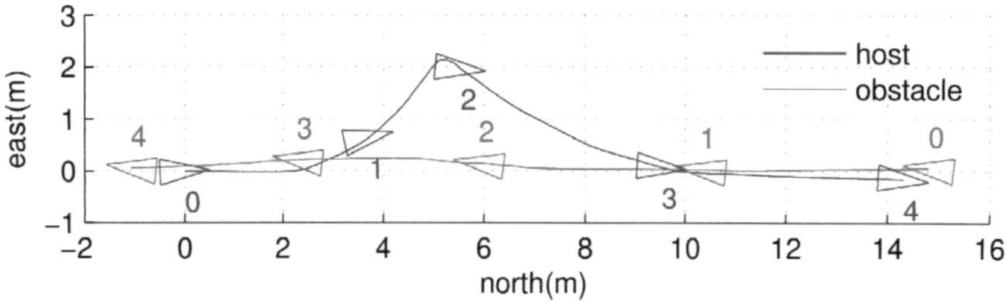

Logged Trajectories of the Host and Obstacle UAV in One Flight Experiment

flight	1	2	3	4	5	6	7	8	9	10
h_dis(m)	2.365	2.14	4.431	6.119	3.523	3.296	3.43	3.25	7.757	3.83
v_dis(m)	0.03	0.45	0.01	0.117	0.032	0.05	0.001	0.013	0.015	0.033

The horizontal and vertical distance between the host and obstacle UAV when closest. 2 meter is the horizontal safety threshold.

H_dis: horizontal distance; v_dis: vertical distance

Both the host and obstacle UAV fly at 1m/s

2/6/2017 30

outline

- Background
- Problem formulation
- Algorithm description
- Experiment results
- Conclusion and future work

An UAV encounters a commercial aircraft

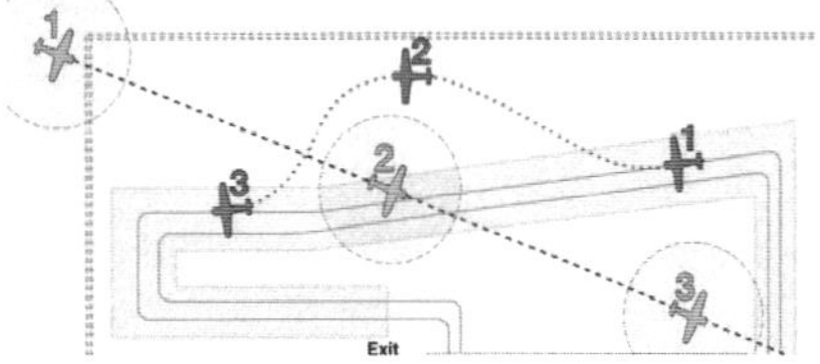

Concept Demo of Path Planning for UAV Collision Avoidance

2/6/2017

31

Conclusion

1. A path planning algorithm based on waypoint sampling and trajectory simulation using the vehicle's closed-loop system is proposed.
2. A collision re-checking and path re-planning strategy is implemented to deal with changes of UAV and obstacles' states.
3. The algorithm is validated in Hardware-in-the-Loop simulation and real flight experiments

Future work

1.Further validate the algorithm using a fixed-wing UAV in a larger area.
2.Integrate the planning algorithm with sensing system to achieve 'Sense-and-Avoid'.

2/6/2017

32

A Functional Safety Management Approach for Certification of Small VTOL sUAS

Dr. Daniel P. Schrage, Professor and Director,
VLRCOE, School of Aerospace Engineering,
Georgia Institute of Technology, Atlanta, GA 30332-0150
e-mail: daniel.schrage@ae.gatech.edu

Abstract

A Functional Safety Management (FSM) Approach for Certification of Small VTOL sUAS into National & Military Airspace will be presented and described in this paper. Civil certification for manned rotorcraft are addressed in FAR Parts 27 and 29, and appropriate advisory circulars which now lead to the accepted use of Aerospace Recommended Practices (ARPs). The standards and guidelines are as follows:

- •AC 27.1B, 27.1309 provides guidance for compliance to FAR 27.1309
- •AC 29-2C, 29.1309 provides guidance for compliance to FAR 29.1309
- •Both ACs recognize SAE-ARP 4761/4754 System Safety Assessment (SSA) and Development processes
- •AC 20-174 for compliance to the new ARP 4754A: Guidelines for Aircraft and Systems Development

However, the FAA manned aircraft approach is basically derived from civil fixed wing aircraft certification standards and guidance and is based almost solely on a transportation Concepts of Operations (ConOps), e.g. moving people or cargo from point A to point B. As a result, certification starts with identifiable aircraft functions which drive the certification requirements, as illustrated in the figure below.

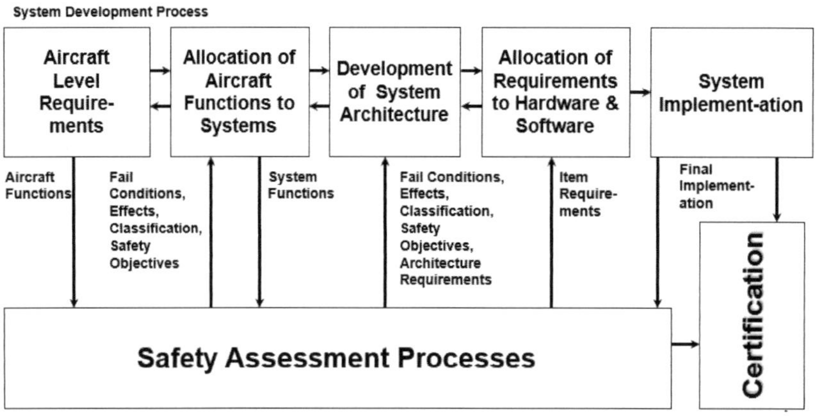

Recommended Certification Process for Manned Aircraft

Since UAS, especially sUAS or Drone ConOps, are not limited to mostly people and cargo transportation missions, there needs to be a certification approach that allows sUAS operators to address their Concept of ConOps and only certify to those number of functions, often limited, that have to be addressed to insure, or minimize, no in-flight collisions and ground impact fatalities, as Target Levels of Safety (TLS). Therefore, sUAS operators need a Functional Safety Management (FSM) approach that is affordable and insures safety for their limited operations. This paper will propose a FSM certification approach for sUAS, To further develop and examine the impact of this approach, a team project in the Georgia Tech Spring 2016 graduate course in *Safety By Design (SBD) and Flight Certification* was used as a case study on how to implement this new FSM approach. This paper will also present the FSM approach using results from this case study.[1]

7th AHS Technical Meeting on VTOL Unmanned Aircraft 24-26 January 2017 Mesa, AZ 85201.

Introduction

Recent advances in small unmanned aerial systems (sUAS) technologies have greatly broadened their potential applications. Proposed applications include videography, surveying, supporting intelligent transportation systems[2], and logistics support[3], among others. Of particular interest from a safety and certification standpoint is logistic support applications related to package delivery. Delivery of goods directly to customers' doors has become ubiquitous and has proliferated due to internet shopping. Companies, such as Amazon and DHL, are considering using sUAS to complete the so called "last mile" of deliveries in order to deliver packages to customers faster than conventional ground based means. In this scenario, packages would be delivered by conventional means to a local distribution center. Then, a sUAS would complete the delivery from the distribution center to the final destination by autonomously traveling to the delivery site, dropping the package and returning to the distribution center.

However, major roadblocks currently exist that prevent such operations from taking place. sUAS need to be safely integrated into the current airspace and air traffic management systems to coordinate flights. Furthermore, there are not clear safety requirements for commercial sUAS, and there is an absence of formal certification processes. In general, establishment of safety requirements and a certification process are necessary to mitigate the risk of collisions both in the air and on the ground that result in property damage and/or injury. Should sUAS be used in a commercial capacity for package delivery, the number of sUAS operations would greatly increase resulting in a corresponding increase in risk.

This paper presents a brief development of a package delivery sUAS from the perspective of safety and certification. A package delivery concept of operations (CONOPS) coupled with an operational risk assessment (ORA) drives a target level of safety (TLS) and reliability of a baseline sUAS architecture. Then, modifications to the baseline architecture are presented and modeled to assess their impact on the vehicle's reliability and meeting the TLS.

sUAS Overview

Numerous sUAS vehicle configurations exist. Configurations include fixed-wing, rotary wing, and hybrid platforms. Among the simplest is a multirotor configuration, often called a "drone", which is a rotary wing vehicle that has two or more rotor pairs that achieve thrust modulation through RPM control. Figure 1 depicts Amazon's package delivery sUAS concept which is a multirotor with four rotor pairs as illustrated[4].

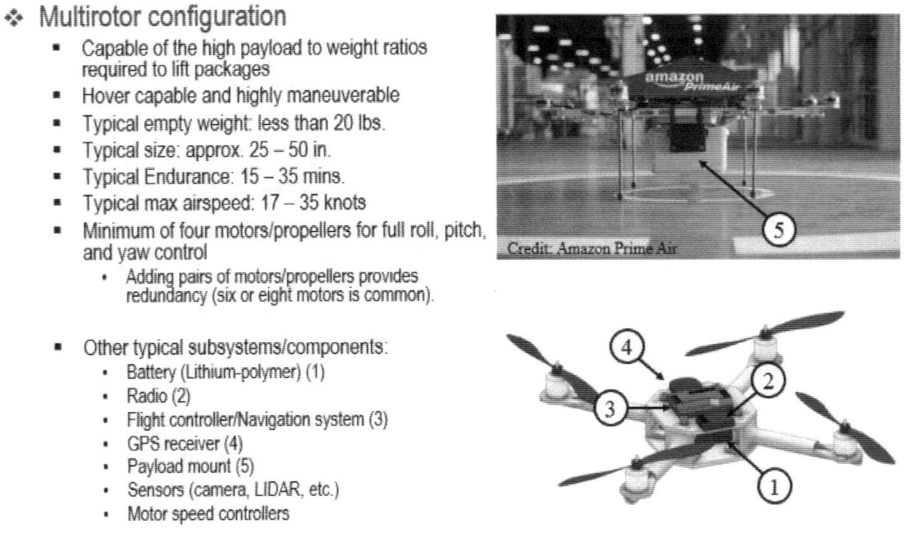

Figure 1. Testbed for VTOL sUAS Certification Approach[4]

Current FAA sUAS Restrictions

Currently, the FAA restricts sUAS operations to conform to the following[7]:

Operational restrictions

- Visual line-of-sight (VLOS) only
- Cannot operate over persons not directly involved in the operation
- Daylight-only
- Must yield right-of-way to all other aircraft
- Maximum airspeed of 100 mph
- Maximum altitude of 500 feet above ground level

- Operations without permission can occur in Class G airspace only
- Preflight inspection by operator

Aircraft restrictions

- Gross weight less than 55 lbs.
- Normal aircraft markings required

These restrictions prohibit package delivery operations which require heavy gross takeoff weights and autonomous beyond line of sight flights. Accordingly, the developments in this paper assume that a less restrictive set of regulations are in place which permit delivery operations. Such a set of regulations was proposed by Amazon, and is discussed in the next section.

Proposed Air Traffic Management Revision

A basic assumption for this paper is that autonomous and beyond line of sight operations are permitted. Furthermore, this paper assumes that a coherent air traffic management system is in place that integrates sUAS operations into the current airspace exists. Amazon proposed one such system where airspace is separated into four sections as illustrated in Figure 2[4]:

1) Below 200 feet reserved for

a. non-transit operations such as surveying, videography and inspection

b. operations of vehicles without sense and avoid technology

2) Between 200 and 400 ft. reserved for vehicle equipped with autonomy and sense and avoid technologies. This air space section is for drone delivery sUAS.

3) No operations between 400 and 500 ft. This space will be reserved for emergency situations.

4) No changes to airspace above 500 ft. This means the sUAS operators has to follow the current regulation

The developments in this paper assume Amazon's proposed airspace segregation scheme represents how the air traffic management system coordinates air traffic over an urban area.

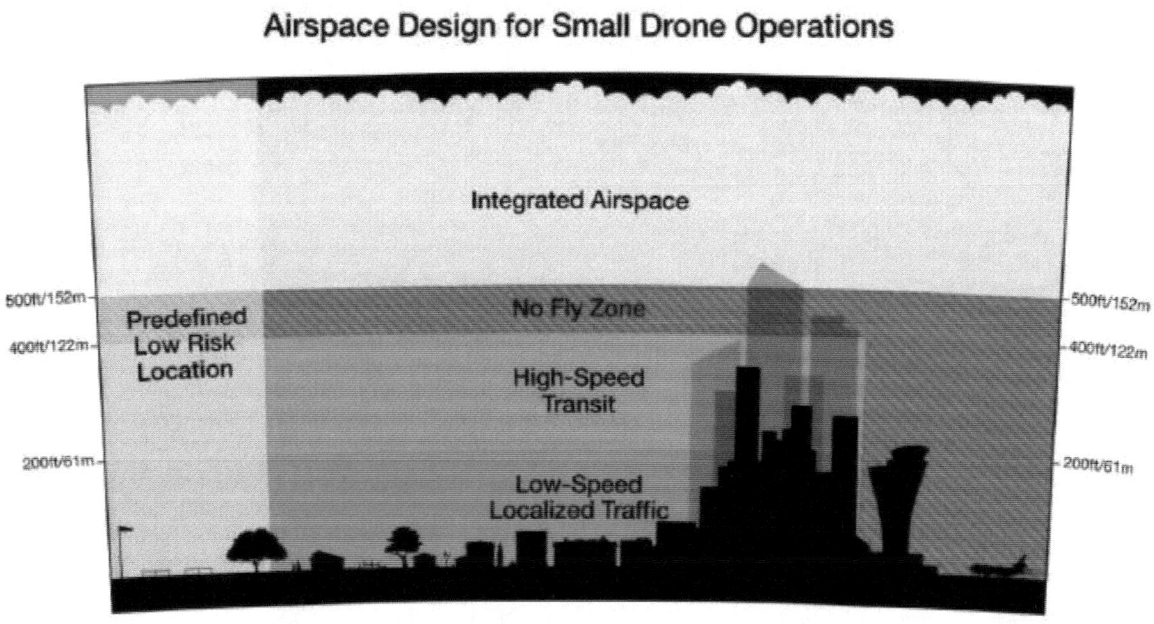

Figure 2: Proposed Air Traffic Management by Amazon[4]

Safety By Design Approach

This paper presents the development of a package delivery sSUAS from a safety and certification standpoint following the approach shown in Figure 3. The approach is based on SAE ARP-4654A, and is modified to include elements necessary for sUAS development. Starting with a baseline vehicle configuration, a CONOPS is defined which is then used to define a mission profile. The mission profile is decomposed into system level functions, which feed into a preliminary aircraft safety assessment (PASA). This includes a system level functional hazard assessment (FHA) and risk assessment. A target level of safety is then determined. At this point, a functional allocation shows which components of the baseline architecture cover which system level functions, which allows for the generation of a system level fault tree diagram. The baseline architecture is further modeled using reliability block diagrams. A Preliminary System Safety Assessment (PSSA) is conducted for the subsystem that has the greatest impact on system reliability. Finally, modifications to the selected subsystem are presented and modeled to show a means of meeting safety requirements.

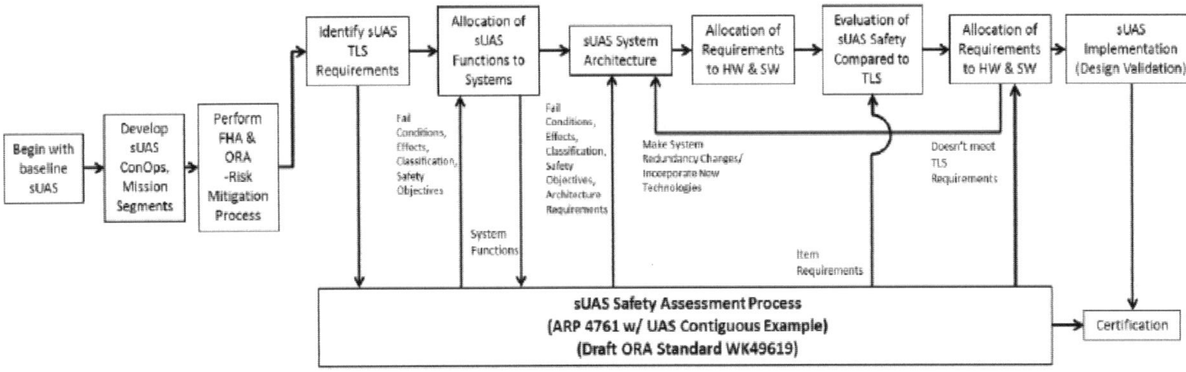

Figure 3: sUAS Safety by Design Certification Approach

The Concept of Operations (CONOPS) used in this paper is from the 2015 AHS Student Design Competition[5], as summarized in the Figure 4 below.

- Notional system CONOPS detailed in the 32nd annual AHS student design competition: "Distributed Logistics in an Urban Setting Using Small Unmanned Aerial Vehicles" [5]
  - Deliver packages in a 50 sq. mi. urban area from one central warehouse
  - 5,000 packages per 10 hour delivery day
  - Uniformly random orders across first 8 hours of the day
  - Package delivery within 2 hours (30 min for Amazon prime air)
  - *60% of packages weight less than 1 lbs.*
  - *85% of packages weight less than 5 lbs.*

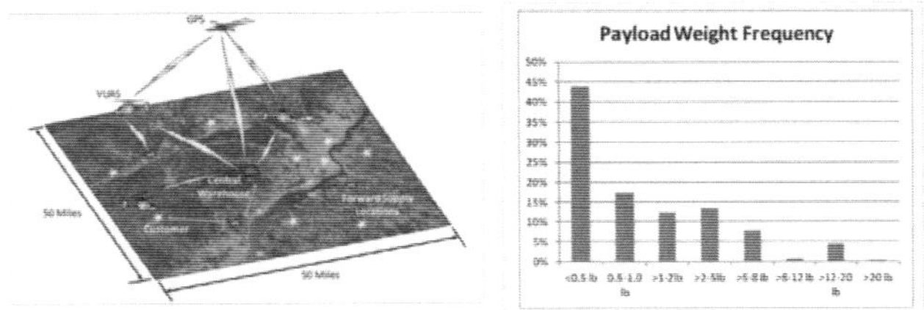

Figure 4. 32nd AHS SDC CONOPS for VTOL sUAS Case Study[5]

Concept of Operations and Mission Segments

A small multi-rotor quad rotor, similar to the Amazon Prime, as illustrated in Figure 1 was used as the sUAS testbed for the AE6372 SBD and FC case study[1]. The paper and presentation will go through the step by step process of the proposed new Certification Approach for VTOL sUAS and recommend on how to further implement it.

The mission profile of the notional sUAS system is based on both Amazon Prime and the AHS competition CONOPS as described above, and is shown in Figure 5. At the start of the mission, the package will be loaded onto the sUAS at central distribution center, the navigation information will be uploaded to the vehicle and batteries will be changed. Then, the vehicle will take off vertically and climbed to an altitude between 200ft and 400ft, which is approved by FAA for sUAS flight [4]. The vehicle will then cruise to the destination autonomously, at which time it will descend below 200 ft to deliver the package. During the package delivery, the sUAS will enter slow flying mode in order to sense and avoid obstacles at the destination and successfully deliver the payload. At the end of the mission, the sUAS will climb back to the cruise altitude and fly back to the distribution center, where it will descend and land. Additional assumptions included in the mission profile include:

- The vehicle shall be capable of carrying a logistics payload of 13 lbs an unrefueled distance of 10 statute miles (mi) from a central distribution center to the delivery point
- The mission will include a 1 min hover at takeoff and an additional 1 min hover at

the delivery point followed by landing and unloading the payload

- Hover shall be assumed to occur 10 ft above the ground, measured from the lowest point of the combined vehicle and payload

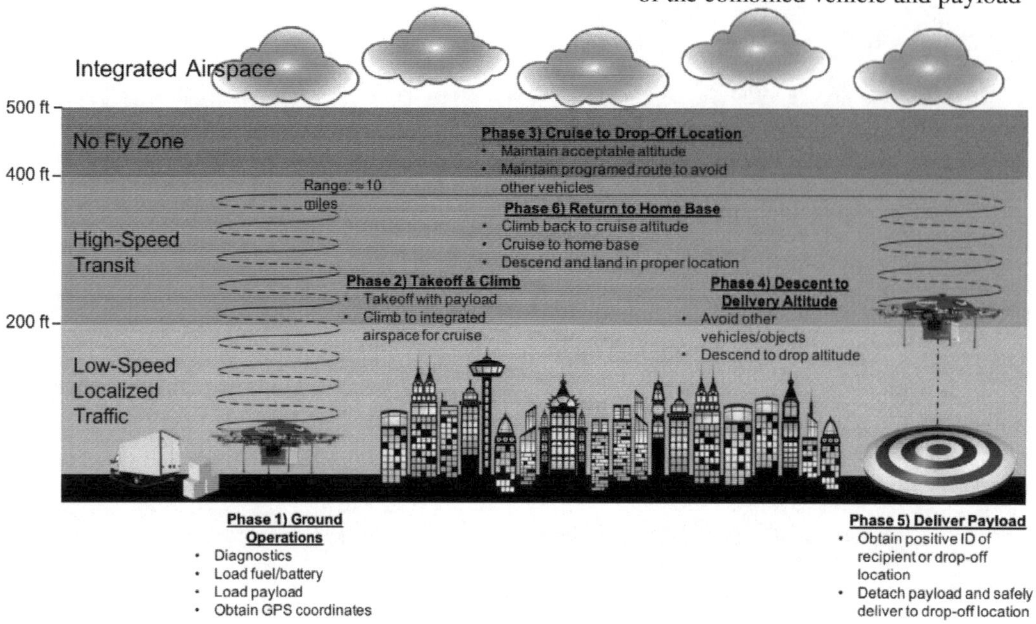

Figure 5. sUAS mission profile based on AHS Logistics CONOPS for package delivery

The mission functional decomposition is based on the mission profile. It delineates all necessary functions to complete a successful mission. At the start of the mission, the sUAS must be loaded and launched from the distribution center. It will climb to the cruise altitude and cruise to the destination. It will then descent and land at the designated location. The payload will then be released. After successful delivery of the payload, the sUAS will return to the distribution center. Beside the mission profile based function, a category of general functions are also listed. They are the necessary function to ensure the safety of the vehicle and the payload, such as simultaneous localization, sense and avoid, generate lift, etc. The functions that a sUAS must perform to successfully complete a package delivery mission are shown in Figure 6. Each function is described in more detail in the subsections of the final report[1].

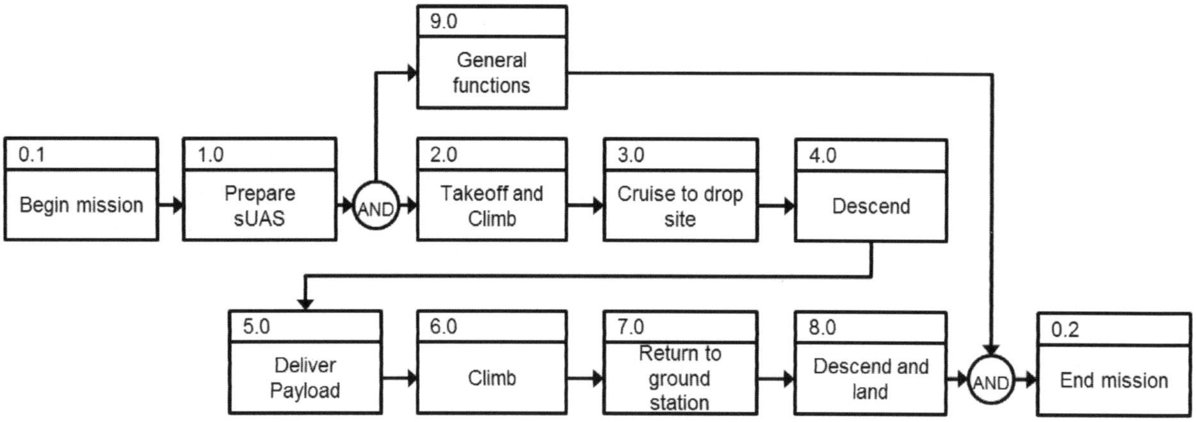

Figure 6. Functional decomposition of sUAS for package delivery mission[1]

Overview of Functional Hazard Assessment (FHA) and Operational Risk Analysis (ORA)

The key purpose of performing a Functional Hazard Assessment (FHA) is to identify and evaluate the hazards associated with aircraft-level functions. This is done by first determining major malfunctions or failures that may occur for each function. After identifying these malfunctions, each will be assessed based on their impact on the mission, including how the malfunction may affect other aircraft or systems, and other persons such as bystanders on the ground. Each hazard will be classified according to severity of their impact using the WK49619 Standard for sUAS[7], presented in Table 1. It is necessary to note that persons are defined as people who are involved with the operation of the sUAS or are affected by its operations directly or indirectly. Along with this, equipment is defined as any object(s) or property *external* to the sUAS, with the exception of persons.

Severity Level	Definition	Value
Catastrophic	• Equipment destroyed (such as electrical transmission lines, substation, water treatment facility, etc) • One or more deaths	5
Hazardous	• Large reduction in safety margins • Large reduction in ability to complete duties accurately • Serious injury to any persons • Major equipment damage	4
Major	• Significant reduction in safety margins • Injury to any persons • Equipment damage	3
Minor	• Nuisance • Operating limitations • Requires use of emergency procedures	2
Negligible	• Little or no consequence	1

Table 1. Hazard severity classification[1]

Along with performing a FHA on the sUAS, an Operational Risk Assessment (ORA) is performed concurrently in order to evaluate the overall risks associated with each function's hazard. A level of risk associated with a hazard is established through assessing the severity of the hazard and the likelihood of the hazard, combined. Hazard likelihood was determined based on the WK49619 Standard for sUAS[6], presented in Table 2.

Likelihood Level	Definition	Value
Frequent	Likely to occur many times	5
Occasional	Likely to occur sometimes	4
Remote	Unlikely, but possible, to occur	3
Improbable	Very unlikely to occur	2
Extremely Improbable	Almost inconceivable to occur	1

Table 2. Hazard Likelihood Definitions[6]

The level of risk is determined by way of calculating a risk score using the values associated with severity and likelihood levels, as shown in Equation 1[7].

$$Risk\ Score = (Severity\ Value) * (Likelihood\ Value)$$

Equation 1

Assumptions Made in Hazard and Risk Assessment

In order to perform a hazard and risk assessment on the sUAS, it was necessary to make a few assumptions regarding the sUAS and its operation. First, it is assumed that mission success relies on a single sUAS, therefore the loss of the sUAS will result in mission failure. Secondly, it is assumed that no backup systems are in place to mitigate risk, as current sUAS's do not have redundant systems. Therefore, any loss of a system or reduction in performance is permanent for the mission, and any system failure will result in loss of functionality of the sUAS. Lastly, all hazards and failures are based on currently state of the art technology to perform a hazard and risk assessment on the sUAS, it was necessary to make a few assumptions regarding to the sUAS and its operation. First, it is assumed that mission success relies on a single sUAS, therefore the loss of the sUAS will result in mission failure. Secondly, it is assumed that no backup systems are in place to mitigate risk, as current sUAS's do not have redundant systems. Therefore, any loss of a system or reduction in performance is permanent for the mission, and any system failure will result in loss of functionality of the sUAS. Lastly, all hazards and failures are based on current state of the art technology.

FHA and ORA Summary

Figure 7 presents the compiled risk matrix. The functions associated with failures that have the highest risk deal with flight control and sense and avoid capabilities. The functions with the highest risk associated with them are listed below. Based on these findings, risk mitigation focus areas should include sense and avoid, as well as guidance, navigation, and control (GNC).

1. Perform pre-flight checks/diagnostics Detect obstructions
2. Execute avoidance maneuvers
3. Update flight plan
4. Reconfigure system/mitigate failures
5. Deliver electrical power
6. Generate lift
7. Update Flight Plan

#	Unacceptable Risks
1.6	Perform pre-flight checks/diagnostics
2.3/3.9	Generate Lift
2.4/3.10	Identify Obstructions
3.8	Deliver electrical power
3.11/4.6	Execute avoidance maneuvers
3.12	Update flight plan
9.7	Reconfigure system/mitigate failures

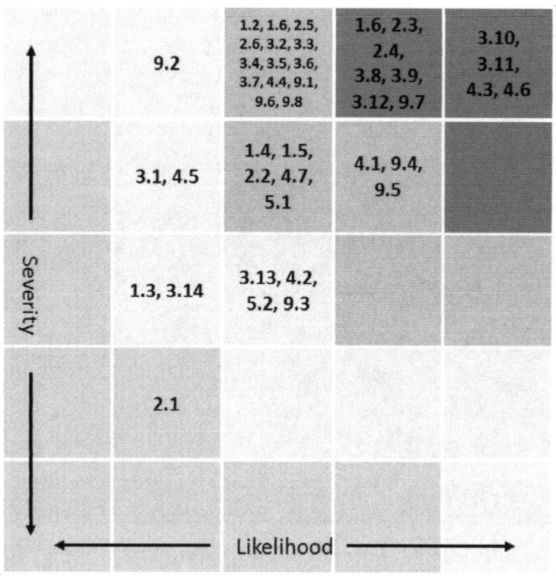

Figure 7. Compiled risk matrix for all phases of the mission[1]

Target Level of Safety (TLS)

In order evaluate the safety of the system in a more quantitative manner (by means of future Fault Tree Analysis, Reliability Block Diagrams, and Abridged Petri Nets), it is necessary to establish a top-level safety target for the sUAS. The target level of safety (TLS) quantifies the tolerable frequency of accidents occurring as measured by some rate, such as accidents per flight hour, or accidents per mission. For manned aircraft, the current TLS is $2 * 10^{-7}$ ground fatalities per hour of operation. This includes operations in all environments. Amazon intends to operate in an urban environment, as well as to frequently deliver packages within this urban area. The worst case scenario considered for the urban environment is if a sUAS were to experience a failure and fall out of the sky, potentially causing a traffic accident or striking a bystander, leading to one or multiple deaths. Due to the nature of the planned operation of the vehicle(s), it is necessary to consider the population density of urban environments, as well as the region of the ground covered by equipment or buildings which, when impacted, could lead to damage to them. Since the UAS is a different kind of vehicle, a Ground Impact Model[8] was adopted to estimate the acceptable vehicle level failure rate. Figure 8 presents the logic behind the ground impact model: if something happens to the vehicle and it is unable to recover, does it lead to an accident? If it does lead to an accident, was there any exposure? If there was exposure, was there an injury or any damage caused? If so, did it cause a fatality or any destruction.

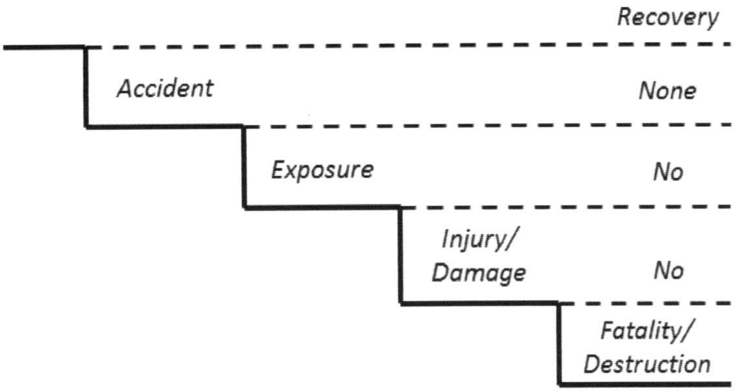

Figure 8: Ground Impact Model[7]

For the vehicle level, the mean time between failure (MTBF) can be measured by Equation 2, where $TLS_{vehicle}$ is the target level of Safety of single vehicle, A_{exp} represents the ground exposed area of vehicle (ft^2), ρ is the population density of operating area, γ is the percent of ground area covered by equipment, and P_{pen} is the probability that debris causes equipment destruction or fatality given exposure[7].

$$MTBF_{vehicle} = \frac{1}{TLS_{vehicle}} A_{exp}(\rho + \gamma)P_{pen}$$

Equation 2

Along with this, the probability that the vehicle fails is given by Equation 3:

$$P_{failed,veh} = \frac{1}{MTBF_{veh}}$$

Equation 3

This probability is the top level safety target to be used in figuring FTA, RBD, and APN analyses. In order to obtain an estimate based on the above equations, it was assumed that the vehicle would be operating above the Atlanta metropolitan area, yielding the following data:

- $TLS_{veh} = 1 \times 10^{-7}$
- $P_{pen} = 0.01$
- $A_e = 9 \text{ ft}^2$
- $\rho = 1.13 \times 10^{-4} \ {}^{\text{ppl}}/_{\text{ft}^2}$
- $\gamma = 0.811 \ (based\ on\ Georgia\ Tech)$

Figure 9 presents the ground mappings of Georgia Tech, where all of the yellow area represents regions where equipment covers the area. This mapping leads to a calculation of 81.1% equipment coverage by area

Figure 9: Georgia Tech's Equipment Ground Coverage[1]

Based on the above data, the following calculations were performed, ultimately yielding a value of $1.37\text{x}10^{-6}$ for the probability of the vehicle failing. For the purposes of flowing this down to future system and sub-system safety analysis, the TLS was taken as $1.0\text{x } 10^{-6}$ system level failure per flight hour.

$$MTBF_{veh} = \frac{9 * (1.13 \times 10^{-4} + 0.811) * 0.01}{1 \times 10^{-7}}$$

$$MTBF_{veh} = 730000 \text{ hours}$$

$$P_{failed,veh} = \frac{1}{730000} = 1.37 \times 10^{-6}$$

Equation 4

sUAS Physical System Architecture

To determine whether the current sUAS meets the required target level of safety, a physical decomposition of the system must be performed to analyze its reliability. First, the key mission functions are allocated to the corresponding subsystem that performs each function, as shown in Figure 10. The subsystems considered in this analysis include: propulsion, guidance navigation and control (GNC), communications, and the structural and payload delivery subsystems. The electrical subsystem was not considered separately, as the electrical components of each subsystem were considered within each of the subsystem architectures.

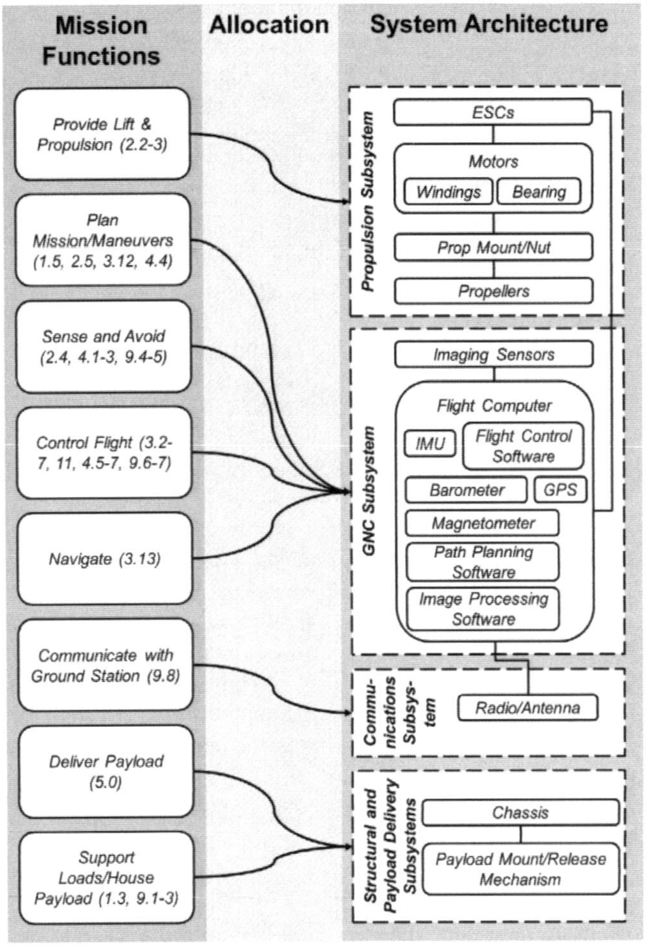

Figure 10: Allocation of functions to different physical subsystems of sUAS, and physical decomposition of these key subsystems[1]

Each subsystem was then physically decomposed so that the reliability of the system could be analyzed at the component level. The physical decomposition of the sUAS is displayed under the system architecture in Figure 10. A literature review was performed to determine the reliability of each component, the results of which are presented in Table 3. Component relatabilities were found for everything except the software and structural components, which required estimations to perform a complete reliability assessment of the entire system. Using this physical decomposition and the component reliabilities, a preliminary aircraft safety assessment (PASA) was performed to determine the current reliability of the sUAS and assess where reliability must be improved to meet the TLS.

Component	Failure Rate (per flight hr)	Source
Accelerometer	2.0E-05	"Reliability, Maintainability, and Risk" (textbook), pp. 298
Printed Circuit Board Connector/Wiring	1.0E-05	Mil-Hdbk-217F, pp. 15-4
Gyroscope	2.5E-04	"Reliability, Maintainability, and Risk" (textbook), pp. 301
Barometer	2.4E-05	"Reliability, Maintainability, and Risk" (textbook), pp. 302
Printed Circuit Board	3.4E-08	"Reliability, Maintainability, and Risk" (textbook), pp. 302
GPS Unit and Signal	3.4E-07	"Reliability, Maintainability, and Risk" (textbook), pp. 302
Magnetometer	2.94893E-06	Mil-Hdbk-217F, pp. 9-3
Camera	2.67426E-07	Mil-Hdbk-217F, pp. 5-7
Electronic Speed Controller (ESC)	8.0E-08	"Reliability, Maintainability, and Risk" (textbook), pp. 301
Motor Bearing	2.5641E-05	Mil-Hdbk-217F, pp. 5-7
Motor Winding	6.25E-07	Mil-Hdbk-217F, pp. 5-7
Encoder/Decoder of Integrated Circuit Board	3.23E-07	Mil-Hdbk-217F, pp. 5-24
Radio	2.9859955E-06	Mil-Hdbk-217F, pp. 15-1

Table 3. Reliability of subsystem component[1]

Preliminary Aircraft Safety Assessment

The preliminary aircraft safety assessment (PASA) assesses the aircraft level functions that were addressed during the functional hazard assessment (FHA) that combine several systems failures that are not studied at the system level. It allows for the identification of the development assurance level (DvAL) for the different aircraft functions and systems. The PASA is also useful in providing a preliminary indication that the anticipated aircraft or system architectures can meet the designated system requirements. Three techniques were employed during the PASA to assess the system level reliability: a fault tree analysis of each subsystem, reliability block diagram analysis of each subsystem, and a system level abridged petri analysis. The results of each analysis are summarized at the overall system level in the following subsection. More details are provided in the final report[1].

Fault Tree Analysis

Fault tree analysis (FTA) is used to determine the possible causes of hazards within the system, which could potentially lead to complete system failure. It provides a graphical design technique to identify and control the most dangerous hazards to the system, in an effort to ultimately achieve the desired target level of safety. The FTA is first presented at the top system level, and is then broken down by the sUAS subsystems. For each subsystem, the FTA provides an expected failure rate, which together provides an estimate for the total system failure rate, which must be compared to the TLS to determine if the system has met the safety requirements set by the FHA analysis.

Overall System Level

The system level FTA is shown in Figure 11. For the sUAS, a system failure is considered any failure that causes the sUAS to impact an object or bystander in the environment causing serious injury, death, or damage. Such a system failure may be the result of a failure in any one of the four subsystems specified in Figure 10. Based on these results, two conclusions can be drawn: (1) the current system architecture is unable to meet the target level of safety required for the sUAS to operate in an urban environment, and (2) the GNC subsystem is the main cause of this unreliability. The fault tree analysis of the individual subsystems provided more insight into why the reliability of the sUAS is unacceptably high.

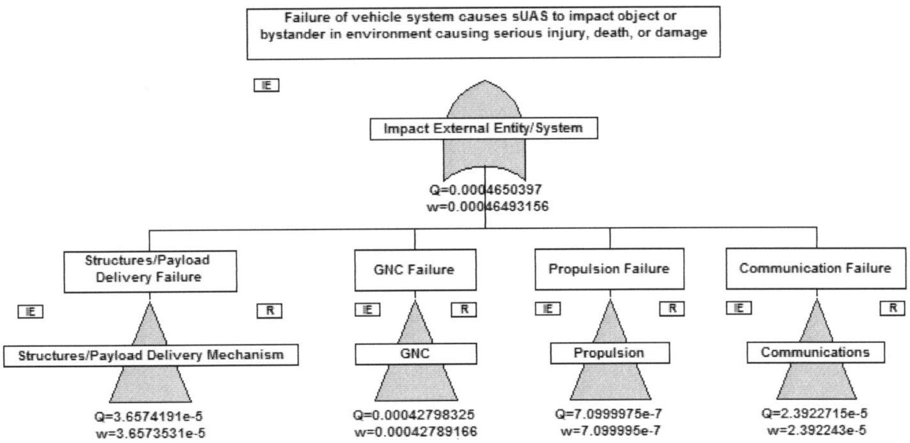

Figure 11. Top level FTA for sUAS. W represents failure rate, while Q represents the unavailability or unreliability of the system[1]

Reliability of Subsystems

The fault tree analysis is a useful technique for identifying the components that are associated with each of the functions that the subsystems must perform. However, through FTA it can be difficult to determine where redundancies should be added to increase the reliability. To assess this, the reliability of each subsystem was assessed through the use of reliability block diagrams (RBD). The use of RBDs is key when identifying single point failures within each of the subsystems.

Overall System RBD

The RBD of the sUAS is provided in Figure 12, showing all subsystems in series. Thus, a failure in any one of the subsystems will result in the failure of the overall system. As expected, the results of the RBD match that of the FTA for each individual subsystem as well as the top level reliability. The subsystem level RBD will be used to identify single point failures to determine where redundancies should be added to increase reliability. This is particularly important for the GNC system, which has the lowest reliability of any system by an order of magnitude.

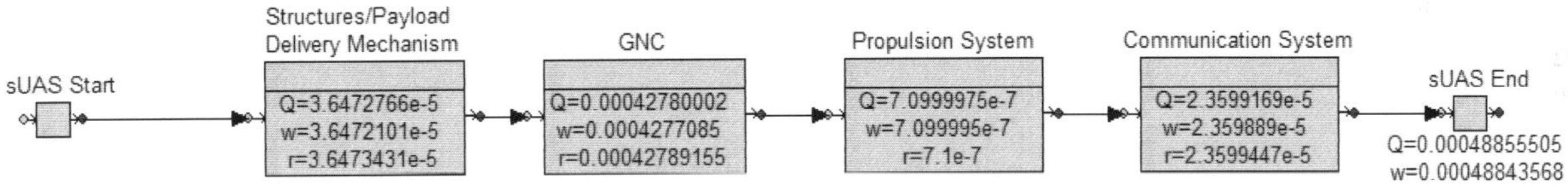

Figure 12. Reliability block diagram of entire sUAS[1]

Guidance, Navigation, and Controls System

By analyzing the baseline GNC subsystem architecture, it becomes clear as to why the reliability of this subsystem is so low. There are many components in the GNC that must work in series, which quickly reduces the reliability of the subsystem even if each component has a high reliability individually due to the high number of single point failures. The baseline GNC architecture is expected to fail at a rate of 0.000428 times per flight hour, which is 428 times larger than the TLS (as shown in Figure 13). Based these results, it is suggested that a new GNC architecture be developed in order to allow the sUAS to achieve the desired level of safety.

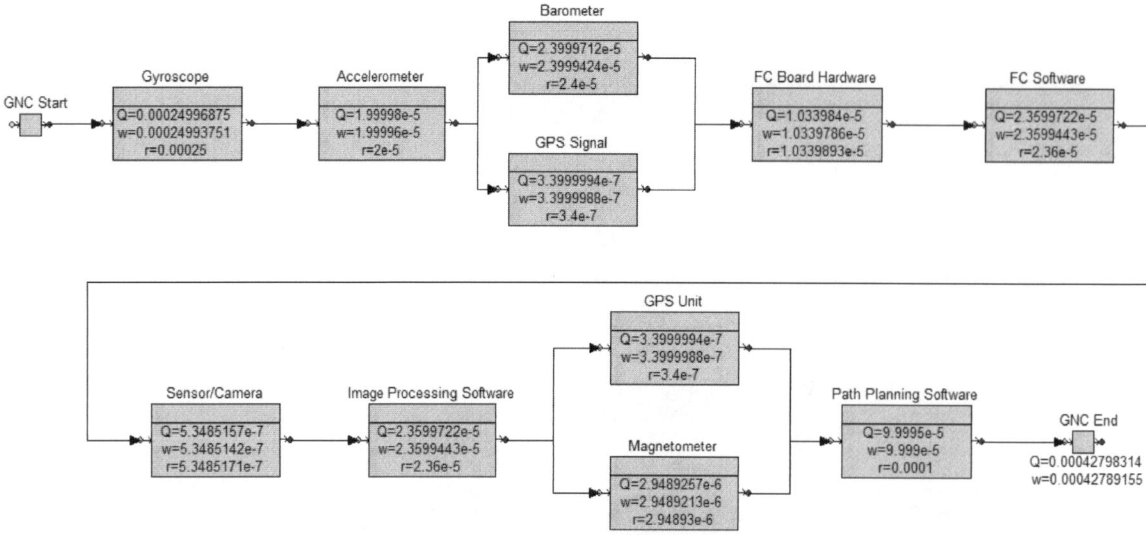

Figure 13: Baseline GNC subsystem architecture[1]

Common Mode Type	Common Mode Sub-Type	Common Mode Source	Common Mode Failure
Concept and Design	Design Architecture	Common External Sources (Power)	Failure of common power source
		Operating Characteristics	Failure to operate in mode or detect current operating mode
		Performance/Failures	Designer failed to predict potential failure and didn't design redundancies or take protective measures
		Equipment Protections	Designer failed to predict potential failure event
	Technology, Materials, & Equipment Type	Component Type	Hardware failure
		Component Software	Software error/failure
		Component Use	Non-redundant component failure
Operation	Staff	Common Staff	Error due to inadequately trained personnel, or distracted operator/monitor
	Procedures	Typical/ Repeated Procedures	Omission of action, incorrect or inadequate action, misdiagnosis
Maintenance	Staff	Common Staff	Error due to inadequately trained personnel, or incorrect human action
	Procedures	Typical/ Repeated Procedures	Failure to follow repair procedures, defective repair procedures, lack of repair procedures

Table 4: Common mode analysis[1]

Common Cause Analysis (CCA)

Static safety analysis methods, such as FTAs and RBDs, don't do a sufficient job of addressing the interactions that can be occur between systems/sub-systems due to common causes. In addition, the traditional CCA methods,

Particular Risk Analysis (PRA), Common Mode Analysis (CMA) and Zonal Safety Analysis (ZSA) that are used for CCA are also static analysis tools. Since this study is looking at incorporating safety into the design process, a Common Mode Analysis was performed at the base level of the vehicle design, operation, and maintenance. The CMA is presented in Table 4. Since CMA is only a static analysis tool, a state-based dynamic analysis (using SPN's) is also performed for the PASA/PSSA in order to develop a more mathematical based CCA.

Abridged Petri Net Analysis

The next step in the PASA is to utilize a dynamic safety analysis method, which attempts to account for the stochasticity of the system and its components, unlike the static FTA and RBD methods previously used. The use of Abridged Petri Nets (APN)[8], a form of Stochastic Petri Nets, allows for the analysis of common cause failures among the components, sub-systems, and systems that form the sUAS. This approach provides a visual and dynamic means of modeling stochastic discrete-event processes that are equally suitable for modeling failure processes (including proper description of component degradation due to various aging mechanisms) and complex logistical scenarios, and coupling among component behaviors during both failure and repair processes. The GUI driven Java™ implementation of this SPN framework is called SPN (Stochastic Petri Nets with Aging Tokens) and was developed by Dr. Vitali Volovoi. Since it had been identified in the FTA that the GNC system is the system which is in need of the most improvement, the SPN approach will be used on the GNC system.

The GNC system is comprised of three sub-systems: the flight controller, computer vision, and navigation system. Any failure in these sub-systems will cause the GNC system to fail. Since the desire is to use SPN to identify key components or sub-systems which fail more often than others, it was necessary to develop a SPN model of the GNC system, down to the component level. Figure 14 presents the full SPN model developed for the GNC system. It was developed by starting with defining the components and their failure rates on the left-hand side. Each component has a transition to either fail, or succeed at performing its function based on its defined failure rate. From here, the tokens then flow into sub-systems and success or failure is determined based on whether the necessary components have failed or succeeded. This logic flows all the way up to the flight controller, computer vision, and navigation system sub-systems. If any of these three sub-systems fails, then the GNC systems is considered to have failed since it requires all three to be functioning. It is also necessary to note that the developed APN appears complex because several enablers have been implemented to ensure that upper level sub-systems and components do not get evaluated until the components at lower levels have already been evaluated. This allows all failures to be modeled correctly. Along with this, additional places, tokens, and enablers are used in order to create success and failure counters for all components, sub-systems, and the top level GNC system. Since only failure rates of the components are known from research for the FTA and RBD's, and the GNC system's components and sub-systems are constantly running, token aging was not used. Due to this, it was still necessary to allow for stochasticity in the simulation. This was done by modeling the transitions with uniform distributions that would be equivalent to the failure rates of each of the bottom level components.

With this APN model developed, a Monte Carlo simulation of the GNC sub-system failures was able to be run. For the purpose of identifying key sub-systems that fail more often than the others, multiple Monte Carlo simulations of 1,000,000 runs were averaged, yielding the results presented in Table 5. As one can see, both the RBD and APN models indicate that the current architecture cannot meet the TLS, therefore requiring redundancies or new technologies to be implemented to improve reliability. Both the RBD and APN models identified that the flight controller sub-system is the key area to focus mitigation efforts

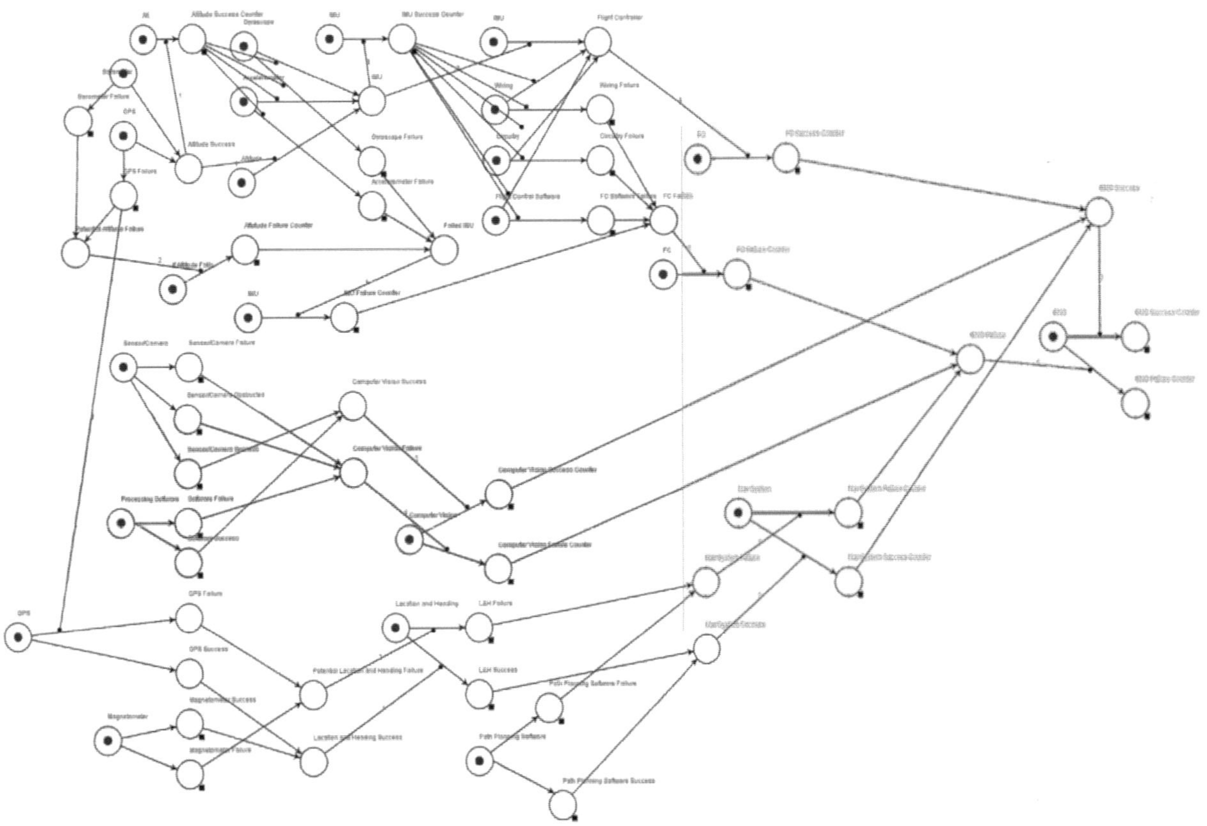

Figure 14. Full APN Model of the GNC System[1]

Event	Failure Probability
FC Failure	0.000507
Computer Vision Failure	0.000011
Nav System Failure	0.000033
GNC Failure	**0.001138**

Table 5: APN Results for GNC Sub-Systems[1]

Preliminary System Safety Analysis (PSSA)

Inspection of Figure 13 shows that in the baseline system architecture, most of the items (all of which are flight critical), are arranged in series. There is little redundancy built into this architecture and it exhibits a multitude of single point failures. For example, if there is a failure of any part of the IMU, an error in any software onboard, or an obstruction/failure of the camera, the result is a system level failure. Based on the model in Figure 7-9, the baseline GNC subsystem architecture is expected to fail at a rate of $f = 0.000428$ times per flight hour, which is 428 times larger than the TLS. Furthermore, as shown in Figure 12, it is the GNC subsystem's failure rate that is driving the system level failure rate. Accordingly, modifications to the GNC subsystem architecture are necessary to achieve a system level failure rate in line with the TLS. This section proposes a triplex flight control architecture and describes its impact on the expected failure rates.

Proposed Architecture for GNC Subsystem

Flight control systems on commercial and military aircraft must conform to strict safety requirements. In order to comply with these stringent regulations, many primary flight control systems have triplex, or triple redundant, flight control computer (FCC) architecture. In this type of architecture, three nearly identical flight control computers are installed on the aircraft. Each contains (nearly) identical software. Each computer receives its own feedback from its own instruments and attempts to control the aircraft's flight based on those feedback signals. The three FCCs are interconnected with cross channel buses, and so each is aware of other two's flight control command.

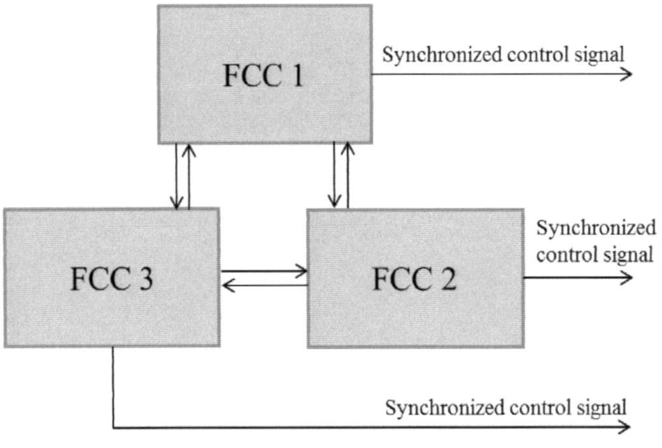

Figure 15. Simplified Depiction of Triplex FCC Architecture[1]

Flight control is achieved by selectively outputting one synchronized control signal output based on the three individual control signals. Criteria for computing the value of the synchronized signal is application specific, and algorithms are often proprietary. Figure 15 depicts a triplex FCC architecture.

The GNC subsystem architecture modification is modeled after a triplex FCC architecture. Three identical FCCs (which include IMUs, barometers, and all the software used on the sUAS) are installed and interconnected. The new architecture is shown in Figure 16. By simply adding two more FCCs, nearly all of the single point failures in the subsystem are eliminated.

The model in Figure 16 predicts a GNC failure rate of $f = 5.35 \times 10^{-7}$: a three order of magnitude improvement over the baseline architecture. With this architecture, the GNC subsystem's failure rate approaches the camera's failure rate, which is the last remaining single point failure. Incorporating camera and sensing technologies that improve the camera's reliability would be required for further improvements to the GNC subsystems' reliability.

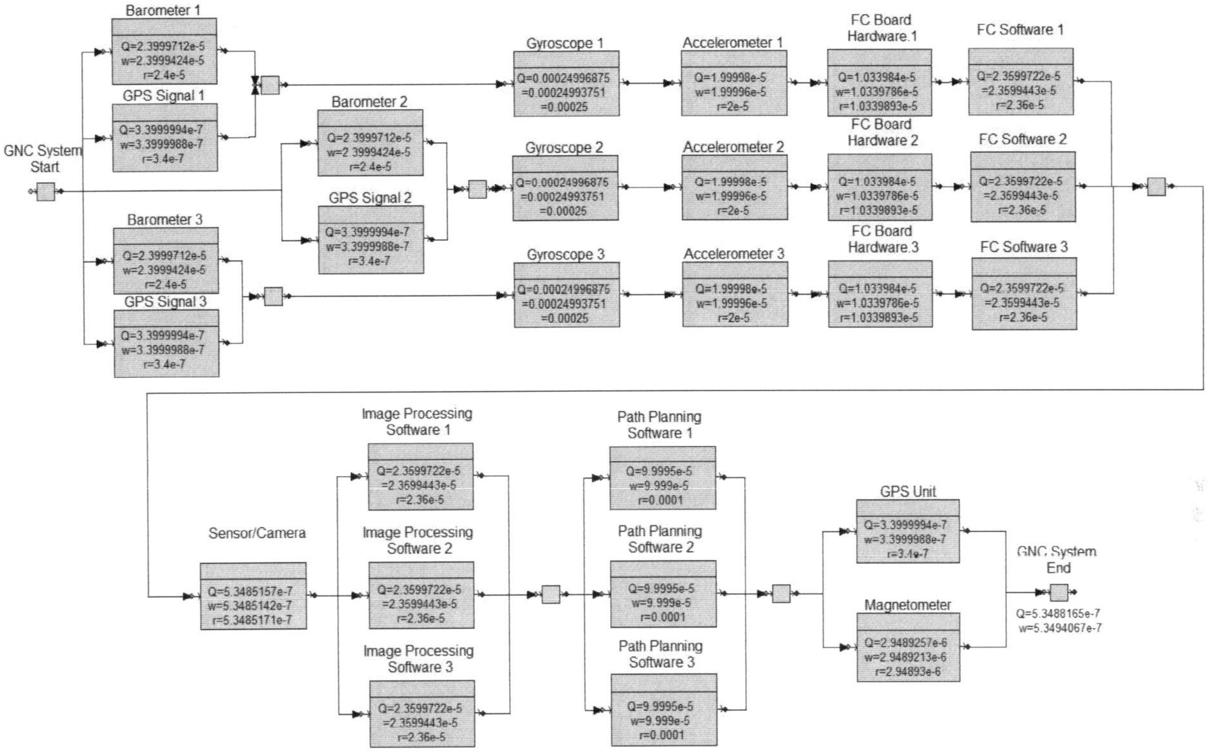

Figure 16: GNC subsystem architecture modification incorporating triplex FCC[1]

Figure 16 shows the impact of the improved GNC architecture at the system level. Before, the system reliability was limited by the GNC subsystem's reliability which was expected to fail 10 times more often than the next most failure-prone subsystem. The system failure rate was therefore well above the established TLS. Implementation of the triplex architecture reduces the GNC subsystem's failure rate so that it is now the most reliable subsystem. Furthermore, the system's failure rate is now approximately 60 times larger than the TLS – a large improvement from the previous figure of 428 times the TLS. Reducing the system's failure rate further can be achieved by modifying the architecture of the structures and/or communications subsystems, which now exhibit the highest failure rates of the four subsystems.

The GNC subsystem architecture modification is modeled after a triplex FCC architecture. Three identical FCCs (which include IMUs, barometers, and all the software used on the sUAS) are installed and interconnected. The new architecture is shown in Figure 17. By simply adding two more FCCs, nearly all of the single point failures in the subsystem are eliminated. The model in Figure 17 predicts a GNC failure rate of $f = 5.35 \times 10^{-7}$: a three order of magnitude improvement over the baseline architecture. With this architecture, the GNC subsystem's failure rate approaches the camera's failure rate, which is the last remaining single point failure. Incorporating camera and sensing technologies that improve the camera's reliability would be required for further improvements to the GNC subsystems' reliability. Figure 17 shows the impact of the improved GNC architecture at the system level. Before, the system reliability was limited by the GNC subsystem's reliability which was expected to fail 10 times more often than the next most failure-prone subsystem. The system failure rate was therefore well above the established TLS. Implementation of the triplex architecture reduces the GNC subsystem's failure rate so that it is now the most reliable subsystem. Furthermore, the system's failure rate is now approximately 60 times larger than the TLS – a large improvement from the previous figure of 428 times the TLS. Reducing the system's failure rate further can be achieved by modifying the architecture of the structures and/or communications subsystems, which now exhibit the highest failure rates of the four subsystems.

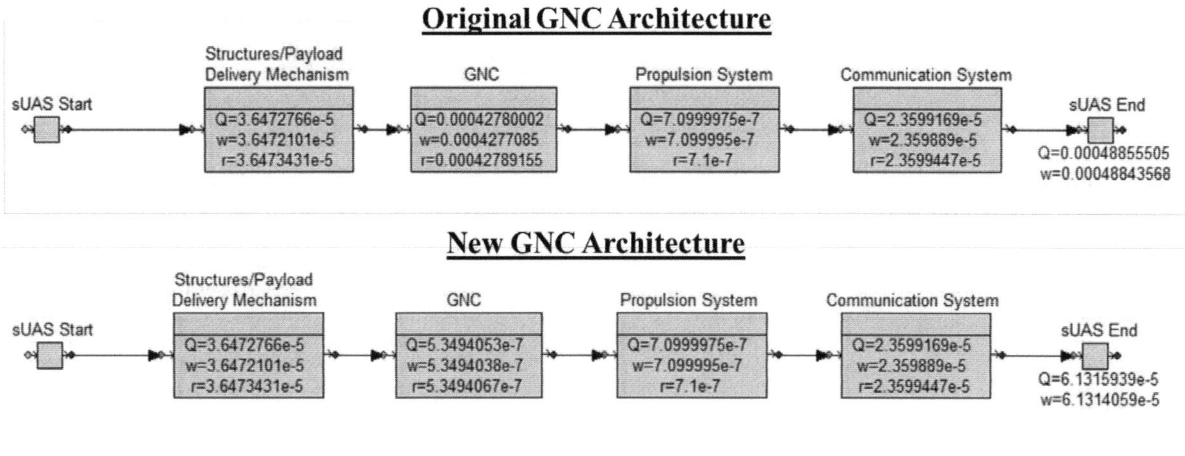

Figure 17. Impact of the triplex FCC architecture on the system level failure rate[1]

Summary and Conclusions

sUAS are relatively small and inexpensive compared to other aircraft. Operators generally will be unable to afford lengthy certification processes and costly testing/V&V cycles. Therefore, certification of a new design should occur in a relatively short period of time (approximately 30 days or less). Unlike commercial transports and larger UAS in general, sUAS are often designed with a specific operation in mind. Each design is optimized to perform that operation. As a consequence of their small scale, each vehicle is relatively inefficient in off-design missions. Therefore, certification should be *operation specific*, meaning that a sUAS is certified along with a defined operation (mission) and the sUAS is only certified to fly when it is performing that mission or a segment of that mission. Achieving such a certification could follow the following steps:

1. Define the operation the operator wishes to certify the SUAS for. This can be accomplished using a mission profile.
2. Identify failure conditions that can lead to property/equipment damage (external to the SUAS) and/or injury. This can be accomplished using a system level FHA. In the future, this set of conditions may be required to contain a set of standard failure conditions that all SUAS are subject to.
3. Demonstrate via ground tests that the SUAS is able to reconfigure and/or mitigate all failures identified in (2). This can be accomplished by performing controlled test cases where faults are intentionally injected into the hardware/software and observing the system's response
4. Demonstrate via flight tests over an unpopulated area that the design is capable of performing the operation (mission profile) defined in (1)
5. Demonstrate via flight tests over an unpopulated area that the design is capable of successfully mitigating the failure conditions defined in (2). This can be accomplished by injecting the same faults from (3) during a flight test.

References

1. Baier, D., Fisher, Z., Schmitt, M. and Yao, Y., "Certification of Small VTOL UAS (sUAS) for Package Delivery", Final Report, AE6362: Safety By Design and Flight Certification, Final Report,April 28, 2017.
2. Xing, Z,"Analysis of UAV Concept of Operations in ITS, University of Minnesota Intelligent Transportation Systems Institute Report, CTS 11-06, Minneapolis, MN, March 2011.
3. Markus, K,"Unmanned Aerial Vehicles in Logistics", DHL, 2014.
4. "Revising the Airspace Model for the Safe Integration of Small Unmanned Aircraft Systems," Amazon Prime Air, July 2015. Web. 16 Feb. 2016. "Amazon Prime Air," Amazon. Web. 17 Feb. 2016.
5. 32nd Annual AHS Student Design Competition,"Distributed Logistics in an Urban Setting Using Small UAVs", 2015.
6. "Small UAS Notice of Proposal Rulemaking", (NPRM), FAA, 15 Feb 2015
7. "*Standard Practice/Guide for Operational Risk Assessment of sUAS.*" Draft Standard WK49619. Harrison Wolf, September 2015. Web. 16 Feb. 2016.
8. Roland E. Weibel, R. J. "Safety Considerations for Operation of Different Classes of UASs" in the NAS". AIAA's 3rd "Unmanned Unlimited" Technical Conference, Workshop and Exhibit (p. 11), 2004. Chicago, Illinois.: AIAA.

Basic Understanding of Airfoil Characteristics at Low Reynolds Numbers

Justin Winslow [1] Hikaru Otsuka [2] Bharath Govindarajan [3] Inderjit Chopra [4]

Alfred Gessow Rotorcraft Center
Department of Aerospace Engineering
University of Maryland, College Park, MD 20742

ABSTRACT

A computational study has been conducted on various airfoils in simulated flows of Reynolds numbers (*Re*) primarily between $10^4 – 10^5$ to provide understanding and guidance for MAV and other low Reynolds number designs. The computational fluid dynamics (CFD) tool utilized in this study was a Reynolds-Averaged Navier-Stokes solver with a Spalart-Allmaras turbulence model and a correlation-based laminar-turbulent boundary layer transition modeling. The airfoils investigated in this study include NACA 0009, NACA 0012 (conventional and reversed configuration), Clark-Y, flat plate airfoils (1, 3, and 5% thickness), and thin cambered plates (3, 6, and 9% camber). Airfoils were examined for lift and drag performance as well as surface pressure and flow field characteristics. In general, it is observed that below Reynolds numbers of 10^6, lift and drag characteristics for most airfoils cannot be assumed to be constant with Reynolds number. Below *Re* of 10^5, cambered plate airfoils are shown to have better lift and drag characteristics than thick conventional airfoils with rounded-leading edges. Flat plate performance is generally invariant to Reynolds number, but performance improves as thickness is decreased for a given Reynolds number.

Introduction

Recent technological advances in compact electronics and power storage combined with the growing need for situational awareness in urban military combat scenarios have provided the impetus for current low Reynolds number (*Re*) design interests. For instance, Micro Air Vehicles (MAVs), which operate in the low *Re* regime because of their relatively small scale, can accomplish a variety of civilian and military based missions that conventional aircraft and large scale Unmanned Aerial Vehicles (UAVs) cannot. MAVs are particularly useful for confined and indoor reconnaissance, especially in hazardous environments. Recent interest has also been generated for Mars exploration UAVs as a fast and mobile alternative to ground-based rovers. Due to the low atmospheric density and temperature on Mars, low *Re* aerodynamics is even more pertinent in aerial vehicle design[1].

In previous studies, most researchers have generally focused on $10^5 < Re < 10^6$ as the low *Re* range. However, this range excludes the lower *Re*, which is of most interest to MAV, Mars UAV, and low speed flight designers. In the present research, chord *Re* between $10^4 – 10^5$ is defined as the low *Re* range. This range is two orders of magnitude smaller than those of large-scale conventional aircraft, which typically operate at *Re* above 10^6. Although large-scale airfoil characteristics, such as high thickness-to-chord (*t/c*), rounded leading edge, and smooth surfaces, have been extensively researched and optimized for decades, these design characteristics may not be directly applicable to the low *Re* regime. Figure 1 shows the variation of maximum sectional lift to drag ratio with Reynolds number and it can be seen that *Re* has a strong effect on conventional airfoil performance. While conventional airfoils perform best at high Reynolds numbers ($10^6 – 10^8$), their performance quickly deteriorates at lower Reynolds number and a geometrically simpler flat-plate out-performs the conventional airfoil. Therefore, it is evident that simply trying to downsize these airfoils to low *Re* scales (*Re* below 10^5) drastically reduces their aerodynamic efficiency in terms of the lift-to-drag ratio[2,3]. Consequently, aerodynamic bodies operating in low *Re* designed with conventional airfoils incur high power requirements because of excessive aerodynamic drag. There is a need to systematically understand the performance of airfoils at these low *Re* numbers and their sensitivities to geometric variations, such as camber and thickness.

[1] Graduate Research Assistant, **jwinslow@umd.edu**
[2] Visiting Ph. D. Student from Tohoku University, **h-otsuka@frl.mech.tohoku.ac.jp**
[3] Assistant Research Scientist, **bharath@umd.edu**
[4] Alfred Gessow Professor and Director, **chopra@umd.edu**
Presented at The 7th AHS Technical Meeting on VTOL Unmanned Aircraft Systems, Mesa, AZ, January 24-26 of 2017.

Flow Separation Physics

Flow field studies, such as particle image velocimetry (PIV), smoke flow visualization, and hot-wire turbulence measurement, have been conducted by previous researchers to provide physical insight into the effect of *Re*, highlighting the difference between low and high *Re* flows. The sensitivity of airfoil performance within high or low *Re* ranges is strongly related to the boundary layer characteristics at a given *Re*.

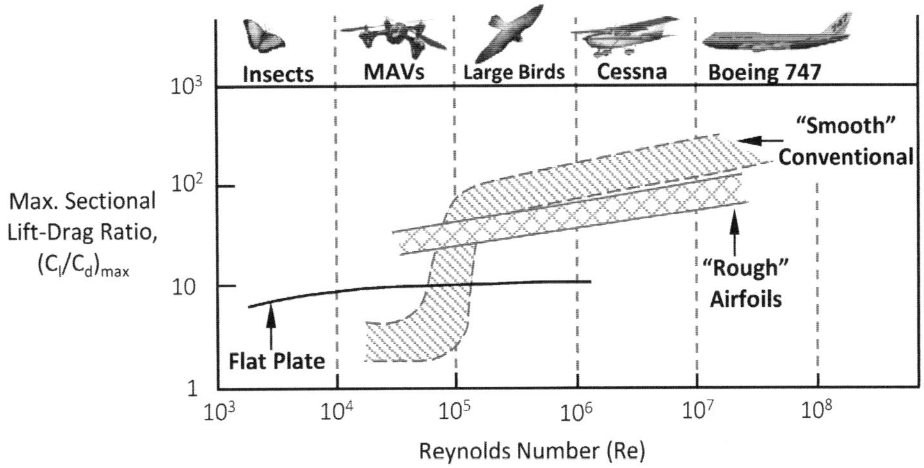

Figure 1: Effect of Reynolds number on airfoil maximum sectional lift-to-drag ratio (Adapted from McMasters[2] and Mueller [3])

Figure 2 shows a schematic of the streamlines over a representative airfoil at various *Re* for a moderate angle of attack. At higher, more conventional, *Re*, represented in Figure 2(c), laminar flow over the upper airfoil surface is subjected to a large adverse pressure gradient close to the leading edge. Because of the inherently lower kinetic energy of the laminar boundary layer, the flow separates as a shear layer. The shear layer gains momentum from the free-stream, and reattaches as a turbulent boundary layer creating a laminar separation bubble (LSB)[4]. Because the turbulent boundary layer is energized from the free-stream, it is much less prone to separation and usually remains attached until the trailing edge. However, at sufficiently high angles of attack (α), the turbulent boundary layer will begin to separate close to the trailing edge resulting in increased pressure drag and decreased lift. Once the turbulent separation reaches the leading edge, the drag rises dramatically, the lift abruptly drops, and the airfoil is considered stalled. This form of separation is known as trailing edge separation and is typically associated with thick, rounded leading edge airfoils.

As *Re* decreases, the separation bubble and turbulent boundary layer thickness both increases in size, resulting in increased parasitic drag. At $5 \times 10^4 < Re < 10^5$, shown in Figure 2(b), the separated shear layer still gains enough momentum from the free-stream to reattach to the airfoil surface as a turbulent boundary layer. However, in this *Re* range, the reattachment point is relatively far back on the airfoil. As α increases, the reattachment point moves towards the trailing edge, creating relatively large separation bubbles (15% to 40% chord) [5]. When the airfoil is close to stall α, the flow fluctuates between reattachment and complete separation, resulting in hysteresis, which can make lift and drag measurements difficult. At $10^4 < Re < 5 \times 10^4$, shown in Figure 2(a), the laminar separation point is delayed until close to the trailing edge, even at very low α [5]. As α increases, the separation point moves towards the leading edge, increasing the separation distance of the shear layer. Unlike at high *Re*, the shear layer is not able to transition and reattach to the airfoil as a turbulent boundary layer, resulting in high pressure drag and poor lift generation[5]. The airfoil is effectively in trailing edge stall for most or all of its operational range. The trailing edge separation increases in size as *Re* is further decreased for a given α, resulting in even lower airfoil efficiency[6].

Previous Studies

The effects of decreasing *Re* on airfoil performance were initially investigated by researchers, such as Schmitz[7], as early as 1930. Schmitz conducted experiments on three airfoils, thin flat plate, thin cambered plate, and a

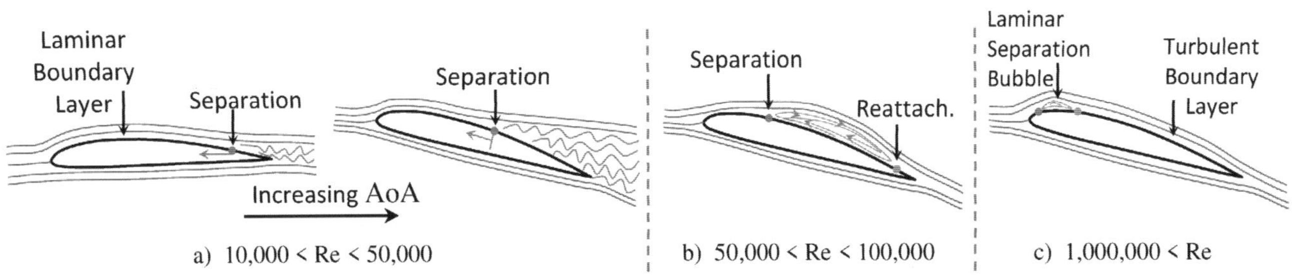

Figure 2: Conventional airfoil characteristics at *Re* < 10^6

conventional N60 airfoil (12.4% *t/c*, 4% camber) within a *Re* range of $2x10^4$ to $2x10^5$. This study resulted in two key conclusions.

1. Thin plate airfoils consistently performed better below *Re* 10^5, and
2. The flow is more susceptible to separation below *Re* 10^5, especially for the thicker N60.

In the 1980's, interest in low *Re* aerodynamics was renewed by researchers such as Mueller, Selig, and Hoerner, whose studies[8] also agreed with those of Schmitz's experiments. Specifically, thin flat and cambered plates are more efficient than conventional airfoils below *Re* of 10^5. Another observation was that cambered plates show a small variation of lift coefficient (C_l) and drag coefficient (C_d) with *Re* while flat plates remain virtually unchanged, which is also evident in Figure 1. Hoerner showed that from $Re = 4x10^4$ to $Re = 1.2$ $x10^5$, cambered plate C_{lmax} rose by 4% whereas N60 C_{lmax} rose by over 180%[8]. Further low *Re* experiments by Selig, Donovan, and Fraser[9] and Mueller and Batill[10] on a range of airfoils also concluded that C_{lmax} consistently increases with higher *Re* and that minimum C_d significantly decreases above *Re* 10^5. Another observation made by Selig was that for the 60 sailplane-type airfoils tested, the drag polar is visually similar and is nearly insensitive to *Re* variation once above *Re* 10^5. But below this *Re*, there is a large degree of non-linearity in the drag polar with *Re* variation[9]. The initial results from each of these studies indicated that low *Re* airfoil performance is very dependent on airfoil geometry and the specific *Re* at which they operate.

While the experimental measurements of previous studies are useful for understanding trends in low *Re* aerodynamics, they also have many limitations. Previous studies on $Re < 10^5$ typically contain lift and drag data only for a single *Re*[11] for each airfoil, which is not sufficient to fully characterize its performance across the *Re* spectrum. Furthermore, the data sets which are below 10^5 contain uncertainties and discrepancies. For instance, C_d measurements taken on the E387 airfoil (9.1% *t/c*, 3.2% camber) at *Re* $6x10^4$ have been shown to vary by 28–68% between independent measurements, which have been taken at different facilities[12]. Whereas at *Re* $2x10^5$, the drag measurements were much more consistent and vary by less than 13%[12]. These discrepancies are due to factors such as the drag measurement method used (wake-momentum deficit[13] or load measurement), inadequate force measurement sensitivity[14], and wind tunnel turbulence levels[15]. Additionally, below *Re* 10^5, even small changes in *Re* have a much more prominent effect on airfoil performance than they would at higher *Re*. It has been shown that for the NACA 0012 airfoil, the lift curve is highly non-linear[16], particularly below *Re* $5x10^4$. And, C_l at *Re* $5x10^4$ can be as much as 3 times higher than C_l at *Re* 10^4 for a given angle of attack[16]. Since many of the airfoils tested below *Re* 10^5 are only tested at one *Re*, there is not enough data to capture each airfoil's performance sensitivity to *Re*.

It may be possible to perform a large number of experiments to assess the performance characteristics of various airfoils at different Reynolds number over a wide range of angle of attack and Mach number. However, these tests can be prohibitively expensive and time consuming making them practically infeasible. On the other hand, computational fluid dynamics (CFD) has been significantly refined over the past three decades and has now been routinely used by academia and industry to the point of accurately predicting both internal and external aerodynamic flows past complex geometries. CFD is being extensively used to design and predict the performance of a wide range of single and multi-element airfoils. Therefore, the goal of the present research is to utilize an in-house computational tool capable of generating reliable lift, drag, and moment data for an arbitrary 2D airfoil at any *Re* above 10^4 to extract the useful trends from the data to: 1) understand the behavior of airfoils at these low Reynolds number, and 2) to provide insight and guidance for low *Re* designers.

Computational Method

The CFD solver used in this study to investigate 2D low *Re* aerodynamics is *Transonic Unsteady Rotor Navier-Stokes 2D* (TURNS2D)[17] fluid dynamics solver developed at the University of Maryland. TURNS2D has been widely used in the past for flows past airfoil and rotor blades at high-Reynolds number with confidence[17,18]. TURNS2D uses a dual volume formulation to solve for the Reynolds-Averaged Navier-Stokes (RANS) equations. The inviscid fluxes uses a third-order MUSCL reconstruction scheme in conjunction with Roe's scheme, while the viscous fluxes are computed using a second-order central difference. For steady problems, an Euler implicit formulation was used for time marching with the matrix inversion performed using a lower-upper symmetric line-Gauss Seidel approach. A key feature in TURNS2D is the inclusion of a laminar-turbulent transition model coupled with the Spalart-Allmaras (SA) turbulence model, which improves the numerical solution by better predicting the flow transition physics[19]. The formulation keeps track of the "intermittency" in the flow, whose value varies between 0 and 1, indicating a fully attached or fully separated flow, respectively. C-meshes were generated around the airfoil of interest and the RANS equations were solved using TURNS2D to generate sectional coefficients of surface pressure (C_p), skin friction (C_f), lift (C_l), drag (C_d), and pitching moment (C_m) for an airfoil under given flow conditions[17].

To generate the following CFD results, a C-mesh was used as the grid type around the airfoil with an outer boundary at 15 chord lengths away from the airfoil surface. A mesh size with 267 grid points in the surface-wrap direction and 123 grid points in the surface-normal direction. Grid spacing within the boundary layer in the normal direction was 0.001% of the chord length. Dual time stepping was implemented with a time step size of 0.1. The freestream Mach number was fixed at 0.1 because the majority of studies did not report an exact Mach number, and the flows were assumed to be within the incompressible range. These parameters were found to be sufficient for solution convergence at the range of low *Re* investigated. Using these parameters, the solution for a single airfoil at a single *Re* over a range of ten angles of attack takes approximately nine hours to converge on a typical desktop computer.

Results and Discussion

To demonstrate the capability of TURNS2D, validation cases were carried out by comparing previous experimental results to the current computational results. The primary validation data used was existing C_l and C_d data versus angle of attack (α). C_p along the chord was also used to validate the TURNS2D results with available test data. Because the NACA 0009 experiment by Lutz, Würz, and Wagner[20] reported both C_l and C_d data and C_p data for low *Re* flows, it was selected as an early validation case. The upper surface pressure coefficient validation for the NACA 0009 airfoil at multiple angles of attack is shown in Figure 3 ($Re = 5 \times 10^4$). It is evident that at low angles of attack ($\alpha \leq 5°$), the C_p results from TURNS2D agrees very well with the experimental data[20]. At higher angles of attack, the agreement is less satisfactory but still follows the trend as angle of attack increases. Figure 4 shows the lift and drag comparison for the same NACA 0009 airfoil at $Re = 5 \times 10^4$. It can be seen that for the majority of the lift and drag curves, the C_l and C_d calculated by TURNS2D compare very well until close to stall due to large separated flow regions, which is a limitation of the S-A turbulence model[19]. The SA turbulence model within RANS is known to overpredict the Eddy viscosity,

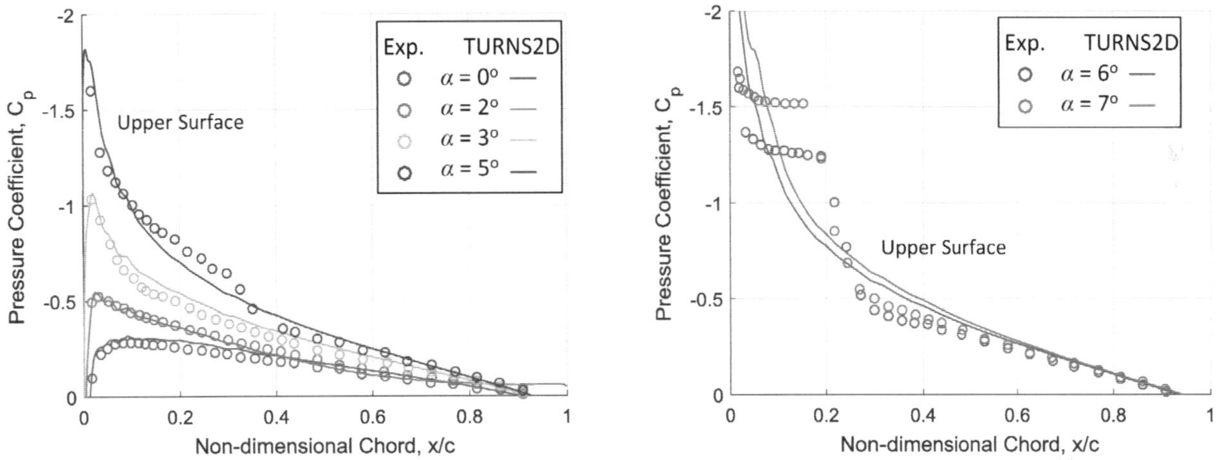

(a) NACA0009: Low angle of attack *(b) NACA0009: High angle of attack*

Figure 3: C_p over the upper surface of a NACA0009 airfoil for various angles of attack at $Re = 5 \times 10^4$ (Experimental data from Lutz, et al.[20])

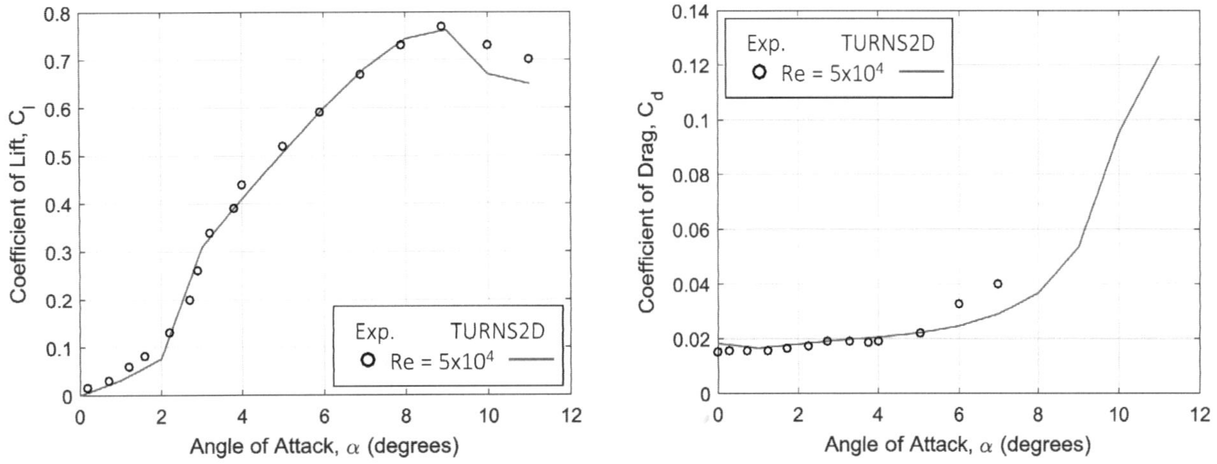

Figure 4: C_l and C_d comparison of TURNS2D CFD code with experiment for NACA0009 at $Re = 5 \times 10^4$ (Experimental data from Lutz, et al.[20])

which tends to lead to high-rates of dissipation in the flow. To better predict the onset of stall and the immediate post-stall characteristics, either a Large Eddy Simulation or a Detached Eddy Simulation may be required, which is beyond the scope of the present study.

Effect of Re on NACA 0012

The NACA 0012 is a common airfoil which has been experimentally tested for a wide range of *Re*. Therefore, the experimental results serve as a useful validation database for both high and low *Re*. The wide range of *Re* also provides insight into how strongly *Re* effects airfoil performance.

Figure 5 shows the validation of C_l and C_d predictions from TURNS2D with the experimental measurements taken at Sandia National Laboratories[21] for the NACA 0012 at high Re ($3\text{x}10^5 < Re < 1.8\text{x}10^6$). The lift and drag correlate well for both $Re = 3\text{x}10^5$ and $Re = 10^6$ for the range of α shown, including the angles around stall. Already, a 23% drop in peak C_l from $Re = 10^6$ to $Re = 3\text{x}10^5$ is observed. The drag also rises sharply at a smaller α as Re is decreased. However, the lift curve slope and minimum drag appear to be very similar.

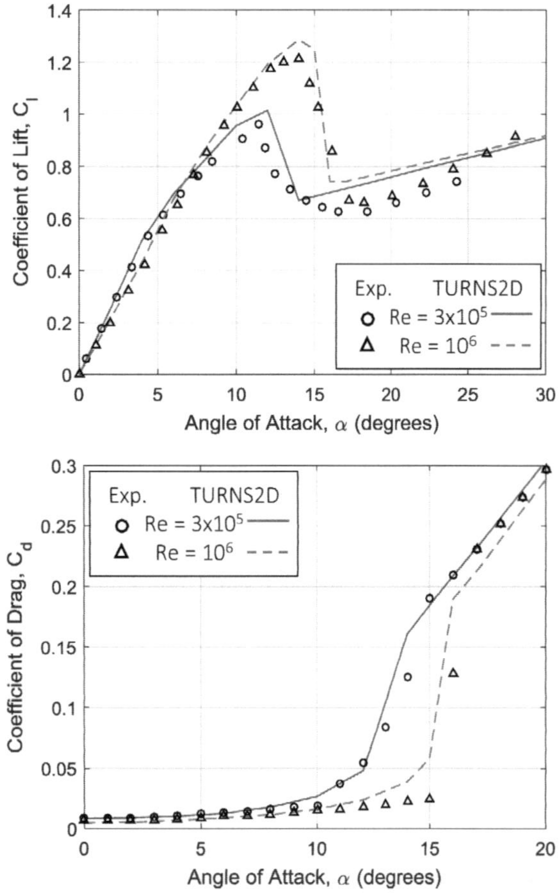

Figure 5: High *Re* C_l and C_d comparison of TURNS2D CFD predictions for NACA0012 at $Re = 3\text{x}10^5$ and $1\text{x}10^6$ (Experimental data from Sheldahl and Klimas[21])

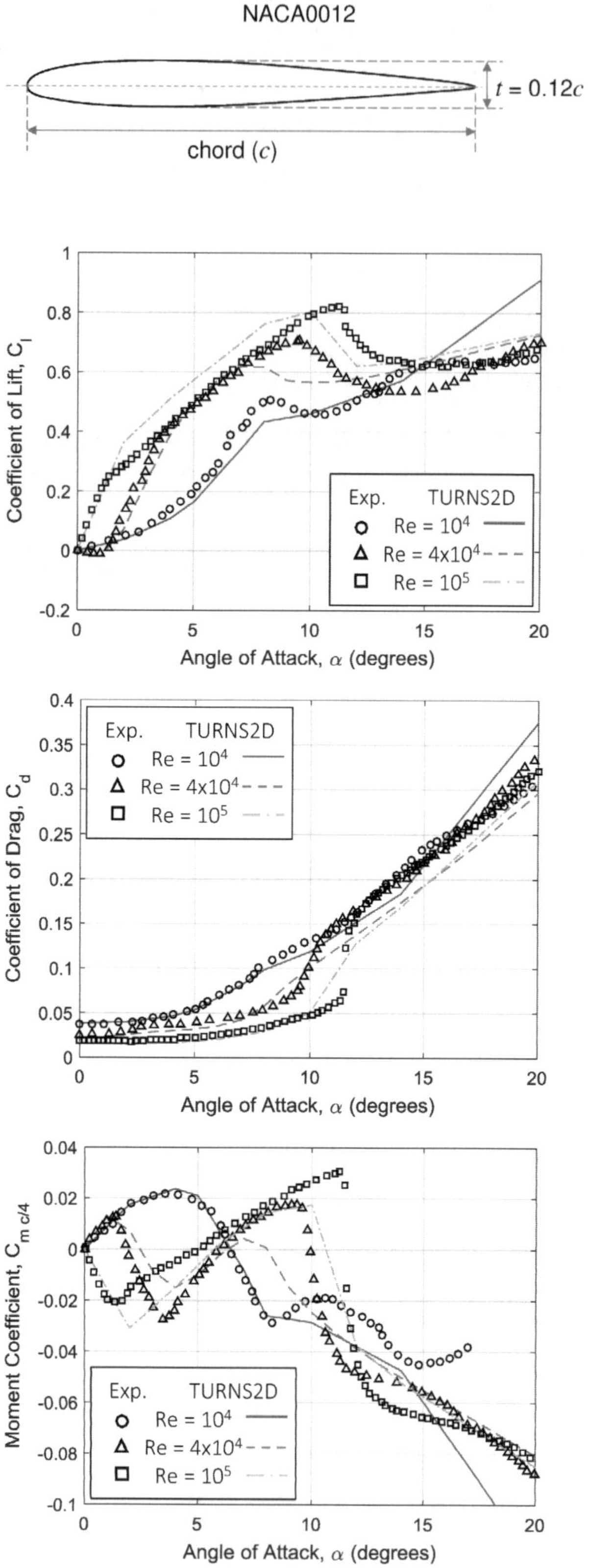

Figure 6: C_l, C_d, and C_m comparison of TURNS2D CFD predictions for NACA0012 at $10^4 < Re < 10^5$ (Experimental data from Ohtake, et al.[16])

The C_l, C_d, and C_m comparisons for a NACA 0012 airfoil at various low *Re* between 10^4 and 10^5, is shown in Figure 6. The results show the extreme sensitivity of airfoil performance to changing *Re* below 10^5 for the NACA 0012. As *Re* decreases, there is an increase in drag, particularly due to shrinking of the drag bucket, and a large decrease in lift. Here it is shown that peak C_l decreases by approximately 46% between $Re = 10^5$ and $Re = 10^4$. Unlike the higher *Re* results in Figure 5, the lift curves are highly non-linear, particularly for $\alpha < 5°$, and a single lift-curve-slope value cannot be assumed. It should also be noted that the drag bucket is significantly smaller at low *Re* and the minimum drag noticeably increases compared to the minimum drag in Figure 5. The general shapes of the moment curves are shown to vary significantly as *Re* changes, but these are still predicted moderately well.

Effect of Thickness on Flat Plates

The specific airfoil characteristics, such as camber and thickness, heavily affect airfoil lift and drag qualities. For example, the effect of thickness (1, 3, and 5%) for simple flat plates at *Re* of 10^4 is shown in Figure 7. Results are shown from both TURNS2D and experimental measurements[22]. It is evident that as flat plate thickness-to-chord (*t/c*) decreases, lift and drag characteristics improved, with the 1% flat plate being the most efficient. For each flat plate, the lift-curve slope is generally constant up to approximately $\alpha < 7°$, where the slope then abruptly drops for $\alpha > 7°$. However, the high lift-curve slope is maintained to higher α as *t/c* is decreased. Flat plate C_d also decreases as *t/c* decreases. C_d appears to decrease with *t/c* between $0° < \alpha < 10°$, then converges to similar values at high α. Minimum *t/c* is shown to be a key characteristic for increased aerodynamic efficiency at low *Re*.

An explanation for the decreased performance as flat plate *t/c* increases is the increased separation of the flow around the airfoil. The thicker the flat plate is, the blunter its leading edge becomes. The increased bluntness of the airfoil leading edge increases the strength of the adverse pressure gradient. Thus, the flow around the airfoil is separated by a greater distance which increases form drag. Furthermore, the increased separation of thicker flat plates reduces the pressure differential between the upper and lower surfaces, thus reducing its lifting force.

Effect of Camber on Thin Plates

In order to understand the performance trends at Low *Re* due to camber, thin plates with varying camber were investigated. C_l and C_d for 1% thick plates with cambers of 3, 6, and 9% at $Re = 10^4$ are shown in Figure 8. Results are shown from TURNS2D with experimental measurements[22] overlaid. The results show that as camber is increased, so does the lift curve and peak lift by a significant margin. However, there was also increased drag as camber was increased, particularly for 9% camber. Figure 9 shows the lift-to-drag ratio of each cambered plate. It is evident that despite the higher C_l of the 9% camber, a moderate camber (6%) yields the highest lift-to-drag. Clearly, camber has a strong effect on low *Re* aerodynamic performance.

The effect of increasing airfoil camber causes a greater change in momentum of the flow around the airfoil, thus increasing lift. However, the trade-off is that increasing thin plate camber also increases drag. When the camber is too great, a large separation region is created on the aft end the airfoil. This separation region effectively increases the profile size of the thin plate, which increases drag. At low *Re*, the camber value at which the drag increases begin to outweigh the lift gains appears to be larger than 6%. For a low *Re* wing design to achieve high lift, approximately 9% camber may be needed. Whereas, 6% camber should be utilized if efficiency is a desired design factor.

Effect of Re on Thin Cambered Airfoils

The effect of camber on thin (3% *t/c*) NACA airfoils was examined with the use of computational results from TURNS2D. The performance results for symmetric

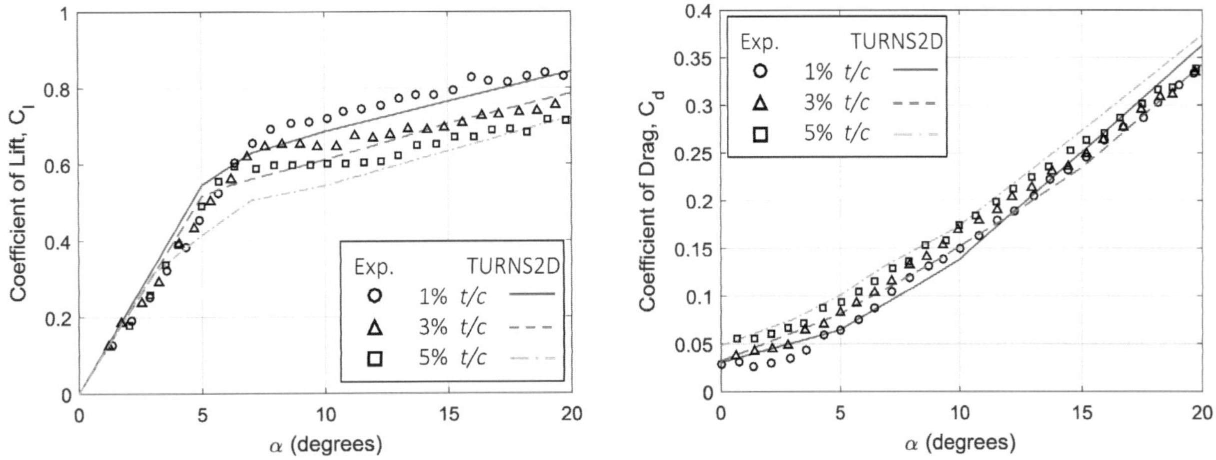

Figure 7: Effect of flat plate thickness on C_l and C_d at $Re = 10^4$ (Experimental data from Okamoto, et al.[22])

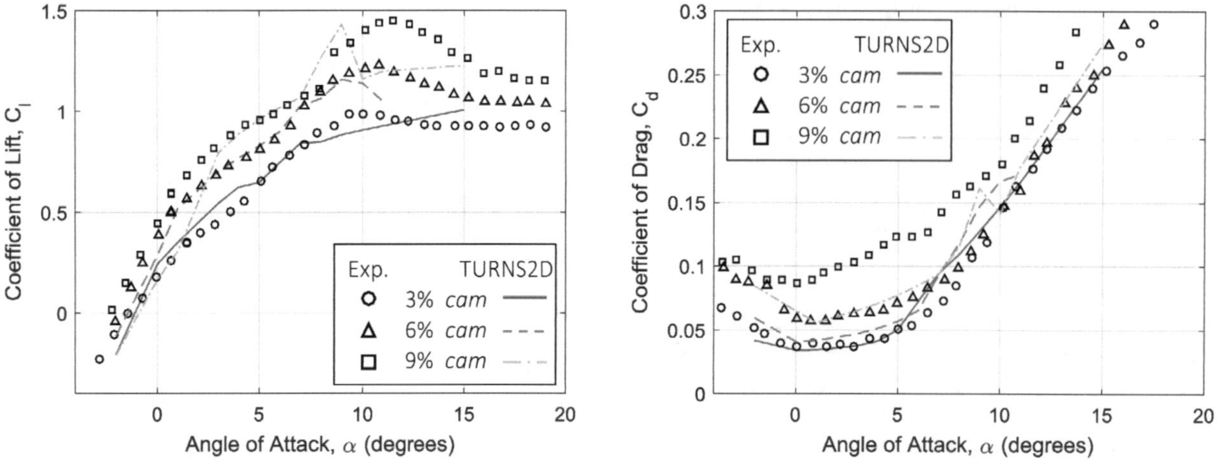

Figure 8: Effect of thin plate (*t/c*=1%) camber on C_l and C_d at $Re = 10^4$ (Experimental data from Okamoto, et al.[22])

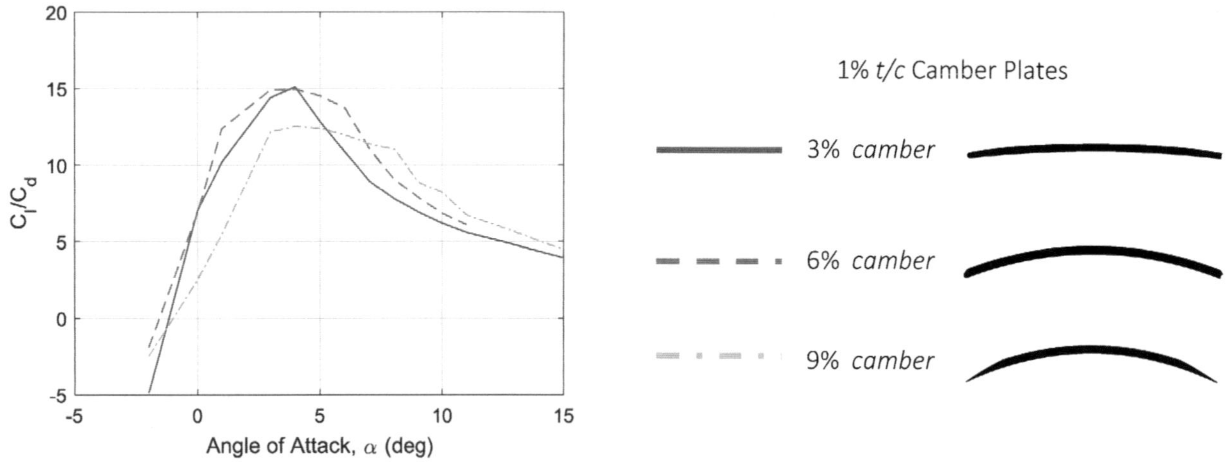

Figure 9: Effect of thin plate (*t/c*=1%) camber on lift-to-drag ratio at $Re = 10^4$

(NACA0003) and 2% (NACA2403), 4% (NACA4403), and 6% (NACA6403) cambered airfoils are shown in Figure 10. Results are shown for two flows, $Re = 2\text{x}10^4$ and $Re = 10^5$.

At both $Re = 2\text{x}10^4$ and $Re = 10^5$, increasing camber increases the C_l over the given range of angle of attack. However the increase in camber also causes a drag penalty by increasing the minimum drag. As a result, the lift-to-drag curves of the cambered airfoils are similar to each other at $Re = 2\text{x}10^4$, but they are slightly higher compared to the symmetric airfoil's lift-to-drag. At $Re = 10^5$, increasing camber has a more pronounced effect on performance. It is seen in Figure 10 that increasing *Re* increases C_l by a larger amount for higher camber airfoils. Additionally, C_d decreases more as *Re* increases for higher camber airfoils, particularly within $4^o < \alpha < 12^o$. As a result the lift-to-drag curves at $Re = 10^5$ are more differentiated than at $Re = 2\text{x}10^4$. The symmetric airfoil performance changes very little due to the increase in *Re*. As camber is increased, the lift-to-drag improves by a greater margin. This result indicates that the degree to which an airfoil's characteristics effect its performance is also dependent on *Re*.

Flat Plate and NACA 0012

It was seen in Figure 1 that simple flat plates outperform conventional airfoils for $Re < 5\text{x}10^4$. To obtain a deeper understanding of this rationale as to why certain airfoil characteristics are better suited to low *Re* flows than others, the flow field solution from TURNS2D are examined. For example, the comparisons between the NACA 0012 airfoil and a 2% *t/c* flat plate at $Re = 2\text{x}10^4$ are shown in Figure 11. The key, distinguishing factor between the thick NACA 0012

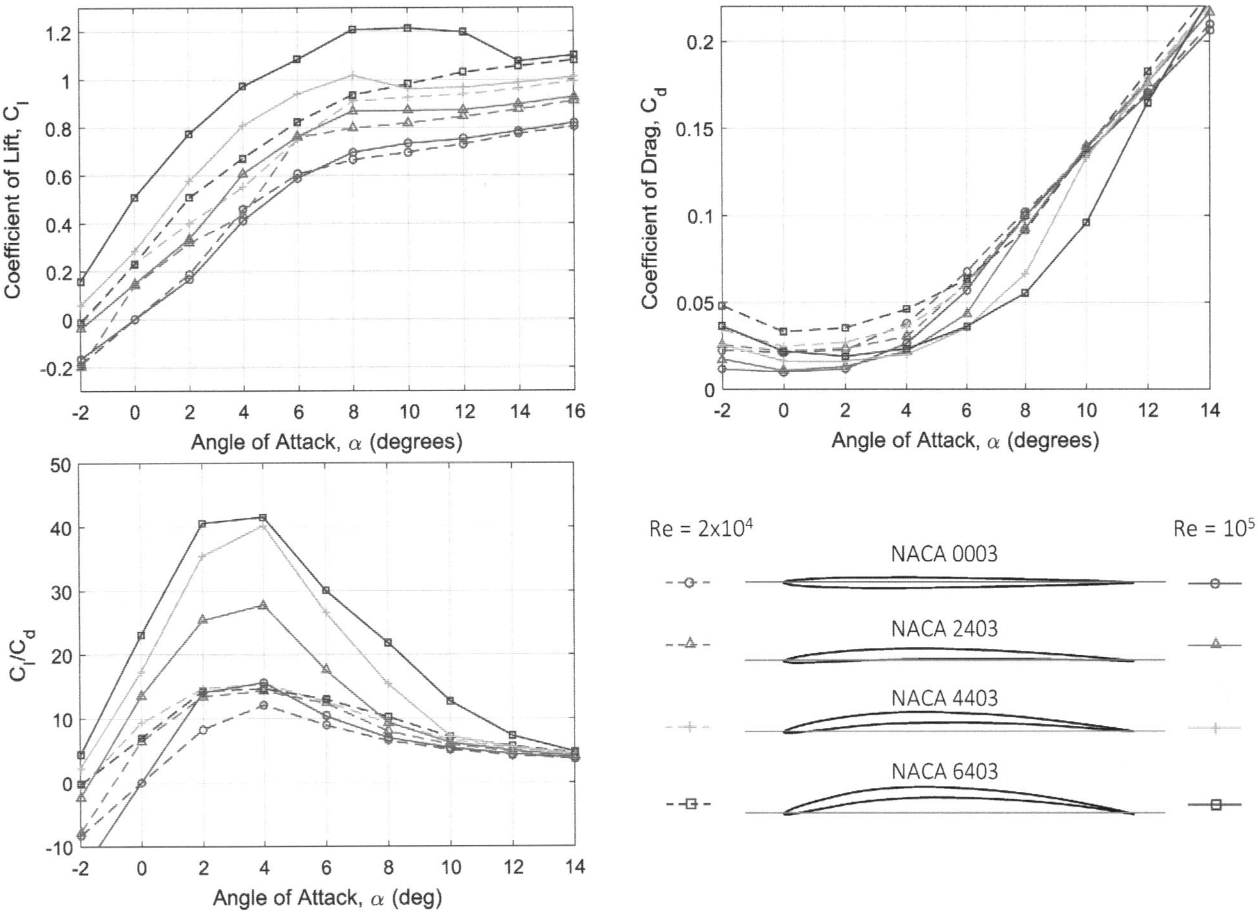

Figure 10: Effect of camber on C_l, C_d, and C_l/C_d for NACA Airfoils of 3% *t/c* at $Re = 2\times10^4$ and 10^5

and thinner flat plate is the difference in boundary layer separation. For the NACA 0012, the upper surface boundary layer stays attached for a majority of the airfoil until it separates close to the trailing edge. This separation characteristic is similar to flow-field results from previous studies[5,6] on thick, conventional airfoils at $Re < 5\times10^4$. For the range of α shown, the flow separates too close to the trailing edge to gain momentum from the free-stream and reattach as

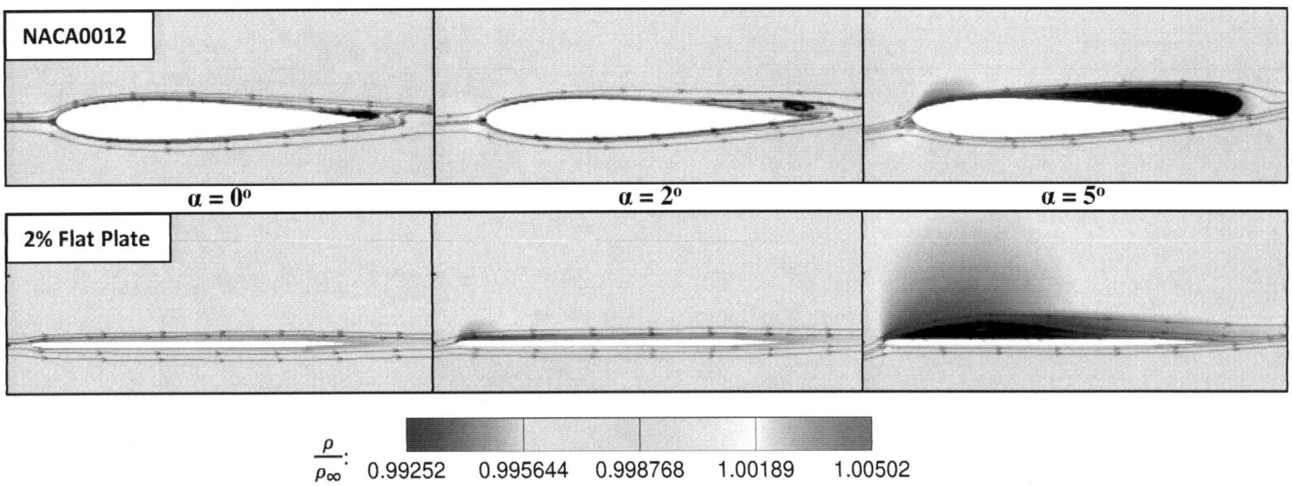

Figure 11: Difference in boundary layer characteristics between NACA 0012 and 2% *t/c* flat plate at $Re = 2\times10^4$ (contours represent non-dimensional density)

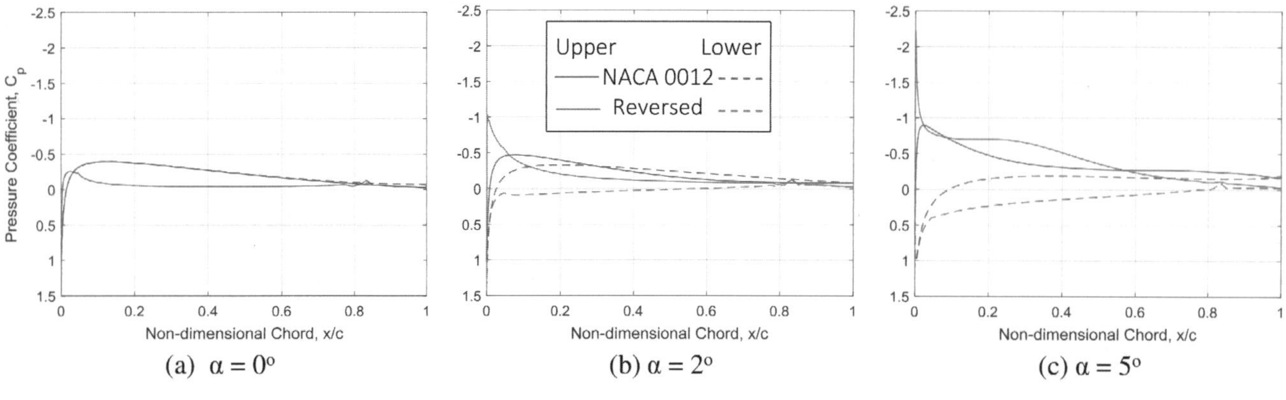

Figure 12: Pressure distribution on upper and lower surfaces of a NACA 0012 and 2% *t/c* flat plate at *Re* = 2x10⁴

a turbulent boundary layer. The resulting large separation region causes high form drag and prevents formation of the beneficial drag bucket around low angles of attack.

The large trailing edge separation results in low lift due to increased pressure on the upper surface. This low *Re* separation effect is particularly influential on the C_l curve for $0° < \alpha < 3°$ and $Re < 5x10^4$. In this range of angle of attack, the separation region can be large enough that the upper surface C_p actually becomes higher than the lower surface pressure C_p for most of the chord. This effect is evident in Figure 12 at $\alpha = 2°$ where the upper and lower surface C_p curves for the NACA 0012 cross at 30% chord. At $Re < 5x10^4$, this pressure difference causes the lift curve slope close to $\alpha = 0°$ to be less steep at other angles. It can even cause slightly negative C_l values as seen in Figure 6 for $Re = 4x10^4$.

The 2% *t/c* flat plate airfoil exhibits significantly different flow characteristics than the NACA 0012 in the same flow regime. As seen in Figure 11, the flat plate does not undergo trailing edge separation. Instead, the laminar flow is tripped early by the sharp leading edge. The separated laminar shear layer gains momentum from the free-stream and transitions to turbulent. Since this transition occurs near the leading edge. There is enough chord length remaining for the turbulent boundary layer to reattach, forming a separation bubble. This bubble is most visible at $\alpha = 5°$ in Figure 11 for the flat plate. The C_p distribution over the upper surface of the flat plate in Figure 12 also indicates the presence of the laminar separation bubble. At $\alpha = 5°$, there is a flattening of the thin plate upper C_p, which is typical of a separation bubble. The reattachment of the upper boundary layer preserves the lower pressure and therefore allow the flat plate to provide more lift than the NACA 0012 at $Re < 5x10^4$.

Conventional and Reversed NACA 0012

The effect of reversing airfoils at low *Re* may actually be beneficial depending on the airfoil. In addition to the maximum thickness of the airfoil, its leading edge (LE) geometry, sharp or rounded, has a strong effect on its performance. By reversing a symmetric airfoil, the LE

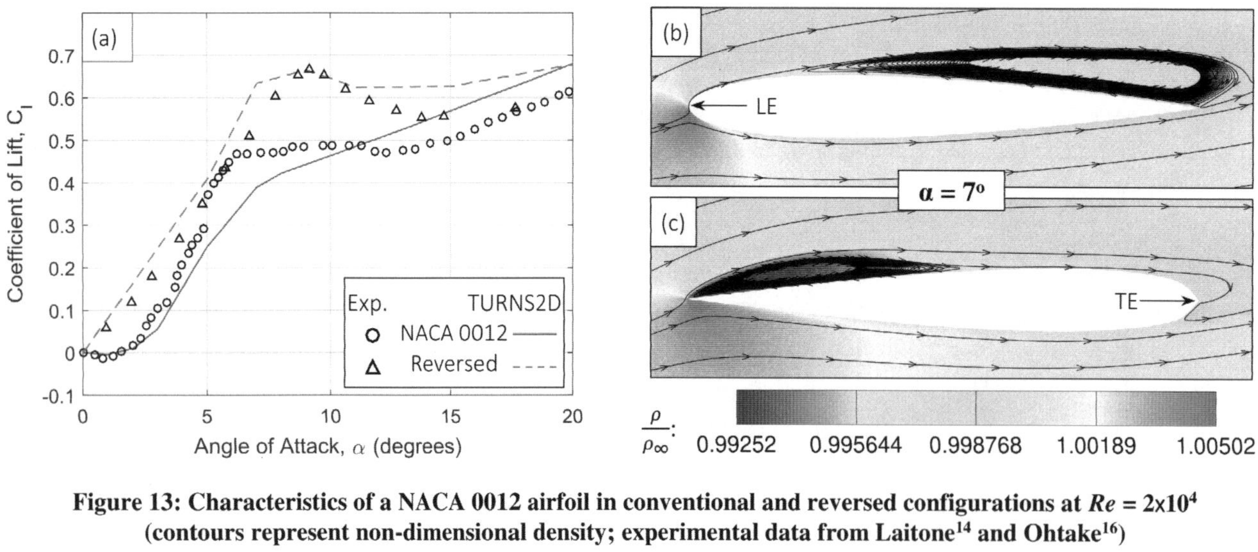

Figure 13: Characteristics of a NACA 0012 airfoil in conventional and reversed configurations at *Re* = 2x10⁴ (contours represent non-dimensional density; experimental data from Laitone[14] and Ohtake[16])

geometry is essentially changed but the maximum thickness remains the same.

Figure 13 shows the characteristics of a NACA 0012 in its conventional configuration (rounded LE) compared to its reversed configuration. In Figure 13(a), both the experimental data[14,16] and TURNS2D predictions show that the reversed configuration yields increased lift at $Re = 2 \times 10^4$. In particular, there is a very strong effect on peak lift between $\alpha = 7^o$–10°. The computational flow fields around the NACA 0012 at $\alpha = 7^o$ provides insight into the difference in lift values. Figure 13(b) shows how the NACA 0012 in its conventional configuration creates a large separation region over 80% of the upper surface. However, when the NACA 0012 is reversed in the same flow conditions, it behaves more like a flat plate. Most notably, the now sharp leading edge creates a similar separation bubble as seen on the 2% *t/c* flat plate in Figure 11. The early separation of the boundary layer on the reversed NACA 0012 allows it reattach as a turbulent boundary layer further downstream. The flow reattachment maintains lower pressure on the upper surface and increases lift. Despite the fact that the NACA 0012 has the same maximum thickness in either configuration, these results indicate the distribution of the thickness along the chord, particularly at the LE, has an effect on airfoil performance at low *Re* that is counter to what is expected at high *Re*.

Overall Re Effects

TURNS2D has been used as to determine other trends in aerodynamic performance of airfoils at low *Re*. For example, Figure 14 shows the effect *Re* of multiple types of airfoils: 2% thick flat plate, 6% cambered plate (2% thickness), NACA 0012, and Clark-Y. This comparison was conducted to obtain an understanding of *Re* effects on vastly different airfoils. The major distinguishing factor between the airfoils is that the more conventional NACA 0012 and Clark-Y are much more sensitive to *Re* than plate-type airfoils. As seen in Figure 14, the peak C_l of the 6% cambered plate only increases from approximately 1.15 at $Re = 2 \times 10^4$ to approximately 1.32 at $Re = 10^6$, and the peak C_l of the flat plat is virtually unchanged. Over the same *Re* range, the peak C_l more than doubles for the NACA 0012 and Clark-Y, increasing from 0.55 to 1.3 for the NACA 0012 and from 0.72 to 1.63 for the Clark-Y.

In Figure 14, the largest change in performance between the four airfoils occurs after $Re = 10^5$. At $Re = 2 \times 10^4$ and 4×10^4, the thin flat and cambered plates clearly outperform the conventional airfoils in terms of lift and drag. But at $Re = 1.2 \times 10^5$, the C_l of the Clark-Y is comparable to that of the 6% cambered plate and the C_l of the NACA 0012 is primarily greater than that of the flat plate for a given C_d. These performance margins increase further at a higher *Re* of 10^6.

It should also be noted that *Re* appears to have a uniform effect on the C_d of each of the four airfoils. Figure 14 shows that as the *Re* increases, the C_l vs C_d curves gradually shift to the left, indicating decreasing drag. The minimum C_d appears to decrease by approximately the same amount for all four

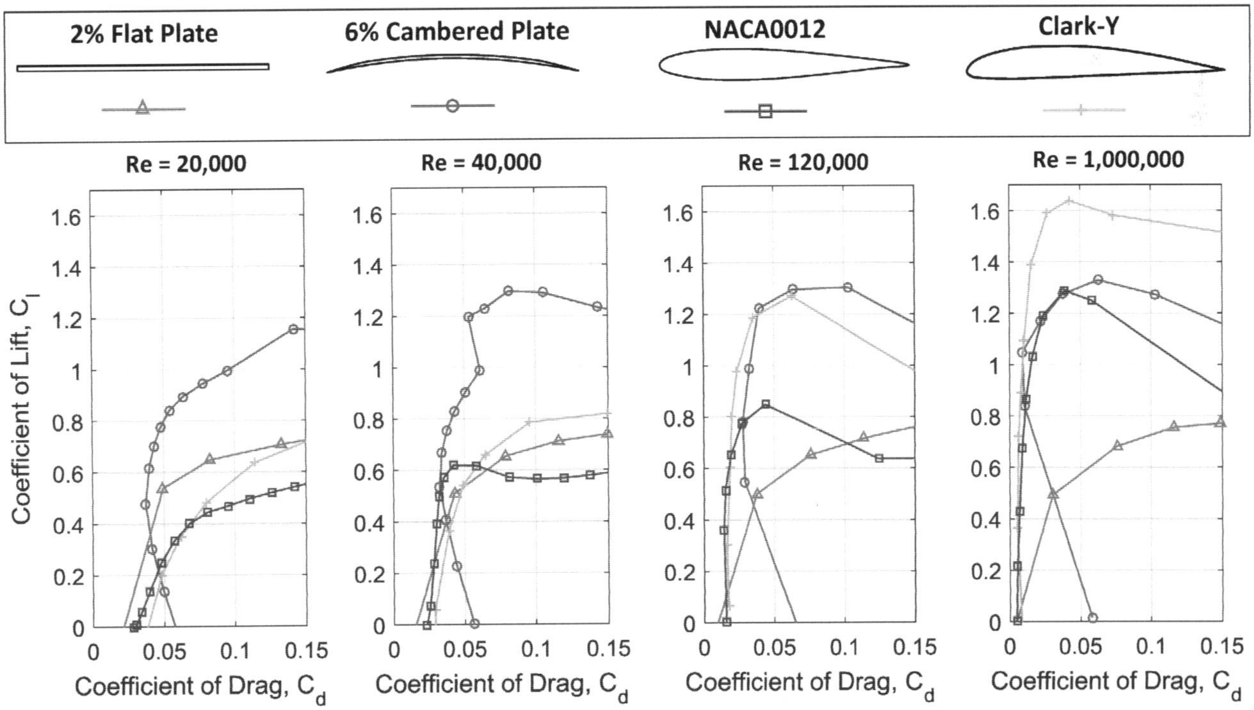

Figure 14: Effect of Reynolds number on multiple airfoils in terms of C_l vs C_d

airfoils. Because the geometric characteristics of the four airfoils are vastly different, the similar decrease in their minimum C_d, is likely attributed to decreasing viscous drag effects as *Re* increases.

Figure 15 shows the effect of *Re* on each airfoil (2% *t/c* flat plate, 6% cambered plate, NACA 0012, and Clark-Y) separately in terms of lift-to-drag (C_l/C_d) as a function of α. Similarly to the results observed in Figure 14, the conventional NACA 0012 and Clark-Y performance is shown to be more sensitive to *Re* compared to the plate-type airfoils. Surprisingly, the 6% cambered plate appears to achieve the same peak lift-to-drag as the Clark-Y at $Re = 10^6$, as seen in Figure 15. However, the Clark-Y maintains a larger envelope of higher C_l/C_d for $\alpha > 5°$.

The primary implications for design drawn from these results are: 1) Below $Re = 10^5$, cambered plate airfoils will have better lift-to-drag characteristics than conventional airfoils 2) Above $Re = 10^5$, thicker airfoils will have better lift to drag characteristics than thin flat or cambered plates 3) Below $Re = 10^6$, lift to drag characteristics for most airfoils cannot be assumed to be constant with *Re* and 4) Flat plate performance is generally invariant to *Re*.

Conclusions

The goal of the present research was to utilize TURNS2D as a means to understand low *Re* aerodynamics and therefore to provide guidelines for MAV and low *Re* designs. The results from TURNS2D are compared with available experimental C_l, C_d, C_m, and C_p data, as means of validation. Overall, the calculated values from TURNS2D agree quite well with the experimental measurements at low *Re*. The C_l, C_d, and C_p curves for the NACA 0009 at $Re = 5x10^4$ correlate very well for low to moderate α. The calculated and experimental C_l, C_d, and C_m curves for the NACA 0012 at various *Re* between 10^4-10^6 are in good agreement. TURNS2D also predicts the performance trends of increasing airfoil camber and thickness at $Re = 10^4$ quite well.

The results from TURNS2D show the extreme sensitivity of airfoil performance to changing *Re* below 10^5, particularly for conventional airfoils like the NACA 0012. As *Re* decreases there is an increase in drag, particularly due to the

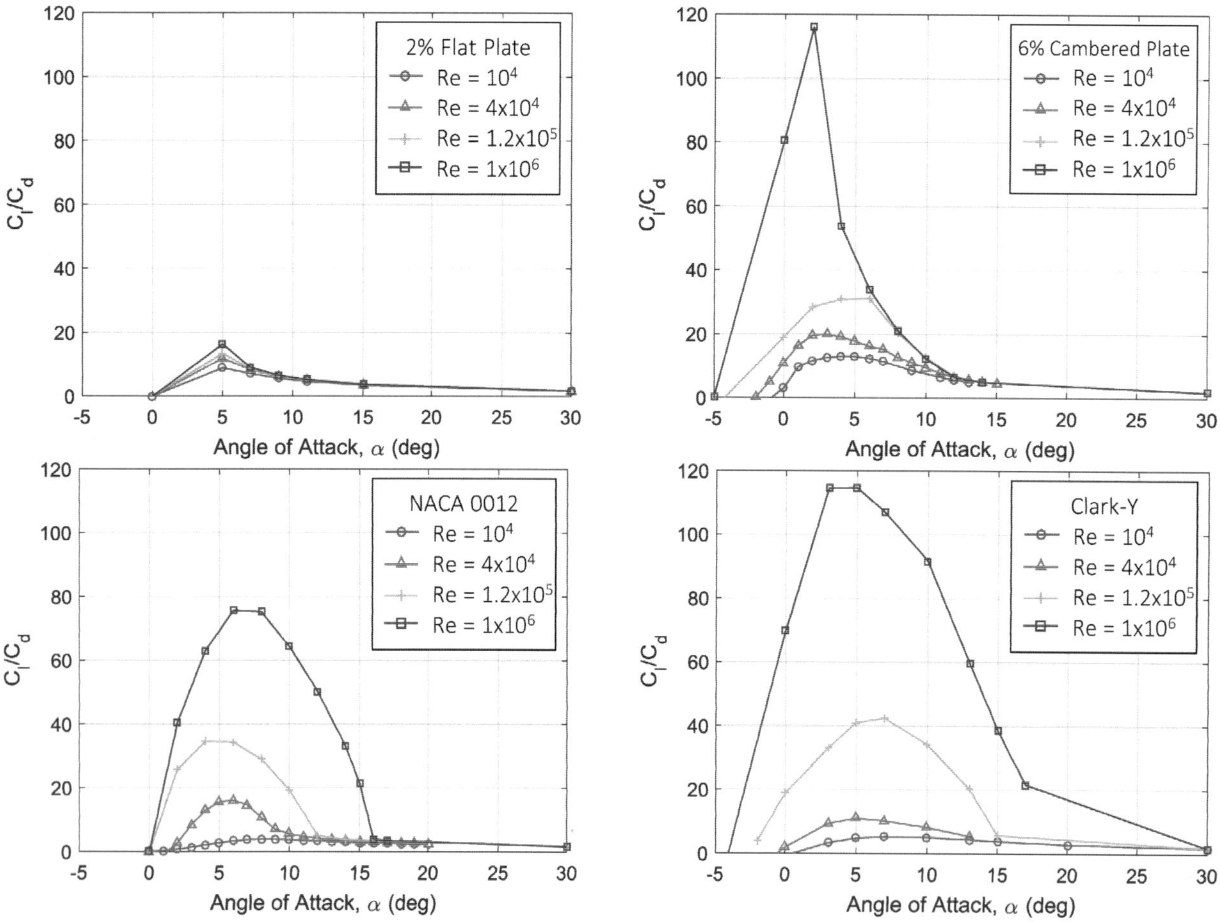

Figure 15: Effect of Reynolds number on multiple airfoils in terms of lift-to-drag vs angle of attack

shrinking of the drag bucket, and a large decrease in lift. C_{lmax} decreases by approximately 46% between $Re = 10^5$ and $Re = 10^4$. Unlike performance results above $Re = 10^5$, the lift curves are highly non-linear, particularly for $\alpha < 5°$, and a single lift-curve-slope value may not be appropriate. It should also be noted that the drag bucket is significantly smaller at $Re < 10^5$ and the minimum drag noticeably increases.

The effect of thickness (1, 3, and 5%) for simple flat plates was examined at $Re = 10^4$. It is evident that as flat plate thickness-to-chord (*t/c*) decreases, lift and drag characteristics improved, with the 1% flat plate being the most efficient.

The effect of camber (3, 6, and 9%) on thin plate airfoils was examined at $Re = 10^4$. As camber is increased, both the C_l and C_d increase and the lift-to-drag is significantly improved compared to flat plate airfoils of the same thickness. For a low *Re* wing design to achieve high lift-to-drag, approximately 6% camber appeared optimum. On the other hand, 9% camber could be utilized if maximum lift is a stronger design factor.

The effect of camber on thin NACA airfoils (NACA 0003, 2403, 4403, 6403) was also examined at $Re = 2x10^4$ and $Re = 10^5$. The high camber airfoils (NACA 4403 and 6403) were shown to have the highest lift-to-drag at both *Re*. Additionally, the lift-to-drag curves are more similar at $Re = 2x10^4$ and more differentiated at $Re = 10^5$.The symmetric NACA 0003 airfoil performance changes very little with an increase in *Re*. As camber is increased, the lift-to-drag improves by a greater margin. This result indicates that the degree to which an airfoil's characteristics effect its performance is also dependent on *Re*.

Examination of the boundary layer at low *Re* shows significant differences between thin plate airfoils and thicker airfoils. The NACA 0012 is more susceptible to trailing edge separation at low *Re* whereas the flat plate trips the laminar boundary layer at the leading edge, allowing it to reattach. The reattachment of the upper boundary layer preserves lower surface pressure and therefore allow the flat plate to provide more lift than the NACA 0012 at $Re < 5x10^4$.

When the NACA 0012 is placed in a reversed configuration at $Re = 2x10^4$, its lift qualities improve. This effect is due to the sharp leading edge creating an early laminar separation bubble similar to the flat plate's. Despite the fact that the NACA 0012 has the same maximum thickness in either configuration, this result indicates that the distribution of the thickness along the chord, particularly at the LE, has an effect on airfoil performance at low *Re* that is counter to what is expected at high *Re*.

Acknowledgment

This work is sponsored by the Army, Navy and NASA VLRCOE partnership. The authors would like to thank Dr. Mahendra Bhagwat as Technical Monitor and Dr. Alex Moodie for his many discussions and comments.

References

[1] Shrestha, R., Benedict, M., Hrishikeshavan, V., and Chopra, I., "Hover Performance of a Small-Scale Helicopter Rotor for Flying on Mars", *Journal of Aircraft*, Vol. 53, No. 4 (2016), pp. 1160-1167.

[2] McMasters, J. H., Henderson, M. L., "Low Speed Single-Element Airfoil Synthesis", 3rd International Symposium on the Science and Technology of Low Speed and Motorless Flight; p 1-31, NASA Langley Research Center, Hampton, Virginia, March 29-30, 1979.

[3] Mueller, T., J., "Aerodynamic Measurements at Low Reynolds Numbers for Fixed Wing Micro-Air Vehicles", RTO AVT/VKI Special Course on Development and Operation of UAVs for Military and Civil Applications, VKI, Belgium, September 13-17, 1999.

[4] Abbott, I., H., and Von Doenhoff, A., E., "Theory of Wing Sections", 2nd ed., Dover Publications, Inc., New York, New York, 1959.

[5] Mueller, T. J., and DeLaurier, J. D., "Aerodynamics of small vehicles", *Annual Review of Fluid Mechanics*, Vol. 35 (2003), 89–111.

[6] Tsuchiya, T., Numata, D., Suwa, T., and Asai. K., "Influence of Turbulence Intensity on Aerodynamic Characteristics of an NACA 0012 at Low Reynolds Numbers", 51st AIAA Aerospace Sciences Meeting including the New Horizons Forum and Aerospace Exposition, Grapevine, Texas. January 7 - 10, 2013.

[7] Schmitz, F. W., "Aerodynamik des Flugmodells", C. J. E. Volckmann Nachf. E. Wette, Berlin-Charlottenburg, 1942.

[8] Hoerner, S.F., "Fluid-Dynamic Lift: Practical Information on Aerodynamic and Hydrodynamic Lift", Second Ed., Published 1985.

[9] Selig, M. S., Donovan, J. F., and Fraser, D. B., "Airfoils at Low Speeds", Publisher: H.A. Stokely, Virginia Beach, Virginia, 1989.

[10] Mueller, T. J., and Batill, S. M., "Experimental studies of separation on a two-dimensional airfoil at low Reynolds numbers", *AIAA Journal*, Vol. 20, No. 4, pp. 457–463, 1982.

[11] Williamson, G. A., McGranahan, B. D., Broughton, B. A., Deters, R. W., Brandt, J. B., and Selig, M. S., "Summary of Low-Speed Airfoil Data", Vol. 5,

Department of Aerospace Engineering, University of Illinois at Urbana-Champaign, 2012.
[12] Lyon, C. A., Broeren, A.P., and Giguère, P., Gopalarathnam, A., Selig, M.S., "Summary of Low-Speed Airfoil Data", Vol. 3, Department of Aerospace Engineering, University of Illinois at Urbana-Champaign, 1997.
[13] Selig, M. S., Lyon, C. A., Giguere, P., Ninham, C. P., AND Guglielmo, J. J., "Summary of Low-Speed Airfoil Data", Vol. 2. Publisher: SoarTech Publications, Virginia Beach, Virginia, 1996.
[14] Laitone, E. V., "Wind tunnel tests of wings at Reynolds numbers below 70000", *Experiments in Fluids*, Vol. 23, No. 5, pp. 405–409, 1997.
[15] McAurthur, J., "Aerodynamics of Wing at Low Reynolds Numbers: Boundary Layer Separation and Reattachment", Ph.D. Dissertation, Department of Aerospace and Mechanical Engineering, University of Southern California, 2008.
[16] Ohtake, T., Nakae., Y., Motohashi, T., "Nonlinearity of the Aerodynamic Characteristics of NACA 0012 Aerofoil at Low Reynolds Numbers", *Japanese Society for Aeronautical and Space Science Papers*, Vol. 55, No. 644, pp. 439–445, 2007.
[17] Srinivasan, G. R., and Baeder, J. D., "TURNS: A Free-Wake Euler/Navier-Stokes Numerical Method for Helicopter Rotors", *AIAA Journal*, Vol. 31, No. 5: Technical Notes, 1992.
[18] Yang, K., "Aerodynamic Analysis of an MAV-Scale Cycloidal Rotor System Using a Structured Overset RANS Solver", M.S. Thesis, Department of Aerospace Engineering, University of Maryland, 2010.
[19] Medida, S. and Baeder, J. D., "Application of the Correlation-based $\gamma - \overline{Re_{\theta t}}$ Transition Model to the Spalart-Allmaras Turbulence Model", 20th AIAA Computational Fluid Dynamics Conference, 27 - 30 June 2011, Honolulu, Hawaii.
[20] Lutz, T., Würz, W., and Wagner, S., "Chapter 9: Numerical Optimization and Wind-Tunnel Testing of Low Reynolds-Number Airfoils", Fixed and Flapping Wing Aerodynamics for Micro Air Vehicle Applications, Mueller, TJ., editor, AIAA, Vol. 195, pp. 169-190, 2001.
[21] Sheldahl, R. E., and Klimas, P. C., "Aerodynamic Characteristics of Seven Symmetrical Airfoil Sections Through 180-Degree Angle of Attack for Use in Aerodynamic Analysis of Vertical Axis Wind Turbines", Sandia National Laboratories, Albuquerque, New Mexico, printed March 1981.
[22] Okamoto, M., Yasuda, K., and Azuma, A., "Aerodynamic Characteristics of the Wings and Body of a Dragonfly", *The Journal of Experimental Biology* No. 199, pp. 281–294 (1996).

Analysis of Flapping Wings from a Rotorcraft Perspective

Camli Badrya* Bharath Govindarajan† James D. Baeder‡

Department of Aerospace Engineering, University of Maryland, MD

ABSTRACT

The aerodynamic flow beneath a flapping wing and the resulting forces and moments resemble to a large extent the behaviour of rotorcraft flows. While there are dissimilarties between the two vehicles, analogies between the two platforms can be drawn in terms of the wake structure, induced velocities and the vehicle performance. Therefore, techniques developed for the field of rotorcraft can be adapted and applied to flapping wings for analyzing steady flight. This paper aims to draw a comparison between flapping-wing and rotary-wing aerodynamics considering the similarities and differences between the nature of the generated forces, aerodynamic simplified models and advance analysis of trim conditions in hover and forward flight. The study relies heavily on a simplified quasi-steady model coupled to a three-dimensional CFD Navier–Stokes solver. This multifidelity analysis has been to shown to be able to drive the flapping wing system to a trim solution and provide valuable insight into the fundamentals of insect flight. Furthermore, the lift-to-power ratio versus average lift was identified as a principal efficiency metric to assess the performance of flapping-wing vehicles for a given geometry and kinematic parameters.

MOTIVATION

Initial micro air vehicle (MAV) design requirements were defined in 1997 by the US Defense Advanced Research Projects Agency (DARPA) through a program called "MAV - project" (Ref. 1). As outlined by DARPA, MAV are required to be: a) Small, where no dimension exceeds 15 cm (6 inches), b) Lightweight, where the gross take-off weight (GTOW) is less than 100 grams, c) Capable of a loiter time at least 60 minutes, and d) Capable of carrying at least 20 grams of payload. The motivation behind developing these vehicles was driven largely by potential military and civilian applications, with possible extension into the research and commercial sectors. These vehicles could be used to investigate dull, dirty and dangerous environments, such as reconnaissance into hostile territory or scouting missions where the risk of endangering human life is high. Therefore, in addition to the DARPA sizing requirements, capabilities such as sustained hover, efficient cruise and high maneuverability are essential for these small-scale platforms operating within confined environments. Although substantial advances have been made in the last decade in this field, the DARPA sizing requirements are still considered a challenge to satisfy. Currently there is no vehicle that can satisfy the sizing requirements. Three concepts (fixed wing, rotating wings and bio-inspired flapping wings) are being considered for application to hovering MAVs. Fixed wings might offer better performance in terms of high lift-to-drag ratio (L/D), but these vehicles suffer from the inability to hover, a crucial requirement. In contrast, rotary-wing and flapping-wing platforms have a significant advantage in their capability to hover and perform relatively complex maneuvers.

Thus far at least, rotary-wing concepts represent the bulk of the vehicles that have been developed for MAV applications, largely driven by the progress in full-scale rotary-

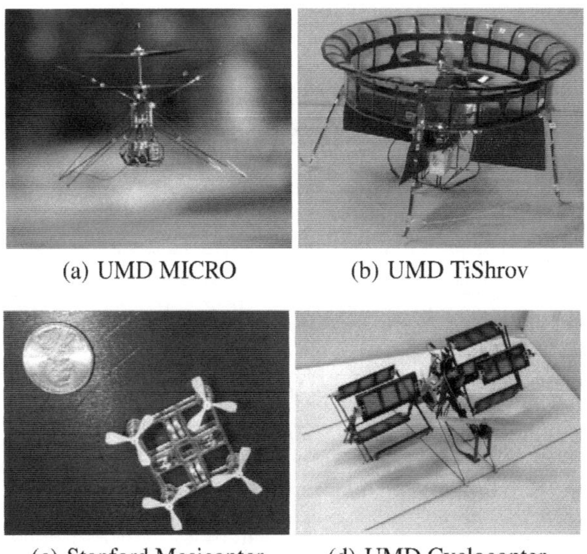

(a) UMD MICRO (b) UMD TiShrov

(c) Stanford Mesicopter (d) UMD Cyclocopter

Fig. 1. Rotary-wing MAVs (Refs. 2,3,4,5).

* Graduate research assistant, E-mail: cbadrya@umd.edu

† Assistant research scientist, E-mail: bharath@umd.edu

‡ Professor, E-mail: baeder@umd.edu

Presented at the 7th UAV Specialists Meeting of the American Helicopter Society, Mesa, AZ, January 24–26, 2017.

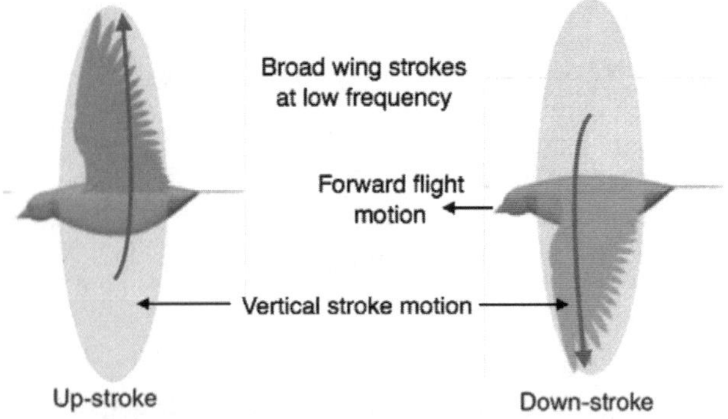

(a) Avian stroke

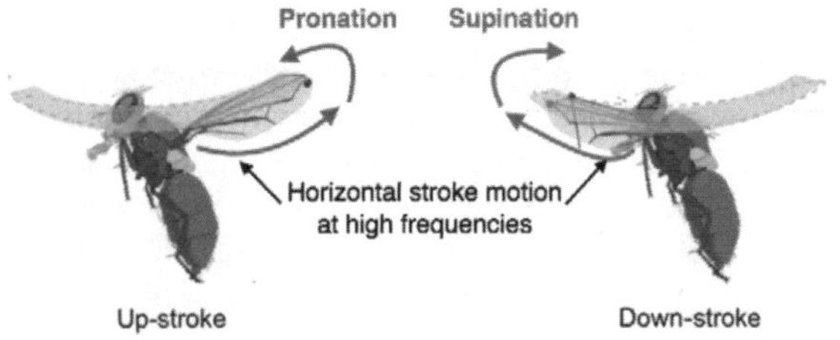

(b) Insect stroke

Fig. 2. Avian and insect stroke, the shaded area represents the flapping stroke path.

wing design and technology. A variety of MAVs have been designed successfully with traditional rotor configurations, such as single main rotor, coaxial rotors and quadrotor (Ref. 4). Non-traditional rotary-wing vehicles, such as the cyclocopter (Ref. 5) with the rotors rotating along a horizontal axis have also been developed and tested to varying degrees of success. Some of these concepts are shown in Fig. 1. These systems show promising progress toward achieving DARPAs MAV requirements in term of weight, dimensions, and performance. A plethora of studies have been performed on optimizing the rotor design to improve vehicle performance, especially in hover (Ref. 6); but the 60 minutes endurance requirement is still a significant technological challenge. These rotary-wing MAVs operate at ultra-low Reynolds numbers $O(10^3$–$10^5)$ because of the size of the vehicles, and consequently, at these low Reynolds numbers the aerodynamic behaviour of airfoil are vastly different resulting in high rotor profile and induced losses, and therefore, poor vehicle performance metrics. Geometrically scaling down full-scale rotary wing concepts to small scales is not effective as the aerodynamic principles do not scale linearly across Reynolds numbers. Therefore, a systematic study is required to understand the fundamental flow physics around airfoils of different shapes at these ultra-low Reynolds numbers; a field of ongoing research (Ref. 7).

Another approach to design MAVs is to draw inspiration from nature and the multitude of "hovering vehicles" in the form of small birds and insects. Flapping-wing based MAVs are divided into two types based on their flight kinematics; avian and insect-like, as shown in Fig. 2. Examples are the

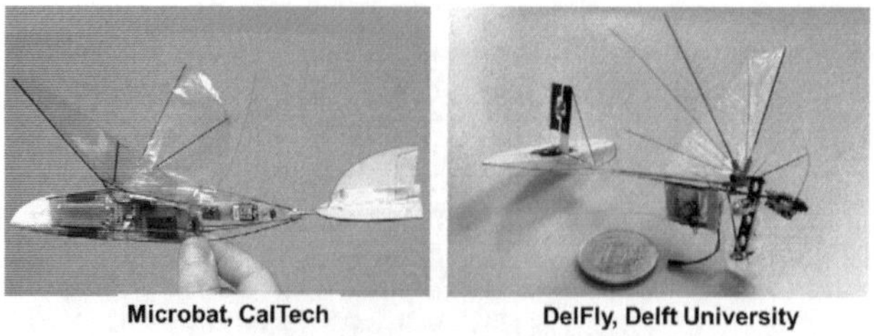

(a) Avian flappers (Refs. 8, 9)

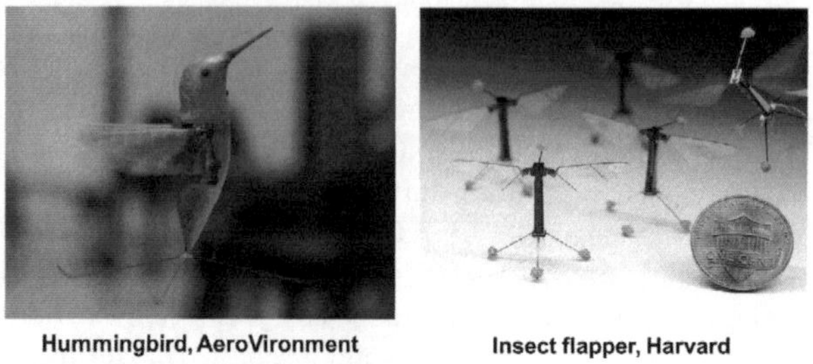

(b) Insect flappers (Refs. 11, 12)

Fig. 3. Different types of laboratory built flappers: (a) Avian flappers, and (b) Insect flappers.

Microbat developed by Caltech labs (Ref. 8) and DelFly developed at Delft University (Ref. 9) shown in Fig. 3(a). Ornithopters are not considered in this study because of their inability to hover. In the first type of avian flappers (Fig. 2(a)), the flapping kinematics are similar to those of a birds' flight, where the flapping-stroke is perpendicular to the travel direction, change in angle of incidence (defined as the angle between the wing chord and the flapping stroke) is relatively small, and the flapping frequency is low (order of magnitude of 10 Hz) (Ref. 10). The second type of flapping wing MAVs are the insect-based flappers, where the kinematics are similar to insects. The flapping stroke is not necessarily perpendicular to the travel direction (Fig. 2(b)), and unlike avian kinematics, the change in angle of incidence (pitch angle) is large and the flapping frequency is high (order of magnitude of 100 Hz (Ref. 10)). Representative examples of insect-based MAV are the AeroVironment Hummingbird (Hummingbirds fly more like insects than birds); see Fig. 3(b). The Hummingbird is equipped with a small video camera for surveillance and reconnaissance purposes and for now operates in the air for up to 11 minutes, with the ability to operate both indoors and outdoors.

Despite the progress that has been achieved especially in understanding the underlying aerodynamic principles behind insect-flight, a fully functional insect-based MAV seems to be a challenging goal to achieve at this point in time and many related projects remain undergoing research. Perhaps the most notable development in this area comes from the Harvard University Microrobotics lab which has developed a bio-inspired robot with insect kinematics and geometric features with a weight of 80 mg, wing span of 3 mm, and wingbeat frequency of 120 Hz (Ref. 12). This flapper was able to demonstrate a controlled vertical flight in hover (Fig. 3(b)) for a few seconds, in part because of battery size limitations. Insect flight is very complex and the flowfield around the wings is highly viscous and unsteady. Flapping wing kinematics are difficult to untangle into independent parameters, and at small scales the flowfield around these wings is both highly viscous and unsteady. Research over more than three decades has contributed much to understanding the aerodynamic basis of insect flight. Progress has been made toward establishing the principles underlying the design of insect wings in isolation, but trimmed and controlled insect-like MAV flight has not yet been achieved.

Philosophically, both flapping-wing and rotary-wing platforms aim to achieve similar capabilities, i.e., efficient hover albeit through different aerodynamic principles to generate the necessary lift. The forces generated by the rotor blades (in hover) are steady with minimal flow unsteadiness, whereas for flapping-wings, the insect relies heavily on unsteady mechanisms to produce lift. There are however, certain similarities in the flow field and forces produced by flapping wings and a rotor in hover, namely the periodicity of the forces (every flapping cycle or every revolution).

The aim of this paper is to draw a comparison between flapping wings and rotorcraft to highlight the similarities and differences along with the corresponding tools required to perform a wholistic vehicle analysis, which includes aerodynamic modeling, identification of efficiency metrics, trim and performance analysis. Can engineers and developers analysis flapping wings using engineering tools that have been used in rotorcraft ? It is recognized that the underlying aerodynamics are different between the two platforms, but the question is how much ground can be covered with conventional tools used in rotorcraft and where do further tools need to be developed. Such an effort is not without precedent as simplified momentum theory analysis (used typically in rotorcraft) has been used to analyze the wake beneath flapping wings and basic performance analysis (Refs. 13, 14).

FLAPPING WING PHYSICS

After three decades of research aimed at understanding insect flight, it is evident that insects rely on the production of unsteady forces to sustain flight and to perform maneuvers. The kinematics of the insect wings behind these forces are quite complex. Insects flap their wings (two wings: fruitfly, bumblebee, hawkmoth, or four wings: dragonfly) at high frequencies and high pitch angles, with the flapping configuration varying between different insect species. The nature of insect wing flapping kinematics results in the generation of unsteady but periodicity forces. Prior experimental and computational studies identified several unsteady flow phenomena that enable the production of sustained lift. The force production mechanisms of insect during hovering flight can be classified into delayed stall, rapid pitch-up and wake capture.

A brief description of each of the mechanisms is presented below. Figure 4 describes the two-wing (*Diptera*) insect stroke, that can be divided into four motions; a translational upstroke and downstroke along with supination and pronation at the end of each of the translational strokes. Specific instances of strong unsteady aerodynamic interactions are annotated in Fig. 4 to highlight the source of delayed stall, rotational lift and wake capture. Additionally Fig. 5 details a quantitative variation of the lift and drag forces experienced by a wing in hover using observations from experiments and numerical predictions from a previous study by the authors (Ref. 16). Figure 5 also contains six distinct events and the vortex structure around the wing at these instances, which further aid in understanding the physical phenomenon behind the production of unsteady forces in flapping wings.

Delayed Stall: A phenomenon predominantly because of the leading edge vortex (LEV). As the wing angle of attack increases as a consequence of flapping kinematics, the flow over the wing tends to separate near the leading edge, especially in the case of insect wings (similar to thin airfoils). However, the flow reattaches before it reaches the trailing edge. In such cases, a leading edge vortex (LEV) forms on the upper surface of the wing. Because the flow reattaches, the fluid continues to flow smoothly from the trailing edge and the Kutta condition is maintained. The presence of the attached leading edge vortex produces high lift coefficients, a phenomenon termed "delayed stall." Delayed stall in insect flight was observed by Maxworthy (Ref. 17), who visualized the leading edge vortex on the model of a flying wing.

Figure 5 shows that during each flapping cycle there are three positive peaks in the lift and drag denoted by A, B, and C in the down-stroke and D, E, and F in the up-stroke. By definition, lift is the component of the total aerodynamic force along the wing-span perpendicular to the translational velocity and drag is the component of the total aerodynamic force parallel to the translational velocity. The formation of the LEV during the flapping flight occurs at instances B and E in Fig. 5 during the downstroke and upstroke, respectively. This LEV is stable during the flapping stroke, which implies that it remains attached to the wing upper surface during the translational periods creating strong suction forces. This vortex can maintain a lift enhancement for 3–4 chords length of wing travel, before the LEV breakdown occurs at the end of the stroke.

Rapid Pitch-Up or Rotational lift: The LEV is a major source of lift enhancement for insect flight during the translational portions of the flap stroke. In addition, the rapid rotation of the wing at the end of the upstroke and downstroke, i.e., pronation and supination (see Fig. 4), also results in lift generation. Kramer (Ref. 18) was first to show that a wing rotating from low to high angles of attack results in a lift coefficient greater than the static stall value, known as the "Kramer effect." The rotation effect in a flapping wing was investigated experimentally by Dickinson et al. (Refs. 15, 19), and computationally by Sun and Tang (Ref. 20). Sun and Tang suggested that the vorticity associated with rapid pitch-up rotation results in lift enhancement. Sane and Dickinson (Ref. 19) explained it as reestablishing the Kutta condition due to the change of the pitch angle during the rotation. Pitch-up rotation is influenced by the kinematics of the flapping wing.

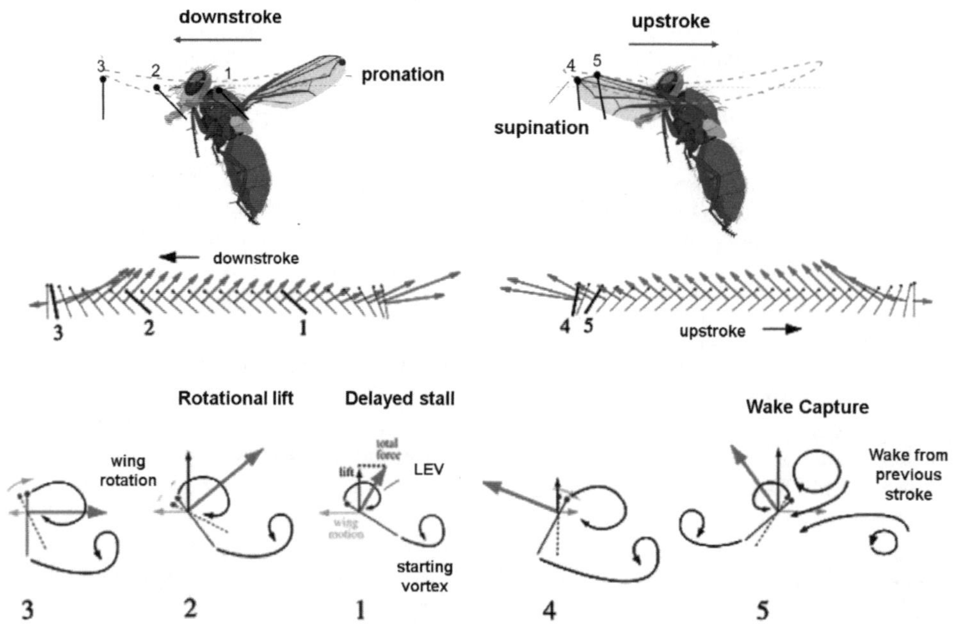

(a) Insect flapping stroke during downstroke (b) Insect flapping stroke during upstroke

Fig. 4. Insect flapping stroke divided into four motions: 1) Downstroke (translational) 2) Upstroke (translational) 3) Pronations and 4) Supination. Blue arrows indicate the direction of wing movement and red arrows the direction and magnitude of the forces generated in the stroke plane, and the associated forces production mechanism (Refs. 10, 15).

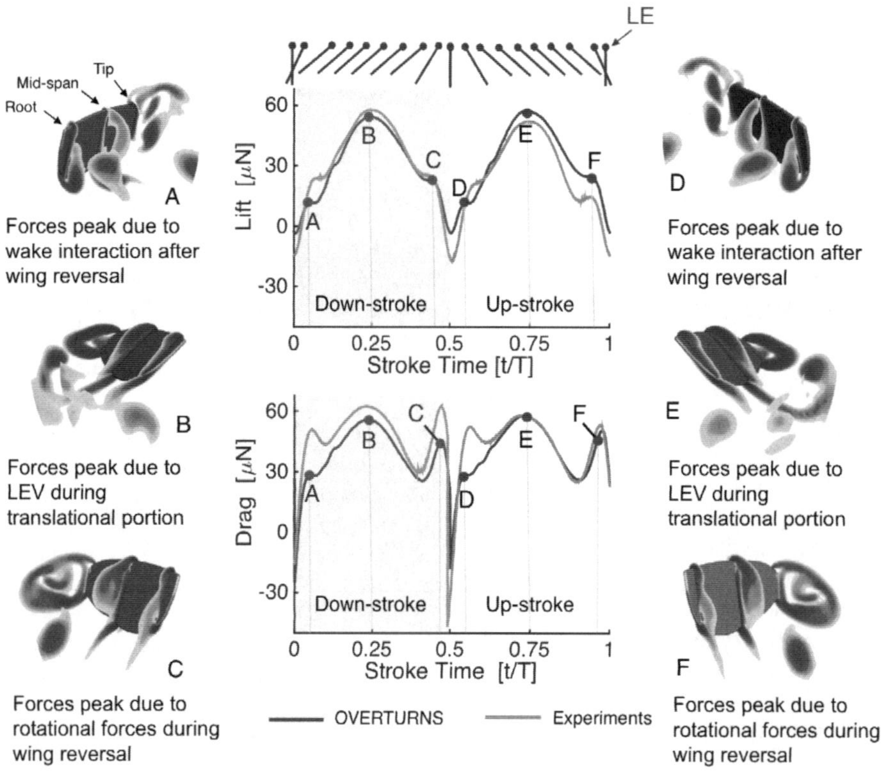

Fig. 5. Experimental and computational lift and drag forces at Reynolds number 346, where the peaks A, B, C, D, E and F are described by the vortex magnitude at three chord-wise sections; root, mid-span and tip during the flapping cycle (Ref. 16).

The peaks C and F in Fig. 5 are a result of this rapid pitch-up mechanism.

Wake capture: Dickinson et al. (Refs. 15,19) observed from the experimental data of the Robofly in hover that two lift peaks (enhancement) were identified. One peak was attributed to the aforementioned rotational forces and the second peak a result of wake capture (Refs. 15,21). This latter lift enhancement occurs when the wing starts the translational motion at the end of each half-stroke (upstroke or downstroke). During this portion of the stroke, the wing interacts with the flow field that consists of the trailed and shed vortices because of the previous supination/pronation, which results in transferring momentum from the flow to the wing, and generates additional lift. In other words, the wing is capable of extracting energy from the wake. Note that there are two "modes" defined in hover for flapping wings in viscous flow: 1. Water-treading mode where the starting pitch angle is zero and increases to the prescribed value, and 2. Normal hovering mode, where the pitch angle starts at 90° and decreases. The consequence is that the wake capture mechanism was observed also for normal-hovering mode (Refs. 22,23) while being absent in water-treading (Refs. 23,24), indicating that the wake capture mechanism, similarly to rapid pitch-up, is affected by the kinematics. The time instants A and D in Fig. 5 show the wake behind the wing during the reversal.

MODELLING OF FLAPPING WINGS

Reliable experimental measurements on millimeter-scale models that flap at high frequencies (≈100 Hz) are challenging to obtain. This difficulty arises from the small magnitudes of the forces involved (resulting in low signal-to-noise ratio), very small length scales and the high sampling rates required to obtain time-resolved information within each flap cycle. Experimental work involves conducting experiments on living insects or scaled up mechanical wings such as robotic wings (Refs. 21, 25, 26, 27, 28), each with its unique set of challenges. Majority of these experiments considered an isolated wing using simplified kinematics to quantify the effects of wing rotation, wing-wake interaction and leading edge vortex formation.

Quasi-Steady Model

The goal of quasi-steady (QS) models is to approximate and represent the various aerodynamic phenomenon of flapping wings through a simplified mathematical model. In certain flight conditions such as hover, these simplified QS models are sufficient to capture performance trends. However, these models are limited in their predictive capability for complex flight conditions such as forward flight or maneuvering flight. Nonetheless, these QS models are essential and can be used for either preliminary design or to aid a higher-fidelity trim process (discussed later in this study). QS models aim at predicting the lift, drag and pitching moment across the flapping stroke at different wing kinematic conditions. These aerodynamic effects are broadly classified into the translational lift, added mass effects or wake capture and rotational circulation. These models are not presented here in detail for reasons of brevity. However, the formulation has been published in earlier works (Refs. 29,30).

While the true kinematics of an insect wing are complex and the wings themselves are flexible and deform, a simplified kinematic representation is used in the present study, which consists of three positional angles as shown in Fig. 6; 1. Flap

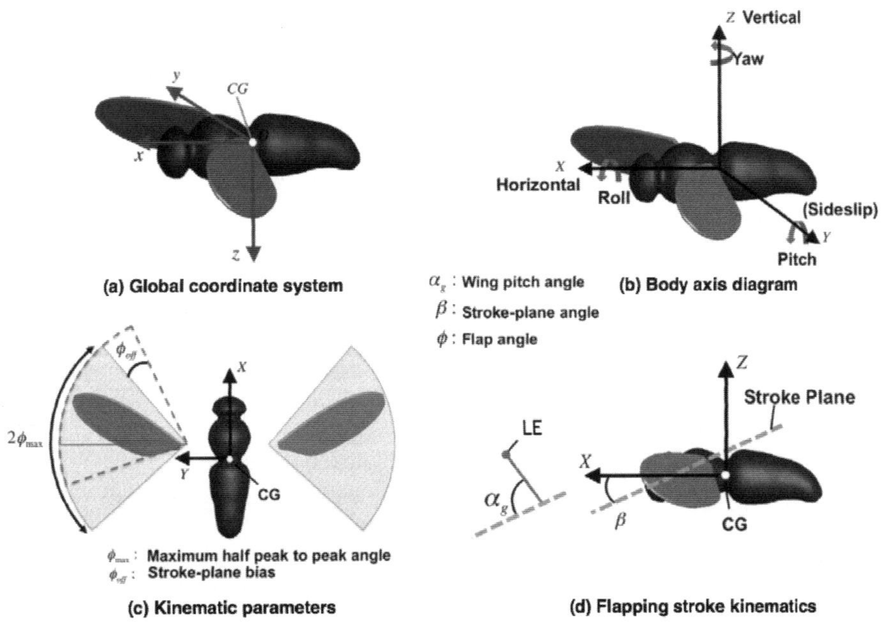

Fig. 6. Inertia coordinate systems and wing kinematic parameters ϕ, α and β.

angle ϕ, the wing position in the flap stroke, 2. Stroke plane angle β, the inclination of the stroke plane with respect to the x axis, and 3. Pitch angle α_g, the angle between the wing chordwise direction and the flap-plane, also referred to as "geometrical angle of attack."

Wing kinematics used in this study are qualitatively similar to the those used by Faruque and Humbert (Ref. 31) for *Dipteran* insect flight dynamics, and by MacFarlane and Humbert (Ref. 32) for *Drosophila* type platforms. The flap stroke velocity increases at constant rate and the maximum velocity occurs at mid-stroke. The azimuthal position of the wing in the flap cycle is given by

$$\phi(t) = -\phi_{\text{max}} \cos(2\pi f t) + \phi_{\text{off}} \quad (1)$$

where ϕ_{max} is the peak-to-peak amplitude and ϕ_{off} is the stroke bias. The geometric pitch angle (Fig. 6) with respect to the flapping stroke plane remains constant except near the stroke reversal during supination and pronation, and is given by

$$\alpha_{\text{g}}(t) = 90 - \alpha_{\text{max}} \tanh(2.7 \sin(2\pi f t)) \quad (2)$$

where α_{max} is the complimentary angle of the maximum geometric pitch at the mid-stroke with respect to flapping stroke plane. This angle remains constant during the translation periods and increases as the wing approaches stroke reversal, as shown in Fig. 4. The wing aspect ratio (AR) is defined as wing span squared (b^2) divided by the wing area ($S = \bar{c}b$), therefore AR is equivalent to $b/\bar{c}$. The wing shape is bio-inspired by blowfly shape and fruitfly shape Fig. 7(a). The simulation parameters are bio-inspired by *drosophila* flight features, i.e., flapping frequency range from 100–300 Hz, wing aspect ratio of 2.8, and Reynolds numbers of $O(10-10^2)$.

Computational Fluid Dynamics

Alternatively, CFD simulation models can help provide a good understanding of the unsteady flow mechanisms and insight into the critical design parameters. The computational approach is to calculate the flow variables by solving the governing equations and associated boundary conditions with a flapping wing numerically. In the last decade, researchers have taken advantage of the advancements in computing facilities and developments in computational techniques to develop solvers that can address the insect flights with varying degrees of success.

The use of computational tools poses two major challenges. The first of these is the computational cost. Simulations needed to resolve insect flight require extensive computational resources to handle viscous flows, accurately capture vortical structure and allow for the complexity of moving grids. The second issue is that CFD studies rely on experimental results for both validation and for the relevant kinematic inputs. Currently, high fidelity solvers can be executed in parallel on high performance computing clusters to shorten the wall clock time needed to obtain well resolved solutions. To address the second concern, collaboration between empirical and computational studies can be employed. Once the CFD solver is verified against experimental results, it can be used to provide insight about more complicated kinematics and trends that experiments find challenging to capture. There are a variety of mathematical models in the literature that can sufficiently address insect flight simulation, such as direct numerical simulations (DNS) and Navier–Stokes solvers.

Navier–Stokes solvers have been used to simulate the flow around flapping-wings by many authors (Refs. 20, 22, 33, 34, 35). While this approach calculates the properties of the entire flowfield and instantaneous forces, there are drawbacks in terms of computational cost, as mentioned previously. Furthermore, a considerable effort is required to construct a suitable mesh that discretizes the flowfield. In the case of a flapping wing, the simulations require moving boundaries, and the most popular grid methods have been: 1. Immersed boundaries, 2. Conforming body mesh (overset) grids, or 3. Overlapping unstructured meshes.

The governing equations used to describe flapping wings physics are the unsteady Navier-Stokes (NS) equations. Progress in computing facilities and new algorithmic capabilities in the last decade now allow for researchers to use CFD tools to simulate flapping wings with their inherent kinematic complexities. Solution to the NS equations provide time accurate flow-field variables (velocity field, density and pressure), which can be used to extract valuable aerodynamic and fluid dynamic quantities. Much computational work has been carried out on insect-wing related kinematics at low Reynolds number (Refs. 20, 33, 34, 36). These works predicted the leading-edge vortical structure, and showed good qualitative agreement with smoke flow visualization and computational predictions were validated against Dickinson's Robofly experiments. At low Reynolds number, a turbulence model is not required and the solution of the NS equations can be considered a direct numerical simulation (DNS).

Lower-fidelity panel methods have also been used to model flapping-wing flight, where the wake geometry is prescribed as a series of vortex rings (Refs. 13, 37, 38). 3D panel methods have been used to study the experimental flapping of the Hawkmoth, and vortex panel methods have been used to study the inviscid lift generation, i.e., Wagner effect in a pitching airfoil. However, these models are limited in their inability to capture the evolution of the wake geometry over time and are not a focus of this study.

Navier–Stokes Solver : OVERTURNS (Overset Transonic Rotor Unsteady Navier-Stokes) is a time-accurate, structured, overset, compressible solver, that solves the three dimensional Navier–Stokes equations (Refs. 39, 40). OVERTURNS was developed at the University of Maryland for rotorcraft applications, and was extended for this work to analyze flapping wing platforms. The solver was modified to simulate a three dimensional *diptera* type vehicle, consisting of two flapping wings and a rigid body in various flight modes (hover, forward flight and coordinated turn). All the computational results presented in this paper were performed using this computational platform.

The governing equations solved in OVERTURNS are the three-dimensional Navier–Stokes equations, which are mathematical representations of the three conservation laws of mass, momentum and energy. These laws are expressed as a system of partial differential equations that model an unsteady, compressible, and viscous flowfield. Additional equations, such as the equation of state, are required to ensure that the system is fully determined. The general formulation is given by:

$$\frac{\partial \mathbf{Q}}{\partial t} + \nabla \cdot (\mathbf{F}_C^M - \mathbf{F}_V) = \mathbf{S} \tag{3}$$

where $\mathbf{Q} = [\rho, \rho u, \rho v, \rho w, p]$ is the vector of conserved variables, $\mathbf{F}_C^M$ and $\mathbf{F}_V$, are the inviscid and viscous convective fluxes, respectively, and $\mathbf{S}$ is the body-force source term. ρ is the fluid density, u, v, w are the components of the velocity vector in the Cartesian coordinate system, and p is the local static pressure. $\mathbf{F}_C^M$ accounts for the grid motion terms and can be expressed as $\mathbf{F}_C^M = \mathbf{F}_C - \mathbf{V}_g\ \mathbf{Q}$, where $\mathbf{F}_C$ is the inviscid fluxes without grid motion and $\mathbf{V}_g$ is the local contravariant velocity of the grid.

In OVERTURNS, the baseline algorithm uses a finite volume approach to solve the unsteady compressible Navier–Stokes equations in space and time. For spatial reconstruction of the inviscid fluxes, a third-order Leers Monotone Upstream-Centered Scheme for Conservation Laws (MUSCL) scheme with Korens (Ref. 41) limiter was used with an option to use higher order schemes such as the Compact-Reconstruction Weighted Essentially Non-Oscillatory (CRWENO) schemes (Ref. 42). The viscous terms were computed using a second order central differencing scheme. Because the flapping-wing simulations are performed at low-Mach numbers O(10^{-3}) and low Reynolds numbers (10–10^2), Roe-Turkel low-Mach preconditioning was applied to improve the convergence and accuracy of the compressible NS solver (Refs. 39, 43, 44). These low Reynolds numbers result in either a purely laminar flow or with eddies large enough that they can be captured such that no turbulence model was required.

Initial conditions were specified by assigning the density, flow velocities and pressure everywhere in the computational region. In forward flight the initial conditions were set such that the density, pressure and flow velocities are free-stream values. In hover the free-stream velocities were set to be zero. There are two types of boundary conditions (BC) that are specified; wall boundary and far-field boundary conditions. Wall boundaries are natural boundaries of the physical domain which arise from the surfaces being exposed to the flow. For a viscous fluid, the relative velocity between the surface and the fluid at the surface is zero (no-slip condition). The second boundary condition is the far-field BC. The direction of the velocity vector and sonic velocity, are used to determine these characteristic boundary conditions.

Overset Grid System: A single structured mesh can not preserve the complexities in the flow-field such as tip vortices, trailing edge vortices and far field wake features. The approach in the present work to account for moving meshes is the use of overset grids. The insect is modeled as a collection of two rigid flapping wings and a rigid body, the meshes of which are overset onto a coarser nested background mesh. Nested Cartesian background meshes are used to capture the far-field effects. For the background grids a simple Cartesian mesh generator is used for this purpose. Two levels of nested Cartesian background are used. The coarser mesh extends

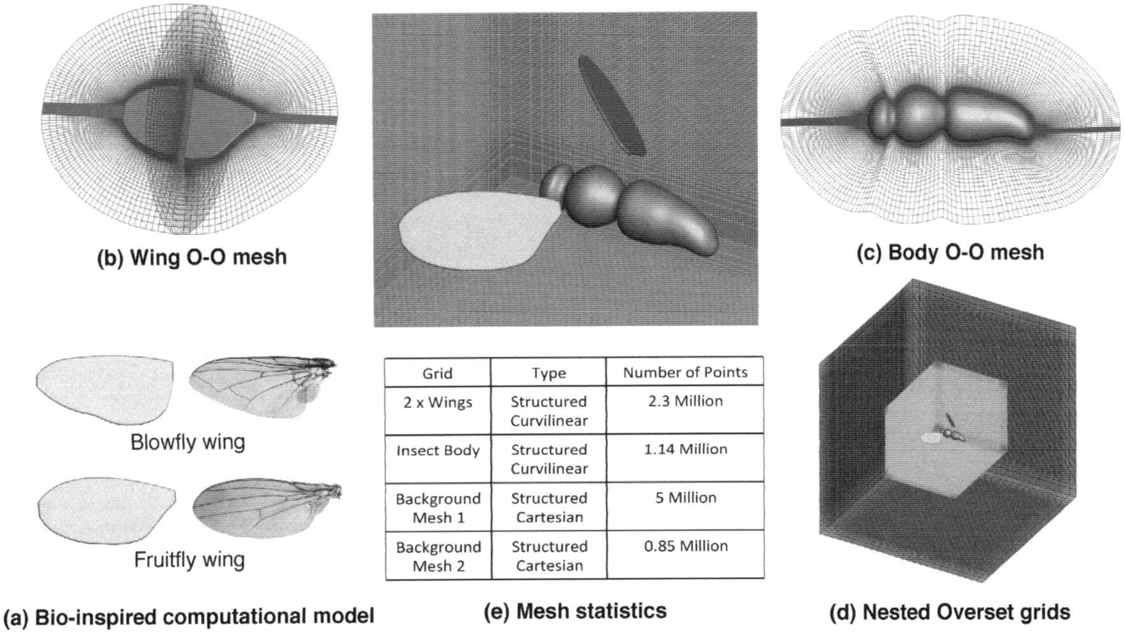

Grid	Type	Number of Points
2 x Wings	Structured Curvilinear	2.3 Million
Insect Body	Structured Curvilinear	1.14 Million
Background Mesh 1	Structured Cartesian	5 Million
Background Mesh 2	Structured Cartesian	0.85 Million

Fig. 7. Nested overset grid system highlighting the near body wing and insect body meshes.

up to 5 wing chords on each side. The nested background meshes are stationary and extended to allow for the motion of the wings. The wing grids translate and rotate according to user-prescribed kinematics while the body mesh is fixed for hover and forward flight modes. Such a setup minimizes the total mesh points required to obtain a detailed resolution of the flowfield, and allows for selectively refining the flow domains of interest. The overall overset grid system has five grids for the *diptera* configuration, as shown in Fig. 7(d), and consists of roughly 10 million mesh points (depending on the simulation type). The important grid parameters are the following: a) spacing near the wall (wing surface) is 0.01% of the mean chord, and the wings grid is extended in the normal direction over 82 planes to resolve the near-wake features b) Wing thickness is 5.8% of the averaged chord c) The inner nested background mesh has equal spacing in all directions and equal to 5% of the mean chord, where the outer background has a length scale of approximately 4 times the inner background cell length. More details of the computational domain and its borders are described in Fig. 7. Overall approximately 9 million mesh points are in the computational domain, and computations are performed on 20 processors in parallel, on high performance computing platforms. The connectivity and information transfer across the overlapping areas is handled by an implicit hole cutting technique (Ref. 39).

Body-fitted, O-O curvilinear grids are used for the wings and the insect body, which essentially capture the near-field flow, i.e., around the wings and body surface. The wing surface is treated as a solid wall and move according to the prescribed kinematics relative to the insect body, which is fixed (Fig. 7(c)). A 3D algebraic mesh generator (Ref. 40) for structured, curvilinear, body-fitted CFD grids was used for the wings and insect body grids. The insect body and wing meshes are finely spaced near the surface to resolve the boundary layer features and associated viscous effects (Ref. 40), as shown in Figs. 7(b) and 7(c).

RESULTS AND DISCUSSION

The results section is broken down into four parts: 1. Aerodynamic modelling, 2. Efficiency metrics, 3. Insect trimmed flight, and 4. Performance. Each section provides an overview of the state of the art and proceeds to discuss the similarities and differences between rotorcraft (revolving wings) and flapping wings. The CFD methodology outlined is also used to provide more insight into the workings of flapping wings.

I. Aerodynamic Modelling

The three primary approaches to explain the physical principles and to estimate the generated aerodynamic forces and power for rotary and flapping wings are: 1. Formulation of a simplified model, 2. Experimental studies, and 3. Computational simulations.

One of the earliest approaches in numerically analysing the unsteady aerodynamics in flapping wing was to verify if the developed simplified aerodynamic theories for fixed and rotary wing were applicable to flapping wings (Refs. 13, 14, 37, 38, 46). It is evident that fixed-wing aerodynamic theory under the assumptions of steady forces and potential flow does not apply to flapping/rotary wings. While the assumption of steady flow is valid for fixed wings, flapping-wing and rotary-wing flows are unsteady and result in time-dependent periodic forces, largely attributed to the underlying time-dependent kinematic behaviour of flapping wings and rotors.

Similarities in the wake can be observed between flapping and rotating wings and a comparison of the flow fields in hover is shown in Fig. 8. The rotor flow field is experimentally obtained (Ref. 45) using particle image velocimetry (PIV) and the flapping wing flow around a *Diptera* is obtained numerically through CFD. Note the presence of coherent tip-vortices and root vortices in the flow fields. The contraction in the wake of the flapping wing is not as pronounced as that of the rotor flow. Nonetheless, sufficient similarities exist that warrant a deeper understanding of the aerodynamics of flapping wings.

For rotary wing, simplified models have been developed successfully such as momentum theory, blade element theory (BET) and combined blade element momentum theory (BEMT); see Chapter 2 of Leishman (Ref. 47). Momentum theory is a first-order estimation of the thrust and the power of a lifting rotor, derived from the integral form of the conservation laws, under the assumptions of one-dimensional, steady, incompressible flow where the rotor is approximated as an infinitesimally thin actuator disk. This analysis has its limitations, such as: 1. Does not provide power and thrust distribution along the blade, 2. Challenging to extend for maneuvering flights, 3. Viscous effects are introduced into the model based on experimental results as a post-dictive model, and 4. Detailed blade design cannot be incorporated. Therefore, although acceptable for rough sizing and required power of a rotating wing, it is not suitable for detailed design. Therefore blade element theory was developed to include the airfoil details, twist and flexibility, where lift and drag coefficients can be determined from experiments, lift-line theory predictions or numerical simulations.

Momentum theory based on actuator disk has been used to analyze flapping wing and simplify the aerodynamic forces calculations (Refs. 13, 14). For this analysis a few assumptions were made: 1. Wing flapping frequency is high enough that the flapping plane can be simplified as an actuator disk; and 2. uniform velocity distribution occurs downstream of the wake. Using this approach, Weis-Fogh (Ref. 14) calculated the induced downwash velocity at the flapping stroke in hover, assuming that the far wake velocity was twice the downwash velocity at the full actuator disk (similar to the result obtained from rotorcraft). Ellington (Ref. 13) calculated the induced power, that was modified to consider a partial actuator disk and stroke plane incline. These results, however, when compared to experimental data resulted in predictions that were far from the observations and did not provide a good estimation.

To understand the shortcomings of this analysis (to provide

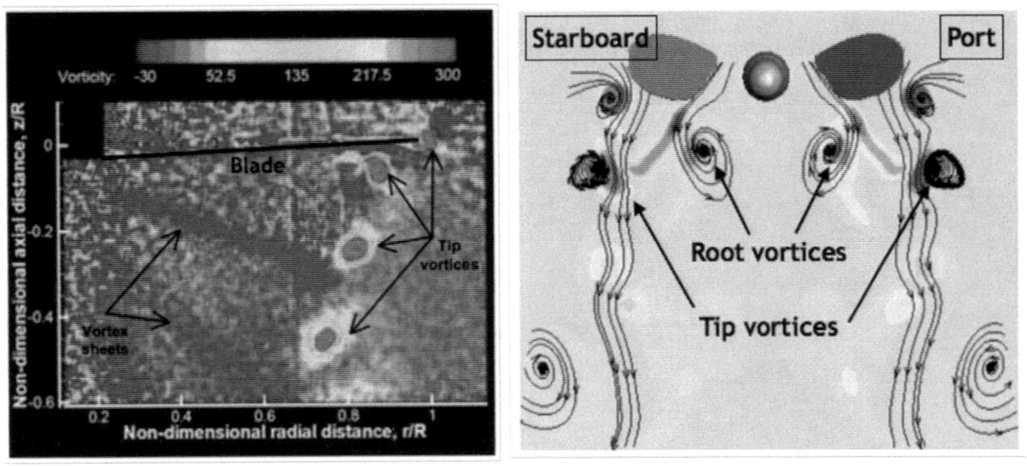

(a) PIV on a hovering rotor

(b) CFD on a hovering flapping wing

Fig. 8. Comparison of flow fields in hover between a: (a) Rotor, and (b) Flapping wing. Source of experimental PIV: Ramasamy et al. (Ref. 45).

even a first-order estimation), a closer look at the flapping wings aerodynamics is required. First, the flowfield around flapping-wings have a three dimensional structure because of the resulting vortical flow along the wing span and around the wings (see Fig. 5), and because of the typical low aspect ratio of the wings the flow can not be approximated as a 1D wake, even in a time-averaged sense. The inability to accurately define a suitable control volume to apply the conservation laws result in mis-predictions of the induced velocity and power. Therefore, the approximation of a thin actuator disk does not hold in this case. Second, the viscous effects are dominant in this low Reynolds number regime and it should be recognized that the modified momentum theory for rotorcraft relies on experimental observation to include the viscous effects in the profile power term. However, the luxury does not yet exist for flapping wings making it harder to predict the effects of this viscous dominated flow.

While steady-state aerodynamic theory does not explain the lift enhancement sources (delayed stall, pitch-up and wake capture) in insect flight, a different approach was adapted and examined by many authors using a quasi-steady model. A quasi-steady aerodynamic model is a reduced order model in which the instantaneous aerodynamic forces and moments are computed solely based on instantaneous velocity, wing geometry and effective angle of attack, i.e. there are no flowfield hysteresis effects such as shed wake. Though this model provides qualitative trends that describe the coupling between wing kinematics and associated loads, it does not capture unsteady aerodynamic effects due to shed vortices, a key phenomenon in flapping wing flight (Ref. 15). Therefore, a higher-fidelity aerodynamic model is necessary to obtain accurate airloads. By comparing the quasi-steady results against experimental data, Ellington et al. (Ref. 36), demonstrated that a quasi-steady model is insufficient to predict wing force magnitudes accurately.

Later, semi-empirical corrections were proposed to augment the quasi-steady models with unsteady effects (Refs. 36, 48, 28). Though such modified quasi-steady models can be tuned to provide good agreement with experimental measurements, such an approach represents "point calibrations" based on data fitting, and must be re-tuned for each geometry and flapping frequency. Unsteady forces (added mass, wake capture) are known to play a significant role in insect flight and are highly influenced by the wing kinematics and wing shape. Therefore, a quasi-steady model, while proving a good tool to obtain rapid estimations of the aerodynamic force-kinematics trends, is insufficient to capture unsteady force mechanisms from first principles. Therefore, using just a quasi-steady aerodynamic model may result in an over-designed or under-designed vehicle. A higher-fidelity model that captures the essential aspects of insect flight is required for accurate performance estimates.

For sake of completeness, it should be noted that a vortex based simplified model presented by Rayner (Refs. 37, 38) suggested modeling small-core coaxial vortex rings for each half flapping wing. Lift and drag coefficients can be determined from this model, but the effects of stroke amplitude and stroke plane angle were not included. Later on, Ellington (Ref. 36), developed a simplified model based on vortex sheets that can also account for the influence of forward speed.

Quasi-steady Forces

The results in this section correspond to the use of the QS model to breakdown the various components of the lift and drag produced by a wing over a flapping cycle. Details regarding the derivation of the various components can be found in other works of the author (Refs. 29, 30, 32).

The QS model and full NS CFD is compared against available experimental data for lift and drag over a flapping cy-

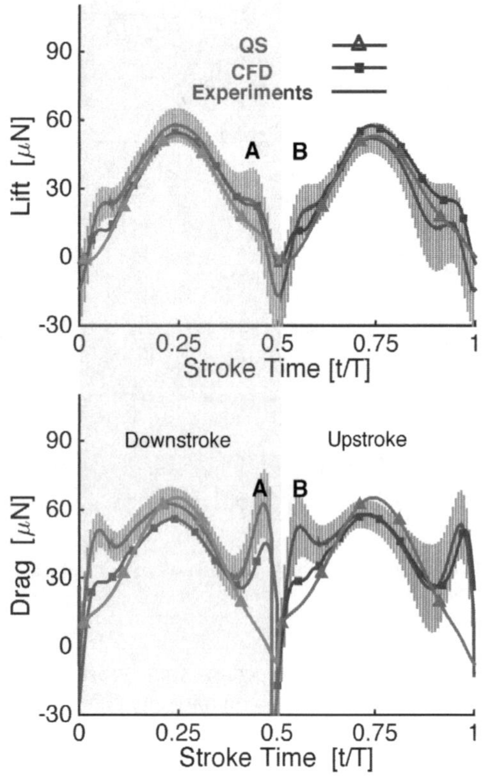

Fig. 9. Comparison of time history lift and drag over a flapping cycle between experiments, quasi-steady model and CFD.

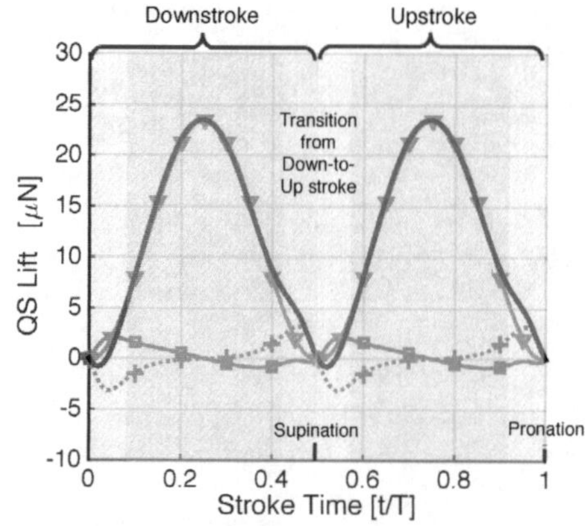

(a) Quasi-steady lift

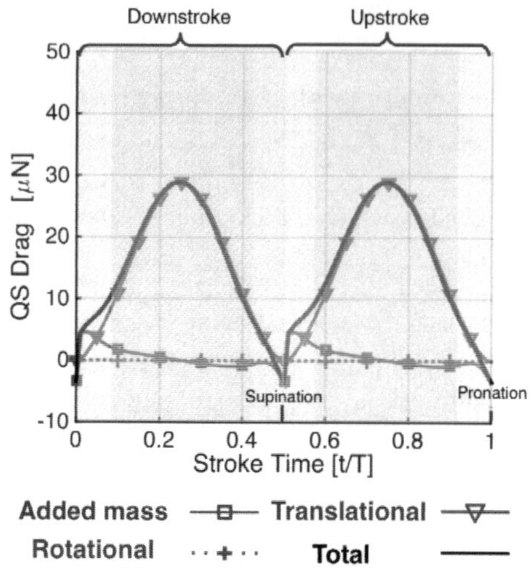

(b) Quasi-steady drag

Fig. 10. Quasi-steady forces and its components over a flapping cycle.

cle. Figure 9 shows the time history of lift and drag of an isolated wing over one time-period. The wing span is 3 mm with a 0.75 mm root cut-out. The flapping frequency is 300 Hz (Reynolds number of 360) and the mid-stroke pitch angle is 45°. The shaded area in Fig. 9 represents the uncertainities in the experimental measurements and is defined as $\pm(4.7+2\sigma)\mu N$, where σ is the standard deviation (Ref. 16). The time-history forces obtained are plotted as a function of the non-dimensional time. The lift is the component of the total aerodynamic force along the wing-span perpendicular to the translational velocity. Drag is the component of the total aerodynamic force parallel to the translational velocity and it is positive when it points in the opposite direction of the translation velocity. Both the QS and CFD predict trends in lift and drag reasonably well when compared against experiments, although the CFD predicts the peaks in forces around $t/T = 0.05,\ 0.5, \text{and} 0.95$ better than QS. These instances in the flapping cycle correspond to the contributions from added mass and rotational circulation and there it is expected that CFD predict the trends better than the simplified QS model. However, for a simple case of hover the QS performs reasonably well.

During the translational portions, lift obtained from a QS model is in good agreement with the CFD results; while for drag prediction, QS model is overpredicted relative to CFD. The QS model does not include an adequate mathematical model for induced inflow that can lead to a reduction in the drag. The significant difference between QS and CFD drag predictions is greatest in the stroke transition regions (up-to-downstroke and down-to-upstroke), represented by the shaded areas marked by A and B in Fig. 9. CFD has dominant peaks, especially the peak in (A) portion which results from the rapid pitch-up effect. The relatively smaller peak in (B) is due to wake interaction. With the added mass and rotational forces effect modeling in QS, QS still underpredicts these peaks in comparison to CFD. This is similar to the observations when comparing QS predictions and experimental measurements.

Figures 10(a) and 10(b) show the quasi-steady (QS) lift and drag variation, respectively, during the flapping stroke of an insect in hover using CFD predictions. The solution was obtained at a flapping frequency of 200 Hz, which has an equivalent Reynolds number of 240. The contribution of the added-mass, translation and rotation forces to the total force is also presented. The total force is mostly contributed by the translational forces. Added mass effect contributes mostly during the stroke reversal (transition from up to down-stroke - supination, or from down to up-stroke - pronation) and less during the translational portions (as shown in Fig. 10(a)). The peaks in the lift due to added mass are not equal in magnitude during the supination and pronation. For example, during the supination there are two peaks due to added mass, one at $t/T = 0.45$ and one at $t/T = 0.55$. The first peak is less in magnitude as can be seen from Fig. 10. The rotational effects are emphasized during stroke reversals. During the transition from downstroke to upstroke, the wing starts to pitch up rapidly and results in a positive lift ($t/T = 0.45$). When the wing flips ($t/T = 0.5$) there is a drop in the rotational force, and when the wing starts the upstroke and pitches down, that results in a negative lift (as shown in Fig. 10(a)). The upstroke and downstroke are symmetric in hover, therefore the same pattern in forces due to added mass and rotational forces is obtained during the transition from up-to-down stroke. The negative peak of the rotational force is in the opposite direction to the added mass contribution, and they eliminate each others effect in the total force ($t/T = 0.05$, 0.55). The positive peak of the rotational force enhances the added mass peak ($t/T = 0.55$, 0.95) in the total force. Added mass and rotational forces have a minor contribution during the translational portions. The QS prediction for total drag is shown in Fig. 10(b). The total drag force consists of the added mass contribution and translational forces. Currently, these is no model for rotational drag.

Induced Inflow

The induced inflow ratio is a non-dimensional quantity that is used in rotorcraft and defined for helicopters as $\lambda_i = v_i/(\Omega R)$, where, v_i is the inflow velocity at the rotor disk normalized by the blade tip speed in hovering flight (ΩR). In rotorcraft, using momentum theory assumptions, analytical relation between the thrust and the induced velocity in hover can be derived, and it is:

$$\lambda_h \equiv \lambda_i = \sqrt{\frac{C_T}{2}} \tag{4}$$

where C_T is the rotor thrust coefficient, and equal to $T/\rho A(\Omega R)^2$. Momentum theory assumes a 1-D flow of an

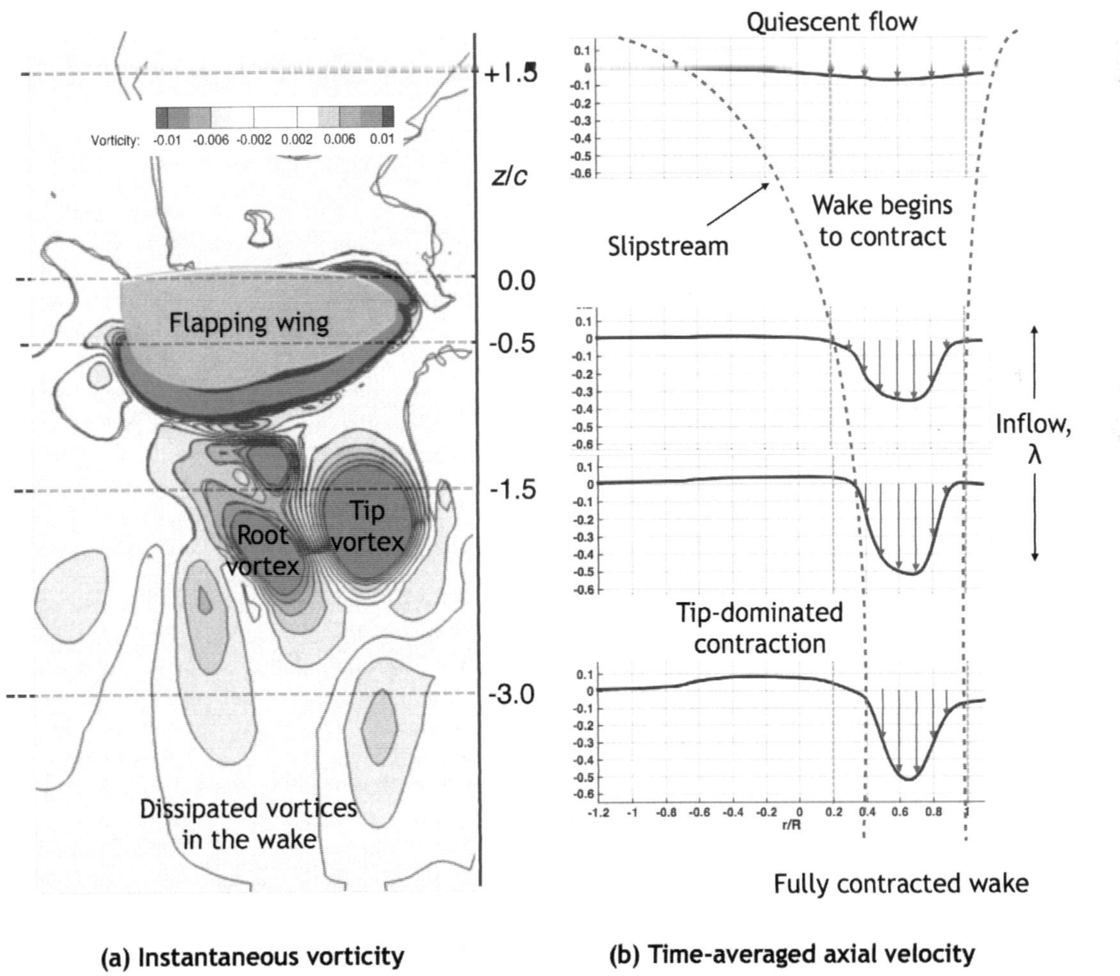

Fig. 11. Schematic diagram of the hovering flap-stroke. Flap-stroke is horizontal.The shed lines describe the planes parallel to flap-stroke at different location in the Z direction below and above the wing. The location is given in non-dimensional quantity (normalized Z by R).

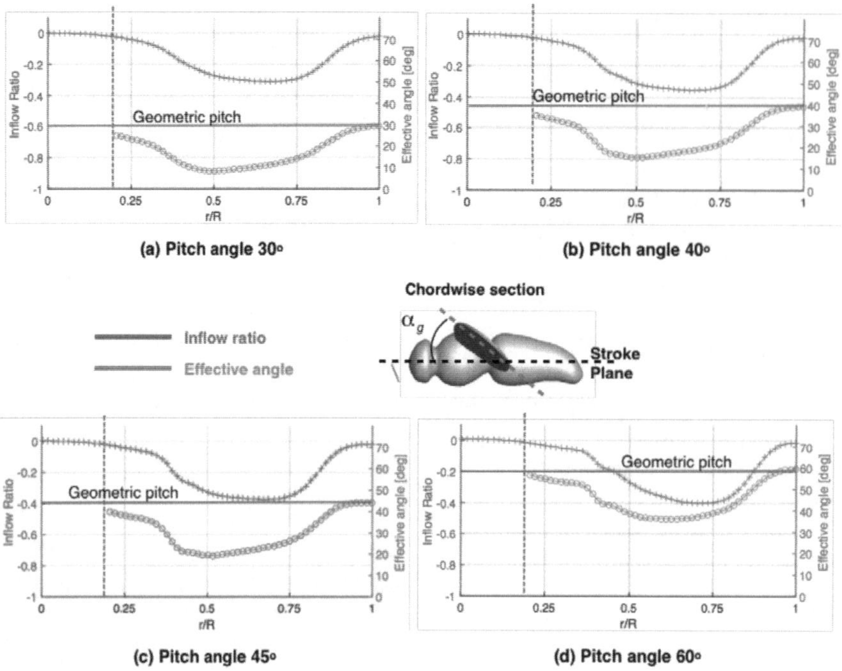

Fig. 12. Schematic diagram of the hovering flap-stroke. Flap-stroke is horizontal.The shed lines describe the planes parallel to flap-stroke at different location in the Z direction below and above the wing. The location is given in non-dimensional quantity (normalized Z by R).

ideal actuator disk, where the rotor is modeled as an infinitely thin disk, with a time-averaged induced constant velocity across the disk. These assumptions do not necessarily hold for flapping-wing insect flight, therefore the relation derived for rotorcraft may not apply here. While recognizing that the expression might not hold one-for-one, quantifying and understanding the induced inflow can provide insights about the wake involved with flapping kinematics. For instance, the inflow helps identify portions of the stroke most affected by the flow from previous wakes, comparisons between different pitch angles and their effect on the wake and induced inflow, and estimating the obtained effective angle of attack, which is instrumental in obtaining the loads on the wing.

Figure 11(a) shows an instantaneous image of the vortices in the wake beneath a flapping wing in hover obtained using CFD. The signature of the tip and root vortex can be seen in the image. The vortices convect down axially and dissipate into the wake. The reason for the dissipation of the vortices are because of the natural viscosity in the flow that causes them to breakdown, and the coarsening of the meshes further away from the wing, which results in added numerical dissipation because of the increasing mesh size. Nonetheless, the signature of the flow is similar to those observed in rotor wakes in hover (see Fig. 8(a)).

Figure 11(b) shows the time-average result of the axial velocity at various longitudinal locations above and beneath the flapping wing ($z/c = 1.5, -0.5, -1.5, \text{and} -3.0$), from CFD over a flapping cycle. These results are shown for the case of a mid-stroke pith angle of 40°. It is interesting to note that similarities between Fig. 11(b) and momentum theory in hover for rotors exist. A quiescent flow upstream of the wing is observed, following a contracting of the wake through the wing, which finally asymptotes to a fully-contracted wake. The downward arrows at each longitudinal section indicate the magnitude of the axial velocity at a given spanwise station. The contracted wake is, however, more tip-dominated, which is expected for both rotors/wings that are aerodynamically loaded towards the outboard sections. While this study does not proceed to derive an appropriate contraction ratio or slipstream velocity relations, the information presented can provide certain clues to do so in future studies.

A time-averaged along upstroke and downstroke over one flapping cycle, inflow distribution normal to the stroke plane is presented in Fig. 12. Inflow at mid-stroke as a function of radial position is plotted for the various pitch angles. The inflow is obtained from CFD predictions by calculating the velocity in the axial direction through a plane that is perpendicular to the flapping axis (wing-stroke plane). Such a technique is common in extracting induced velocities as the velocity on the surface would be zero (no-slip boundary condition). In this study, the wing stroke plane is defined as the plane normal to z-direction that cuts through the middle the of wing during the flapping cycle at ($z = -0.5\bar{c}$). At mid-stroke, a roughly linear increase in inflow magnitude radially up to a peak in induced inflow ratio at approximately $r/R = 0.7$ was observed. The maximum peak location at mid-stroke is approximately the same for all of the considered pitch angles, but the peak magnitude varies with the pitch angle, where a larger peak is obtained for the pitch angle of 60° case with value of 0.4 (downward) and a peak of 0.3 for 30° case; for 40° and 45° cases the induced inflow ratio reaches a peak of around 0.35. The effective angle of attack can be approximated as:

$$\alpha_{eff} = \alpha_g - \tan^{-1}\left(\frac{v_i}{\omega(t)r}\right) \quad (5)$$

The local velocity $\omega(t)r$ also increases linearly with r, thus the induced angle of attack change is nearly constant for the inboard 70% of the wing, before sharply decreasing outboard. Also, this induced angle of attack change continues during the translational portions (before the reversal portions of the stoke). Figure 12 shows the variation of the effective angle distribution along the wing span at mid-stroke for the cases with various geometrical pitch angles. The reduction in the pitch angle is nonuniform along the span due to varying rotational velocity and induced velocity along the span. The minimum effective pitch angle occurs in the wing inboard portion, between $r/R = 0.4 - 0.8$. The maximum reduction magnitude is about 20°–25°.

II. Efficiency Metrics

A key parameter revolving around flapping wing flight is the quantification of flight efficiency, especially in a scenario where kinematics and aerodynamic loads are coupled. Efficiency metrics are an essential tool to compare various kinematics and geometrical parameters for design. It should be noted that the word 'efficiency' has been used in flapping wing literature with little consistency. In the context of the present work, efficiency is related to aerodynamic efficiency and not mechanical or muscular efficiency. For a better understanding of the physiology of insects and the quantification of mechanical efficiency of the muscles, see Weis-Fogh and Jensen (Ref. 49). Conventional fixed-wing flight has relied on lift-to-drag (L/D) as the primary measure of aerodynamic efficiency. Rotary wing flight uses either figure of merit (purely a hovering metric) or a more general thrust-to-power ratio (T/P) for forward flight or rotorcraft with multiple interacting rotor systems. Thrust-to-power ratio is an absolute metric and quantifies the amount of power that needs to be supplied to a system to generate a given quantity of thrust. These common metrics of efficiency of flight are applied to flapping wings in an effort to propose a suitable quantity. The results shown in this section were obtained based on the analysis of CFD predictions of an isolated wing for various pitch angles and Reynolds numbers (through manipulation of the flapping frequencies), totalling close to fifty converged numerical simulations. The results are averaged over one flapping cycle after convergence (i.e., periodicity) in the solution was observed.

Lift-to-Drag ratio (L/D)

For insect flight, most of the lift is obtained during the translational portion because of the sustained formation of the leading edge vortex (LEV); see Fig. 10(a). The resultant force from the LEV is perpendicular to the wing and has both lift and drag components. These components are a function of the pitch angle, Reynolds number and the corresponding LEV size and strength. In Fig. 13(a), the averaged lift during a flapping cycle is shown as function of flapping frequency, for various pitch angles (30°, 40°, 45°, 60°). At a flapping frequency of 100 Hz (Re= 115), the averaged lift produced is similar in magnitude irrespective of the pitch angle. With an increase in flapping frequency (larger Reynolds number), the influence of the pitch angle on the produced lift is more apparent. For a given flapping frequency, the average lift increases to a maximum at a pitch angle of 45° before decreasing when the pitch angle was further increased to 60°.

The flapping cycle averaged drag is shown in Fig. 13(b), with larger differences between the various pitch angles when compared to the lift force. However with increasing in flapping frequency the discrepancy between the drag values increases. A monotonic increase in drag at a particular flapping frequency was observed as the pitch angle was increased, with 60° showing the highest drag values. This increase in drag can be directly attributed to the higher angle of attack.

Figure 13(c) shows the combined lift-to-drag ratio (L/D) for the isolated flapping wing, and it is apparent that the L/D across all cases tested was less than 1. One reason for this low L/D value is the prescribed kinematics. The wing starts the rotational stroke with a pitch angle of 90° which leads to high drag and decreases until it reaches the required mid-stroke pitch angle (geometrical angle of attack). From past studies of the author, it was observed that for pitch angles larger than 45°, the drag component is larger than the lift, which is reflected in the average values of L/D. The lowest L/D wass obtained for 60° pitch angle (Fig. 13(c)) across all flapping frequencies. The maximum L/D is obtained for the wing with pitch angle of 30° at the maximum tested flapping frequency of 300 Hz. This trend in L/D can be explained through the lift and drag curves, Figs. 13(a) and 13(b), respectively. While the lift force is roughly a constant across pitch angles, there is a substantial variation in drag. Therefore, the kinematics of the flapping wing are crucial to efficient flight. However, a higher L/D does not necessarily imply a lower drag. For instance, while a pitch angle of 30° produces higher values of L/D, the lift produced by a pitch angle of 30° is less than that produced by the 45° or 40° cases. Therefore, the L/D ratio in flapping wings provide only a limited understanding about the insect performance and other quantities need to be investigated to obtain a wholistic view of the efficiency of flapping wings.

Power Loading (L/P)

To investigate the efficiency of the kinematics, the lift-to-power ratio (L/P) was calculated across the four mid-stroke pitch angles. It is again noted that the power is aerodynamic power. For fixed wings, where the incoming velocity is constant and power is the product of integrated drag and velocity, L/D is equivalent to L/P. However, this relation does not

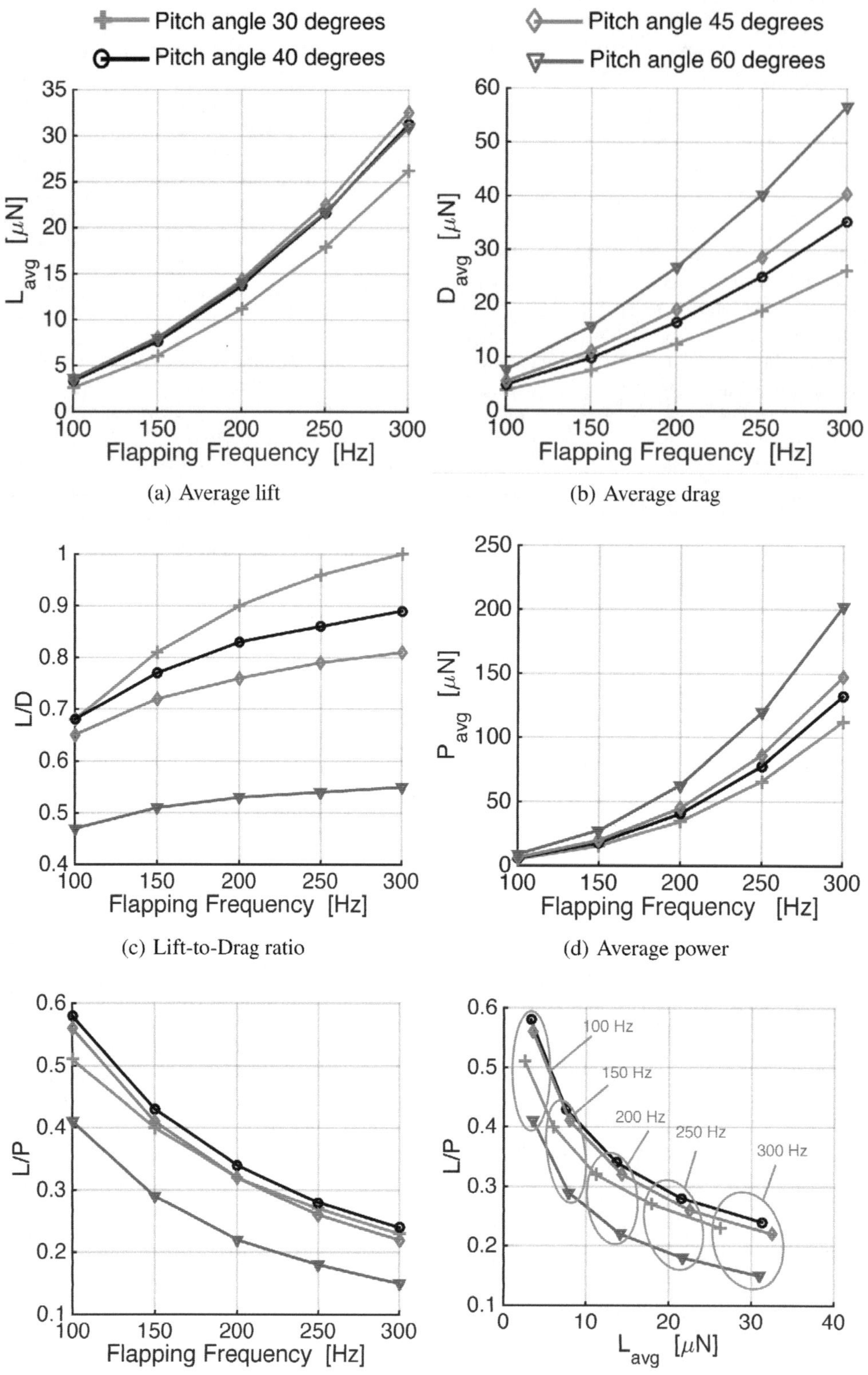

(a) Average lift

(b) Average drag

(c) Lift-to-Drag ratio

(d) Average power

(e) Lift-to-Power ratio

(f) Lift-to-Power ratio

Fig. 13. CFD flapping-cycle average lift, drag and required aerodynamic power of wing with 3 mm span and 20% root cut-off (0.75 mm). The results are shown for different mid-stroke pitch angles: 30,40, 45 and 60 degrees, Reynolds numbers range from 115 ($f = 100$) to 345 ($f = 300$).

hold for flapping or rotating wings where the velocity is time-varying and non-uniform along the wing span. Therefore, to compare across flapping wings with different kinematics or geometrical parameters as well as to other micro-air vehicles lift-to-power ratio is different from lift-to-drag ratio and may be able to provide a more valuable insight into the efficiency of the vehicle.

Aerodynamic power is computed by integrating the dot product of the local aerodynamic force and local wing velocity along the wing-span. The power incurred in biological and mechanical flight was not investigated in this study. The resultant aerodynamic power varies based on the pitch angle, wing-length and flapping frequency. Figure 13(d) shows the variation of aerodynamic power for different flight conditions. The power requirements increase as the mid stroke pitch angle increases at the same Reynolds number and increases with an increase in Reynolds number as well. The difference between the mean aerodynamic power curve for the various pitch angles increases with increasing flapping frequency, i.e., more power is required for larger Reynolds number. Furthermore, an increase in power requirements was observed at higher pitch angles, part of which can be attributed to the increase in wing drag.

Figure 13(e) shows the variation of L/P as a function of flapping frequency (f), to be compared and contrasted against L/D shown in Fig. 13(c). While L/D presented a monotonic behaviour with pitch angle, L/P shows a peak in efficiency at a pitch angle of 40°. The pitch angle of 60° has the lowest L/P ratio when compared to pitch angles of 30°, 40° and 45°. The ratio of L/P decreases with increasing frequency for each pitch angle as the power increases for a decrease in Reynolds number; a consqeuence of increasing drag efficient at lower Reynolds number. Power loading values for 40° and 45° are almost identical with a small advantage to pitch angle 40°. The pitch angle 30° and 45° cases are similar for frequencies greater than 150 Hz.

The efficiency metrics can be inverted into a design question, i.e., what are the optimum kinematics to generate a given lift. Figure 13(f) shows the variation of L/P against average lift for different pitch angles and flapping frequencies. A pitch angle of 40° performs well across the entire spectrum, although a pitch angle of 45° results in similar efficiencies. There are many aspects to consider when evaluating flapping wing efficiency metrics (i.e., maximum lift, minimum drag, minimum required power, flapping frequency and pitch angle). The flapping-cycle averaged power-loading L/P versus flapping-cycle averaged lift (L_{avg}) includes the above factors and is, therefore, a more appropriate efficiency metric than the simpler L/D ratio. The proposed metric properly accounts for the required aerodynamic power and vertical thrust as a function of prescribed kinematics and can be used to determine the performance of a flapping wing vehicle. Similar conclusions were reached by Zheng et. al. (Ref. 50), where lift and power were identified as the most objective metrics to quantify the performance of rotating-wings and flapping-wings in hover.

III. Vehicle Trim Analysis

A plethora of studies exist where the kinematics are prescribed and the resulting forces and moments are studied, i.e., the direct problem. However, the interest in this section (and for practical designers) is to identify the kinematics that will result in the desired forces and moments (trim solution), i.e., the inverse problem. Solution to a trim problem requires the formulation of a Jacobian (sensitivities of the forces and moments to kinematic inputs), which is in effect a new converged CFD solution, making the process prohibitively expensive. Therefore, studies in the past have obtained trim solutions only for simplified aerodynamic models with very low computational cost. However, these simplified models do not contain the key unsteady phenomenon associated with flapping wings and as a consequence, the wing kinematic parameters are incorrect or a trial and error approach is used to only trim the lift.There is a clear need to incorporate three-dimensional CFD simulations to capture the physics of flapping wing flight. This section provides the details of the flight dynamic model, trim condition, trim variables and the loose-coupling algorithm, an idea borrowed from the world of rotorcraft where similar constraints are placed on the computational expense of computing a Jacobian.

Vehicle trim is defined as the flight state when the linear and angular accelerations are zero. Within the context of flapping-wing MAVs, trim is said to be achieved if the steady-state forces and moments are periodic and do not change over successive flapping cycles, i.e., the accelerations are zero in a cycle-averaged sense. This cycle variation in the forces are a result of the flap kinematics and therefore, the loads at the center of gravity of a flapping wing MAV are oscillatory in nature, fluctuating about a non-zero mean. The overall objectives of the coupled (Quasi steady/CFD) trim methodology are: 1. Generalize the flapping-wing trim framework and CFD solver to analyze steady flight including hover, forward flight and coordinated turn, 2. Determine kinematics-driven limitations for the flight envelope in level turn, and 3. Apply CFD coupling to characterize the flight limits, i.e., maximum forward speed) for a particular platform.

Flight Dynamics: The flight dynamics model is obtained by expressing the fundamental equations of force and moment

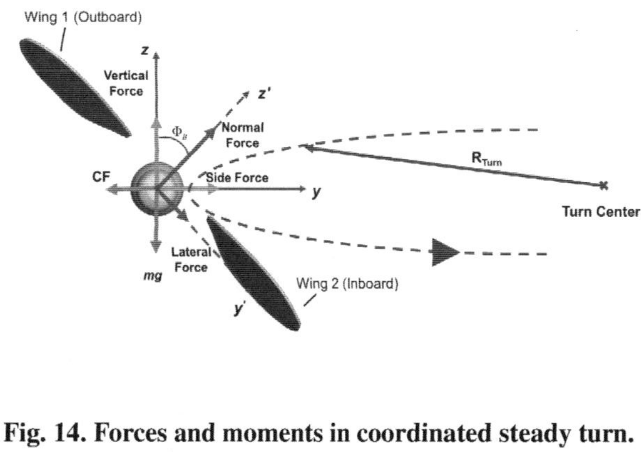

Fig. 14. Forces and moments in coordinated steady turn.

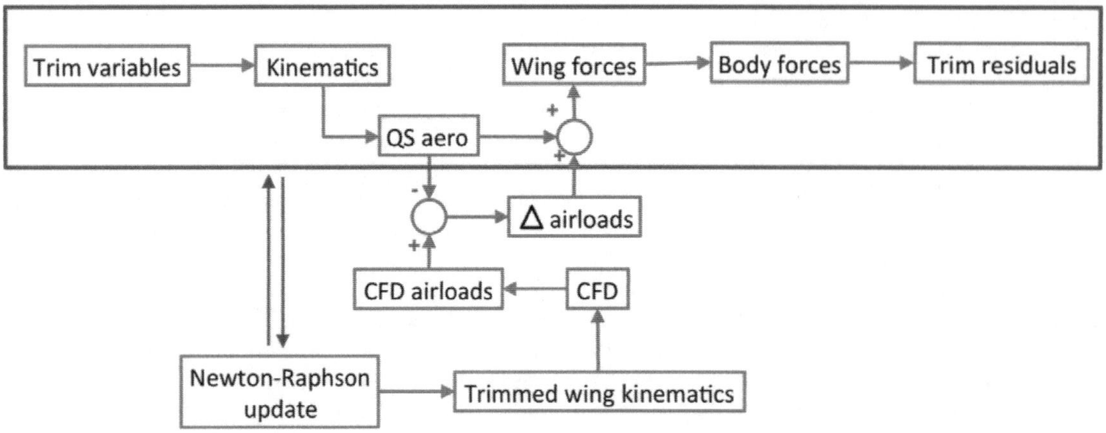

Fig. 15. Schematic of the loose coupling algorithm for high-fidelity insect trim.

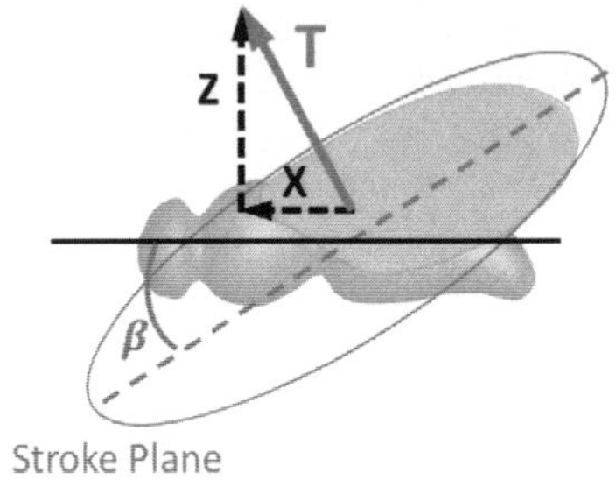

(a) Flapping wing stroke plane

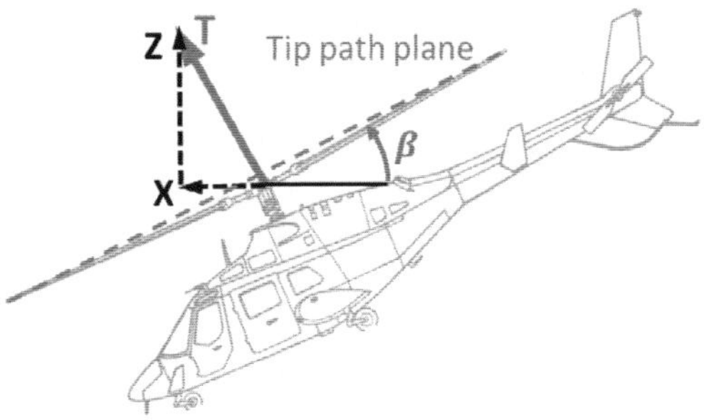

(b) Helicopter tip-path-plane

Fig. 16. Free-body diagram showing analogy between flapping wing and rotary-wing: resultant thrust, time-averaged over one cycle, is nominally perpendicular to the stroke plane (insect) or tip-path-plane (rotor).

equilibrium along the inertial coordinate system, shown in Fig. 6(a), for a general flight condition. The body of the insect is assumed to be rigid and the dynamics are given by

$$\mathbf{F} = \left(\frac{\partial}{\partial t} + \omega\times\right) m\mathbf{V} + m\mathbf{g} \tag{6}$$

$$\mathbf{M} = \left(\frac{\partial}{\partial t} + \omega\times\right) \mathbf{I}\,\omega \tag{7}$$

where **F** and **M** represent, respectively, the external forces and moments acting at the center of gravity due to body drag, and wing aerodynamic and inertial loads. m is the total mass of the insect. At these length scales, the mass of the wings are negligible compared to body mass. **g** is the acceleration due to gravity and ω is the body angular velocity vector. **I** represents the inertia tensor with body-fixed axes as reference. Wing inertial loads are ignored for trim since they are guaranteed to have a mean value of zero for periodic flapping.

Steady flight trim represents the condition where the insect center of gravity experiences zero translational and rotational accelerations measured along the body-fixed axes. Mathematically, this condition is represented using the following equations

$$\dot{u} = \dot{v} = \dot{w} = 0 \tag{8}$$

$$\dot{p} = \dot{q} = \dot{r} = 0 \tag{9}$$

where u, v, w and p, q, r are the translation velocities and body angular rates along the three coordinate directions, respectively. Substituting the translational and angular velocity in Eqs. 6 and 7 yields the following trim equations for force and moment balance

$$\overline{F}_x = 0 \tag{10}$$

$$\overline{F}_y = mV_\infty \dot{\psi}_B \cos\phi_B - mg\sin\phi_B \tag{11}$$

$$\overline{F}_z = -mV_\infty \dot{\psi}_B \sin\phi_B - mg\cos\phi_B \tag{12}$$

$$\overline{M}_x = -(I_{yy} - I_{zz})\dot{\psi}_B^2 \sin\phi_B \cos\phi_B \tag{13}$$

$$\overline{M}_y = 0 \tag{14}$$

$$\overline{M}_z = 0 \tag{15}$$

where $\dot{\psi}_B$ is the turn rate, ϕ_B is the body pitch as described in Fig. 14, which also describes the aerodynamic forces and moments, including the body angles. It should be noted that V_∞ is aligned with the body longitudinal axis, i.e., the insect flies with zero sideslip.

In obtaining the moment balance equations, the cross-products of inertia were assumed to be zero. Here, $\overline{F}_x$ represents the component of stroke-averaged wing aerodynamic force along the body longitudinal axis. A similar notation has been followed for the other loads. In a coordinated turn, the body roll attitude is prescribed so that the lateral component of the stroke-averaged wing aerodynamic force (in the inertial reference) counters the centrifugal force, i.e., the aerodynamic force along the body lateral direction vanishes ($\overline{F}_y = 0$). This condition can be used to solve for the roll attitude as

$$\phi_{\text{B}} = \tan^{-1}\frac{V_\infty \dot{\psi}_{\text{B}}}{g} \tag{16}$$

Therefore, the trim equations for force balance reduce further to

$$\overline{F}_x = 0 \tag{17}$$

$$\overline{F}_y = 0 \tag{18}$$

$$\overline{F}_z = -\sqrt{(mg)^2 + (mV_\infty \dot{\psi}_{\text{B}})^2} = -mg\sec\phi_{\text{B}} \tag{19}$$

where $\sec\phi_{\text{B}}$ is the normal load factor produced by the wings in the vertical direction. The negative sign indicates that the wing force component is upwards, since the coordinate direction for body Z-axis points downwards.

In steady forward flight and hover, symmetry can be assumed with respect to the body X-Z plane. Turn rate and body roll angle is zero, and average side force is zero (no centrifugal force). Further, wing inertial loads are ignored because the inertial loads vanish when time-averaged over one flap cycle for periodic kinematics. Therefore, the flight dynamics reduce to the following longitudinal trim equations:

$$\overline{F}_x = 0 \tag{20}$$

$$\overline{F}_z = -mg \tag{21}$$

$$\overline{M}_y = 0 \tag{22}$$

The results presented in this work do not discuss trim with respect to maneuvering flight and are only limited to hover and forward flight for the sake of brevity. However, additional results for maneuvering flight can be found by previous works of the author (Ref. 51).

Trim Algorithm (Loose Coupling): While CFD can provide detailed numerical representations of the flow-field around the insect body and flapping wings, the time taken for CFD analysis is several orders of magnitude larger than the quasi-steady aerodynamic model. Using CFD directly in the trim process is therefore infeasible because of the associated computational cost. The approach adopted for this work uses CFD to sporadically *correct* the quasi-steady model at select stages of the trim process, instead of replacing it entirely. For this work, the *delta air-loads approach* used commonly in rotorcraft analysis (Refs. 52, 53) is adopted, schematically shown in Fig. 15. While the quasi-steady model cannot accurately predict the forces and moments for a given wing kinematic condition, it captures the trends well. Therefore, the quasi-steady model is used to compute the Jacobian (which is of negligible computational cost) and the results from CFD are used to augment the forces and moments from the quasi-steady mode

$$\mathbf{F}_{\text{corrected}} = \mathbf{F}_{\text{QS}} + \mathbf{F}_{\Delta} \tag{23}$$

where $\mathbf{F}_{\Delta} = \mathbf{F}_{\text{CFD}} - \mathbf{F}_{\text{QS}}$. Using this approach, the temporal resolution of the unsteady air-loads generated by CFD are judiciously introduced into the trim framework without incurring a large computational cost. The flight dynamic model with built-in quasi-steady wing aerodynamics is used to obtain the trimmed wing kinematics using Newton-Raphson iterations. For each iteration, the Jacobian (sensitivity of cycle-averaged force and moments to perturbations in the individual trim variables) is recalculated using forward differences about the current trim guess. For the vehicle parameters considered, 5–10 iterations are required until trim is achieved. As noted previously, the quasi-steady aerodynamic model is very inexpensive, and requires a few seconds to achieve trim. Note that while trim solutions are obtained using a simplified aerodynamic model with very low computational cost, three-dimensional CFD simulations are key to capturing all associated unsteady phenomena and providing fundamental insight into the nature of the flow-field.

Three flapping kinematics parameters for each wing are allowed to vary in this study to achieve the forces and moments required to maintain steady flight. These parameters are identical between the port and starboard wings in steady level forward flight, and are referred to as "trim variables"

1. Half peak-to-peak stroke amplitude (ϕ_{max}), which predominantly affects the vertical thrust.

2. Stroke flapping bias (ϕ_{off}), which primarily affects the center of lift during the flapping stroke, and hence the body pitching moment.

3. Stroke plane angle (β), which produces a forward tilt of the resultant thrust vector, used to counteract drag by generating positive propulsive force.

The trim variables, moments, and forces with respect to the global inertial (space-fixed) coordinate system are depicted in Fig. 6. Within each flap cycle, the aerodynamic forces on the wing are time-varying. Therefore, the insect body experiences non-zero accelerations when flying at constant speed and steady non-accelerating flight is obtained in an *average* sense over one flap cycle. A schematic illustration depicting the analogy between flapping wings and rotary-wings is shown in Fig. 16.

Trim Controls: The variation of stroke amplitude with forward speed is shown in Fig. 17(a). As the forward speed increases, vehicle drag increases primarily because of the increasing wing drag. The total force required from the flapping wing increases significantly beyond 0.5 m/s, and results in a subsequent increase in the stroke amplitude. The variation of stroke phase offset is shown in Fig. 17(b). This parameter

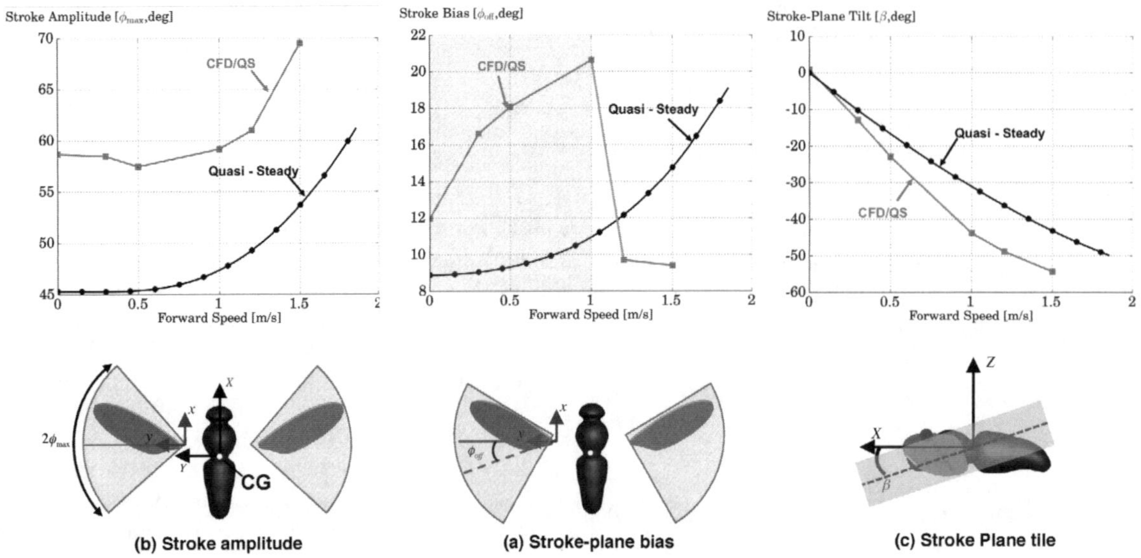

Fig. 17. Variation of trim variables with forward speed: quasi-steady aerodynamics vs. CFD.

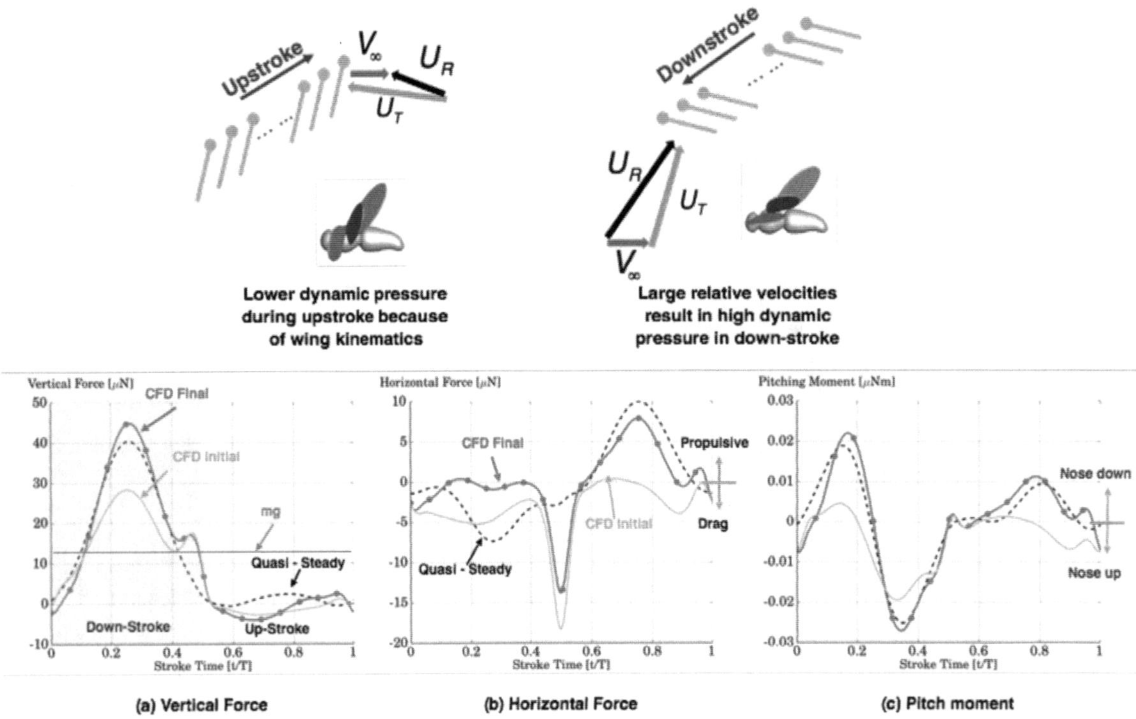

Fig. 18. Time histories of aerodynamic loads during a flapping cycle in trimmed forward flight at 1 m/s: comparison of quasi-steady predictions with initial and final CFD.

affects the longitudinal bias of the center of lift ahead of the vehicle CG, shifting the mid-point of the flap cycle towards the tail for forward speed less than 1 m/s. For speeds higher than 1 m/s, an opposite trend is observed where the mid-point of the flap cycle starts to shift forward. The phase offset ϕ_{off} has a dominant effect on the pitching moment, and must be analyzed in conjunction with the stroke plane tilt β, shown in Fig. 17(c). The reason for the significant change in the bias angle is the increase in the stroke plane tilt for speeds larger than 1 m/s. For instance, at a forward speed of 1.2 m/s the flapping plane tilt is almost 50°, and at a forward speed of 1.5 m/s it is approximately 55°.

The stroke plane is the flapping-wing analog of the rotor tip-path-plane, in that the resultant thrust produced by the wings over one cycle is nominally perpendicular to the plane containing the motion of the lifting surfaces. As vehicle speed increases, the stroke plane and hence the resultant thrust vector (time-averaged over one cycle) tilts forward, as shown in Fig. 16). Similar trends have been observed in the literature, notably by David et al. (Ref. 54) and Willmott and Ellington (Ref. 25). In forward flight, the stroke plane tilt was also observed to significantly delay stall, as demonstrated by Nagai (Ref. 55). This observation emphasizes the need for accurate prediction of trim kinematics to capture dominant flow phenomena that are key to overall vehicle analysis and performance. Therefore, utilizing either incorrect kinematics or an aerodynamic model with insufficient capabilities would result in these characteristics being missed entirely. The present work addresses this requirement by incorporating high-fidelity CFD airloads into the trim process.

Forces and Moments: Equilibrium in forces and moments in level forward flight for a speed of 1 m/s is obtained for the following trim variables, stroke amplitude (ϕ_{max}) of 59.2°, stroke bias (ϕ_{off}) of 20.6° shifted from midpoint towards the tail and a stroke-plane tilt (β) equal to 43.8° tilted forward. In trimmed forward flight, the total velocity vector has components from the induced velocity, rotational velocity and freestream velocity. Due to the flapping-stroke kinematics, the wings experience different resultant velocities during the downstroke and the upstroke. Figure 18 shows the resultant velocity U_R from the vector summation of U_T (perpendicular component of the rotational velocity) and V_{inf} (free stream velocity). The resultant LEV and circulatory forces are a function of the effective angle of attack and the resultant velocity in the flapping plane. The wings have a larger resultant velocity in downstroke than in upstroke as is illustrated in Fig. 18. This increase in dynamic pressure on the wings during the downstroke compared to the upstroke is reflected in the magnitude of the produced aerodynamic forces.

Figure 18 plots the time histories of the total forces over one stroke in trimmed forward flight of 1 m/s. The total forces are contributed by the two wings (effect of insect body forces are negligible compared to those from the wings) and computed in the inertial coordinate system. As in hover, the average quantities must satisfy the trim conditions, i.e., the vertical thrust (Z) is equal to the insect MAV weight and average

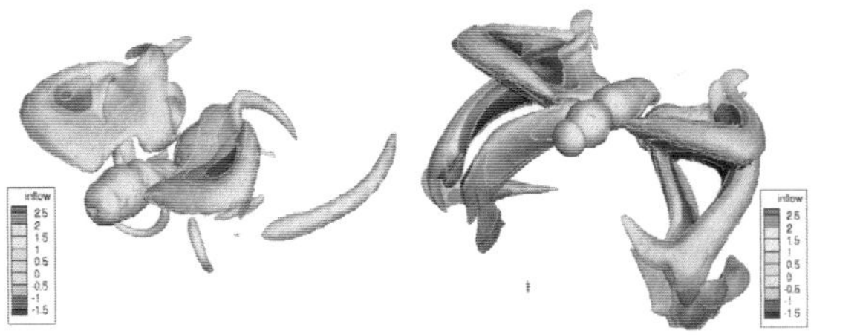

(a) Forward flight (b) Hover

Fig. 19. Iso-surface vorticity magnitude colored by the vertical velocity component (induced velocity), in trimmed forward flight and hover in downstroke.

X and Y forces are zero. The majority of the vertical thrust is obtained during the downstroke. The downstroke and the upstroke are no longer similar due to the tilt of the plane forward and the different resulting velocity magnitude. The averaged X force is approximately zero. Most of the negative X force is obtained during the supination (with a negative peak), with a positive horizontal force obtained during the upstroke. Note that the obtained forces from the QS model are quite different compared to those obtained from CFD, which will affect the power predictions. More results from the trim analysis and the algorithm are presented in previous works (Refs. 16, 56, 57).

Wake: To compare the wake structure in forward flight and the wake in hover, iso-surfaces of vorticity magnitude are shown in Fig. 19, where the vorticity is colored by the vertical velocity component. In hover, where there is no freestream velocity component, the shed vorticity convects below the wings creating distinguishable shed tip and root vortices. In forward flight the shed vortices convect backward due to the free stream velocity and the plane tilt, and downwards due to the induced inflow. The vortical structure are more visible in hover than in forward flight, which in part could be the inability of the coarser meshes behind the wings to accurately preserve the trailed/shed vortex structures.

IV. Performance

The next logical step in the study of flapping wings is to evaluate its performance for a given flight condition after the insect has been trimmed, which is also an important outcome of the trim analysis. For a given vehicle weight, the mean aerodynamic power and its components can be determined at hover and different forward flight speeds. The insect mass was 1.11 mg in all cases. The presented results were obtained using CFD after converged QS-CFD loose coupling trim iterations. This reference trim condition allows for a comparison of insect flight with different wing kinematics.

Figure 20(a) presents the CFD predictions of flapping cycle averaged aerodynamic power in trimmed forward flight as a function of forward speed up to 1.5 m/s. The circles represent CFD solution and the dashed line is a least square curve-fit to better elucidate the trend. Note that the aerodynamic power initially decreases with increasing forward

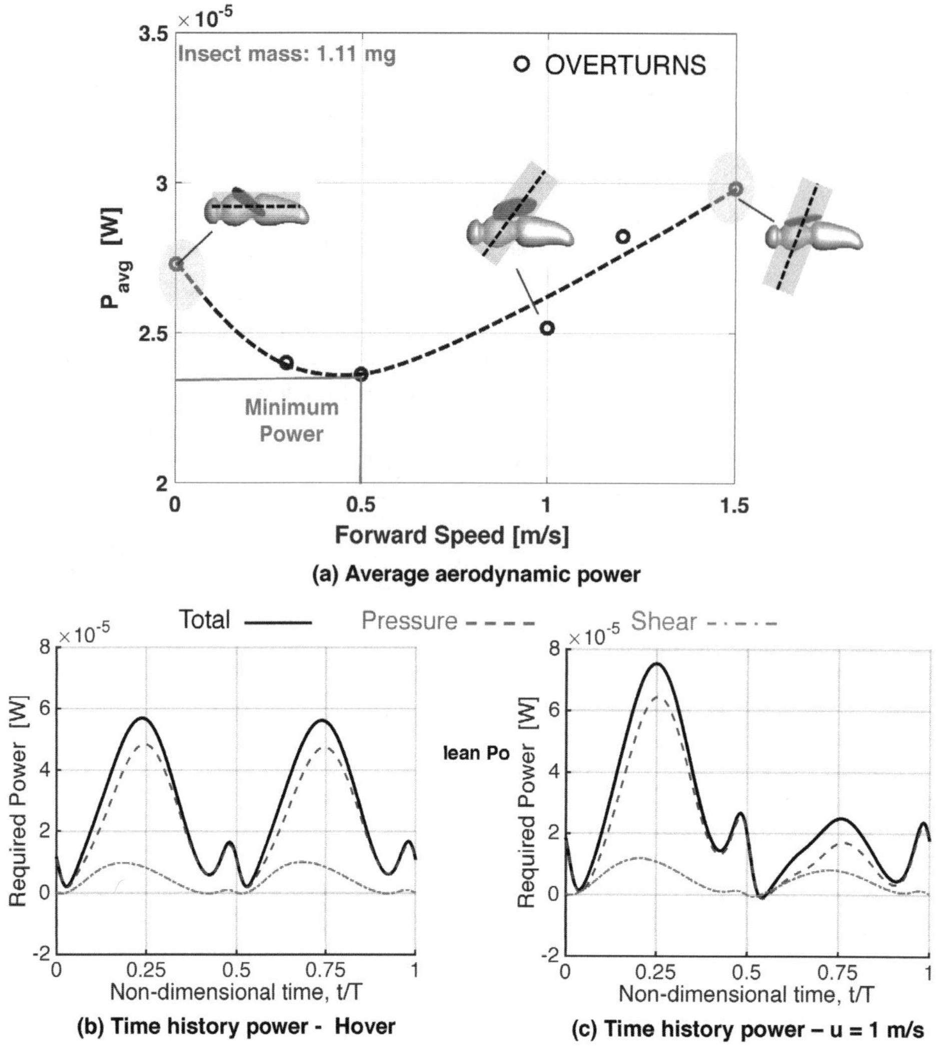

Fig. 20. CFD predictions of flapping wings MAV aerodynamic power mean and time history in hover and trimmed forward flight of 1 m/s.

speed, reaching a minimum at 0.5 m/s before increasing again for higher forward speeds. As flight speed increases, there is an increase in the flapping stroke plane tilt (β). The angle is close to zero in hover and increases to a value of 55° at the maximum tested forward flight speed. This trend in power is similar to those obtained from a rotor in trimmed flight. Figure 21 shows the contribution of the induced power (decreases with forward speed), profile power (increases as a square of the forward speed) and parasitic power (increases as a cube of forward speed) to the total power for typical helicopter. Similarities in the trend between flapping wings and edgewise rotors is apparent in a reduction in power with forward speed before increasing.

Through a CFD analysis it is challenging to isolate the components of power into induced, profile and parasitic. In the present study, power is computed through a summation of the dot product of the local force and local body velocity for all surface mesh elements. While the estimation of power is based on aerodynamic loads obtained from the CFD simulations, real insect flight (mechanical or biological) will experience losses due to inertial forces and mechanical losses in addition to aerodynamics. This aerodynamic power is decomposed into the pressure and viscous components. The physical flow around the flapping wing is highly viscous (low Reynolds number), resulting in a large boundary layer and increased shear drag, which can be referred to as profile drag. Even for low Reynolds numbers, where shear forces can not be neglected, pressure forces still dominate shear forces.

Figure 20(b) shows the variation in shear, pressure and total power over the flapping cycle for an insect in hover. The power losses due to the pressure drag are much larger than the power losses due to the viscous profile drag through the hover stroke. The peak in pressure power losses occurs at the middle of the downstroke and upstroke. The viscous power losses

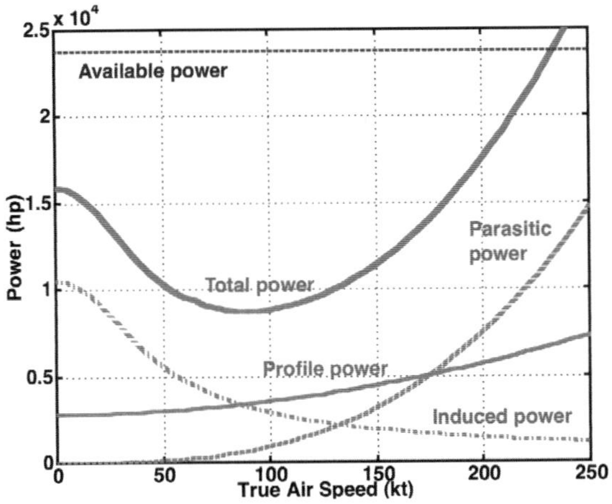

Fig. 21. Main rotor power in forward flight.

increase at the beginning of the downstroke (and upstroke), reaching the peak around $t/T = 0.2$ (upstroke $t/T = 0.7$) and then there is a reduction until the end of the stroke, where there is a relatively small peak due to the rotational forces.

Figure 20(c) shows the time history of required power during one flapping cycle in trimmed forward flight. The larger component of total power is due to the pressure, which can roughly be assumed to be "induced power." The pressure contribution to power is much larger than the shear contribution during the flapping stroke. The symmetry between the upstoke and downstroke seen in hover is no longer true for the insect in forward flight. The key feature in steady forward flight is the forward tilt of the flapping stroke plane, which leads to asymmetry in the forces produced and the power required during during the downstroke and the upstroke. The majority of the vertical thrust is produced during the downstroke, while the positive propulsive force is produced during the upstroke. The required aerodynamic power is larger during the downstroke, mostly due to the pressure drag resulting from the presence of a relatively stable LEV on the upper surface of the wing. Large interactions between the wing and the wake from the downstroke results in a smaller pressure peak on the upstroke. The power required due to the shear forces is nearly zero during the supination and pronation because the LEV detaches and the flapping velocity slows down to zero. Shear forces are non-zero during the translational portions because of the presence of the boundary layer.

It is clear that the power is dominated by both the viscous and non-viscous components of drag. The viscous drag contribution towards power is relatively small in comparison to total required power, and this quantity depends on Reynolds numbers. The non-viscous drag sources in the flapping wing case are attributed to lift-induced drag, vortex-drag and is referred to as the "added mass" effect. Aerodynamic power in flapping wings are mostly dominated by the non-viscous drag, also referred to as "pressure drag."

in the shown figures, and it is Reynolds number independent (discuss with Dr. Baeder and Bharath), where kinematics and formation of vortical structure on the wing and downwash seems to affect the trend and magnitude of pressure power (i.e. in forward flight the pressure power during up-stroke is very small where LEV is significantly small in comparison to the formation of LEV during downstroke).

In comparison to rotating wing at Reynolds number range similar to the cases studied ($O(10-10^2)$), Zheng et. al. (Ref. 50) compared between rotating wing and flapping wing of hawkmoth shape, and they conclude that aerodynamic power of revolving wing at these Reynolds number range (100–5,000) is dominated by viscous drag, which leads that for decreasing Reynolds numbers the required power increasing for rotating wing, in contrast for flapping wings power is dominated by non-viscous effects as well shown in this study, where the viscous drag seems to be small and constant for these Reynolds number, therefore power didn't increases with decreasing Reynolds numbers. So in terms of efficiency (power loading) the flapping wings have a clear advantage over rotating-wing at low Reynolds numbers.

SUMMARY

An unsteady, three-dimensional, Navier–Stokes solver (OVERTURNS), was used to simulate insect flight with two flapping wings (*diptera* configuration). Investigated were the aerodynamics around flapping wings, measure of efficiency metrics, analysis of trim and the performance of the insect. The principle aim of this study was to recognize similarities between rotorcraft and flapping wings, both in terms of the results obtained and the methodology used in rotorcraft that could be transferred to the flapping wing community. Specific conclusions from this study are

1. The flow field beneath flapping wings is inctdicate and laden with vortices, which interact with the wing resulting in complex aerodynamic phenomenon. Simplified models must account for the effects of added mass and rotational circulation along with the translational lift to accurately predict the forces and moments with varying kinematics. These effects are more predominant in forward flight as compared to hover. Flapping wings in hover show similarities to hovering rotor through the presence of axially convecting tip and root vortices. In a time-averaged sense, the wake beneath flapping wings was found to resemble the momentum theory formulation of helicopters. The contraction was more tip-domination and a loose control volume could be defined through the slipstream boundaries.

2. For fixed wings, where the incoming velocity is uniform and power is the product of integrated drag and velocity, lift-to-drag (L/D) is equivalent to lift-to-power (L/P, power loading). However, this relation does not hold for flapping or rotating wings where the velocity is time-varying and non-uniform along the wing span. It can therefore be concluded that to compare between flapping wings with different kinematics or geometrical parameters as well as to other micro-air vehicles, power loading

is a more suitable metric compared to lift-to-drag. Furthermore, power loading as a function of averaged lift (L/P vs. L_{avg}) is more appropriate as an efficiency metric than is the L/D ratio. The proposed metric properly accounts for the required aerodynamic power and vertical thrust as a function of prescribed kinematics. A similar ratio in rotorcraft, i.e., the thrust-to-power ratio is known to be a suitable metric across different platforms and in forward flight and is preferred over more specialized metrics such as figure of merit, which is best suitable only for isolated rotors in hover out of ground effect.

3. A multi-fidelity approach for obtaining CFD-coupled trim solutions was adapted from rotary-wing analysis and applied to a flapping wing platform. Using a judicious number of CFD calculations to correct the quasi-steady aerodynamics to obtained the desired trim state, the detailed flowfield models can be incorporated into the analysis without incurring prohibitive run times. This approach where trim is the anchor point, can be used to compare different configurations for example in hover for a given weight between rotary-wing vehicle and flapping wing, or with different kinematics pattern, or to optimize specific parameters such as pitch angle. While the importance of the QS cannot be understated in this analysis and is key in fast-evaluation of the trim Jacobian, this simplified model over-predicts the wing lift and under-predicts power required, and cannot capture unsteady effects during wing rotation between the up-stroke and down-stroke. CFD is therefore essential for the quantitatively accurate predictions of aerodynamic power and trim kinematics that are crucial for selection of wing geometry, battery, and determination of the actuator torque required and the vehicle flight envelope restrictions. It should be noted that this loose-coupling procedure does converge and provides a stable numerical scheme.

4. The required power of flapping wing in forward flight is similar to rotorcraft, in that there is a decrease from hover to a minimum power before increasing with forward flight speed. The ability to trim with CFD is key in obtaining insights into the mechanics of flight. For instance, the forces generated by the wings in forward flight are asymmmetric between the up-stroke and the down-stroke, with the down-stroke responsible for the vertical force and up-stroke responsible for the propulsive force. It is recognized that this division of forces in forward flight is true for a *Diptera* configuration, and might vary in the presence of insects with four wings.

ACKNOWLEDGMENTS

This work was supported by the Micro Autonomous Systems and Technology Collaborative Technology Alliance MAST CTA, contract number W911NF-08-2-0004. Thanks to Kenneth MacFarlane and J. Sean Humbert, University of Maryland, for providing the quasi-study solver used in this study. The authors would also like to acknowledge the University of Maryland supercomputing resources (http://www.it.umd.edu/hpcc) made available in conducting the research reported in this paper.

REFERENCES

[1]McMichael, J. M. and Francis, Col. M. S., "Micro Air Vehicles Towards a New Dimension in Flight," Technical Report TTO document, USAF, DARPA, July 1996.

[2]Chopra, I., "Rotor-Based Micro Air Vehicles: Challenges & Opportunities," Presentation at First US-Asian Demonstration & Assessment of Micro-Aerial & Unmanned Ground Vehicle Technology, Agra, India, March 10–15, 2008.

[3]Grasmeyer, M. J., Keennon, T. M., and Aerovironment Inc, "Design and Development of a Quad Shrouded-Rotor Micro Air Vehicle," American Helicopter Society 65_{th} Annual Forum, Grapevine, Texas, 2009.

[4]Prinz, I., "The Mesicopter: A Meso-Scale Flight Vehicle," Available at http://aero.stanford.edu/mesicopter/, Accessed on July 14, 2016.

[5]Benedict, M., *Fundamental Understanding of the Cycloidal-Rotor Concept for Micro Air Vehicle Applications*, Doctor of philosophy, Department of Aerospace Engineering, University of Maryland, College Park, MD, 2010.

[6]Benedict, M., Jarugumilli, T. and Chopra, I., "Effect of Rotor Geometry and Blade Kinematics on Cycloidal Rotor Hover Performance," *Journal of Aircraft*, Vol. 50, (5), Sep., 2013, pp. 1340–1352.

[7]Winslow, J., Otsuka, H., Govindarajan, B., and Chopra, I., "Basic Understanding of Airfoil Characteristics at Low Reynolds Numbers," 7th UAV Specialists Meeting of the American Helicopter Society, Mesa, AZ, January 24–26, 2017.

[8]Pornsin-Sisirak, T., Lee, S.W., Nassef, H., Grasmeyer, J.,Tai, Y.C., Ho, C.M., and Keennon, M., "MEMS Wing Technology for A Battery-Powered Ornithopter," Thirteenth IEEE International Conference on Micro Electro Mechanical Systems, Miyazaki, Japan, January 2000.

[9]De Clercq, K. M. E.,de Kat, R., Remes, B. D. W., van Oudheusden, B. W. and Bijl, H., "Aerodynamic Experiments on DelFly II: Unsteady Lift Enhancement," *International Journal of Micro Air Vehicles*, Vol. 1, (4), 2009, pp. 255–262.

[10]Shyy, W., Lian, Y., Tang, J., Viieru, D. and Liu, H. , *Aerodynamics of Low Reynolds Number Flyers*, Cambridge University Press, New York, USA, 2008.

[11]Keennon, M., Klingebiel, K., and Won, H., "Development of the Nano Hummingbird: A Tailless Flapping Wing Micro Air Vehicle," 50th AIAA Aerospace Sciences Meeting including the New Horizons Forum and Aerospace Exposition, Nashville, TN, January 09 - 12, 2012.

[12]"Controlled Flight of a Robotic Insect," Available at http://wyss.harvard.edu/viewpage/428/, Accessed on July 2016.

[13]Ellington, C. P., "The Aerodynamics of Hovering Insect Flight," *Philosophical Transactions of the Royal Society of London. Series B, Biological Science*, Vol. 305, (1122), Feb. 1984, pp. 1–181.

[14]Weis-Fogh, T., "Energetic of Hovering Flight in Hummingbirds and in Drosophila," *Journal of Experimental Biology*, Vol. 56, 1972, pp. 79–104.

[15]Dickinson, M. H.,Lehmann, F. H and Sane, S. P., "Wing Rotation and the Aerodynamic Basis of Insect Flight," *Science Journal*, Vol. 305, Jan. 1999, pp. 1954–1960.

[16]Badrya, C., Harrington, A., Baeder, J. D., Kroninger, C. M., Govindarajan, B. M., "Prediction and Validation of CFD-Coupled Trim Analysis for a Flapping Wing MAV," American Helicopter Society Specialists Meeting on Aeromechanics Design for Vertical Lift,, SanFrancisco, CA, Jan 20–22, 2016.

[17]Maxworthy, T., "Experiments on the Weis-Fogh Mechanism of Lift Generation by Insects in Hovering Flight I. Dynamics of the Fling," *Journal of Fluid Mechanics*, Vol. 93, (01), 1979, pp. 47–63.

[18]Kramer, M., "Die Zunahme des Maximalauftriebes von Tragfluglen bei plotzlicher Anstellwinkelvergrosserung (Boeneffect)," *Zeitschrift fur Flugtechnik und Motorluftschiffahrt*, Vol. 23, (7), 1932, pp. 185–189.

[19]Sane S. P , T. and Dickinson M. H, " The Aerodynamic Effects of Wing Rotation and a Revised Quasi-steady Model of Flapping Flight," *The Journal of Experimental Biology*, Vol. 205, 2002, pp. 1087–1096.

[20]Sun, T. and Tang, J., "Unsteady Aerodynamic Force Generation by a Model Fruit Fly Wing in Flapping Motion," *The Journal of Experimental Biology*, Vol. 205, Jan. 2002, pp. 55–70.

[21]Birch, J. M., and Dickinson, M. H., "The Influence of Wing Wake Interactions on the Production of Aerodynamic Forces in Flapping Flight," *The Journal of Experimental Biology*, Vol. 206, 1998, pp. 124–156.

[22]Wang, Z. J.,Birch, J. M., and Dickinson, M. H., "Unsteady Forces and Flows in Low Reynolds Number Hovering Flight: Two-Dimensional Computations vs Robotic Experiments," *Journal of Experimental Biology*, Vol. 207, 2007, pp. 449–460.

[23]Tang, J.,Viieru, D., and Shyy, W., "Effects of Reynolds Number and Flapping Kinematics on Hovering Aerodynamics," *AIAA Journal*, Vol. 46, (4), 2008, pp. 967–976.

[24]Freymuth, P., "Propulsive vortical signature of plunging and pitching airfoils," *AIAA Journal*, Vol. 26, (7), 1988, pp. 881–883.

[25]Willmott, A. P. and Ellington, C. P., "The Mechanics of Flight in the Hawkmoth Manduca sexta. I. Kinematics of Hovering and Forward Flight," *The Journal of Experimental Biology*, Vol. 200, 1997, pp. 2705–2722.

[26]Dickinson, M. H, and Gotz, K. G., "Unsteady Aerodynamic Performance of Model Wings at Low Reynolds Numbers," *The Journal of Experimental Biology*, Vol. 174, Jan. 1993, pp. 45–64.

[27]Birch, J. M., and Dickinson, M. H., "Spanwise Flow and the Attachment of the Leading-Edge Vortex on Insect Wings," *Nature Journal*, Vol. 412, 2001, pp. 729–733.

[28]Dickson, W. B. and Dickinson, M. H., "The Effect of Advance Ratio on the Aerodynamics of Revolving Wings," *The Journal of Experimental Biology*, Vol. 207, 2004, pp. 4269–4281.

[29]Badrya, C.,Baeder, J. D., Amiraux, M., Kroninger, C. M., "Parametric Study of Insect Flapping Performance Using CFD," fifth decennial specialists meeting on Aeromechanics of the American Helicopter Society Forum,, SanFrancisco, CA, Jan 22–24, 2014.

[30]Badrya, C., *CFD/Quasi-Steady Coupled Trim Analysis of Diptera-type Flapping Wing MAV in Steady Flight*, Doctor of philosophy, Department of Aerospace Engineering, University of Maryland, College Park, MD, 2016.

[31]Faruque, I. and Humbert, J.S., "Dipteran Insect Flight Dynamics. Part 1: Longitudinal Motion About Hover," *The Journal of Theoretical Biology*, Vol. 264, (02), 2010, pp. 538–552.

[32]MacFarlane, K. et al., "Quasi-Steady and Computational Aerodynamics Applied to Hovering *Drosophila* Dynamics," AIAA Applied Aerodynamics Conference, Honolulu, HI, June 27 - 30, 2011.

[33]Liu, H. and Kawachi, K., "A Numerical Study of Insect Flight," *Journal of Computational Physics*, Vol. 2146, March 2003, pp. 2257–2272.

[34]Ramamurti, R. and Sandberg, W, "Simulation of Flow about Flapping Airfoils using Finite Element Incompressible Flow Solver," *AIAA Journal*, Vol. 39, (2), Feb., 2001, pp. 253–260.

[35]Ramasamy, M., Leishman, J. G., and Lee, T. E., "Flow Field of Rotating Wing MAV," 62nd Annual National Forum Proceedings of the American Helicopter Society, Phoenix, AZ, May 9-11, 2006.

[36]Ellington, C. P., van den Berg, C., Willmott, A. P., and Thomas, A. L. R., "Leading-Edge Vortices in Insect Flight," *Nature*, Vol. 384, Dec. 1996, pp. 626–630.

[37]Rayner, J. M. V., "A New Approach to Animal Flight Mechanics," *Journal of Experimental Biology*, Vol. 80, 1979a, pp. 17–54.

[38]Rayner, J. M. V., "A Vortex Theory of Animal Flight. Part 1. The Vortex Wake of a Hovering Animal," *Journal of Fluid Mechanics*, Vol. 91, 1979b, pp. 697–730.

[39]Lakshminarayan, V. K., *Computational Investigation of Micro-Scale Coaxial Rotor Aerodynamics in Hover*, Doctor of philosophy, Department of Aerospace Engineering, University of Maryland, College Park, MD, 2009.

[40]Amiraux, M., *Numerical Simulation and Validation of Helicopter Blade-Vortex Interaction Using Coupled CFD/CSD and The Levels of Aerodynamic Modeling*, Doctor of philosophy, Department of Aerospace Engineering, University of Maryland, College Park, MD, 2014.

[41]Koren, B., , "Upwind Schemes, Multigrid and Defect Correction for the Steady Navier-Stokes Equations," The 11th International Conference on Numerical Methods in Fluid Dynamics, Willansburg, VA, June 1988.

[42]Ghosh, D., Medida, S. and Baeder, J. D., "Compact-Reconstruction Weighted Essentially Non-Oscillatory Schemes for Unsteady Navier-Stokes Equations," 42nd AIAA Fluid Dynamics Conference and Exhibit, New Orleans, LA, June 25-28, 2012.

[43]Buelow, P. E. O., Schwer, D. A., Freng, J. and Markle, T. E., "A Preconditioned Dual-Time, Diagonalized ADI scheme for Unsteady Computations," 13th AIAA Computational Fluid Dynamics Conference, Snowmass Village, CO, July 1997.

[44]Pandya, S. A., Venkateswaran, S. and Pulliam, T. H., "Implementation of Preconditioned Dual-Time Procedures in OVERFLOW," 41*st* AIAA Aerospace Sciences Meeting and Exhibit, Reno, NV, January 2003.

[45]Ramasamy, M., Gold, N. P., and Bhagwat, M., "Rotor Hover Performance and Flowfield Measurements with Untwisted and Highly-Twisted Blades," 36th European Rotorcraft Forum, Paris, France, September 7–9, 2010.

[46]Singh, B. and Chopra, I., "Insect-Based Hover-Capable Flapping Wings for Micro Air Vehicles: Experiments and Analysis," *AIAA Journal*, Vol. 46, (9), 2008, pp. 2115–2135.

[47]Leishman, J. G., *Principles of Helicopter Aerodynamics*, Cambridge University Press, New York, NY, 2006.

[48]Sane S. P , T. and Dickinson M. H, "The Control of Flight Force by a Flapping Wing: Lift and Drag Production," *The Journal of Experimental Biology*, Vol. 204, 2001, pp. 2607–2626.

[49]Weis-Fogh, T. and Jensen, M., "Biology and Physics of Locust Flight. I. Basic Principles in Insect Flight. A critical Review," *Philosophical Transactions of the Royal Society of London*, Vol. Series B 239, 1956, pp. 415–418.

[50]Zheng, L., Hedrick, T., and Mittal, R., "A comparative study of the hovering efficiency of flapping and revolving wings," *Bioinspiration & Biomimetics Journal*, Vol. 8, 2013, pp. 1–13.

[51]Badrya, C., Sridharan, A., and Baeder, J., "CFD-Coupled Trim Analysis of Millimeter-Scale Flapping Wing MAV in Steady Turning Flight," 34th AIAA Applied Aerodynamics Conference, Washington, DC, June, 2016.

[52]Tung, C., Cardonna, F.X., and Johnson, W.R., "The Prediction of Transonic Flows on an Advancing Rotor," *Journal of the American Helicopter Society*, Vol. 31, (3), 1986, pp. 4–9.

[53]Datta, A.,Sitaraman, J.,Chopra, I. and Baeder, D. J., "CFD/CSD Prediction of Rotor Vibratory Loads in High-Speed Flight," *Journal of Aircraft*, Vol. 43, (6), 2006, pp. 1698–1709.

[54]David, C. T., "The Relationship Between Body Angle and Flight Speed in Free-Flying Drosophila," *Physiol. Entomol.*, Vol. 13, 1978, pp. 191–195.

[55]Nagai,H., Isogai, K. , Fujimoto, T. and Hayase, T., "Experimental and Numerical Study of Forward Flight Aerodynamics of Insect Flapping Wing," *AIAA Journal*, Vol. 47, (3), 2009, pp. 730–742.

[56]Badrya, C., MacFarlane, K., Baeder, J. D., Humbert, J. S., "Insect Kinematics in Trimmed Flight at Low Reynolds Numbers Using CFD," 2015 AIAA SciTech Conference and Exhibit, Kissimmee, FL, Jan 5-9, 2015.

[57]Badrya, C., Baeder, J. D., "CFD/Quasi-Steady Coupling for Insects in Trimmed Forward Flight," AHS Specialists' Meeting on Unmanned Rotorcraft and Network-Centric Operations Program, Chandler, Arizona, Jan 20-22, 2015.

Performance and Sizing Tool for Quadrotor-Biplane-Tailsitter UAS

Marius Strom Graduate Research Assistant

James Baeder Professor

Alfred Gessow Rotorcraft Center
Department of Aerospace Engineering
University of Maryland, College Park, MD, USA

ABSTRACT

To support a growing off-shore wind energy industry, this project proposes a new sizing tool for a quadrotor-biplane-tailsitter aircraft: a UAS capable of both long-range, high-speed cruise and hovering flight. This paper presents the initial work done on this tool, including a robust and flexible weight buildup methodology, various aircraft sizing modules, and a modular integration framework. Utilizing both on- and offshore wind turbine inspection missions for preliminary design explorations, this tool can provide a survey of potential designs for a quadrotor-biplane-tailsitter aircraft. This initial study shows the potential for the new configuration for such long-range, high-endurance missions in comparison with pure quadrotor designs sized by the same tool. The integration framework designed for this tool will enable coupling with a free-vortex and panel method code that will allow for interference effects in the rotor/wing/body system to be intelligently accounted for in the sizing process.

NOTATION

C_T	Thrust coefficient
σ	Rotor solidity
P_{reqd}	Maximum power required per motor
E_{reqd}	Total stored energy required
W_{motor}	Motor weight
W_{ESC}	Motor controller weight
$W_{battery}$	Battery weight
$Imax_{motor}$	Motor maximum current draw
$Imax_{esc}$	ESC maximum current draw
$Imax_{battery}$	Battery maximum discharge rate
mAh	Battery capacity in mAh

INTRODUCTION

At present, the methods of inspecting an on- or off-shore wind turbine blade include the use of roped inspectors (positioned on the turbines themselves), a ground station equipped with various sensors, manned aerial inspection platforms, and short-range unmanned aerial systems (UAS) that must be carried within range of individual turbines by ground vehicle or sea vessel. However, each of these methods carries with it a certain cost, be it in terms of financial investment, increased inspection time, or susceptibility to metocean conditions as noted in (Ref. 1), (Ref. 2), (Ref. 3), (Ref. 4), (Ref. 5), (Ref. 6), and (Ref. 7). Furthermore, each of these methods necessitates a human expedition to the turbine itself. On-shore, this is a trivial matter, but, for off-shore turbines, this requirement dramatically reduces turbine accessibility and provides a source of risk to the personnel involved.

A design previously explored in (Ref. 8), (Ref. 9), (Ref. 10), and (Ref. 11), could alleviate most of these issues. Utilizing a Quadrotor-Biplane-Tailsitter (QBT) configuration, this aircraft could, potentially, perform inspection missions in high wind conditions without requiring local transport or human intervention, decreasing risk and increasing turbine accessibility. Able to leverage both helicopter and biplane flight modes via a unique "tilt-body" configuration, this aircraft is capable of high speed cruise followed by extended loiter time in hover; all while preserving the simplicity of a traditional quadrotor UAS. This aircraft would not necessarily need to be transported to an individual turbine, but could, potentially, fly directly to it from a remote location.

However, current QBT performance analysis and sizing studies at the University of Maryland are limited in that the methodologies employed are only capable of considering the various components in isolation. As the aircraft's proprotors, wings, and body are all necessarily in close proximity with the rotor wake impinging on the aircraft in all normal operating modes, the extent of the mutual aerodynamic interactions of these bodies should be taken into account when sizing the aircraft. Furthermore, as only two examples of this aircraft exist (to the author's knowledge) near to the desired scale, creating an accurate weight prediction methodology is of great concern. This study presents a new design tool that is being built for this joint purpose. While the aerodynamic model required to account for these interactions is not fully implemented, the present work provides the weight model, initial sizing studies, and integration framework that will enable use of the aforementioned model as a "drop-in" addition to the existing tool.

Presented at the 7th AHS Technical Meeting on VTOL Unmanned Aircraft Systems, Mesa, Arizona, USA, January 24–26, 2017.

METHODOLOGY

The design space is defined by bounds of wing loading, disc loading, wing aspect ratio, and rotor solidity. Based on these and a desired mission, the code proceeds through the steps shown in Figure 1. As it is of interest to compare variations on the QBT design, such as an octorotor-biplane-tailsitter, and then compare to more conventional designs, such as a typical quadcopter or octocopter, this framework has been made such that it can be applied to all four configurations.

The different modules of this process are explicated as follows. The modules are individually written in Fortran 95 and are connected via a framework written in Python 2.7.6.

Initial Performance Estimation

In the initial module, an aircraft is sized (as possible) for each combination of disc loading, wing loading, rotor solidity, and wing aspect ratio. This sizing utilizes a combination of momentum theory for rotor aerodynamics from Leishman in (Ref. 12), lifting line theory from Anderson in (Ref. 13), and parasite drag approximations from Raymer in (Ref. 14) and Hoerner in (Ref. 15). It is assumed for purposes of simplicity that the aircraft has equal rectangular wings with no decalage or stagger.

While these simplified models do not account for the complex aerodynamic interactions present, some approximations have been adopted. First, to account for proprotor/wing interactions, an experimental investigation carried out by Witkowski in (Ref. 16) is utilized as a guide. In this and other studies, such as conducted in (Ref. 17), (Ref. 8), and (Ref. 18), it is shown that a significant performance increase can be realized by operating a propeller in a tractor configuration in front of a wing across a wide range of Reynolds numbers. As a first approximation, a plot from (Ref. 16) (shown in Figure 2) is utilized as a model. This plot shows the lift and drag coefficients for a semi-span wing operating at a Reynolds number of 470,000 at Mach 0.1 with a tractor propeller positioned one blade length from the wingtip. This Reynolds number is within the expected operating range for all vehicles sized by this tool. Also, while it has been shown that these effects can be modeled via panel methods, such a methodology will be included upon integration of the design tool with the code developed from (Ref. 19) and (Ref. 20).

It should be mentioned that, while the QBT is expected to have a similar operating Reynolds number to the wing in the study, it is also expected to have a much lower propeller advance ratio (on the order of 0.65) as well as a lower wing aspect ratio. However, as mentioned in both (Ref. 16) and (Ref. 17), this favorable induced drag effect is intensified at lower aspect ratios. Hence, the values given by (Ref. 16) are taken as a "best-case" scenario and is fit with an analytic relationship as shown in Equation 1. This fit was exceedingly good, with a correlation coefficient of 0.998.

$$C_{d_{wings}} = 0.056615(C_L^{3.0530}) + 0.0043454 \qquad (1)$$

In fact, it is mentioned by both (Ref. 16) and (Ref. 17) that as the propeller (in this case a proprotor) is moved closer to the wingtip, the favorable performance effects increase due to the effect of the propeller wake on the wingtip vortex, laminar-turbulent transition, and spanwise lift distribution. As seen, these effects are quite significant with respect to the aircraft's cruising performance. In its hovering flight mode, (Ref. 11) provides some empirical evidence to suggest that, when there is an edgewise freestream, the interference effects are diminished. As this flight condition is expected in this particular sizing study, proprotor/wing interference effects in hovering flight will be ignored in the preliminary design module. As noted, they will be captured in the final design module upon integration with a more realistic aerodynamic model.

It should also be mentioned that Munk provides an approximation for the mutual repulsion of the biplane wings in (Ref. 21). The result of this expression is added to the expected wing airloads and is on the order of 10% of those loads.

Geometric Considerations

As shown in (Ref. 22), the interference effects between two tandem rotors operating on the same plane is minimized when their axes are placed 1.1 rotor diameters apart. Accordingly, the sizing module automatically places rotors such that their rotation axes are separated by at least 1.1 rotor diameters. Rotor placement is also varied by fuselage radius as the code enforces a minimum fuselage/rotor axis separation of 1.1 rotor radii.

In this framework, the aircraft was assumed to utilize either 4 or 8 rotors that attach to either to struts or wings, depending on the configuration. As, in the initial sizing phase, these rotors are sized by disc loading and solidity, it is possible to return converged design solution but with a nonphysical blade (e.g., a blade incapable of carrying flight loads). Hence, following design convergence, the code automatically screens designs based on blade load capacity utilizing the material properties given for use in the weight model and disallowing any blade with an aspect ratio less than 6. The former screening is accomplished by assuming a steady triangular lift distribution across the blade and computing the resultant von Mises stress at the blade root.

For the biplane configurations, a set of 4 struts attaches the wings to the fuselage. The fuselage is assumed to be an ellipsoid of a user-specified aspect ratio and is physically sized by both payload geometry, which is assumed to a be sphere, and the battery volume. Likewise, the struts are assumed to be slim, ellipsoidal prisms extending diagonally from the fuselage body to the wing surfaces.

Surface grids for both the wing and fuselage are produced for use in future analysis at the conclusion of the initial design module. Piecewise monotone cubic interpolation splines are utilized to generate a wing surface according to various stretching functions to produce a desirable distribution of grid points. These splines utilize the interpolation scheme are described by (Ref. 23) with a weighted slope correction given

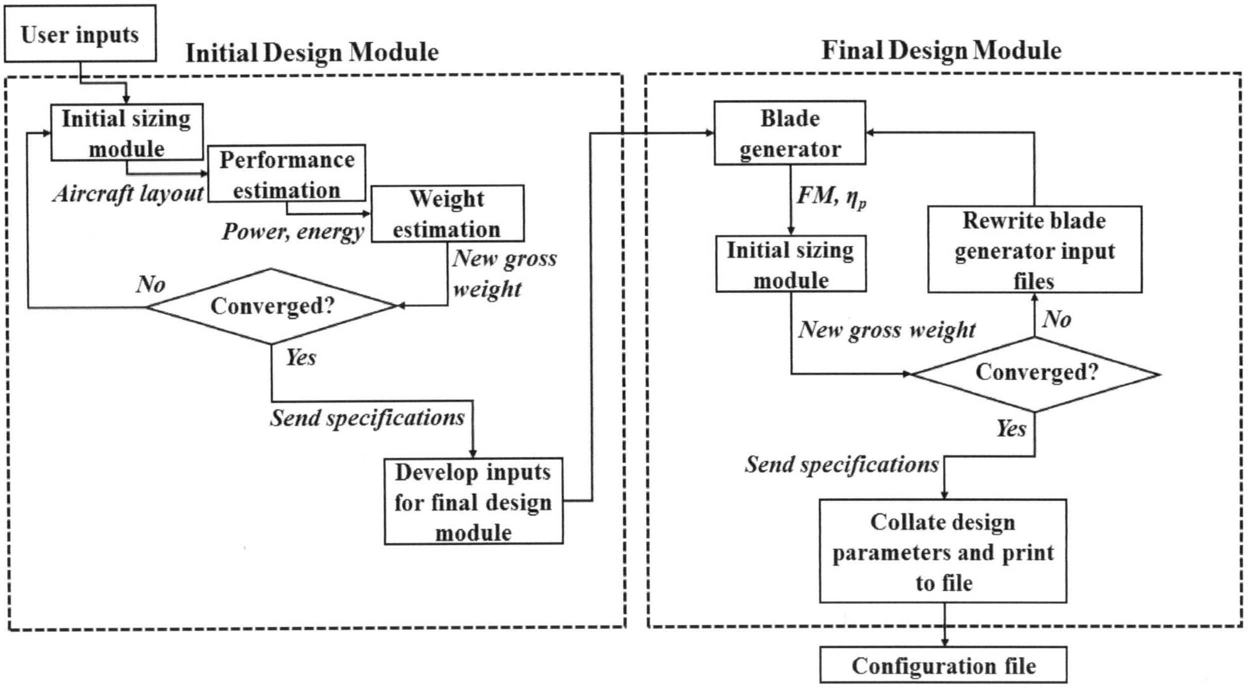

Fig. 1. Design Flowchart

by (Ref. 24), or, as is more easily applied for certain aspects of the wing and fuselage geometry, a Bézier spline. An appropriate grid point distribution is obtained through use of various stretching methodologies.

Weight Model

As this configuration is new, little historical information exists from which to generate trends on the vehicle level. However, on the component level, more can be said. The radio-controlled aircraft market has produced a variety of motors, speed controllers, and batteries and a variety of manufacturers readily provide specifications for these items. Given this, a database has been built, trends established, and the required work flow of the weight build up specified (as seen in Figure 3). The relationships established for this study are as follows with the corresponding correlation coefficient for the nonlinear fit. As the propellers of interest to this study can easily exceed the dimensions of those readily available on the market (no diameters greater than 24 inches are typical), commercially available propellers have not been included in this database. For reference, the motors captured in this database are all brushless, and the electronic speed controllers (ESC) have been collected accordingly.

$$W_{motor} = 0.00084659(P_{reqd}^{0.98831}) \qquad (R^2 = 0.989) \qquad (2)$$

$$I_{max_{motor}} = 0.069163[log(W_{motor})]^{0.98831}) \qquad (R^2 = 0.899) \qquad (3)$$

$$W_{ESC} = 0.0030939(I_{max_{ESC}}^{0.91428}) \qquad (R^2 = 0.899) \qquad (4)$$

$$W_{battery} = 0.000014842(E_{reqd}^{0.91566}) \qquad (R^2 = 0.991) \qquad (5)$$

$$I_{max_{battery}} = 0.000043159(mAh^{1.7971}) \qquad (R^2 = 0.864) \qquad (6)$$

These relationships are employed in the motor, ESC, and battery weight modules. Of particular interest here is the battery module, wherein battery weight is determined by 4 factors: total energy required, maximum energy per pack, maximum motor current, and pack discharge rating. The code first sizes a pack according to the predicted energy needs of the aircraft: if the required energy exceeds that which can be stored by a pack in the battery database, multiple packs of the largest size possible are utilized. However, if additional packs are required in order to accommodate the necessary current draw (as determined by the motor's rated maximum current), then the number of packs required are determined by the discharge requirement instead.

Data concerning structural elements is far more elusive, requiring a different methodology. This study utilizes a technique previously employed by (Ref. 25), wherein the designer assumes a particular structural layout and material properties for a given component, such as a wing. As this aircraft is not only designed for a particular industry but is being developed in cooperation with a plastics injection molding firm, a candidate material has been independently selected. The methodology is as follows, beginning with the wings.

In this tool, the wing is assumed to be a uniform, semi-monocoque structure, comprising of a single, vertical member at the quarter-chord and a thin, load-bearing skin. Its sizing depends on 3 items: load capabilities, aeroelastic flutter, and

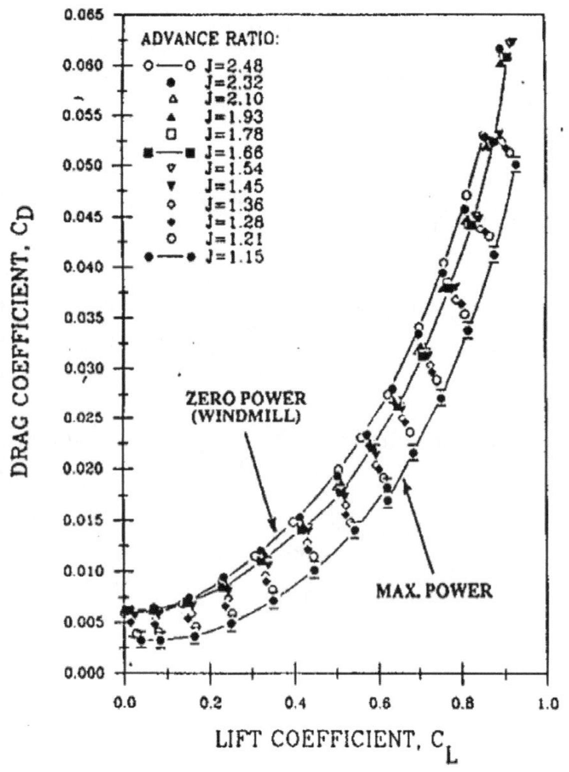

Fig. 2. Drag polar for a semi-span wing/propeller combination at various power settings (wing aspect ratio is 8.25)

whirl flutter. These can be counteracted by either modifying the mass distribution within the wing, strengthening the wing, or moving the wing's supports (the struts). For simplicity, this study only considers the latter two methods. Loads are computed by von Mises equivalent stress and the structure is thickened or thinned appropriately to allow for a suitable load factor. The minimum wall thickness is determined by the confluence of manufacturing limitations and the minimum thickness required to avoid flutter.

The aeroelastic flutter boundary is estimated utilizing the *k*-method as explicated by Scanlan and Rosenbaum in (Ref. 26) and solved by Hodges and Pierce in (Ref. 27) with more modern techniques. The wing's natural frequencies are computed utilizing a finite element formulation of the Rayleigh-Ritz method as shown in (Ref. 28). These frequencies are also compared to the operating frequencies of the motors to determine the wing's susceptibility to whirl flutter. In order to counter either eventuality (failure due to aeroelastic or whirl flutter), the wing strut connection points are moved to various test positions from the inner-most motor location, 0.55 rotor diameters from the aircraft centerline, to the wingtips. The struts attach to the wing at the quarter-chord and are assumed to not exert any pitching or heaving moments on the wing in this analysis. Similarly, the strut attachment points are assumed to be pinned in heaving and free in torsion. Motor, ESC, and rotor inertia are considered as "lumped" masses at the leading edge of the wing and are treated accordingly in the finite element formulation.

The struts, a set of ellipsoidal prisms extending from the fuselage to the wings, are sized similarly, although, for the sake of computational efficiency, they are considered to be stiff in torsion and so insusceptible to aeroelastic flutter. However, as an analytic shape, their natural frequencies are simple to generate and then compare against the operational frequencies of the motors. The struts are then sized accordingly: avoiding coincidence with these forcing frequencies while resisting the anticipated loads.

The fuselage is similarly approximated as a hollow structure with a cylindrical midsection and conical ends. This structure is sized to accommodate the loads and physical size of both batteries and payload. As this results in what is easily the stiffest structure on the aircraft, it is considered to be insusceptible to flutter. The struts are assumed to connect to the fuselage at its center of gravity and extend to the wing's quarter-chord at a spanwise location determined as previously described.

Blade Design

Once the initial design module has completed, the output designs are screened based on blade loading, retaining only those designs with a $C_T/\sigma \leq 0.14$, can resist the anticipated bending loads, and have an aspect ratio of at least 6. Bending loads are determined by assuming a triangular lift distribution on a rectangular, solid blade with a NACA 0012 cross section. With this in hand, the von Mises equivalent stress is computed at the root of the blade and compared to the material properties previously supplied.

Once screening is completed, the initial design module imposes a constraint on the aircraft's maximum dimension. For purposes of this initial investigation, it has been set to 8 feet. Furthermore, the module only passes on the design with the lowest combination of disc loading and installed power. Additional constraints can be added as desired. After this screening, the basic rotor geometry and the most stringent power requirements in both edgewise and axial flight modes for these designs are then passed to the blade design module.

This module is constructed around the linear inflow model of Pitt and Peters as described in (Ref. 12) and uses tabulated lift and drag data for a NACA 0012 airfoil, the proprotor cross-section which is assumed, at present, for the remainder of the design. While the symmetric cross section is not desirable in a production aircraft, it is here selected due as airfoil tables for the section, including angles of attack of +/- 180°. Utilizing these, the code will optimize the rotor twist, taper, collective pitch, and tip speed for minimum energy usage with respect to both of the flight modes considered. As the Pareto frontier emerges from the analysis, the module automatically selects the design point that provides the greatest energy savings for the flight conditions analyzed without exceeding the airfoil's stall boundary.

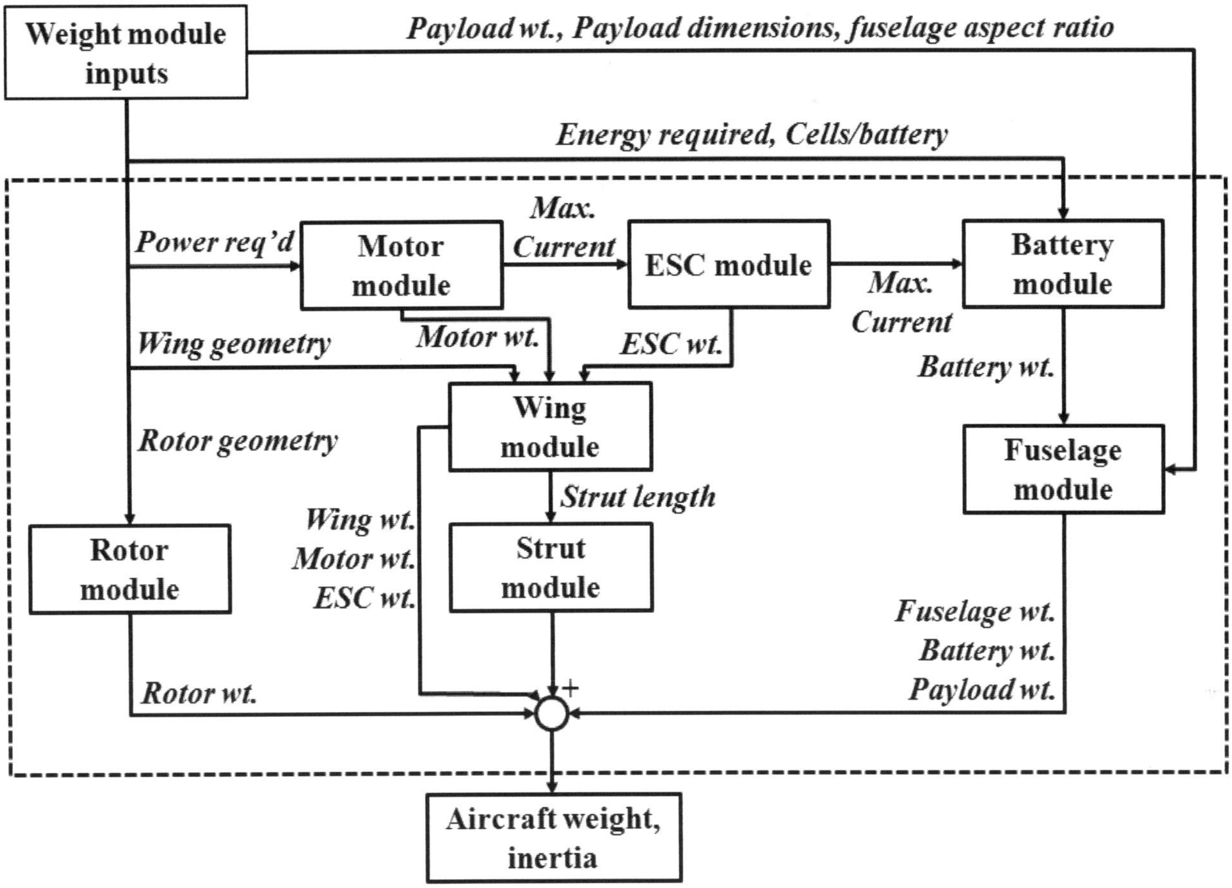

Fig. 3. Weight Module Flowchart

Furthermore, based on the designer's choice, the module can operate in both fixed-pitch and variable-pitch modes, wherein it will generate proprotors optimized for fixed- and variable-pitch blades, respectively. While the benefit of a variable collective is clearly evident in the representative plots shown in Figures 4 and 5 (albeit for a higher Reynolds number), it may be precluded by scale and manufacturing limitations. However, based upon flight condition, it may be impossible to achieve a desirable propulsive efficiency and figure of merit through a fixed pitch blade that is trimmed utilizing tipspeed alone.

This consideration, as it reflects a design decision (particularly on this scale) and does not necessarily follow from the development of the design itself, will be left as an input. As noted, however, it is much easier to design a high-efficiency rotor for both hover and axial flight if variable collective is allowable. The results of this method are printed to input files for the aforementioned free-vortex method.

In the current design framework, this module returns an updated estimate for both figure of merit and propulsive efficiency to the initial sizing module. This is only done for the optimal (i.e., minimal) configuration as designated by the initial design module.

PrasadUM

As the aircraft's proprotors, wings, and body are all necessarily in close proximity, with rotor wake impinging the aircraft in all normal operating modes, aerodynamic interactions must be taken into account in order to properly size the aircraft. To accomplish this, an inviscid, incompressible, time-accurate Lagrangian free-vortex method will be utilized. This methodology has already been implemented in the Parallelized Rotorcraft Analysis for Simulation and Design tool (PrasadUM) developed at the University of Maryland and previously validated against flight test data for an aircraft both trimmed and maneuvering flight (Ref. 19).

This tool places Langrangian markers along a continuous vortex filament emanating from a rotor tip, allowing the tip vortex to be modeled via a piecewise linear reconstruction. Assuming inviscid flow outside of the vortex core, these filament segments convect as material lines under the influence of local flow velocities and, thus, can be tracked via applying the linear advection equation to the Langrangian markers. This method utilizes a 2nd order, 5-point central discretization in space and a 2nd order, 2-step backward discretization in time. Mutual influences of the vortex filaments are deter-

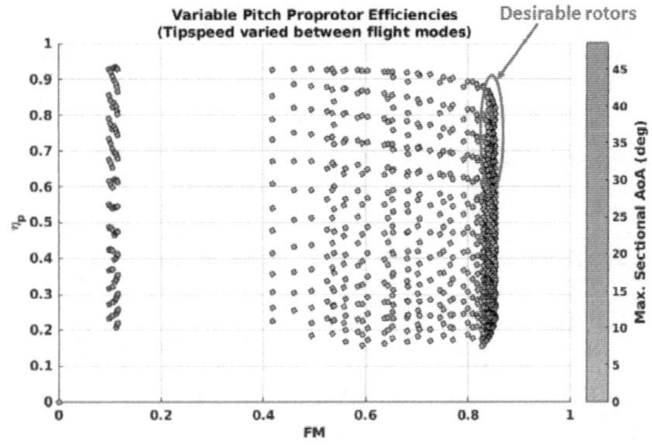

Fig. 4. Rotors of varying taper, twist, and tipspeed: trimmed via collective pitch and tipspeed

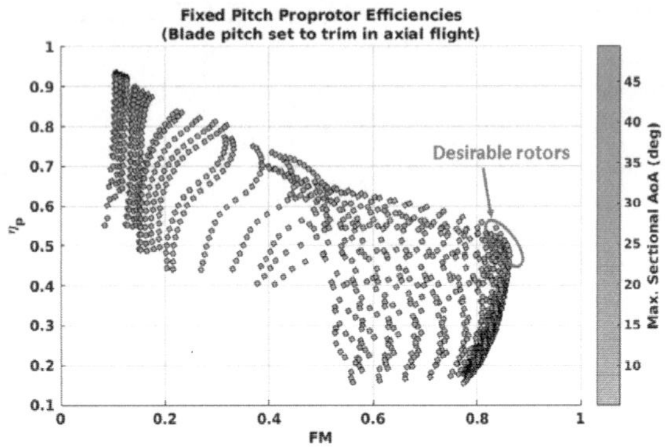

Fig. 5. Rotors of varying taper, twist, and tipspeed (collective set in axial flight): trimmed via tipspeed

mined by the Biot-Savart law. These aerodynamics have been implemented utilizing Fortran, although the code itself builds upon a Python framework that communicates with a CUDA-C module which enables parallelization of the solution using a graphics processing unit.

At present, by placing source terms on a discretized surface, the code also has the capability to compute the induced drag on bluff bodies as shown in (Ref. 20). This capability must now be expanded to lifting bodies as well for the present study, but this work is currently in progress and is not incorporated into this paper. When it is prepared, the design framework has been organized so as to facilitate a drop-in replacement of the blade design module shown previously in Figure 1. The final design flowchart can be seen below in Figure 6. Instead of the brute-force optimization used here (wherein 48,000 unique combinations of rotor geometry and flight mode are analyzed in under 15 minutes), a genetic algorithm will be utilized to decrease the computational expense of a far more complex solver.

Sizing Missions

As no QBT over a gross weight of 8 lbf has yet been manufactured at the University, a relatively small mission will be considered as a first exercise in scaling: an on-shore wind turbine blade inspection. In an effort to demonstrate the QBT's capacity for high-speed forward flight, the mission will entail flight from one row of turbines to another, simulating a transit from an inspected turbine to an uninspected turbine as shown in Figure 7. The prevailing winds are assumed to be a constant headwind of 25 kts, a desirable hub height wind speed that is assumed to be true at cruising altitude in this case. A Vestas V90-3MW wind turbine was utilized for this case.

In this mission, the aircraft takes off from the ground as a tail-sitter, transitions into its biplane flight mode, and cruises in its biplane flight mode at an altitude of 50 ft for 0.55 mi. Towards the end of this cruise, it begins to climb at its estimated best ROC until it arrives at hub height. As it approaches hub height, the aircraft transitions to hovering flight and begins blade inspections. These inspections proceed at a rate of 0.6 ft/s along the blade (defined by the resolution of typical high-definition video and thermal cameras), passing all spanwise locations on each blade twice. Once complete, the aircraft transitions back into its biplane flight mode and descends to the base of the turbine, where it lands. For the purposes of initial sizing, it is assumed that 1% of the battery is consumed per transition.

The offshore mission is similar, although it requires that the aircraft not land at the base of the inspected turbine but instead travel back to its takeoff location. This is taken to be a secondary platform on the farm, such as are used to accommodate crew quarters or electrical substations. To define the mission, the layout of an existing wind farm was utilized, namely a Round 1 farm built at Robin Rigg in the United Kingdom as recorded in (Ref. 29). Range was determined as the maximum distance from substation to turbine for the given farm, requiring 2.29 mi radius of action. As the farm was built with Vestas V90-3MW turbines, the turbine dimensions remain the same from the previous case. Representative atmospheric conditions were found by analysis of metocean weather data from a National Oceanic and Atmospheric Administration weather buoy in the Maryland Wind Energy Area (WEA), as recorded by NOAA buoy 44009 over the past 15 years and available via the National Oceanic and Atmospheric Administration in (Ref. 30). For this mission, the winds were taken to be 95% of the maximum windspeed recorded by the buoy over that time period. At cruising altitude, this results in a 35.9 kts headwind. Adjusting for the atmospheric boundary layer via the power law given in (Ref. 31), this results in a 45.6 kts wind at hub height, which is assumed to remain constant over the diameter of the turbine rotor disc. This mission is illustrated in Figure 8.

RESULTS

For this mission, a range of 400 quadrotor configurations and 17,600 QBT configurations were considered based on variations in wing loading, wing aspect ratio, rotor solidity, disc loading, number of rotors, and number of wings. All configurations utilized the same ranges of solidity and disc loading and were restricted to only 2 blades per rotor.

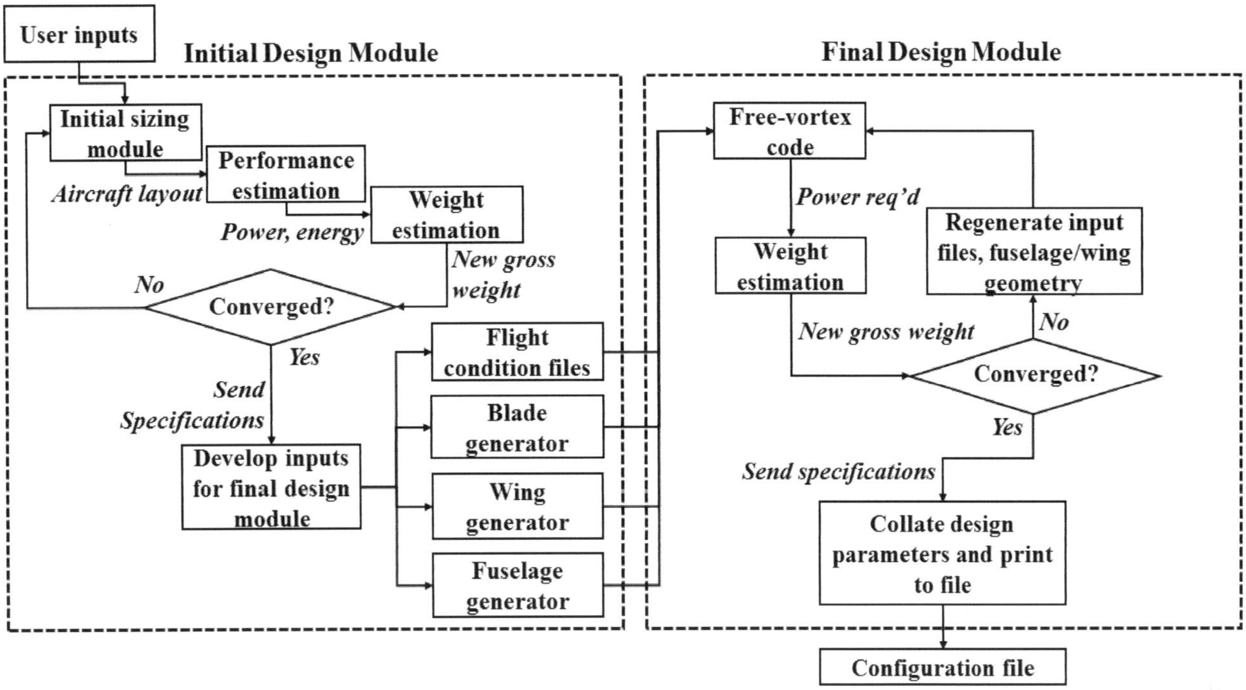

Fig. 6. Design Flowchart

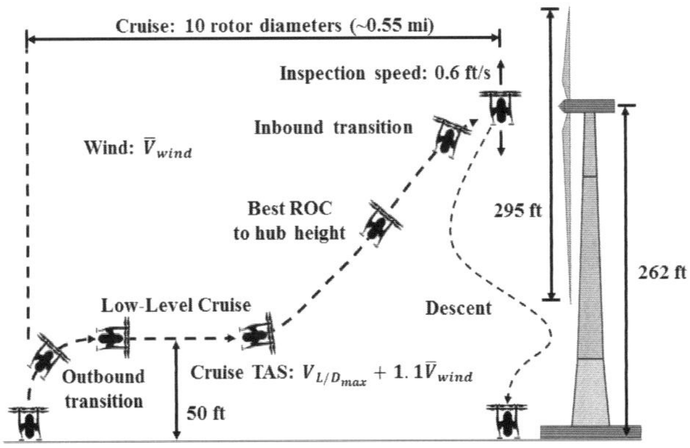

Fig. 7. On-shore turbine inspection mission

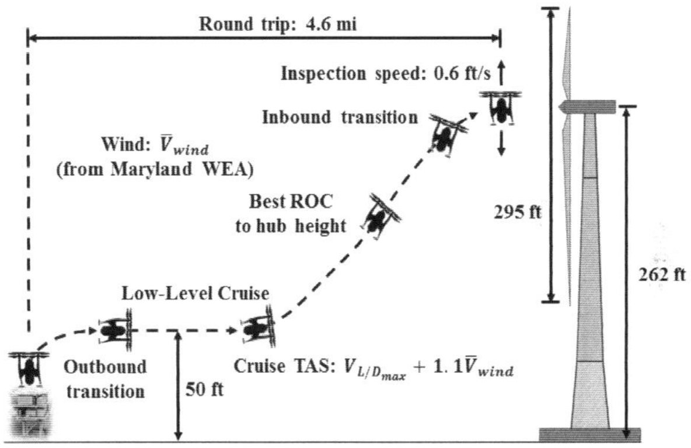

Fig. 8. Offshore turbine inspection mission

For clarity, a selection of the results from the initial design module will be presented first. These include the onshore mission sizing results, see in Figure , and the offshore mission sizing results, seen in Figure .

While limited in scope, these plot does present several general trends concerning multirotor-biplane aircraft with respect to their more conventional multicopter cousins. First, when considering the two onshore cases presented, it is immediately evident that the QBT designs can be built at a similar or power loading and gross weight as the quadcopter design for the same mission. It should be said that this is, obviously, not the norm: the mean power loading for a QBT design is markedly lower than that of the pure quadcopter, which speaks to what each design is intended for: the quadcopter for hover, and the QBT for forward flight.

This difference also indicates that the necessary cruise component of even this onshore mission makes a significant difference in the sizing outcome. While the total endurance for this mission is under a half-hour, well within the normal flight endurance of a typical quadcopter, the winds and the required range are likely more intense than most commercial quadcopters are designed for. While the QBT does incur some extra weight penalty by dint of requiring wings, the benefits, at least in this simplified analysis, appear to be quite significant. This appears to be confirmed by the difference in battery weight fraction between the two aircraft, wherein the maxi-

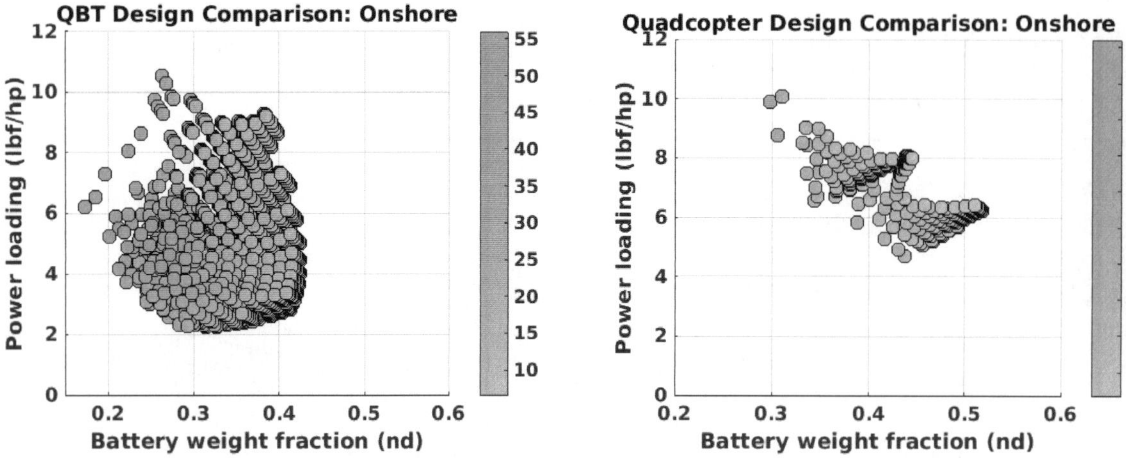

Fig. 9. Initial Design Output - Onshore Mission

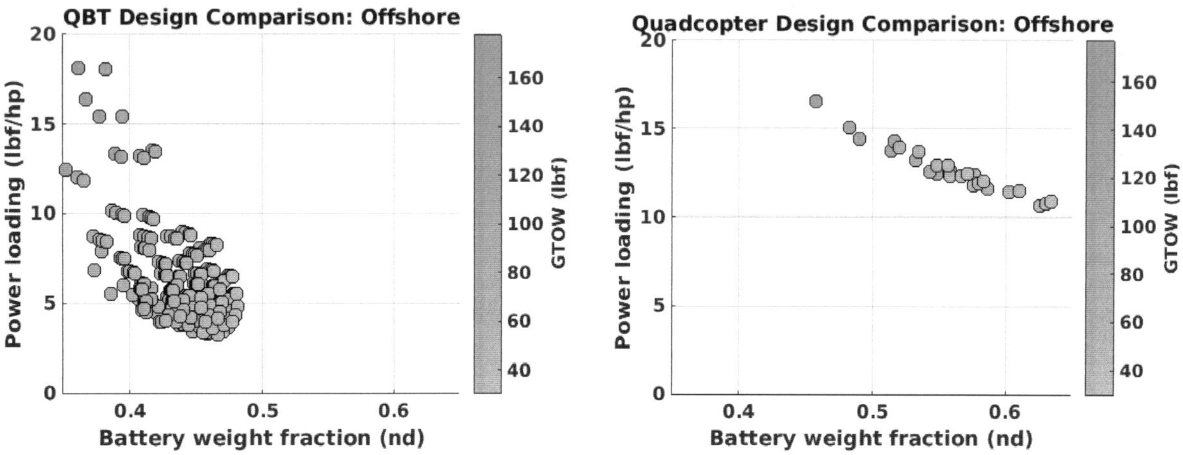

Fig. 10. Initial Design Output - Offshore Mission

mum battery weight fraction of the pure quadcopter is nearly 10% more than that of the QBT design, even at similar gross weights. This can also be seen in their mutual weight breakdowns, given in Tables 1 and 2, respectively.

This effect is exaggerated for the offshore mission, where wind speed increases by 42.6% at cruising altitude and range is increased by 836% while the hover requirement stays the same. The quadcopter designs immediately see a jump in battery weight fraction, increasing by 12.6% on average, and a corresponding jump in power loading. Simultaneously, the minimum gross weight, previously below that of the lightest QBT, is now above that of the lightest QBT design. As power loading is, essentially, a measure of hover efficiency, these plots indicate that hover capabilities are, definitively, not the defining component of this particular sizing mission. Hence, while the QBT design suffers from that, its capabilities to contend with the demands of the increase cruise requirement more than compensate for it.

The effect the slipstreams from multiple proprotors does not appear to be well-documented in the literature, although mul-

Table 1. Quadcopter: Onshore Mission

	Item	Value
Takeoff weight	9.58	lbf
Empty weight	4.39	lbf
Payload weight	1.00	lbf
—	—	—
Battery weight	4.19	lbf
Fuselage weight	0.60	lbf
ESC weight	0.02	lbf/each
Motor weight	0.08	lbf/each
Strut weight	0.09	lbf/each
Rotor weight	0.03	lbf/each
Systems weight	0.30	lbf
—	—	—
Rotor diameter	3.90	ft
Disc loading	0.2	lbf/ft^2
Solidity	0.015	nd
Cruise speed	32.52	kts

tiple methods utilizing panel methods, lifting line theory, are other approximations have appeared to better estimate the effectiveness of wings operating in propwash. While it is undesirable to add such codes to the initial design module of this project, it further emphasizes the need for such a tool as PrasadUM to further analyze any design passed out of the code.

In summary, the code returned the following QBT (shown in Table 2), following sizing of an appropriate rotor in the final design module. However, as the Reynolds number for this rotor (as measured of 0.75R) varies from 80,000 to 260,000 depending on flight mode, the airfoil tables provided for this section are insufficient. However, the module has demonstrated its capabilities, provided an appropriate set of tables can be formulated for it, and will serve, in the future, to initialize a more detailed analysis code.

Table 2. QBT: Onshore Mission

	Item	Value
Takeoff weight	14.89	lbf
Empty weight	9.30	lbf
Payload weight	1.00	lbf
—	—	—
Battery weight	4.60	lbf
Wing weight	0.98	lbf/wing
Fuselage weight	0.71	lbf
ESC weight	0.03	lbf/each
Motor weight	0.13	lbf/each
Strut weight	0.09	lbf/each
Rotor weight	0.03	lbf/each
Systems weight	0.30	lbf
—	—	—
Rotor diameter	3.44	ft
Disc loading	0.4	lbf/ft^2
Solidity	0.020	nd
Propulsive efficiency	0.51	nd
Figure of merit	0.71	nd
Wing loading	4.0	lbf/ft^2
Cruise speed	41.8	kts
Stall speed	35.7	kts

CONCLUSIONS

Motivated by a need for a cheap and reliable inspection solution in the offshore wind energy industry, this paper has presented a code purpose-built for sizing quadrotor-biplane-tailsitter aircraft. At present, this code includes an initial design module utilizing a simplified aerodynamic model conjoined to a comprehensive weight model for preliminary sizing studies. This weight module, while utilizing a database of commercial off-the-shelf components, also takes into account aerodynamic, aeroelastic, and structural requirements for custom parts, such as the wing, fuselage, and struts, allowing greater flexibility in application. Output from this initial design module is screened and a limited selection of designs are passed through a second design iteration wherein a rotor blade design module is coupled to the design module. Utilizing the Pitt-Peters linear inflow model, this blade definition allows for further refinement of a select few designs.

Specific conclusions drawn from this initial work are as follows:

1. QBT designs are not as suited for hovering flight as pure multicopters, leading to a reduction in efficiency for short-range missions in comparison to these aircraft. This study confirms the usefulness of pure multicopter designs for those missions.

2. The study also notes the utility of the QBT configuration in more extended missions. While some weight penalty is incurred, it is offset by the aerodynamic efficiency of the design in its preferred flight mode. This is dramatically confirmed upon analysis of a demanding offshore wind inspection mission.

3. This study has yet to include a more detailed analysis to intelligently incorporate interference effects. Utilizing a more comprehensive aerodynamic model, such as detailed previously, should be considered as the next step for this code.

While presently limited by its simplified aerodynamic models, the code provides a methodology into which PrasadUM, a free-vortex code with integrated panel method, can be integrated. As it has been shown that panel methods are sufficient to provide more intelligent and flexible estimates of these interactions, the implementation of such methods in the context of this code should be considered to be the next expansion of this project.

Author contact: Marius Strom[1] mstrom@terpmail.umd.edu, James Baeder[2] baeder@umd.edu

ACKNOWLEDGMENTS

The authors would like to thank Luke Chow, Vikram Hrishikeshavan, V.T. Nagaraj, Ananth Sridharan, Bharath Govindarajan, and Jean-Pierre Moreau for their invaluable work and support in the continuing development of this project.

REFERENCES

[1]Sarker, B. and Fiaz, T., "Minimizing maintenance cost for offshore wind turbines following multi-level opportunistic preventive strategy," *Renewable Energy*, Vol. 85, 2016, pp. 104–113.

[2]Shafiee, M., Finkelstein, M., and Bérenguer, C., "An opportunistic condition-based maintenance policy for offshore wind turbine blades subjected to degradation and environmental shocks," *Reliability Engineering & System Safety*, Vol. 142, Oct 2015, pp. 463–471.

[3]Kandukuri, S., Klausen, A., Karimi, H., and Robbersmyr, K., "A review of diagnostics and prognostics of low-speed machinery towards wind turbine farm-level health management," *Renewable and Sustainable Energy Reviews*, Vol. 53, Jan 2016, pp. 697–708.

[4]Snyder, B. and Kaiser, M. J., "Ecological and economic cost-benefit analysis of offshore wind energy," *Renewable Energy*, Vol. 34, (6), Jun 2009, pp. 1567–1578.

[5]Musial, W., Butterfield, S., and Ram, B., "Energy from Offshore Wind," Offshore Technology Conference, Houston, TX, 2006.

[6]Båserud, L., Flügge, M., Bhandari, A., and Reuder, J., "Characterization of the SUMO turbulence measurement system for wind turbine wake assessment," *Energy Procedia*, Vol. 53, 2014, pp. 173–183.

[7]Reuder, J. and Jonassen, M., "First results of turbulence measurements in a wind park with the Small Unmanned Meteorological Observer SUMO," *Energy Procedia*, Vol. 24, 2012, pp. 176–185.

[8]Hrishikeshavan, V., Bogdanowicz, C., and Inderjit, C., "Experimental Investigation of Performance of a Wing-Propeller System for a Quad-Rotor-Biplane Micro Air Vehicle," 54th Structures, Structural Dynamics, and Materials Conference, Boston, MA, April 2013.

[9]Hrishikeshavan, V., Bogdanowicz, C., and Inderjit, C., "Design, Performance and Testing of a Quad Rotor Biplane Micro Air Vehicle for Multi Role Missions," *International Journal of Micro Air Vehicles*, Vol. 6, (3), September 2014, pp. 155–173.

[10]Hrishikeshavan, V. and Inderjit, C., "Refined Lightweight Inertial Navigation System for Micro Air Vehicle Applications," Specialists Meeting on Unmanned Rotorcraft and Network Centric Operations of the American Helicopter Society, Pheonix, AZ, January 2015.

[11]Yeo, D., Hrishikeshavan, V., and Inderjit, C., "Gust Detection and Mitigation for a Quad Rotor Biplane," Paper AIAA 2016-1531, AIAA Atmospheric Flight Mechanics Conference, San Diego, CA, January 2016.

[12]Leishman, G., *Principles of Helicopter Aerodynamics*, Cambridge University Press, New York, NY, 2006.

[13]Anderson, J., *Fundamentals of Aerodynamics*, McGraw-Hill Education, New York, NY, 2011.

[14]Raymer, D. P., *Aircraft Design: A Conceptual Approach*, American Institute of Aeronautics and Astronautics, 2011.

[15]Hoerner, D. F., *Fluid-Dynamic Drag*, New York, NY, 1965.

[16]Witkowski, D. P., Johnston, R. T., and Sullivan, J. P., "Propeller/Wing Interaction," Paper AIAA-89-0535, 27th AIAA Aerospace Sciences Meeting, Reno, NV, January 1989.

[17]Veldhuis, L. L. M., *Propeller Wing Aerodynamic Interference*, Ph.D. thesis, Delft University of Technology, 2005.

[18]Ananda, G. K., Deters, R. W., and Selig, M. S., "Propeller-Induced Flow Effects on Wings of Varying Aspect Ratio at Low Reynolds Numbers," 32nd AIAA Applied Aerodynamics Conference, Atlanta, GA, June 2014.

[19]Rubenstein, G., Moy, D. M., Sridharan, A., and Chopra, I., "A Python-based Framework for Real-time Simulation using Comprehensive Analysis," 72nd Annual Forum of the American Helicopter Society, West Palm Beach, FL, May 2016.

[20]Govindarajan, B., *Contributions Toward Understanding the Effects of Rotor and Airframe Configurations on Brownout Dust Clouds*, Ph.D. thesis, Dept. of Aero. Eng., University of Maryland, 2014.

[21]Munk, M. M., "General Biplane Theory," NACA TR 151, January 1923.

[22]Ramasamy, M., "Measurements Comparing Hover Performance of Single, Coaxial, Tandem, and Tilt-Rotor Configurations," 69th Annual Forum of the American Helicopter Society, Phoenix, AZ, May 2013.

[23]Fritsch, F. N. and Carlson, R. E., "Monotone Piecewise Cubic Interpolation," *Society for Industrial and Applied Mathematics*, Vol. 17, (2), 1980, pp. 238–246.

[24]Aràndiga, F., "On the Order of Nonuniform Monotone Cubic Hermite Interpolants," *Society for Industrial and Applied Mathematics*, Vol. 51, (5), 2013, pp. 2613–2633.

[25]Chambers, M. C., Ardema, M. D., Patron, A. P., Hahn, A. S., Miura, H., and Moore, M. D., "Analytical Fuselage and Wing Weight Estimation of Transport Aircraft," NASA TM 110392, 1996.

[26]Scanlan, R. G. and Rosenbaum, R. A., *Introduction to the Study of Aircraft Vibration and Flutter*, MacMillan, New York, NY, 1951.

[27]Hodges, D. H. and Pierce, A. G., *Introduction to Structural Dynamics and Aeroelasticity*, Cambridge University Press, New York, NY, 2011.

[28]Bauchau, O. A., , and Craig, J. I., *Structural Analysis*, Springer, New York, NY, 2009.

[29]"Robin Rigg Offshore Wind Farm (Navigation and Fishing) (Scotland) Bill," , 2003.

[30]"National Oceanic and Atmospheric Administration's National Buoy Data Center: Station 44009," http://www.ndbc.noaa.gov/station_page.php?station=44009.

[31]Manwell, J. F., McGowan, J. G., and Rogers, A. L., *Wind Energy Explained: Theory, Design, and Application*, John Wiley and Sons, Chippenham, Wiltshire, second edition, 2010.

Visual Odometry Onboard a Micro Air Vehicle Using Snapdragon Flight

Eric Solomon,* Cyrus Vorwald,† Vikram Hrishikeshavan,‡
Inderjit Chopra §
Alfred Gessow Rotorcraft Center, Department of Aerospace Engineering,
University of Maryland, College Park, MD 20742

ABSTRACT

Visual odometry has been demonstrated with success in drones through onboard flight computers. However, these methods often suffer from significant drift and are limited by the computational power of the onboard processors. In 2014, the Snapdragon Flight was released by Qualcomm as a robust flight computing platform. It couples fast, low-powered processors with a well-featured sensor array and a variety of options for usage with peripherals. This paper details a method of using the Snapdragon Flight in parallel with a smaller flight computer to create an extensible flight computing platform. In this process, visual inertial odometry is implemented on the Snapdragon Flight and is experimentally verified. The results show that simple uni- and bi-axial position tracking is feasible. Data of more complex scenarios is tested, but is not shown, as the results are not well understood at this point.

*

Nomenclature

ABI	Application binary interface
API	Application program interface
ARM	Advanced RISC Machine
DSP	Digital signals processor
ELKA	Enhanced Lightweight Kinematic Autopilot
ESC	Electronic Speed Controller
Hexagon	Snapdragon 801 digital signals processor
IDL	Interface Description Language
IPC	Inter-process communication
Krait	Snapdragon 801 apps processor
mAH	milliAmp hours
OpenCV	Open Computer Vision
PX4	Open source flight stack
QAIC	Qualcomm Interface Compiler
QURT	Qualcomm Real-time Operating System
RAM	Random Access Memory
ROS	Robot Operating System
SDK	Software development kit
SoC	System on Chip
UART	Universal asynchronous receiver/transmitter

INTRODUCTION

There has been growing interest in use of small scale multicopters and helicopters to conduct autonomous operations for a wide range of missions such as: navigation through small openings (windows, pipes), disaster scenarios involving collapsed buildings and logistical support through delivery of items to unknown locations. In order to accomplish these, information relating the position and orientation to an outside environment is a major requirement for autonomous operation of unmanned air vehicles (UAVs). To estimate relative position and orientation in potentially GPS-denied environments, the UAV needs to know information about its environment. At the same time, the UAV needs to have an estimate of its pose in order to describe the environment. Major breakthroughs to solve this problem of simultaneous localization and mapping (SLAM) have occurred since the turn of the century (Refs. 1, 2). The ability to describe an environment based on image data allows for accurate position and orientation estimation. However, image data consists of very large sets of pixel information that can be computationally expensive to analyze. Powerful computers with large sensors were capable of providing accurate information to solve this problem in real-time. Because of the complexity of camera-based localization, computation was generally carried out using an offline computer.

In recent years, due to the rapid advancement in microelectronics, the focus has shifted to autonomous localization on systems with minimal processing speed and smaller, less accurate sensors. There has been widespread research on using vision aided algorithms to help in object tracking and odometry on mobile devices (Refs. 3–7). It has therefore become possible to perform simultaneous localization and mapping possible in real-time onboard a quadrotor drone and has already been attempted by researchers (Refs. 7, 8) (Fig. 1).

However, many of these research algorithms have been developed with proprietary collaboration with industry and are therefore not open to other research groups. As a result, UAV autonomy research has generally been based on cus-

*Graduate Research Assistant, esolomon1221@gmail.com
†Underraduate Research Assistant, vgt085@gmail.com
‡Assistant Research Scientist, vikramh@umd.edu
§Alfred Gessow Professor and Distinguished University Professor, chopra@umd.edu
7th AHS Technical Meeting on VTOL Unmanned Aircraft Systems and Autonomy, Jan 24-26, 2017, Mesa, AZ

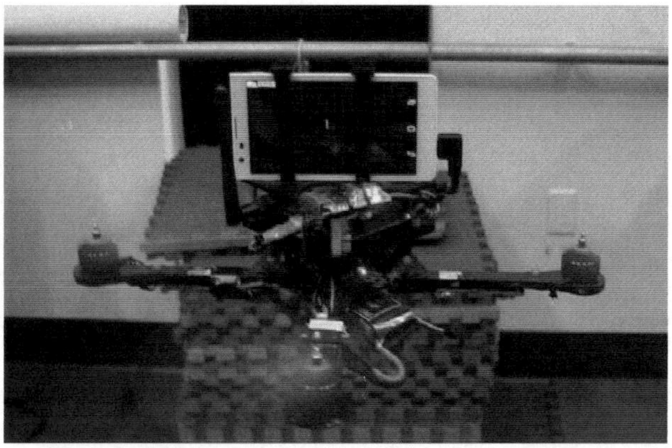

Fig. 1. Visual odometry onboard a 500 gram quadrotor using Android device (Ref. 8)

tom hardware and open source algorithms and development tools (Refs. 12, 13, 16, 29). As can be seen from Fig. 2, the weight allocation for these components is sub-optimal for incorporation with a size and weight constrained micro air vehicle. In order to optimize the hardware choice with computing and power requirements, Qualcomm has released a low-cost ($400) hardware development platform - the Snapdragon Flight (Ref. 34). This has all the hardware and software components required to develop a complete vision based autonomous drone. However the development of this platform is still at a nascent stage and currently can only be used with Qualcomm's custom electronic speed controller board.

It is therefore very desirable to improve upon the functionality and versatility of this development platform. Therefore, the purpose of this paper is two fold: (1) describe the hardware architecture of the Snapdragon Flight and its implementation in a resource constrained localization system that is capable of being used onboard a micro air vehicle, and (2) enhance the class of drones that can be operated with the Snapdragon Flight by integrating it with a custom embedded lightweight kinematic autopilot (ELKA) (Ref. 11) that was developed at the University of Maryland. In this manner, the vision tasks are performed by the Snapdragon Flight and all vehicle flight control tasks are performed by ELKA (Figure 3).

BACKGROUND AND METHODOLOGY

As mentioned before, the purpose is to realize a localization hardware solution that can be applied to mini drones. Fig. 4 shows the hardware architecture of a typical mini-quadrotor with onboard localization. ELKA controls the attitude of the drone at a rate of 1000 Hz. It receives inputs about the drone position from the Snapdragon Flight board at an expected rate of about 20 Hz. This information can then be interpreted towards position feedback control and path planning purposes. The end result of this platform is that the drone software is not tightly coupled to a specific drone hardware.

Next, we discuss the usage of the Snapdragon Flight, often referred to as the Flight, as a feasible drone platform for

Fig. 2. Typical off-the-shelf components used for custom solution for autonomy onboard a UAV - not weight and power optimized

Fig. 3. (a) ELKA and nickel coin (b) Snapdragon Flight, quadrotor drone.

completing visual inertial odometry. The Flight platform is introduced in terms of hardware and firmware. This paper details the Flight's limitations of communication as well as a proposed solution. Then, the paper discusses visual inertial odometry; first by introducing the general state of visual inertial odometry today as it applies to small onboard computers, and followed by a description of visual inertial odometry as it fits into our specific drone schematic. The paper ends with the procedure and discussion of a simple experiment, which demonstrates the capabilities of the Flight for onboard visual inertial odometry.

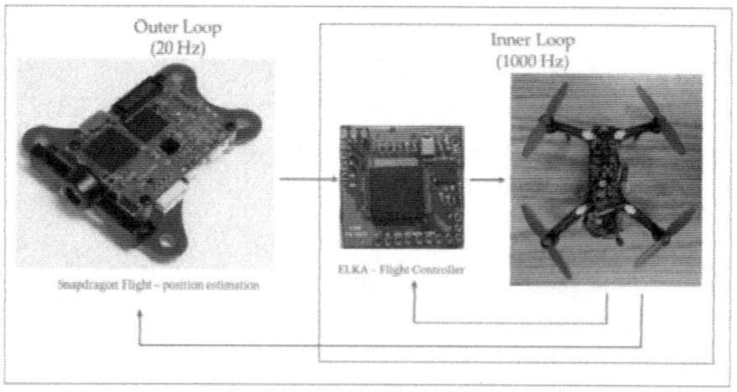

Fig. 4. Hardware architecture

SNAPDRAGON FLIGHT DESCRIPTION

Hardware

The Flight uses the Snapdragon 801 System on Chip composed of a quad-core Krait 400 processor, a Hexagon QDSP6v5 digital signal processor, and an Adreno 330 GPU.

The Krait processor is similar to an ARM Cortex-A processor. The Hexagon DSP is low-power, and efficiently built. It is capable of simulating three independent computing cores, which are useful for parallel code execution. Qualcomm does not make the Adreno 300 series GPU readily accessible to outside developers, so it is thus not used in the project (Ref. 34).

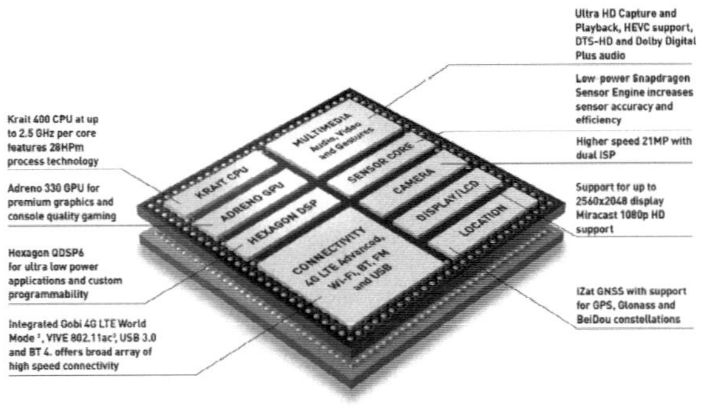

Fig. 5. Snapdragon 801 processor (Ref. 30).

The Flight board has a mass of 26g and consumes 5V DC power through a 2 to 6 series cell (2S-6S) external battery. The input voltage is regulated to 5V through a power adapter. The Flight contains 2GB random access memory (RAM) and 32GB of embedded MultiMediaCard (eMMC) persistent memory. The forward facing camera can capture 4k video at 30 frames per second. The downward facing camera, which performs optic flow velocity sensing, can capture 1080p video. Optic flow analyzes information from consecutive images to estimate vehicle velocity. The Flight has access to 2G/5G WiFi, Bluetooth, and GPS radio communication. Users may attach larger antennae to each radio communication module if desired. Input/output (I/O) is available through USB 3.0, micro SD, serial board connection, gimbal connection, and electronic speed controller (ESC) connections. Digital I/O may be done through protocols including Universal Asynchronous Receiver/Transmitter (UART), Inter-integrated Circuit (I2C), Serial Peripheral Interface (SPI). A 9-axis inertial measurement unit (IMU) may be accessed through SPI, and a barometric pressure sensor may be accessed through I2C (Ref. 34)(Fig. 6).

Firmware

The Flight's Krait and Hexagon processors utilize efficient communication methods to maximize parallelism. The Krait runs QRLinux, a best-effort scheduling operating system that includes Ubuntu 13.04 pre-built binaries (Ref. 24). The Hexagon DSP runs Qualcomm Real-Time OS (QURT), which

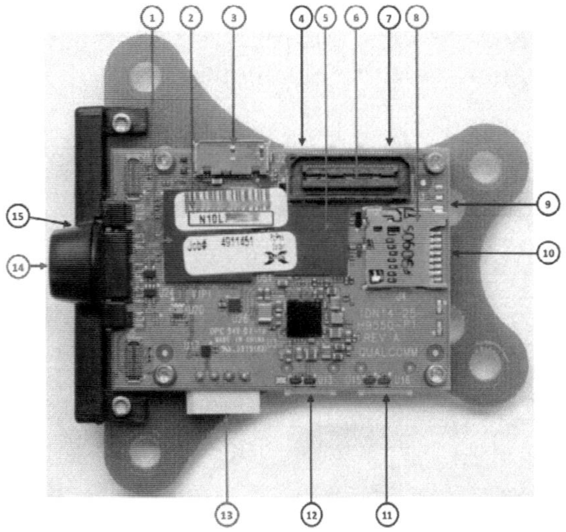

Fig. 6. Description of Flight - (1.) RAM pop memory on top of Snapdragon 801 SoC, (2.) Serial number, (3.) Micro USB 3.0 (Type AB)(OTG)port, (4.) Remote receiver connector, (5.) eMMC flash, (6.) Serial board connector, (7.) IMU connector, (8.) Micro SD card slot, (9.) GPS antenna connector, (10.) WLAN antenna connectors, (11.) Gimbal connector, (12.) ESC connector, (13.) Power connector, (14.) 4K camera, (15.) Optic flow camera (Ref. 28)

is able to efficiently map software processes onto hardware threads without resource overlap. The Hexagon DSP has access to all of the devices attached to the Snapdragon 801 SoC. The the Hexagon DSP is used to access devices and delegate tasks based on hardware efficiency and software support. Visual odometry occurs on the Krait processor, while serial communication tasks occur on the Hexagon DSP. Fig. 7 shows a simple diagram of how different necessary portions of this system are accessed.

Serial communication

As shown in Fig 8, serial communication extends from the Hexagon DSP. Serial communication is a device stream available as '/dev/tty-1...6'. These streams may perform UART communication using either two or four wires, depending on their configuration. Here, two-wire communication is used. The communication packet structure is fairly simple for this program, but is subject to change as the needs of the program change. It consists of a start byte, a control byte, up to 253 data bytes, then an end byte. The packet loading protocol in the program is as yet undefined, but it may be prudent to use a ring buffer data structure to load packets. This will ensure that writing to the packet loading queues will never result in blocking (unnecessary pauses in program) and that only sufficiently recent packets will be sent via UART to ELKA.

VISUAL INERTIAL ODOMETRY

Using computer vision, it is possible to perform simultaneous localization and mapping in real-time onboard a UAV. The

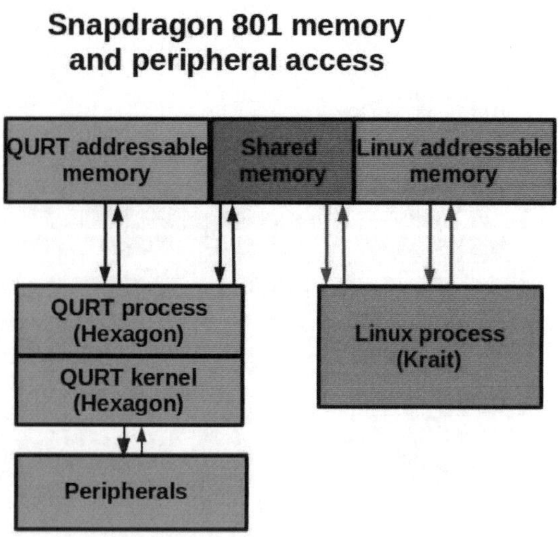

Fig. 7. The division of addressable memory and corresponding addressing processors for the Snapdragon 801. Peripherals are accessed via the Hexagon DSP.

ability to describe an environment based on image data allows for accurate position and orientation estimation. Because of the complexity of camera-based localization, computation is generally carried out using an offboard computer (Ref. 10). Refer to the 'Results and Discussion' section to see that the Snapdragon Flight is capable of localizing using onboard sensors and online processing.

Simultaneous Localization and Mapping (SLAM)

A typical onboard SLAM approach consists of bundle adjustment using a monocular camera and scalar repositioning using an IMU (Ref. 16). The camera frame is a rigid body driven by translation and rotation parameters, as well as estimates of translational and angular velocity. Bundle adjustment requires strong feature correspondence across keyframes with an initial estimate of keyframe poses. The IMU frame measures linear acceleration and angular rates, aiding in the estimation of feature location and keyframe poses. Sensor fusion is typically performed with an error-state Kalman Filter to smooth out the trajectory estimated by the vision algorithm with the IMU (Refs. 15, 36). A working solution that can be used to fuse gyroscope, accelerometer and pose measurements is described in (Ref. 27). The experiment in this paper uses Qualcomm's Machine Vision SDK, for which the source code is not publicly available, in order to show the feasibility of a resource constrained localization system that is capable of being used onboard a micro air vehicle.

Software Architecture

The Krait processor runs a Robotics Operating System (ROS) network to handle multi-processing and asynchronous mes-

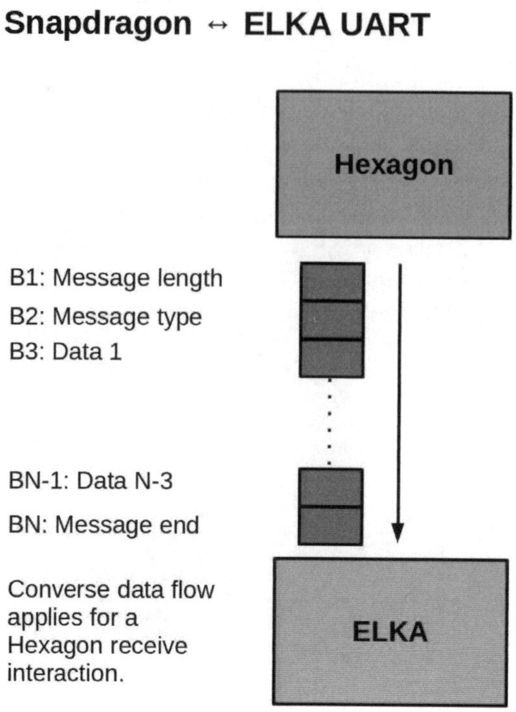

Fig. 8. Snapdragon and ELKA send UART messages back and forth. The format (bytewise) is {start,control,data1,...,datan,end}. Implicity, the max message length is 256 bytes.

saging. ROS is the industry standard for middleware open-source communication between robotic systems. ROS nodes perform three important functions: visual odometry, inter-process communication, and access to additional sensors attached to the Flight. The software control layout is shown in Fig. 9. The visual odometry node uses Qualcomm's Machine Vision SDK. ROS interfaces with PX4, an open-source flight stack, through a single ROS node. PX4 on the Krait processor relays between ROS and Hexagon DSP processes. PX4 on the Hexagon DSP relays between Krait processes, the Flights sensors, and the serial driver. The serial driver relays between PX4 and ELKA, communicating via 2- wire UART. ELKA runs FreeRTOS, which is a simple, multi- threaded real-time operating system. It is capable of receiving motor commands and sending stabilizing control signals to electronic speed controllers attached to motors.

Experimental Measurements

Qualcomm's Visual-Inertial Simultaneous Localization and Mapping package (VISlam) is used for the Flight's onboard tracking and Vicon is used offboard as ground truth data for comparison. Vicon is a motion capture system that consists of cameras with IR light emitters that detect markers that reflect the IR light. The Vicon-460 camera is shown by Windolf, Gotzen, Morlock to be accurate on average between $65 - 129\mu m$ using four cameras. The setup used in this test

Software control layout

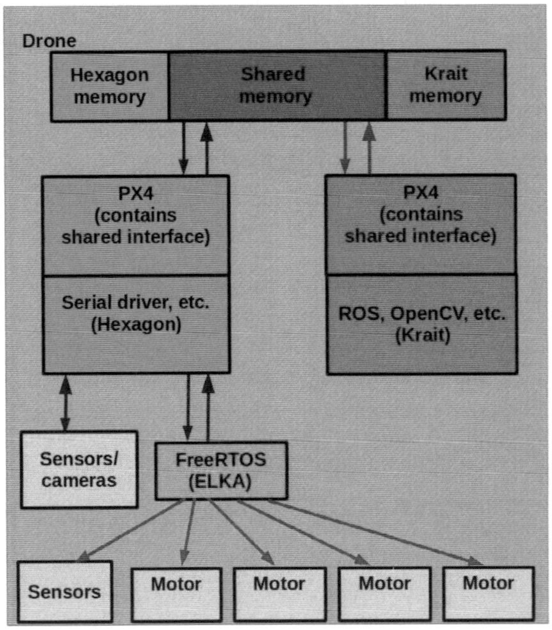

Fig. 9. Software control layout containing the major processes of each processor: ROS, OpenCV, PX4, Serial driver, ELKA.

has seven cameras, although not all of them will track each marker at all times (Ref. 37).

VISlam capabilities for odometry are tested in four characteristic motions: along a single axis, along two axes, along two axes with in-plane rotation, and in arbitrary motion. For the first three test types, the Flight is attached to a rod mounted to a cart, which is rolled around reference lines on the ground. For the third tracking test, the Flight is detached from the cart and moved around at arm's length (Fig. 10, 11). This is the least conservative case for the Flight tracking actual drone motion. This case is not shown, as the results are not well understood at this point.

The Flight is controlled via serial access with Android Debug Bridge (ADB) or via WiFi with Secure Shell (SSH). Flight tracking data due to VISlam is captured onboard. Vicon data is captured on a desktop computer.

Results and Discussion

We present an investigation into the usability of a visual-inertial system that estimates 6 degrees of freedom (DOF) pose in unknown environments. The Flight's output pose ran at an average of 25.189 hertz, which is sufficient for real-time. Results are plotted in the section entitled 'Visual Inertial Odometry Test Data'.

Fig. 12- 14 show position data along simple uni-axial test paths. Fig. 15, 16 show position data along bi-axial test paths. It is shown that the onboard visual inertial odometry results agree well with the Vicon ground truth data. All results show

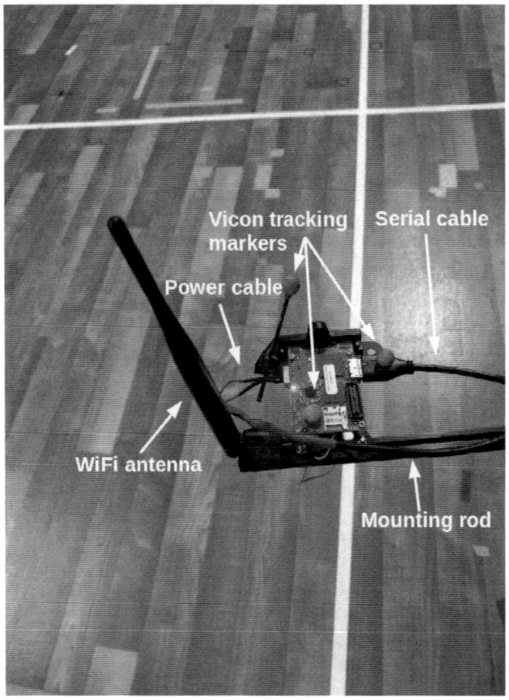

Fig. 10. Snapdragon flight mounted on a rod showing WiFi antenna, power cable, serial cable, and Vicon tracking markers.

a small initial deviation between VISlam and Vicon test data, which does not increase for the duration of the tests. Note that the graph orientations of VISlam data sets are slightly skewed due to the difference between the known Vicon axes orientations and the unknown VISlam axes orientations.

The tests shown do not comprehensively show the Flight's capabilities for visual inertial odometry. The Flight is tracked for approximately 3m in 10s with at most one turn. They are, however, an encouraging starting point and show enormous potential for tracking a complex environment for onboard autonomous navigation.

CONCLUSIONS

The Snapdragon Flight is a strong platform for consideration of onboard autonomous navigation. Its lightweight sensor array supports state-of-the-art navigation techniques. The results show that the Snapdragon Flight is capable of successfully tracking its location relative to a starting point in real-time with low error.

Future work

Currently, UART loopback has not been successfully tested on the Snapdragon Flight. However, the techniques used for

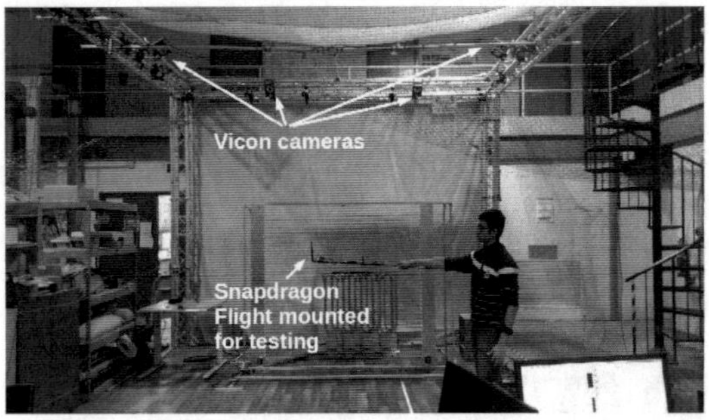

Fig. 11. Visual inertial odometry test setup showing Snapdragon Flight mounted to rod for testing. Tracking due to the Vicon cameras was compared with tracking due to onboard visual inertial odometry.

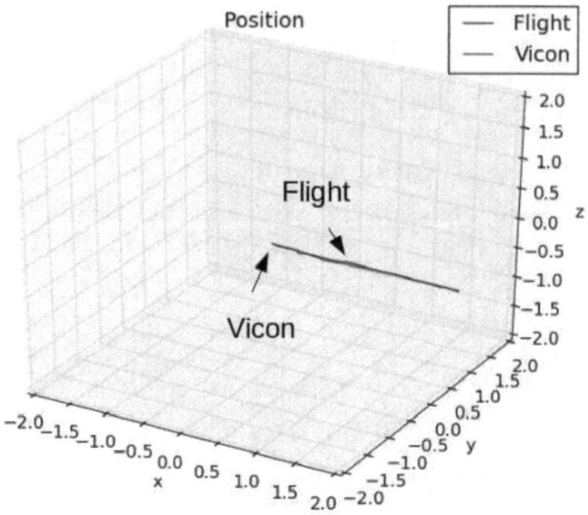

Fig. 12. Positive x position tracking.

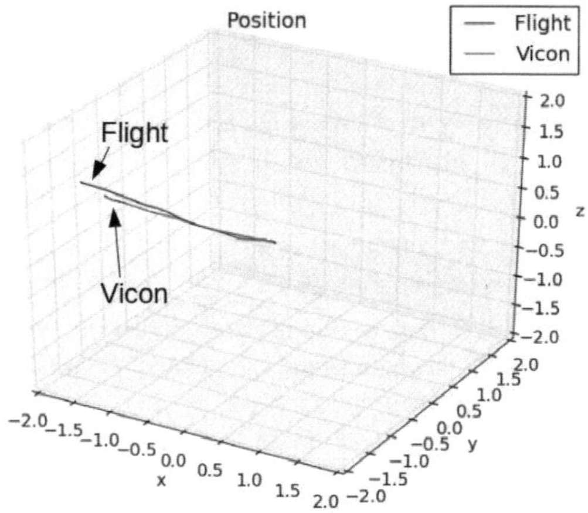

Fig. 13. Negative x position tracking.

UART have been successful on similar embedded systems. Possible this discrepancy arises as a result of peculiarities related to the Flight, such as the serial driver or kernel implementations. These problems will probably dissipate assumming that surrounding community support and internal support by Qualcomm for the Flight continues remains in an active state.

After these problems are resolved, the Snapdragon Flight will be set up to show preliminary capabilities controlling a drone during flight. The outputs of VISlam will be fed into ELKA to simply demonstrate the capabilities of the two of them combined for flight.

This system will fly drones such as the AirEZ developed at the University of Maryland by Hrishekeshavan, Phillips, Rand, Chopra (Ref. 17).

In parallel, visual inertial odometry will be verified and extended by using custom implementations of open-source algorithms for the Flight. This way important internal metrics of visual inertial odometry can be quantitatively measured for a variety of algorithms: including feature detection, loop closure, sensor drift, and pose estimation rate. Doing so will also allow the open-source community to test experimental algorithms onboard a very current and relatively high powered onboard flight computing platform.

VISUAL INERTIAL ODOMETRY TEST DATA

All axes referenced are defined by the Vicon camera system. This uses forward-left-up axes as defined from the perspective in Fig 11 for this experiment. All positions are displayed in meters from the center of the Vicon test area.

Linear Movement Tracking

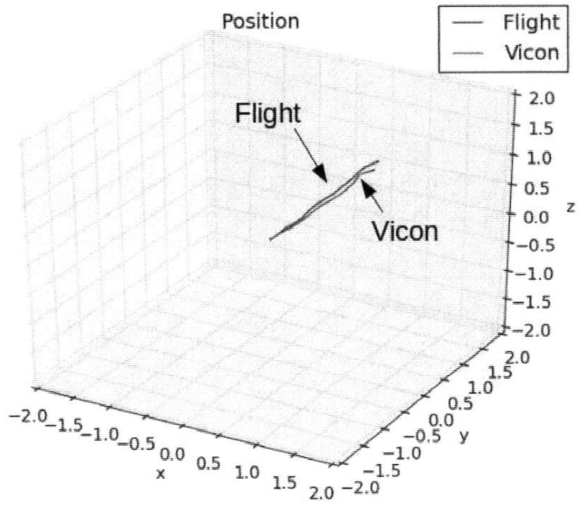

Fig. 14. Negative y position tracking.

Bi-directional Movement Tracking

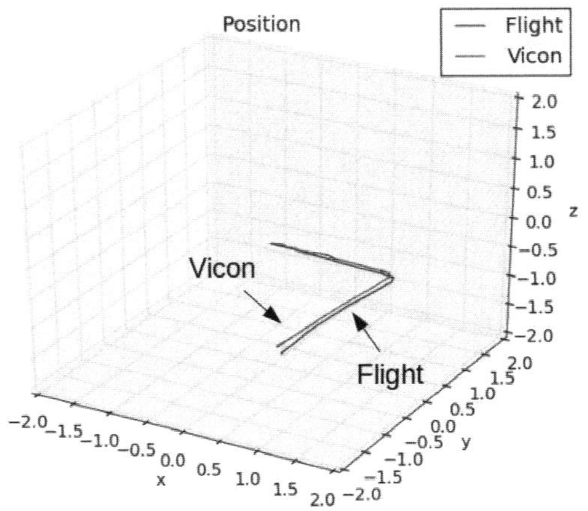

Fig. 15. Positive x, positive y position tracking.

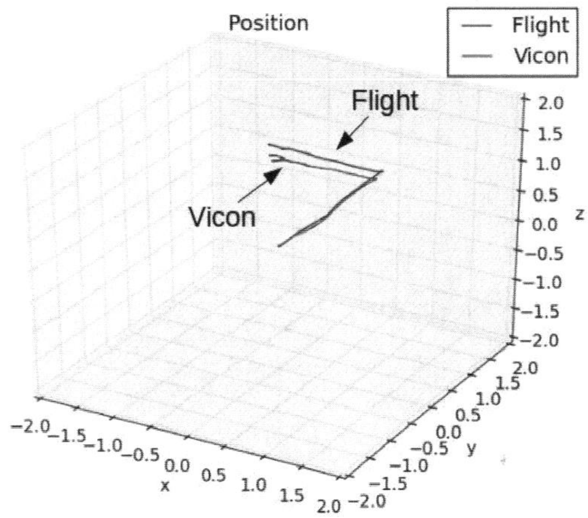

Fig. 16. Negative x, negative y position tracking.

APPENDIX

APPENDIX-1: INTER-PROCESS COMMUNICATION

The Hexagon DSP and the Krait processor handle inter- process communication (IPC) by shared memory. On the Snapdragon 801 SoC, the user controls shared memory through the Qualcomm Interface Compiler Interface Defini- tion Language (QAIC IDL). With the QAIC IDL, the user can declare function prototypes that can be called from the Krait processor and then defined and executed on the Hexagon DSP. The QAIC then performs two major actions: it creates the ap- propriate function stubs (essentially function definitions) for each processor, and it creates libraries that map the interface functions to shared memory. This way, function calls are not different to the user versus how they normally appear, with the exception that only the functions defined in the interface can be used to communicate between processors (Fig 17).

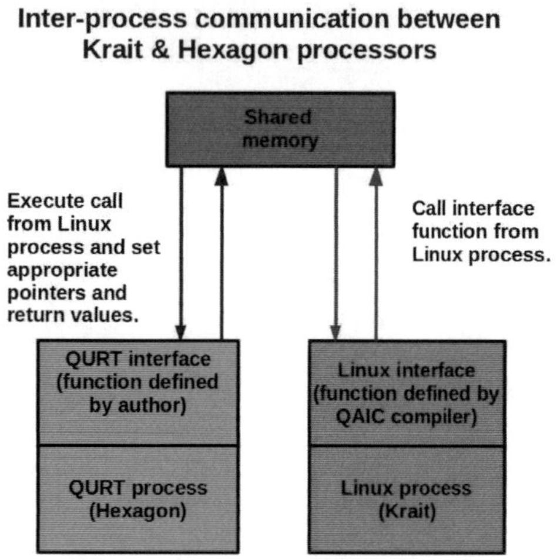

Fig. 17. Snapdragon 801 uses two types of processors almost entirely independently. They communicate with each other thru an inter-process communication scheme defined by the QAIC IDL.

APPENDIX-2: ISSUES AND SUPPORT

Refer to the Qualcomm developer forums (Ref. 7), Intrinsyc support support.intrinsyc.com, ROS forums, and the PX4 discussion forums (Ref. 8) for support on any issues that are encountered dealing with these platforms. Refer to the PX4 github.com/px4and ATLFlight github.com/atlflight Github pages and the ROS wiki (Ref. 9) for updates on program development and issues. In the following table is a list of useful tools for programming with the Flight.

Table 1. Useful tools for programming with the Flight

Name	Use	Reference
PX4	Open soure flight stack for Snapdragon Flight and various other drone platforms	dev.px4.io
ROS	Robot operating system for Ubuntu Linux flavors and some derivatives	wiki.ros.org
Hexagon SDK & Hexagon Tools	Developer kit for programming with Hexagon DSP (includes documentation)	developer.qualcomm.com/software/hexagon-dsp-sdk
Intrinsyc support	Support and documentation for Intrinsyc/Qualcomm products	support.intrinsyc.com
Machine Vision SDK	Machine vision for Snapdragon Flight	developer.qualcomm.com/hardware/snapdragon-flight
CMake	Automated Makefile generation	cmake.org
cmake_hexagon	Cmake build tools for Snapdragon Flight	github.com/atlflight/cmake_hexagon
DSPAL	Hexagon Digital Signal Processor Abstraction Layer	github.com/atlflight/dspal

REFERENCES

[1]Klein, G., and Murray,D., "Parallel tracking and mapping for small AR workspaces," Proc. of the International Symposium on Mixed and Augmented Reality , Nara, Japan, No. 2007.

[2]Whyte, H.D., and Bailey, T., "Simultaneous localization and mapping: Part I," *Robotics and Automation Magazine*, Vol. 13, No. 2, pp.99110, 2006.

[3]Mourikis, A.I., and Roumeliotis.S.I.,"A multi-state constraint kalman filter for visionaided inertial navigation," Proceedings of the IEEE International Conference on Robotics and Automation, pp. 35653572, Roma, Italy, April 2007.

[4]Li,M., Kim,B.H., and Mourikis,A.I., "Real-time motion estimation on a cellphone using inertial sensing and a rolling-shutter camera," Proceedings of the IEEE International Conference on Robotics and Automation, pp. 46974704, Karlsruhe, Germany, May 2013.

[5]Engel. J., Sturm, J., and Cremers, D., "Semi-dense visual odometry for a monocular camera," IEEE International Conference on Computer Vision, pp. 14491456, Dec 2013.

[6]Engel. J., Sturm, J., and Cremers, D., "Scale-Aware Navigation of a Low Cost Quadrocopter with a Monocular Camera," *Robotics and Autonomous Systems* (RAS), volume 62, 2014.

[7]Weiss, S., Achtelik, M. W., Lynen, S., Achtelik, M. C., Kneip, L., Chli, M., and Siegwart, R., " Monocular Vision for Longterm Micro Aerial Vehicle State Estimation: A Compendium,' *Journal of Field Robotics*, 30(5), 803-831. 2013

[8]Loianno, G., and Kumar, V., "Smart phones and flying robots,", Robotics Science and Systems Conference, Berkeley, USA, July 2014.

[9]200qx drone frame, www.horizonhobby.com/product/multirotor/multirotor-aircraft/bind-n-fly-15087--1/200-qx-bnf-with-safe-technology-blh7780, 2017.

[10]Engel J., Sturm J., and Cremers D., "Camera-based navigation of a low-cost quadrocopter In Intelligent Robots and Systems (IROS)", 2012 IEEE/RSJ International Conference, October 2012, pp. 2815-2821.

[11]Hrishikeshavan V, Chopra I. Refined lightweight inertial navigation system for micro air vehicle applications American Helicopter Society International - 6th Ahs International Specialists' Meeting On Unmanned Rotorcraft System 2015: Platform Design, Autonomy, Operator Workload Reduction and Network Centric Operations. 200-210.

[12]ETH Zuric Autonomous Systems Lab, Sensor Fusion Framework. Retrieved: http://bit.ly/2jtVDft

[13]Gazebo: Robot simulation made easy, gazebosim.org, 2017.

[14]Intrinsyc: Qualcomm Snapdragon Flight, https://www.intrinsys.com/vertical-development-platforms/qualcomm-snapdragon-flight/, 2017.

[15]Madyastha, V.K., Ravindra, V.C., Mallikarjunan, S., and Gopalratnam, G., "Extended Kalman Filter vs. Error State Kalman Filter for Aircraft Attitude Estimation," in Proceedings of the AIAA Guidance, Navigation and Control Conference, Portland, Oregon, USA, Aug. 2011.

[16]R. Mur-Artal, J. M. M. Montiel and J. D. Tards, "ORB-SLAM: A Versatile and Accurate Monocular SLAM System," in IEEE Transactions on Robotics, vol. 31, no. 5, pp. 1147-1163, Oct. 2015.

[17]Phillips, B., Hrishikeshavan, V., Rand, O., and Chopra, I., Design and Development of a Scaled Quadorotor Biplane with Variable Pitch Proprotors for Rapid Paylaod Delivery, American Helicopter Society 72nd Annual Forum,West Palm Beach, FL, May 2016.

[18]PX4 Autopilot: The professional autopilot, dev.px4.io+, 2017

[19]PX4 discussion forum, discuss.px4.io, 2017.

[20]Qualcomm Developer Foru", https://developer.qualcomm.com/forum, 2017.

[21]Qualcomm Hexagon, https://developer.qualcomm.com/software/hexagon-dsp-sdk/dsp-processor, 2017.

[22]Hexagon SDK 3.0, Made available by Qualcomm with download of Hexagon 3.0 SDK, 2016.

[23]Qualcomm Snapdragon, https://en.wikipedia.org/wiki/Qualcomm*snapdragon*, 2017.

[24]QR-Linux Overview, https://wiki.codeaurora.org/xwiki/bin/QR+Linux/WebHome, 2016.

[25]Robot Operating System, wiki.ros.org, 2017.

[26]ROS Answers, answers.ros.org, 2017.

[27]Trawny, N. and Roumeliotis, S.I., "Indirect Kalman filter for 3D attitude estimation, " University of Minnesota, Dept. of Comp. Sci. Eng., Technical Report.

[28]Intrinsyc IoT, Qualcomm, Snapdragon Flight Kit (Developer's Edition) Quick Start Guide, 2014.

[29]Michael Montemerlo, Sebastian Thrun, Daphne Koller, and Ben Wegbreit. 2002. FastSLAM: a factored solution to the simultaneous localization and mapping problem. In Eighteenth national conference on Artificial intelligence, Rina

Dechter, Michael Kearns, and Rich Sutton (Eds.). American Association for Artificial Intelligence, Menlo Park, CA, USA, 593-598.

[30]Snapdragon 801 SoC Image, http://i-cdn.phonearena.com/images/articles/114441-thumb/snapdragon-801-soc-image-update.png, 2017.

[31]Qualcomm Snapdragon Flight Developer Guide, October, 2016.

[32]Snapdragon Flight - Board Pin-outs, 2015.

[33]Qualcomm Snapdragon Flight Reference Platform User Guide, January, 2016.

[34]Snapdragon Flight Board Specifications, https://developer.qualcomm.com/hardware/snapdragon-flight/board-specs, January, 2017.

[35]Qualcomm Snapdragon Flight User Guide, October, 2016.

[36]Sola, J., "Quaternion Kinematics for Error State Kalman Filter,", 2012. Retrieved: www.iri.upc.edu/people/jsola/JoanSola/objectes/notes/kinematics.pdf

[37]Windolf, M., Gotzen, N., Morlock, M., Systematic accuracy and precision analysis of video motion capturing systems – exemplified on the Vicon-460 systems.

Controllability Analysis of a Quadrotor in Hover using Empirical Gramians

Vikram Hrishikeshavan,* Elena Shrestha,† Inderjit Chopra ‡
University of Maryland, College Park, MD 20742
J. Sean Humbert,§
University of Colorado, Boulder, CO

ABSTRACT

Controllability gramians have been used to estimate the reachability sets (states attainable by a system for a given control input) of micro air vehicles using prior knowledge of the vehicle dynamics. The dynamics are typically linearized and estimated using elaborate and involved experimental system identification techniques. This paper seeks to validate an alternative approach of estimating the controllability gramians through measurement of system response to impulses along different degrees of freedom. This not only significantly simplifies the procedure for determining the gramians, but also does not rely on any linearity assumption of the system dynamics. The advantage is that any input-to-state or state-to-output nonlinearities in the system can be systematically assessed. In this study, the linear system dynamics of a QAV-180 Luminier quad rotor was initially identified in the hover condition. System responses for impulse inputs ranging from 30% to 50% of maximum control input were then measured. Finally, the controllability gramians generated from linear approximation and from empirical impuslse response data were compared. For low impulse inputs, of up to 30% of maximum, there was satisfactory comparison between the Frobenius norms (a measure of controllability gramian) of the linear and empirical gramians. However, at higher input magnitudes (50%), there was about a 20-30% reduction in reachability sets for lateral and longitudinal degrees of freedom, suggesting some presence of non-linearities. Yaw and heave degrees of freedom did not exhibit significant deviation. The empirical gramian approach presented in this paper represents one of the few experimental verifications of this technique applied to flying air vehicles. It can provide insights on degree of nonlinearity present in the physical system and also opportunities to modify the feedback control system to tailor the nonlinearity to desirable levels.

*

Nomenclature

Δt	Impulse duration (ms)
ϕ	Roll attitude (rad)
θ	Pitch attitude (rad)
c_m	Impulse magnitude (fraction of maximum)
p	Roll attitude rate (rad/s)
q	Pitch attitude rate (rad/s)
r	Yaw attitude rate (rad/s)
u	Longitudinal velocity (m/s)
v	Lateral velocity (m/s)
w	Heave velocity (m/s)
W_C	Controllability gramian
DOF	Degree of Freedom

*Assistant Research Scientist, Aerospace Engineering Department, vikramh@umd.edu
†Graduate Research Assistant, Aerospace Engineering Department, eshresco@gmail.com
‡Alfred Gessow Professor and Distinguished University Professor , and Director of the Alfred Gessow Rotorcraft Center,Aerospace Engineering Department, chopra@umd.edu
§McLagan Endowed Associate Professor, Mechanical Engineering Department, seanh.humbert@colorado.edu
AHS Specialists' Meeting on Unmanned Rotorcraft and Network Centric Operations.

INTRODUCTION

With the rapid progress in microsystem technologies, small scale air vehicles have received significant interest in the past few years. For many of their missions, they require the fidelity and capability to operate in confined spaces and cluttered environments. A number of potential vehicle systems such as quadrotors, coaxial rotors and cyclocopters may be used, each having different degrees of gust tolerance and maneuverability. It is therefore important to develop quantitiative metrics to understand and compare the platform capabilites so that they may be appropriately mapped to different scenarios and missions (Fig. 1).

In the past, systematic time and frequency domain system identification techniques (Refs. 1–3) were applied to study the flight dynamics of these platforms (Refs. 4–8). Within this framework, linear time-invariant (LTI) models were developed to characterize the essential physics of these aerial microsystems. A control-theoretic methodology was formulated which relies on the aforementioned models to extract key maneuverability and disturbance rejection characteristics of these platforms in a bare airframe (open-loop) condition (Refs. 5,7,9). This enabled a comprehensive quantitative comparison of the capabilities inherent to the bare airframe of varions platforms without the need to accound for the feedback control strategy.

There is a great demand for these platforms to fly in highly cluttered and gusty conditions. With the requirement to operate under large control and disturbance inputs, the platforms may exhibit nonlinearities in their input-to-state as well state-to-output descriptions. Therefore it becomes important to quantify the degree of non-linearity. There have been different measures to quantify nonlinearity, which have been proposed in the past, such as differences between values of state transition matrix at a deviated state and at a trim state (Ref. 11). Other approaches, such as empirical gramians, which are estimated from system impulse response data, have been used for several years for model reduction of nonlinear systems (Refs. 12, 13). The idea is that empirical gramian matrices capture the nonlinear behavior within the region of operation and can be used for nonlinearity quantification when compared to gramians of a linear system. One advantage of this method is that it determines how relevant the nonlinearities are to the control of the system (Ref. 15). Recently, these were also used for controllability and observability analysis (Refs. 14, 15) in addition to nonlinearity quantification.

Even though these numerical procedures exist, there have been limited comparisons between linear and empirical gramians from experimental data and there appears to be no reported literature of its application to aerial flight vehicles. Taking a quadrotor vehicle as a representative case, this paper will address two issues: (1) Do the empirical gramians reduce to gramians obtained from a linear model assumption for small input magnitudes, and (2) How severe is the non-linearity at higher input magnitudes. The dynamics of the quadrotor is studied in hover, and inherent nonlinearities are quantified by comparing the gramians obtained from impulse response (empirical controllability gramian) with that obtained from a linearized assumption at the steady state operating point (linear controllability gramian). This method has the added benefit of utilizing the impulse response data directly to determine unit input energy reachability sets without having to generate a model of the vehicle dynamics. In this manner, changes to the feedback control strucutre can also be quickly analyzed through simple flight experiments as opposed to more involved system identification techniques.

BACKGROUND AND METHODOLOGY

Controllability gramian is a measure that can be used to determine easiest directions to move in state space (movements that require the least amount of energy). In (Refs. 5, 7) certian properties of this gramian were extracted from the linear time invariant dynamics to provide more useful insights on the characteristics of micro air vehicles. A controllability gramian can be visualized in the form of an ellipsoid as shown in Fig. 2(a). By observation, one can determine directions of motion that require the least energy (the most controllable state). A further utility of this ellipsoid approach is that one can also estimate which vehicle is more maneuverable than another. For example, vehicle 2 (Fig. 2(c)) can be

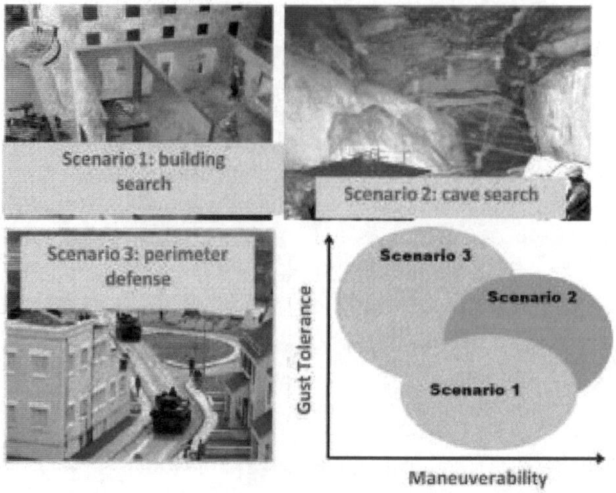

Fig. 1. Examples of MAV operations in aerial missions which require platforms with different degrees of gust tolerance and maneuverability

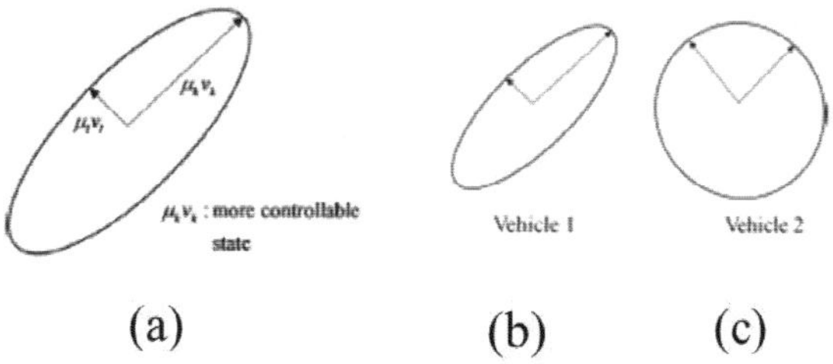

Fig. 2. Principal axes of a controllability ellipsoid and vehicles with different ellipsoid structures

said to be more maneuverable than vehicle 1 (Fig. 2(b)) due to the larger volume of its ellipsoid. However these analyses are based on the assumption of linearity in the dynamics about a certian operating point (say, hover). If we consider a generalized micro air vehicle such as a quadrotor, the linear assumption about hover may be acceptable for small inputs. However, if more aggressive manuevers about hover are desired, non-linearities in the system may become significant. The sources of non-linearity could be many, such as: motor dynamics under instantaneous torque accelerations, aerodynamics of the rotor when it instantaneously climbs or falls etc. It is therefore of desire to estimate the limits of such a linearized assumption on the properties of the controllability ellipsoid. From a procedural standpoint, it may also be useful to estimate the controllability gramians without going through cumbersome system identification techniques.

Both of these aspects can be addressed if one obtains the flight data generated from impulse responses. Gramians computed from these are generally referred to as *Empirical* gramians (Ref. 15). This paper focuses on input-to-state behavior, and therefore, empirical *controllability* gramians are the topic of investigation.

For a linear system of the form

$$\dot{x} = Ax + Bu \tag{1}$$
$$y = Cx + Du \tag{2}$$

the linear controllability gramian is given by,

$$W_{C_{Linear}} = \int_0^\infty e^{At} BB^T e^{A^T t} dt \tag{3}$$

For nonlinear systems,

$$\dot{x} = f(x) + g(x,u) \tag{4}$$
$$y = h(x,u) \tag{5}$$

the controllability gramian can be determined through system impulse response as shown below and is described in further detail in (Ref. 15),

$$W_{C_{empirical}} = \Sigma_{l=1}^{r} \Sigma_{m=1}^{s} \frac{1}{rsc_m^2} \Sigma_{k=0}^{q} \Phi_k^{lm} \Delta t_k \tag{6}$$

where $\Phi_k^{lm} = (x_k - x_{ss})(x_k - x_{ss})^T$, x_k is the state of the system at time k, x_{ss} is the steady state operating point. The system is excited by impulse input $u_k = c_m \delta_{k=0} + u_{ss}$, where c_m is the magnitude of the impulse input, δ_k is the Dirac delta function and u_{ss} is the input to maintain steady state x_{ss}. The gramian for the entire system can be computed by exciting each input direction multiple times and averaging the response (Eq. 6). Here, r is the number of input directions and s is the number of input magnitudes.

These gramians can be geometrically expressed as n dimensional ellipsoids, where n is the degree of A. One way of establishing a metric for this controllability ellipsoid is the Frobenius norm,

$$||W_c||_F = \sqrt{trace((W_C)^{\frac{1}{2}}(W_C)^{\frac{1}{2}})} \tag{7}$$

which is the square root of the sum of squares of the axes lengths of the ellipsoid shown in Fig. 2.

Example

This approach can be illustrated with a simple example. Consider the following linear time invariant system,

$$\dot{x} = \begin{bmatrix} 0 & 1 \\ -1 & -1 \end{bmatrix} x + \begin{bmatrix} 0 \\ 1 \end{bmatrix} u \tag{8}$$

The linear controllability gramian ($W_{C_{Linear}}$) can be obtained using Lyapunov equation solvers such as $lyap(A, BB')$. The empirical gramian can be obtained by first measuring the impulse response of the system from Eq. 8. These measured states can then be recast through Eq. 6 to obtain $W_{C_{empirical}}$. For the present example, $lsim$ is used to simulate impulse response with an impulse magnitude of 1 (i.e. c_m = 1) and time step of 0.01 s. The computed gramians for the system shown in Eq. 8 are,

$$W_{C_{Linear}} = \begin{bmatrix} 0.5 & 0 \\ 0 & 0.5 \end{bmatrix}$$
$$W_{C_{empirical}} = \begin{bmatrix} 0.499 & -8.3 \times 10^{-6} \\ -8.3 \times 10^{-6} & 0.505 \end{bmatrix} \tag{9}$$

It can be seen that when the system is linear, the empirical controllability gramian is equivalent to the linear controllability gramian. The slight differences seen above are due to the choice of time step (difference reduces as time step is made finer). Note that the above result is the same for any choice of c_m. Now, consider addition of input-to-state nonlinearity to Eq. 8,

$$\dot{x} = \begin{bmatrix} 0 & 1 \\ -1 & -1 \end{bmatrix} x + \begin{bmatrix} 0 \\ 1 \end{bmatrix} u + \alpha \begin{bmatrix} 0 \\ 1 \end{bmatrix} u^2 \tag{10}$$

For $\alpha = 0.01$, following are the empirical gramians for different impulse magnitudes,

$$c_m = 0.01, W_{C_{empirical}} = \begin{bmatrix} 0.52 & 2 \times 10^{-5} \\ 2 \times 10^{-5} & 0.54 \end{bmatrix}$$
$$c_m = 0.1, W_{C_{empirical}} = \begin{bmatrix} 0.72 & 2.7 \times 10^{-5} \\ 2.7 \times 10^{-5} & 0.74 \end{bmatrix} \tag{11}$$
$$c_m = 0.5, W_{C_{empirical}} = \begin{bmatrix} 2 & 7.5 \times 10^{-5} \\ 7.5 \times 10^{-5} & 2.07 \end{bmatrix}$$

Clearly the differences between the linear and empirical controllability gramians increase with impulse input magnitudes as may be expected. And this difference is exacerbated with larger values of α.

From the above illustrative example, two things can be stated: (1) The controllability gramians estimated from linear model and empirical impulse response methods are indisitnguishable when the system is linear, (2) These differences increase with additional input-to-state nonlinearity as well as increased impulse magnitudes. For many systems, identification of such non-linearities in the form shown in Eq. 10 may not be possible. However their presence can nevertheless be studied through the empirical gramian approach. In addition, this approach can be used to generate comparative ellipsoids such as in Fig. 2 in a much simpler manner as opposed to having to first identify the linear time invariant system dynamics of the micro air vehicle.

This study seeks to generate such ellipsoids (from $W_{C_{Linear}}$ and $W_{C_{empirical}}$) and compare the reachability sets and Frobenius norms computed from the linear model and impulse responses for a candidate micro air vehicle such as a mini-quadrotor (Fig. 3). It should be noted that $W_{C_{empirical}}$ is defined for stable and control affine systems. Since most micro air vehicles such as flybarless helicopters, flappers and multirotors are inherently unstable, feedback stabilization has to be incorporated to render them control-affine. Therefore the closed loop dynamics of the vehicle is first identified (the dashed box in Fig. 4. Following this, impulse responses are collected and

Fig. 3. Luminier 180-QAV racing quadrotor (Weight - 250 grams).

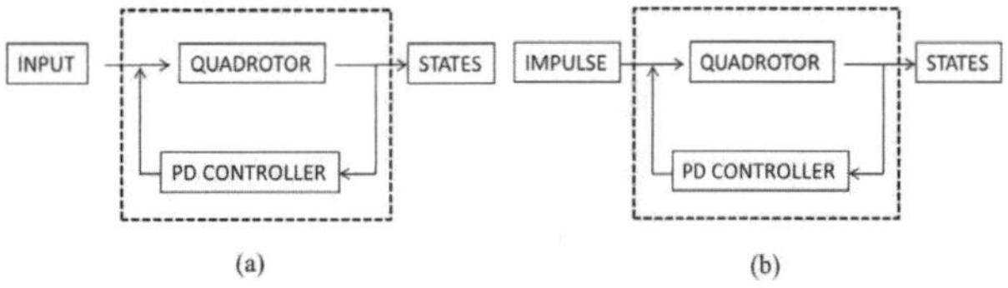

Fig. 4. (a) identification of closed loop dynamics (dashed box), (b) extract impulse response of closed loop system

subsequent comparisons between linear controllability gramians and empirical gramians are made. A motion capture system shown in Fig. 5 in conjunction with an onboard autopilot (Ref. 10) was used for flight data collection.

EXPERIMENTAL MEASUREMENTS

Linear Time Invariant Parameter Identification

Sample flight data charts collected for different input profiles and degrees of freedom are shown in Fig. 6. The data was sampled at 50 Hz. The proportional-derivative (PD) gains for roll, pitch and yaw as a fraction of the maximum input are shown in Table 2. These gains were chosen to be sufficient to render stable flight testing procedures without introducing

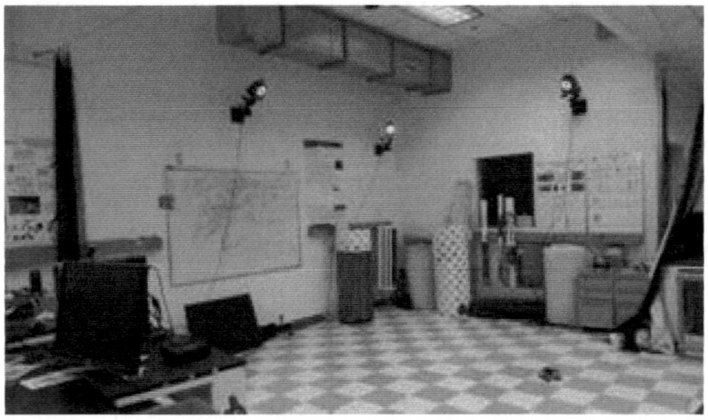

Fig. 5. Motion capture setup for input/output measurements

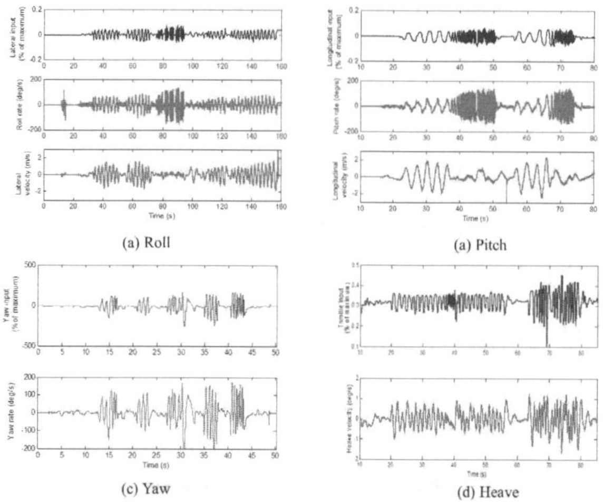

Fig. 6. Input/output data for different input profiles and degrees of freedom

$$\begin{pmatrix}\dot{u}\\ \dot{v}\\ \dot{p}\\ \dot{q}\\ \dot{\phi}\\ \dot{\theta}\\ \dot{w}\\ \dot{r}\end{pmatrix} = \begin{bmatrix} X_u & 0 & 0 & 0 & 0 & -g & 0 & 0\\ 0 & Y_v & 0 & 0 & g & 0 & 0 & 0\\ 0 & L_v & L_p & 0 & L_\phi & 0 & 0 & 0\\ M_u & 0 & 0 & M_q & 0 & M_\theta & 0 & 0\\ 0 & 0 & 1 & 0 & 0 & 0 & 0 & 0\\ 0 & 0 & 0 & 1 & 0 & 0 & 0 & 0\\ 0 & 0 & 0 & 0 & 0 & 0 & Z_w & 0\\ 0 & 0 & 0 & 0 & 0 & 0 & 0 & N_r \end{bmatrix}\begin{pmatrix}u\\ v\\ p\\ q\\ \phi\\ \theta\\ w\\ r\end{pmatrix} + \begin{bmatrix} 0 & 0 & 0 & 0\\ 0 & 0 & 0 & 0\\ 0 & L_{lat} & 0 & 0\\ 0 & 0 & M_{lon} & 0\\ 0 & 0 & 0 & 0\\ 0 & 0 & 0 & 0\\ Z_{thr} & 0 & 0 & 0\\ 0 & 0 & 0 & N_{rud} \end{bmatrix}\begin{pmatrix}\delta_{throttle}\\ \delta_{lat}\\ \delta_{lon}\\ \delta_{rud}\end{pmatrix}$$

Fig. 7. Closed loop model structure of a quadrotor in hover

excessive stiffness and oscillations from perturbations. Due to the presence of a PD feedback controller, the closed loop dynamics model structure of the quadrotor can be expressed as shown in Fig 7. The lateral, longitudinal, yaw and heave dynamics are assumed to be decoupled in hover.

The stability and control derivatives are assumed to be time invariant and were determined using output error forumlation in SIDPAC (Ref. 2). All inputs were scaled with their maximum inputs and expressed between [-1 1]. Table 1 lists the identified derivatives with their respective standard errors. The R^2 fit between model and flight data were above 80% for all cases as can be seen in Fig. 8. It should be noted that with different set of PD gains, the identified stability derivatives will change. However for a default set of gains, the gramians extracted from the linear system as well as impulse response can be compared. In this manner, the effect of gains can be assessed later.

Impulse Response Measurements

Because empirical gramians are by their nature dependent on discrete-time generation of impulses and corresponding measurement of system response, it is important to arrive at schemes that can closely approximate idealized impulses. In theory, an impulse is represented as a Dirac delta function

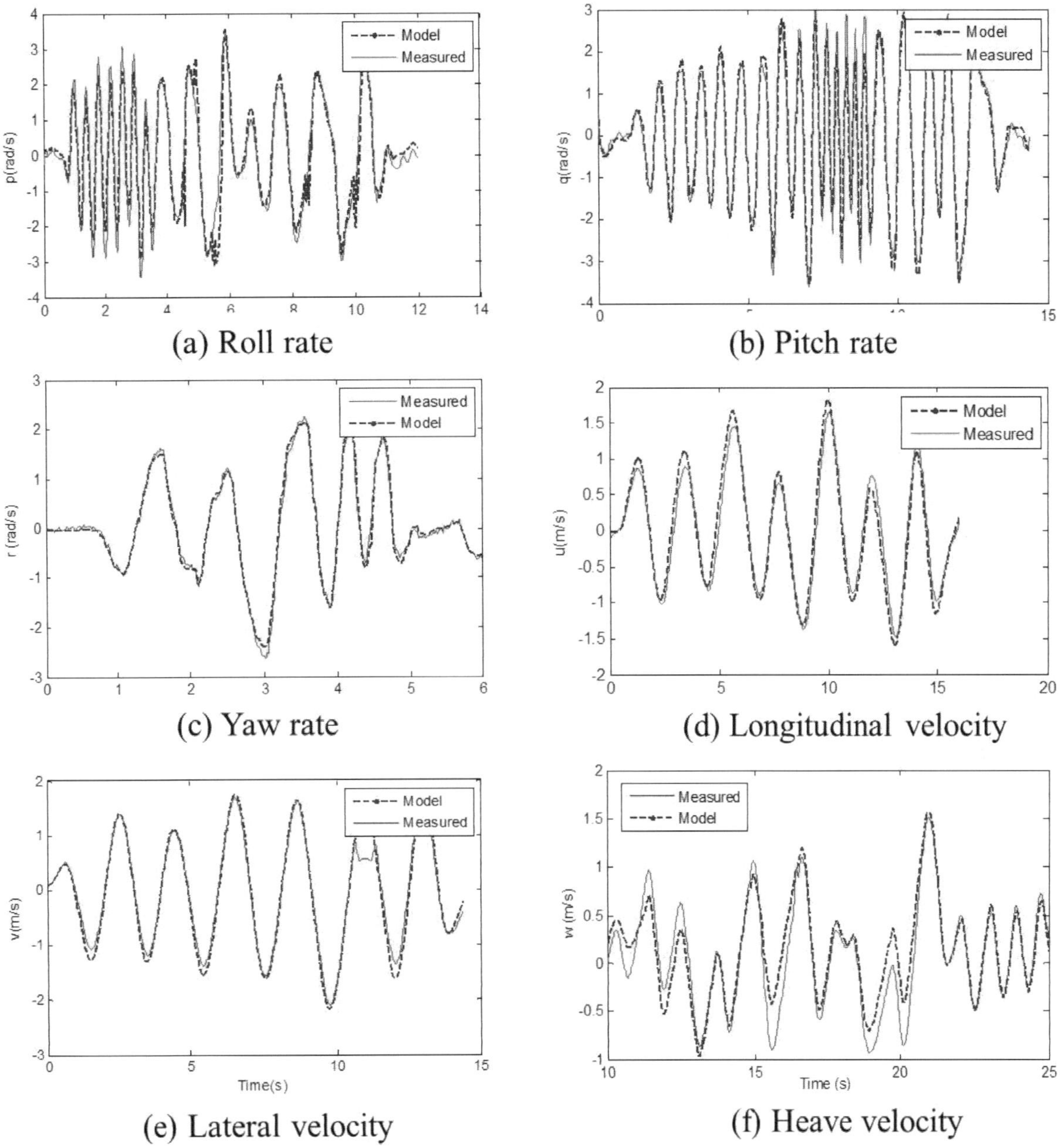

Fig. 8. Comparison between identified linear time invariant model prediction with flight data

Table 1. Identified parameters for Luminier Quadrotor in hover with PD controller

Param	Value	Std. Dev	Param	Value	Std. Dev
Y_v	-0.62	0.17	X_u	-0.776	0.2
L_v	-12.83	1.8	M_u	4.63	1.4
L_p	-54.2	6.9	M_q	-42.2	6.39
L_ϕ	163.63	29.2	M_θ	-129.3	24
L_{lat}	1400	180	M_{lon}	1090	138
N_r	-58.5	6.7	Z_w	-0.7	0.15
N_{rud}	337	43	Z_{throt}	23.93	0.98

Table 2. PD gains used for system identification and impulse response measurements as a fraction of maximum control input

	Pitch	Roll	Yaw
Proportional (/rad)	0.057	0.057	-
Derivative (/rad/s)	0.02	0.02	0.15

with an appropriately scaled magnitude. For numerical simulations, Dirac delta functions are approximated as pulse signals with a width equal to the discrete time step and with an area of one. For experiments demanded by this study, the best impulse that can be accomplished by a pilot during flight testing is schematically shown in Fig. 9. It can be noticed that this is an approximate replication of the idealized discrete impulse. Additionally, the empirical gramian computation requires knowledge of the area of the impulse. Obtaining this information from impulses such as ones shown in Fig. 9 become cumbersome and non-repetitive since it is hard for a human to replicate an impulse input.

To overcome this difficulty, an electronic impulse with a controlled duration and magnitude was setup to be generated by the autopilot, which can be issued on command by a ground operator. This setup is schematically shown in Fig. 10. Since the sampling rate of the data acquisition setup was 50 Hz, a minimum impulse duration of 30 ms was chosen. Figure 11 shows typical impulse responses of the various quadrotor degrees of freedom. It should be noted that for every degree of freedom, the off-axis response is not shown since they were negligible and not necessary based on the decoupled nature of the system dynamics. For empirical gramian computation, data slices are considered up to the time taken for all vehicle

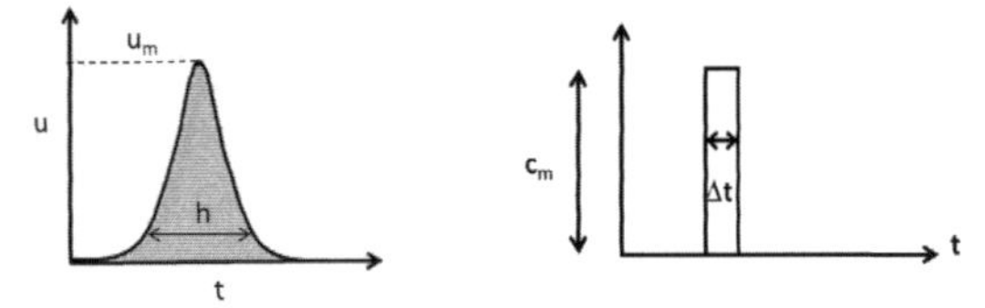

(a) Impulse generated by human pilot

(b) Impulse generated by digital autopilot

Fig. 9. Impulse inputs generated by human operator vs. digital autopilot. h is the duration of impulse. Ideally h $\rightarrow \Delta t$ where Δt is sampling period

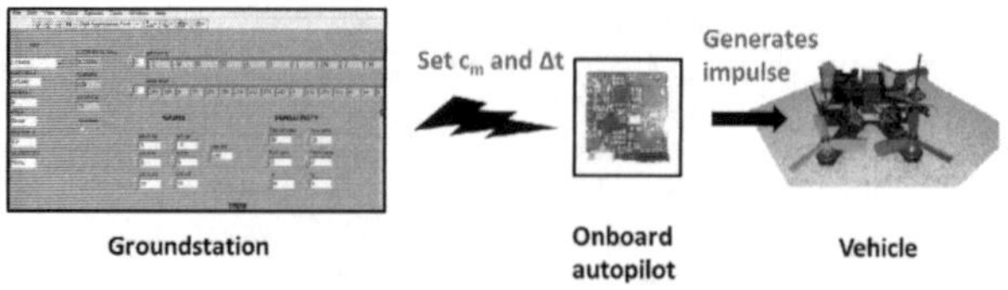

Fig. 10. Experimental setup to generate impulses

Table 3. Frobenius norm computed from linearized system dynamics

Degree of Freedom	$\|\|W_{C_{Linear}}\|\|_F$
Lateral	143.3 ± 9.2
Longitudinal	129.4 ± 8.3
Yaw	31.23 ± 5.97
Heave	20.8 ± 2.8

states to return to steady state.

If A, x and Δt represent the area ($c_m \Delta t$) of the digital impulse, system state response and the sampling period, then the empirical gramian from Eq. 6 for a given impulse profile can be recast as,

$$W_{C_{empirical}} = \frac{x^T x \Delta t}{A^2} \tag{12}$$

RESULTS AND DISCUSSION

The experimental measurements helped provide a linearized model of the quadrotor in hover as well as a data set of impulse responses. Using the methodology described earlier, reachability sets computed from the linear controllability and empirical gramians can be compared. This section will also discuss how the empirical gramians scale with feedback controller gains through a separate set of impulse responses collected at higher gains. This will be shown using the yaw DOF as a simplifying case.

Reachability sets estimated from linearized system

In order to simplify the analysis, the system dynamics is broken down into its mutually decoupled states - lateral, longitudinal, heave and yaw. Based on the identified parameters, the linear controllability gramians (Eq. 3) for these degrees of freedom can be calculated using Lyapunov solvers such as *'lyap'* as mentioned previously. Table 3 lists the Frobenius norms for the various degrees of freedom that were computed using Eqs. 3, 6, 7 and Table 1

The controllability gramian can also be expressed in ellipsoid form (reachability sets). Since the yaw and heave gramians are ellipsoids of degree one, the lateral and longitudinal ellipsoids are shown in Fig. 12 for clarity and visualization purposes. These represent the maximum states attainable by the system with maximum lateral and longitudinal control inputs, which are determined by the linear approximation of the

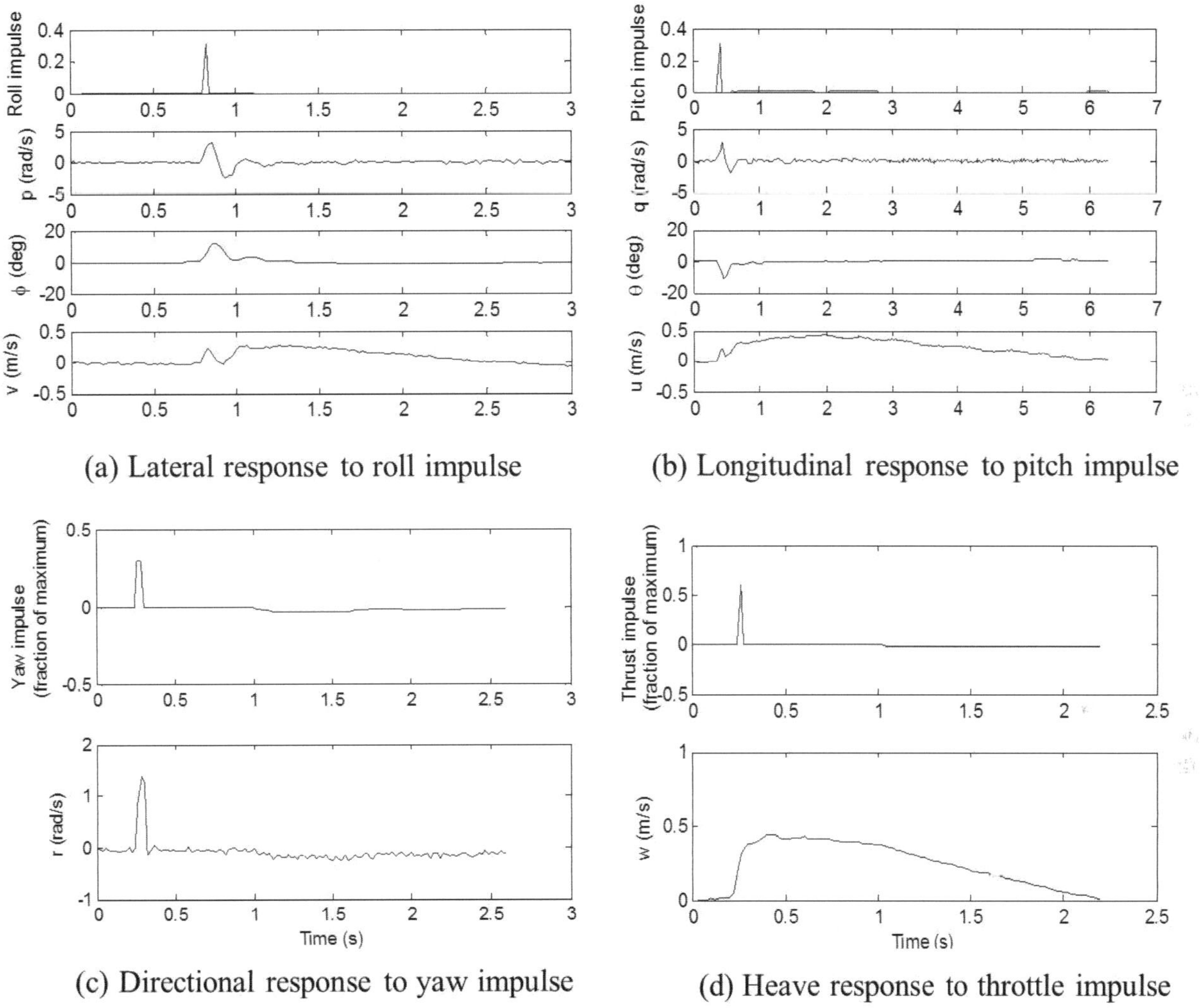

(a) Lateral response to roll impulse

(b) Longitudinal response to pitch impulse

(c) Directional response to yaw impulse

(d) Heave response to throttle impulse

Fig. 11. Response of quadrotor to impulses in each degree of freedom. Impulse magnitude represented as fraction of maximum allowable. Off-axis response for each degree of freedom not shown since they were negligible.

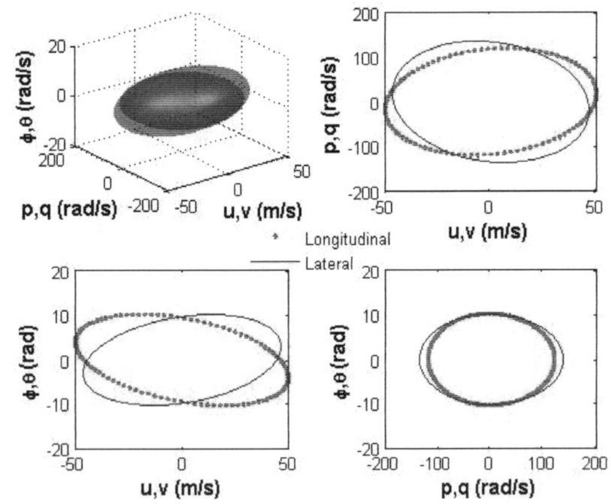

Fig. 12. Sample impulse input for lateral degree of freedom

system. A unique design property of the chosen quadrotor (Fig. 3) is that the motor offsets along the pitch axis is about 20% larger than along the roll axis. Considering that the moments of inertia along these axes do not scale similarly, the angular velocities generated along roll axis is expected to be higher than pitch. Interestingly, this hypothesis is partially reflected in the reachability set shown in the fourth graph in Fig. 12 where the roll rates reachable are higher than the pitch rates.

Assessment of system nonlinearities

The reachability sets from the impulse responses can now be compared with those estimated by the linear system (Table 3). These are calculated using Eq. 12. Instead of using the simulated impulse response as illustrated in a previous example (Eq. 10), the measured impulse response time histories of the following states are used,

$$\begin{aligned} x_{lateral} &= [v, p, \phi] \\ x_{longitudinal} &= [u, q, \theta] \\ x_{yaw} &= [r] \\ x_{heave} &= [w] \end{aligned} \tag{13}$$

Table 4 lists the empirical Frobenius norms calculated from this data for impulses with varying magnitude (c_m = 0.3, 0.5) and duration (Δt = 30 ms, 50 ms). Consider the baseline case of an impulse with c_m = 0.3 and Δt = 50 ms represented by the first entry for each DOF. It can be seen that $||W_{C_{empirical}}||_F$ for this impulse case compares satisfactorily with $||W_{C_{Linear}}||_F$ from Table 3. The ellipsoid for the lateral DOF for the baseline impulse is compared with the linear controllability ellipsoid in Fig. 13(a). It can be seen that the volume and principal axes lengths for the linear and impulse response cases are comparable. However, the nonlinearity becomes apparent when the impulse magnitude is increased to

Table 4. Frobenius norm of empirical gramians from impulse response data

Degree of Freedom	Impulse mag (fraction of max)	Impulse duration (ms)	$\|\|W_{C_{empirical}}\|\|_F$
Lateral	0.3	30	142 ± 7
	0.5	30	90.95 ± 7.4
	0.3	50	135.9 ± 8.1
Longitudinal	0.3	30	132 ± 12
	0.5	30	97.3 ± 5
	0.3	50	127 ± 10.6
Yaw	0.3	30	33.4 ± 2
	0.5	30	29.5 ± 2
	0.3	50	30.15 ± 3.85
Heave	0.3	30	21.05 ± 2
	0.6	30	22.4 ± 1.85
	0.6	50	22.7 ± 3

50% of the maximum allowable input (Fig. 13(b)) where the Frobenius norm decreases by about 20-30%. Similar observations were seen for the longitudinal degree of freedom as well. However, the effect of impulse magnitude on nonlinearity was not observed for the yaw and heave DOF (see cases for c_m = 0.5 and 0.6 respectively in Table 4). In the future, systematic variation of impulse magnitudes will be conducted to carefully assess its effect on nonlinearity for the various DOF. One interesting observation is that increasing the impulse duration (from 30 ms to 50 ms) slightly decreased the Frobenius norms for the lateral and longitudinal DOF, but had negligible effect for the heave and yaw DOF.

In addition to the ellipsoids, the time histories for impulse responses predicted from a linear model (Fig. 7) were also compared with flight data. Representative comparisons are made for yaw DOF (c_m = 0.3) and lateral DOF (c_m = 0.3 and 0.5) which are shown in Figs. 14 and 15. It can be seen that the linear model satisfactorily predicts response within margin of error for yaw DOF. However as observed with the ellipsoids, the linear model overpredicts response for the higher impulse magnitude case (Fig. 15(b)) in lateral DOF, thus possibly indicating the influence of nonlinearities at these impulse input magnitudes.

Effect of feedback control gains

As mentioned previously, impulse response analysis is only applicable to control-affine systems and it is undefined for unstable systems. As a result, the feedback controller is embedded within the empirical gramian analysis. Here we seek to understand if these gains affect the reachability sets and/or inject nonlinearity into the system. For the sake of simplicity we study yaw DOF since it has just a single feedback state variable. Consider the linearized yaw dynamics below,

$$\dot{r} = N_r r + N_{rud}\delta_{rud} \tag{14}$$

Assuming negligible aerodynamic damping in yaw, we can say that N_r primarily reflects the yaw feedback gain ($K_{d_{yaw}}$). Hence, if the gains are scaled by a factor α, we can assume that N_r also scales similarly. Using this assumption and prior measurements of N_r with baseline feedback gain values (from

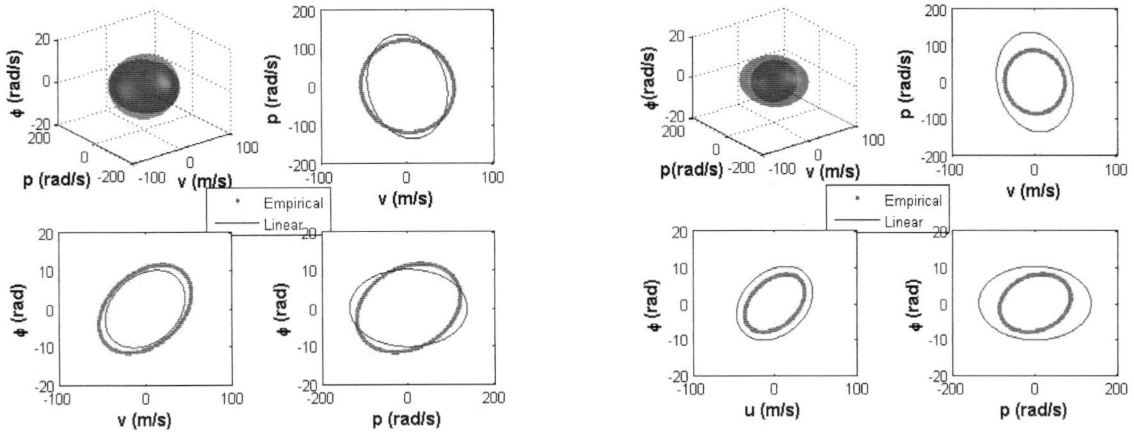

(a) $c_m = 0.3$ (b) $c_m = 0.5$

Fig. 13. Comparison between linear and empirical controllability ellipsoids for two different impulse magnitudes in lateral DOF

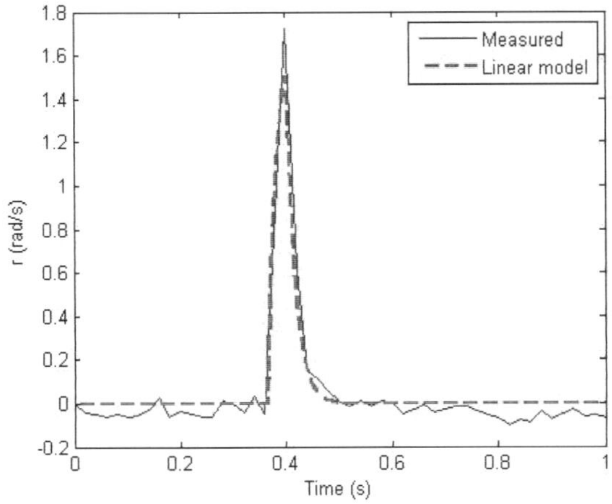

Fig. 14. Comparison between impulse response from flight data and model prediction in yaw DOF for c_m = 0.3.

Table 2, various impulse response measurements were conducted for a given impulse profile ($c_m = 0.3$, $\Delta t = 50$) and three different gains $[K_{d_{yaw}}/2, K_{d_{yaw}}, 2K_{d_{yaw}}]$ where $K_{d_{yaw}}$ represents the baseline value. The corresponding Frobenius norms are shown in Table 5.

It can be seen that the Frobenius norm from impulse data is well in agreement with estimation from linearized dynamics

Table 5. Frobenius norm computed for different yaw feedback gains for impulse profile $[c_m, \Delta t] = [0.3, 50]$

$K_{d_{yaw}}(s/rad)$	Empirical	Linear
0.15 (baseline)	29.5 ± 2	31.23
0.3	21.2 ± 0.3	22 (estimated)
0.075	48.2 ± 2	44 (estimated)

although there is some difference at very low gains. However, by making the system very stiff (as seen with the higher gains), there is very little deviation from a linearized system prediction. This could have some implications towards tailoring nonlinearities by manipulating the control system architecture as opposed to the physical system.

CONCLUSIONS

Controllability gramians have been used to study the reachability sets of micro air vehicles (states attainable when maximum control input is given). These are typically determined from the linear time invariant (LTI) dynamics identified through system identification techniques. These gramians have been recently used to quantify and compare the maneuverability of different vehicles. In this paper, experimental validation of an alternative approach of determining controllability gramians through measurement of impulse response is conducted. Gramians estimated through this approach are referred to as empirical gramians and offer two main advantages over linear controllability gramians - (1) there is no assumption of linearity in system dynamics, and (2) reachability sets can be obtained through direct empirical impulse response methods. These are significantly simpler than having to necessarily identify the LTI model of the system that are often cumbersome and time-consuming to perform. As a part of validation, the linear system dynamics of a typical quadrotor (QAV-180 Luminier) was initially identified in the hover condition. Impulse responses for impulses ranging from 30% to 50% of maximum control input were then measured. Finally the controllability gramians generated from linear approximation and from empirical data were compared. Using Frobenius norm as one of the metrics of quantifying the gramian, following are some of the key observations from the study:

1. There was satisfactory agreement between the linear and empirical Frobenius norms of the controllability gramian

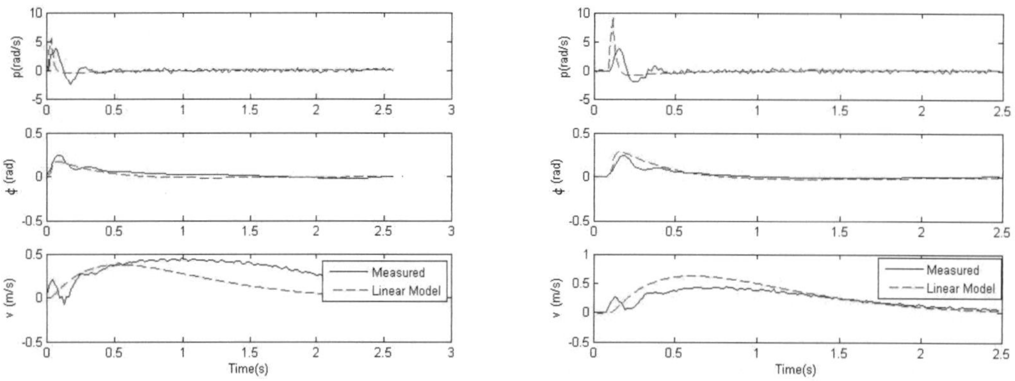

(a) $c_m = 0.3$ (b) $c_m = 0.5$

Fig. 15. Comparison between impulse response from flight data and model prediction in lateral DOF

for all degrees of freedom for impulses up to 30% of maximum input. This suggests mild to non-existent nonlinearities at these input levels. It also validates the fact that empirical gramians agree with linear gramians when system is linear.

2. When the impulse was increased to 50%, there was a 20-30% decrease in Frobenius norms for lateral and longitudinal axes. This suggests presence of input-to-state nonlinearities in these degrees of freedom (DOF). However, differences between linear and empirical Frobenius norms for yaw and heave DOF were minimal and within margin of error.

3. The duration of the impulse had a very mild effect. There was about 5% difference in Frobenius norm for lateral, longitudinal and yaw DOF and no change in heave DOF when the duration of the impulse was increased from 30 ms to 50 ms.

4. Effect of feedback gains were studied for yaw DOF. Increasing gains had the effect of reducing the attainable yaw rates as expected. There was good agreement between the Frobenius norms, although there was a 10% difference at the lowest gains tested. This could have some implications towards tailoring nonlinearities by manipulating the control system architecture as opposed to modifying the physical system itself.

The empirical gramian approach presented in this paper represents one of the few experimental verifications of this technique using flying air vehicles. It provides a simple method of extracting reachability sets of stable micro air vehicles without the requirement to identify the system itself. It can provide insights on degree of nonlinearity for various DOF and also opportunities to tune the gains appropriately. Future work will involve a systematic study of impulse responses for a wider range of impulse magnitudes to more accurately assess system nonlinearities.

ACKNOWLEDGEMENTS

This research was supported by the Army's MAST CTA Center for Microsystem Mechanics with Christopher Kroninger and Brett Piekarski as technical monitors.

REFERENCES

[1]Tischler, M., and Remple, R., "Aircraft and Rotorcraft System Identification: Engineering Methods with Flight Test Examples," AIAA Education Series, AIAA, Reston, VA, 2006

[2]Morelli, E., "Low-Order Equivalent System Identification for the Tu-144LL Supersonic Aircraft," *Journal of Guidance, Control and Dynamics*, Vol. 26, No. 2, Mar-Apr 2003, pp. 354-362.

[3]Maine, R., and Illif, K., "Application of Parameter Estimation to Aircraft Stability and Control: The Output-Error Approach," NASA TR RP 1168, 1986

[4]Gremillion, G., and Humbert, J.S., "System Identificaiton of a Quadrotor Micro Air Vehicle," AIAA Conference on Atmospheric Flight Mechanics, AIAA-2010-7644, Toronto, Canada, 2010.

[5]Hrishikeshavan, V., Humbert, J.S, and Chopra, I., "Gramian Analysis of a Shrouded Rotor Micro Air Vehicle in Hover," *Journal of Guidance, Dynamics and Control*, Vol. 37, No. 5,Sep. 2014.

[6]Hrishikeshavan, V., Benedict, M., and Chopra, I., "Identification of Dynamics of a Cyclocopter Micro Air Vehicle in Hover," *Journal of Aircraft*, Vol.52, pp. 116-129, 2015

[7]Gardner, R., Humbert, J.S., and Chopra, I., "Maneuverability comparison of flybar and flybarless micro-helicopters," American Helicopter Society Specialists' Meeting on Unmanned Rotorcraft, Scottsdale, AZ, Jan. 2013.

[8]Faruque, I., Samuel, P., and Humbert, J.S., "Yaw Dynamics Identification of an Insect Inspired Flapping Wing Micro Air Vehicle," AIAA Science and Technology Forum, National Harbor, MD, Jan. 2014.

[9]Humbert, J.S, and Faruque, I., "Analysis of Insect-Inspired Wingstroke Kinematic Perturbations for Longitudinal Control," *Journal of Guidance, Control and Dynamics*, Vol. 34, (2), 2011, pp.618-622.

[10]Hrishikeshavan, V., and Chopra, I., "Refined Lightweight Inertial Navigation System for Micro Air Vehicle Applications," American Helicopter Society Specialists' Meeting on Unmanned Rotorcraft and Network Centric Operations, Chandler, AZ, January 20-22, 2015.

[11]Junkins, J. L., and Puneet, S., "How nonlinear is it? A tutorial on nonlinearity of orbit and attitude dynamics." Proceedings of the John L. Junkins Astrodynamics Symposium, No. AAS 03-286, AAS Publications, College Station, TX, May 2003.

[12]Lall, S., Marsden, J.E., Glavaski, S., "A Subspace APproach to Balanced Truncation for Model Reduction of Nonlinear Control Systems," *Int. Journal of Robust and Nonlinear Control*, Vol. 12. No. 5, pp.519-535, 2002.

[13]Hahn, J., and Edgar, T.F., "An Improved Method for Nonlinear Model Reduction Using Balancing of Empirical Gramians," *Computers and Chemical Engineering*, Vol. 26, No. 10, pp. 1379-1397, 2002.

[14]Singh, A.K., and Hahn, J., "On the Use of Empirical Gramians for Controllability and Observability Analysis," American Control Conference, Portland, OR, Jun. 8-10, 2006

[15]Hahn, J., and Edgar, T.F., "A Gramian Based Approach to Nonlinearity Quantification and Model Classification,"*Industrial engineering chemistry research* Vol. 40, No. 24, pp. 5724-5731, 2001.

Development of MAV-scale Quad-Cyclocopter Capable of Aerial and Terrestrial Locomotion

Elena Shrestha
Graduate Research Assistant
eshresco@umd.edu

Benjamin Williams
Undergraduate Research Assistant
bjw19@umd.edu

Vikram Hrishikeshavan
Assistant Research Scientist
vikramh@umd.edu

Inderjit Chopra
Alfred Gessow Professor and Distinguished University Professor
Director of Alfred Gessow Rotorcraft Center
chopra@umd.edu

Alfred Gessow Rotorcraft Center, Department of Aerospace Engineering,
University of Maryland, College Park, MD 20740

ABSTRACT

The capability to transition between different modes of locomotion improves micro air vehicle (MAV) application in missions such as reconnaissance where avoiding detection is critical to the mission's success and potentially increases endurance by transitioning to operational modes with lower power consumption. This paper describes the design and development of a 900-gram quad-cyclocopter MAV capable of efficient aerial and terrestrial locomotion solely using combination of thrust vectoring and rotational speed of all four cycloidal rotors (cyclorotors). During flight, independent control of four rotational speeds and thrust vectors were used to efficiently navigate and maintain stable hover. Pitch control is provided by modulating the RPM of the fore and aft pairs of cyclorotor, roll control is achieved by differentially varying rotational speeds of the side pairs, and finally yaw control is possible by differentially rotating the thrust vectors of the side pairs. During operation on land, the quad-cyclocopter maneuvers using motor RPM. Reversible electronic speed controllers permit the forward pair of the cyclorotors to rotate in any direction, enabling the vehicle to translate forward and backward. Both control strategies are implemented using a custom autopilot, which is also used for attitude stabilization in hover. Transition between operational modes is enabled using a retractable landing gear system in conjunction with carbon fiber wheels that are integrated with the cyclorotor endplates. Bench-top experiments shows that the wheels operate around 200 RPM compared to the 1600 RPM required in hover. As a result, terrestrial locomotion requires 44 Watts compared to 280 Watts in aerial mode, which is a 84% reduction in power consumption. The developed quad-cyclocopter is one of the few MAVs to achieve both aerial and terrestrial locomotion.

INTRODUCTION

Mission scenarios ranging from search-and-rescue (SAR) to precision package delivery require robust micro air vehicle (MAV) platforms capable of vertical take-off and landing (VTOL) as well as efficient forward flight, high endurance, portability, and high maneuverability. SAR missions often require MAVs to navigate between multiple terrains, including confined spaces in an urban setting. One of the challenges faced by rotary MAVs is limited endurance, which can be attributed to low Renold's number aerodynamics and limitations in motor efficiency and battery capacity from commercially available sources. Many of the current rotary-based MAVs rely on conventional rotor blade designs that experience degraded performance at low Reynold's numbers (10,000–50,000). Reduced aerodynamic efficiency of blades at the MAV-scale is caused by high viscous forces, which leads to increased boundary layer thickness, increased profile drag, and large induced losses (Ref. 1).

In order to increase endurance, it may be advantageous for an aerial vehicle to transition to another form of locomotion that requires less power. The capability to transition between different modes of locomotion also improves MAV application in missions such as reconnaissance where avoiding detection is critical to the mission's success. Previously, researchers at the University of Tokyo developed a multi-field universal wheel for air-land (MUWA) vehicle that is capable of aerial, terrestrial, and aquatic locomotion (Ref. 2). The

Presented at 7th American Helicopter Society Technical Meeting on VTOL Unmanned Aircraft, Mesa, AZ, January 24-26, 2017.

Fig. 1. MUWA quadrotor developed by the University of Tokyo (Ref. 2).

quadrotor platform used variable pitch propellers that are actuated with rotary servos. In addition, each rotor is attached to individual motors to provide additional means of varying thrust during different modes. For terrestrial and aquatic modes, the quadrotor relied on a ring-shaped polystyrene foam body that is rigidly attached to the vehicle and also functions as a landing gear (Fig.1). The vehicle, along with its foam outer-ring, weighed 2100 grams and spanned 0.91 meters in diameter and 0.12 in body thickness. In terrestrial mode, the MUWA platform acrobatically flipped onto its foam body and transformed into a mono-wheel while relying on motor torque for rotational control. For aquatic locomotion, the foam body simply provided buoyancy. Overall, to function in multiple modes of locomotion, the MUWA platform used eight different actuators and also relied on an external foam body.

Researchers at the Illinois Institute of Technology also developed a hybrid terrestrial and aerial quadrotor (HyTAQ) by suspending the vehicle inside a rolling cage (Ref. 3). HyTAQ uses fixed-pitch propellers and commands thrust by modulating rotational speeds of four motors (Fig.2). The quadrotor, along with its cage, weighs 570 grams and spans 0.20 in diameter. In order to move forward, the entire vehicle pitches down at a 45° angle such that it uses only a fraction of the total generated lift as propulsive thrust. While the design is mechanically simple, a large percentage of available lift is not being utilized for terrestrial locomotion.

Furthermore, both groups used a quadcopter configuration that relies on a conventional edgewise rotor design, which may not be the optimal design at the MAV-scale as previously discussed. An alternative rotary-winged configuration is the cyclocopter (Fig.3), which uses cycloidal rotors (cyclorotor). Blades on the cyclorotor rotate about a horizontal axis such that the blades span parallel to the axis of rotation and perpendicular to the direction of flight (Fig.4). In order to produce a net resultant lift, each blade pitch angle is cyclically varied using a four-bar linkage mechanism that simultaneously results in a positive geometric pitch angle at the top and bottom points of the circular trajectory in hover (Ref. 4). The magnitude of the lift vector is controlled by modulating the cyclorotor rotational speed and the direction can be instantaneously varied by phasing the cyclic blade pitch through the four-bar linkage mechanism.

Using cyclorotors as main source of propulsion offers numerous advantages in terms of aerodynamic efficiency, high-

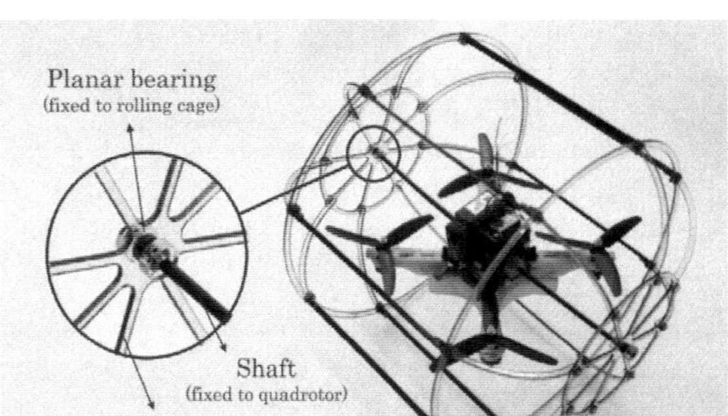

Fig. 2. HyTAQ quadrotor developed by the Illinois Institute of Technology (Ref. 3).

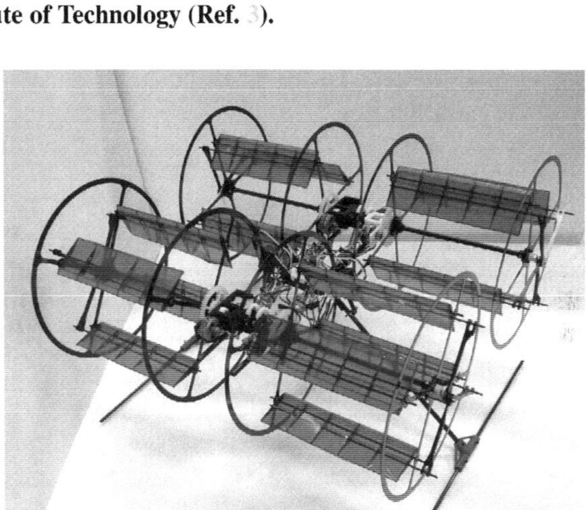

Fig. 3. Quad-cyclocopter.

speed forward flight, maneuverability, and gust tolerance. Previous research showed that an optimized cyclorotor has a higher power loading (thrust/power) in hover compared to a conventional edgewise rotor at the same disk loading (thrust/disk area). The observed increase in aerodynamic efficiency can be attributed to multiple sources such as blade design and beneficial contribution from unsteady aerodynamic

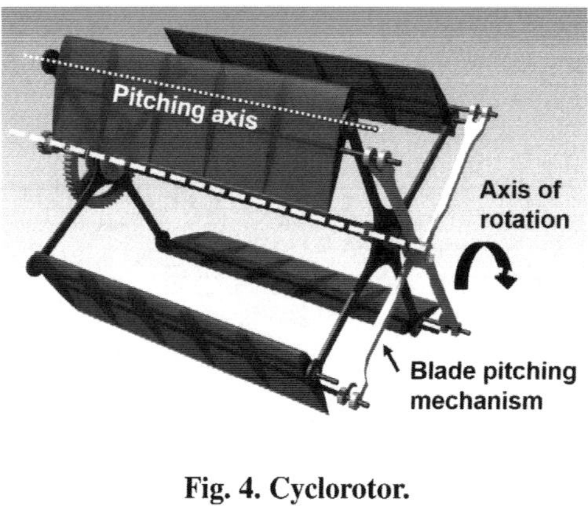

Fig. 4. Cyclorotor.

forces. Since all spanwise blade sections of the cyclorotor operate under similar aerodynamic conditions (flow velocity, angle of incidence, Reynold's number, etc.), the blades can be optimized to achieve higher aerodynamic efficiency. In addition, unsteady aerodynamic mechanisms such as leading edge vortices delay the onset of blade stall and enable the blades to operate at higher pitch angles (Ref. 5).

Results from wind tunnel studies showed that the power required for a cyclorotor to generate adequate propulsive force and lift for steady 13 m/s forward flight drops by 36% compared to power in hover (Ref. 6). Because the cyclorotor is able to instantaneously rotate the lift vector by phasing the cyclic blade pitch, a cyclocopter MAV can maintain level pitch attitude in forward flight, which reduces parasitic drag. Unlike other rotary-winged MAVs, the cyclocopter does not undergo any mechanical changes to transition from hover to forward flight. Previously, a twin-cyclocopter using two cyclorotors and a horizontal anti-torque nose rotor demonstrated level forward flight up to 5 m/s (Ref. 7). Remarkably, the cyclocopter was able to achieve steady 5 m/s flight from hover within two seconds and also immediately return to hover from forward flight solely by thrust vectoring. In a recent wind tunnel study, the cyclocopter was shown to tolerate up to 6 m/s in gust perturbation along the longitudinal axis using the thrust vectoring control strategy (Ref. 8).

The goal of the present work is to develop a cyclocopter capable of efficient aerial and terrestrial locomotion. Unlike the previously mentioned quadcopter platforms, the cyclocopter will only rely on its four cyclorotors to achieve both modes of locomotion. During flight, independent control of four rotational speeds and thrust vectors are used to efficiently navigate and maintain stable hover. During operation on land, the quad-cyclocopter maneuvers using differential motor RPM. In addition to the blade pitching mechanism and thrust vectoring components necessary for flight, the vehicle also requires wheels and retractable landing gear system to transition between the operational modes. Therefore, the main challenge is overcoming the high empty-weight fraction penalty in order to achieve both modes of locomotion. The paper will first present a detailed overview of blade design, rotor performance, closed-loop feedback controls system with custom in-house autopilot, and landing gear system design. Next, the different control strategies used for both modes of locomotion will be discussed in details. Finally, flight testing results will show a stable, sustained hover along with attitude data from the onboard processor.

ALL-TERRAIN CYCLOCOPTER VEHICLE DESIGN

The current vehicle uses four cyclorotors configured into two pairs at the fore and aft of the vehicle (H-configuration) (Fig.3). Both pairs rotate in opposite directions in order to counteract the resulting torque. The vehicle spans 0.51 meters (20 inches) in width, 0.43 meters (17 inches) in length, and 0.30 meters (12 inches) in height. The quad-cyclocopter

Table 1. Weight distribution of the quad-cyclocopter.

System	Weight (g)	% Total
Cyclorotors	333	37
Fuselage	160	18
Transmission	142	16
Batteries	115	13
Landing Gear	100	11
Rotary Servos	45	5
Avionics	5	<1
Total	**900 grams**	**100%**

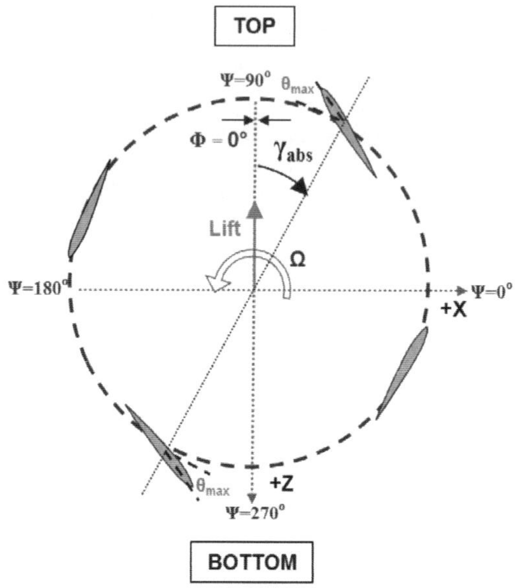

Fig. 5. Coordinates in hover where $\gamma_{abs} > 0°$ and Φ=0°.

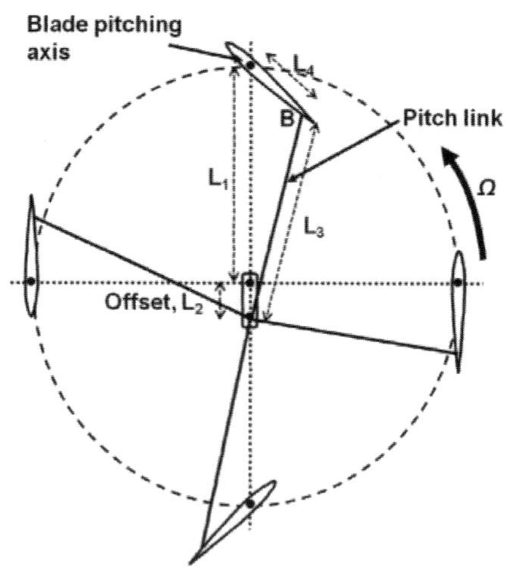

Fig. 6. Four-bar linkage pitching mechanism.

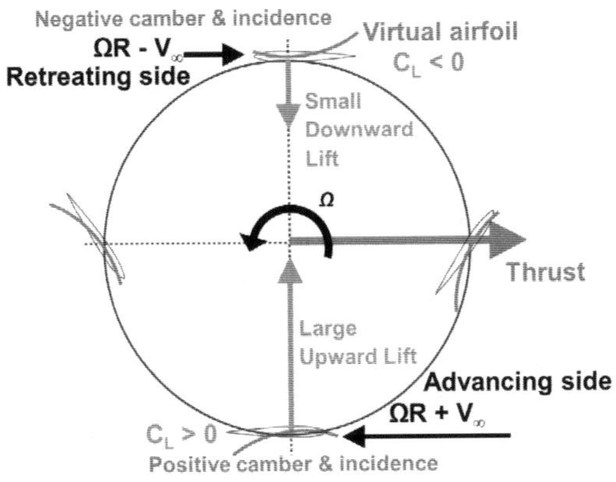

Fig. 7. Lift production based on virtual camber and virtual incidence.

weighs approximately 900 grams with the cyclorotors contributing up to 37% of the total gross take-off weight. Table 1 provides a breakdown of the vehicle's weight distribution.

The cyclorotors are independently driven by a 75 watts Axi out-runner motor through a 7.5:1 single-stage gear transmission. Thrust vectoring for each rotor is actuated using a rotary BMS-373MG servo that controls the blade pitching mechanism. Overall, there are eight actuators that command lift and propulsive thrust in both aerial and terrestrial modes of locomotion. The motors are powered by a 3s 11.1V LiPo battery with 1300 mah capacity while the servos are powered by a single-cell 3.7 V LiPo battery with 500 mah capacity. The following sections elaborate on the operating principle of cyclorotors, structural component designs of the wheels and landing gear system, avionics, and the closed-loop feedback system implemented on the vehicle.

Cyclorotor Kinematics

The key operating principle of the cyclorotor is the passive blade pitching kinematics which is accomplished using a four-bar linkage pitching mechanism. The blade pitching kinematics results in a net positive thrust after a full rotor revolution (Fig.5). In the coordinate frame, the +X-axis is towards the right and +Z-axis is towards the bottom and coheres with the vehicle's body frame. The azimuthal position of the blade (Ψ) is measured counter-clockwise along the axis of rotation and initializes from the +X-axis. The blade pitch angle (θ) is the geometric angle between the blade chord line and line tangent to the circular trajectory. The phasing input (γ_{abs}) is used to rotate the thrust vector by the same angle. In hover, γ_{abs} = 0° and the maximum blade pitch amplitude occurs at the top (Ψ = 90°) and bottom (Ψ = 270°) points of the circular trajectory. Subsequently, the resultant thrust vector only has a vertical lift component. By phasing the blade pitch cyclic, the γ_{abs} increases and the point of maximum pitch amplitude shifts clockwise along the trajectory such that the resultant thrust vector has a propulsive thrust component. Since the blade pitch amplitude is kept at a constant 45°, the magnitude

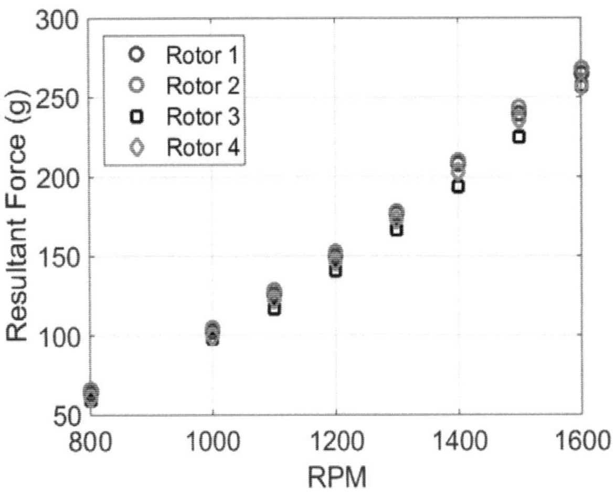

Fig. 8. Averaged thrust vs. RPM for cyclorotors.

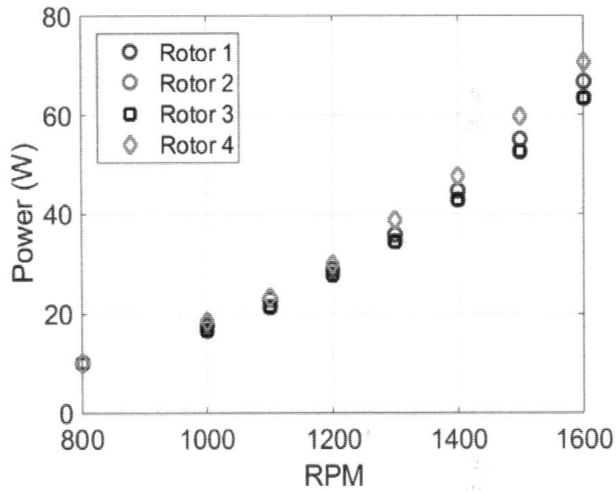

Fig. 9. Averaged power vs. RPM for cyclorotors.

of the resultant thrust vector is instead adjusted by varying the rotational speed of the cyclorotors.

Previous experimental parametric studies varied pitching amplitudes and studied their effect on rotor performance (Ref. 9). For symmetric pitching, the ±45°pitching amplitude achieved a higher maximum thrust at the tested rotational speeds. However, the optimum power loading was achieved with a ±40° blade pitching amplitude followed by ±45°. The reason for performance differences when comparing thrust and power loading is because power loading varies inversely with rotational speed. Therefore, it is more efficient to increase thrust by increasing blade pitching amplitude so the present cyclorotor is designed with a ±45° symmetric blade pitching.

Blade Pitching Mechanism

The schematic of the blade pitching mechanism is depicted in Fig. 6 where the four bars of the linkage system are labeled L_1, L_2, L_3 and L_4. L_1, also referred to as rotor radius, is the distance between the blade pitching axis and the horizontal axis of rotation. The pitch links (of length L_3) are connected to the end of the offset link on one end and the other end is connected to point B which is at a distance L_4 behind the pitching axis. The connections at both ends of the pitch link are through pin joints to allow the rotational degree of freedom. With this arrangement, as the rotor rotates, the blades automatically pitch cyclically. Because the pitching amplitude depends on the offset length, L_2, the offset length is kept at a constant value to achieve $\pm 45°$ pitching amplitude. The rotation of the offset link changes the phasing of the cyclic pitching and thereby changes the direction of the thrust vector. With a fixed blade pitching amplitude, the magnitude of thrust is modulated during flight by varying the rotational speed.

Virtual Camber and Incidence Effect

The blades of the cyclorotor span parallel to the rotating axis and therefore experience a curvilinear flowfield. The curvilinear flow introduces a chordwise variation of velocity distribution along the blade, which causes the symmetric NACA 0015 blade to behave as a virtually cambered airfoil at a virtual incidence in a rectilinear flow (Ref. 10). Due to blade kinematics, blades experience a negative virtual camber and incidence at the top of the trajectory ($C_l < 0$) and a positive camber and incidence at the bottom portion of the trajectory ($C_l > 0$) (Fig. 7). The direction of rotation does not affect rotor performance in hover, but is very critical in forward flight because the advancing side of the blades experience an increase in net tangential velocity.

Optimized Cyclorotor Design

Previous study found that virtual camber and virtual incidence are dependent on the chord-to-radius (c/R) ratio and the location of the blade pitching axis (Ref. 10). Results from systematic parametric experiments at various solidities and at 40° pitching amplitude showed that cyclorotors with c/R of 0.83 were more aerodynamically efficient. However, the increase in performance did not drastically change for c/R between 0.6 to 0.8. Structurally, c/R of 0.83 would result in heavy blades with low aspect ratios and lead to increased centrifugal loads on the blades. Therefore, the optimized cyclorotor on the quad-cyclocopter has a c/R of 0.67.

Aerodynamics The c/R of 0.67 is achieved with cyclorotor radius of 0.076 meters (3 inches) and blade chord of 0.051 meters (2 inches). The blades span 0.171 meters (6.75 inches) and use a NACA 0015 airfoil profile and a symmetric pitching amplitude of 45°. A linear approximation (Ref. 11) shows that the virtual chamber is approximately 7% while the virtual incidence is 8°. The pitching axis locations (x/c) of the

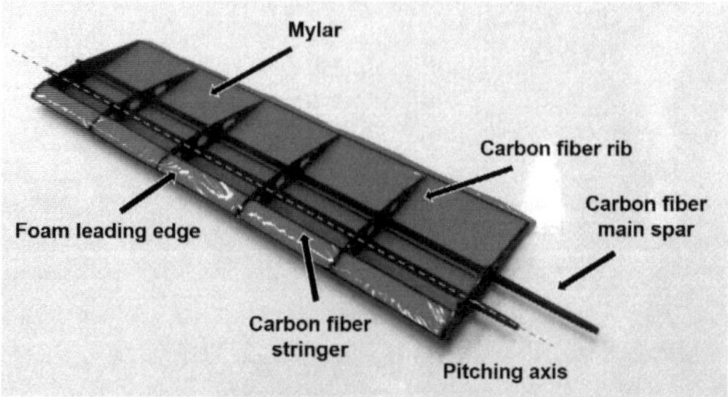

Fig. 10. Blade design.

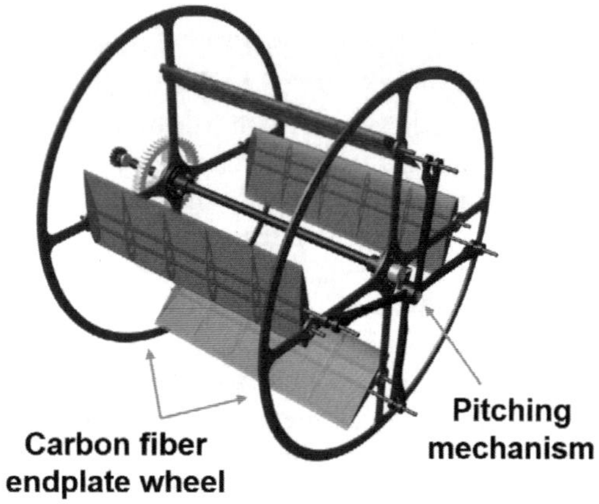

Fig. 11. Integrated composite wheels.

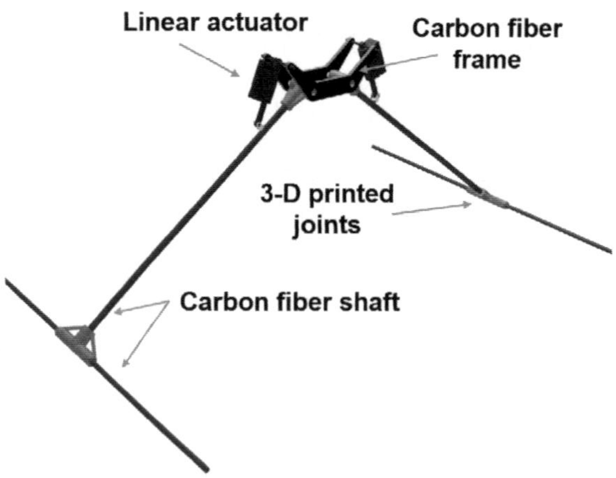

Fig. 12. Landing gear design.

blades were centered around the chordwise center of gravity (CG) location, which resulted in x/c of 0.25. The pitching axis aligned with the chordwise location of blade CG in order to to reduce the pitching moments on the blade caused by the centrifugal force. In addition, it reduces the loads on the pitching mechanism and ultimately on the rotary servos. Both thrust and power were systematically measured for each rotor from 800 to 1600 RPM. Figures 8 and 9 shows that each cyclorotor produces 270 grams (total of 1080 grams) while drawing 65 watts at the operating speed of 1600 RPM.

Structure Since cyclorotors contribute up to 37% of total vehicle weight, a lot of consideration was placed on redesigning the blade structure to achieve maximum lift at minimal weight. The optimized cyclorotor blades are composed of a foam leading edge, carbon fiber ribs, spars, and stringers (Fig. 10). The composite blade is then wrapped with a thin layer of Mylar. The carbon fiber stringers are evenly distributed along the blade chord to prevent the Mylar layer from forming cavities between the ribs. Bending and torsional stiffness is provided by carbon fiber spars along the pitching axis and connected to the linkage system. Additional bending stiffness is contributed from the closed airfoil cross-section and torsional stiffness from the Mylar cover. Overall, the optimized cyclorotor uses four blades with each blade weighing approximately 5.8 grams. Along with other structural components, the optimized cyclorotor weighs 83 grams and has a thrust-to-weight (T/W) ratio of approximately 3.25 at 1600 RPM.

Wheel and Landing Gear

The critical components for terrestrial locomotion are the wheels and the retractable landing gear system. In order to reduce overall weight, the carbon fiber wheels are integrated with the cyclorotor endplates (Fig. 11). The 0.23 meters (9.25 inches) diameter wheels are directly controlled by motor RPM through a 7.5:1 single-stage gear transmission. Unlike previous cyclorotor designs (Refs. 12–14), there is an additional steel spar inside the main carbon fiber shaft that connects the pitching mechanism to a rotary servo. The main shaft and wheel endplates rotate with the blades while the inner shaft is stationary and controlled by servos when thrust vectoring. Currently, the wheels have been lined with rubber coating to improve traction. Bench-top experiments showed that the wheels operate around 200 RPM compared to the 1600 RPM required in hover. As a result, terrestrial locomotion requires 44 Watts compared to 280 Watts needed in hover, which is a 84% reduction in power consumption.

Transition between aerial and terrestrial modes are enabled using a retractable landing gear system that is driven by two Firgelli PQ-12 micro linear servos (Fig. 12). The landing gear offers 0.10 meters (4 inches) of clearance for vertical take-off while completely retracts until the wheels contact the ground. The entire landing gear structure weighs 100 grams and is rigidly connected to the carbon fiber shafts that run across the fuselage frame. During operation, the pilot is able to retract and deploy the landing gear through a switch on the transmitter.

Fig. 13. Custom embedded lightweight kinematic autopilot (ELKA-R).

Avionics

Previous system identification studies showed that the cyclocopter is inherently an unstable system due to limited aerodynamic damping (Ref. 15). Therefore, onboard attitude stabilization was implemented using a revised version of a custom embedded lightweight kinematic autopilot (ELKA-R) (Fig. 13) (Ref. 16). Main components of ELKA-R include a STM32 microprocessor with a 32-bit ARM Cortex-M4 core, MPU-9150 inertial measurement unit (IMU), and nRF-based wireless transceiver. The STM32 microprocessor has a clock speed of 168 MHz, includes 24 ADC channels, and supports USART/UART, SPI, and I2C communication protocols. The MPU-9150 IMU includes tri-axial gyroscopes, accelerometers, and magnetometers, which are updated at 1000 Hz. Wireless communication with a nearby ground control station is enabled by an onboard nRF24L01 chip, a low-power 2.4 GHz RF transceiver. Overall, the board weighs only 1.3 grams and is powered by a single 1-cell 3.7 volt 125mAh Li-Po battery.

The ground control station communicates with ELKA through a wireless IEEE 802.15.4 data link. Onboard IMU data is streamed to a LabVIEW interface, which permits data logging and ability to directly modify closed-loop feedback system gains. All the data processing, switches between control strategies for different locomotion modes, and feedback control calculations are performed onboard by ELKA. A separate connection between the Spectrum receiver and transmitter ensures that the pilot still has command over the vehicle if connection is lost between the ground station and the onboard processor.

Closed-loop Feedback System

Inputs to the feedback are the four individual motor rotational speeds and four thrust vectoring positions of the servos. The feedback states are aircraft pitch and roll Euler angles (ϕ, θ) and the attitude rates (p, q and r), which are recorded by ELKA-R. The pitch (q), roll (p), and yaw (r) attitude rates are extracted from the tri-axial gyroscopes and combined with the data provided by the accelerometer (Fig. 14). The integration of gyro measurements with time has been known to cause

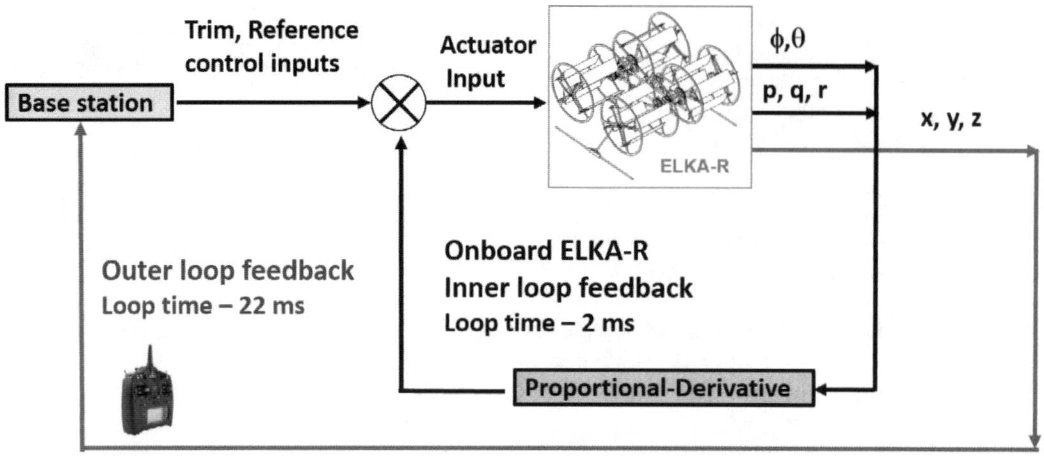

Fig. 14. Overview of closed-loop feedback system.

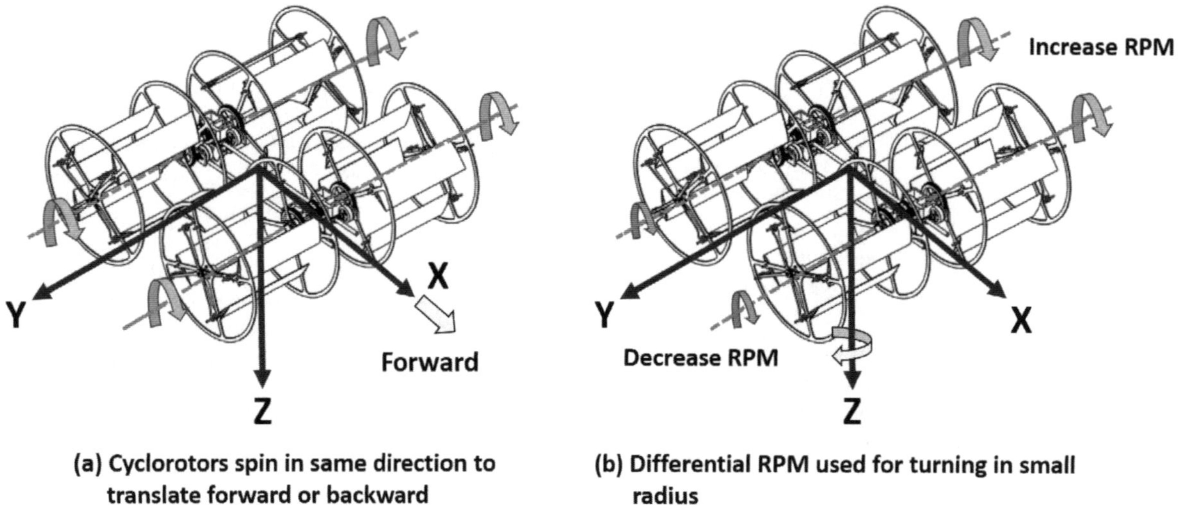

Fig. 15. Control strategy for terrestrial mode.

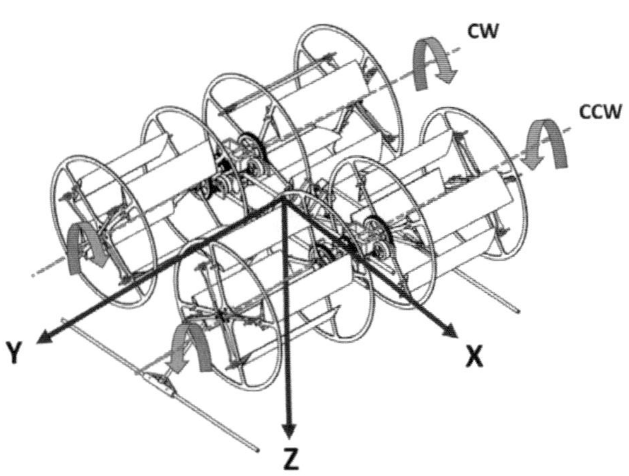

(a) Rows of counter-rotating cyclorotors used to balance resultant torque.

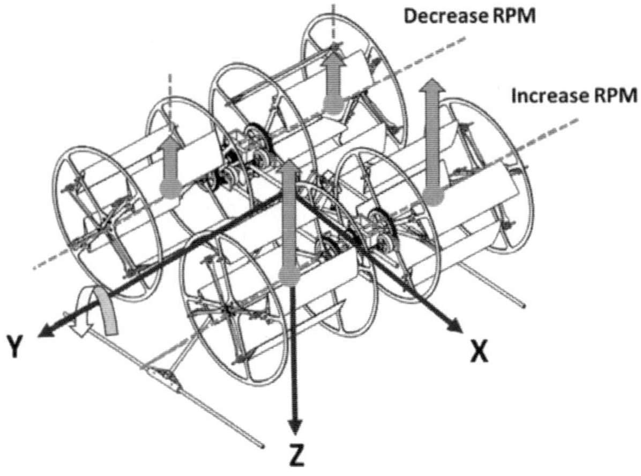

(b) Pitch control achieved through differential RPM of fore and aft rows.

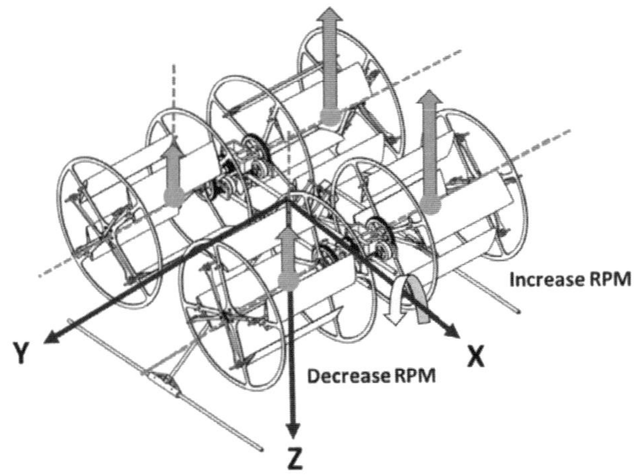

(c) Roll control achieved through differential RPM of lateral cyclorotor pairs.

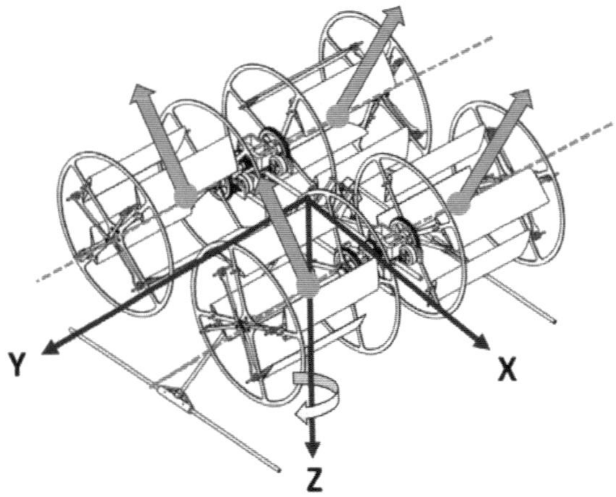

(d) Yaw control achieved by thrust vectoring.

Fig. 16. Control strategy for aerial mode.

Fig. 17. Demonstration of terrestrial locomotion and stable hover.

drift in attitude estimation (Refs. 17, 18). To account for the drift, a high pass complementary filter is used with a 4 Hz cut-off frequency for the gyro measurements (Ref. 19). Vibrations from the rotary systems affected accelerometer measurements, which are processed through a low pass complementary filter with a 6 Hz cut-off to reject disturbances from the rotor vibrations occurring at significantly higher frequencies than the body dynamics. The inner-loop attitude stabilization uses a proportional-derivative (PD) controller which is sufficiently able to stabilize the vehicle and reject the disturbances encountered during indoor operations.

CONTROL STRATEGY

Terrestrial and aerial modes of locomotion require two different control strategies. The terrestrial mode uses differential drive while aerial mode relies on both rotational speed and thrust vectoring to maneuver. Switching actuator controls mixing between both operational modes is enabled through the ground control station and also through the pilot's transmitter.

Terrestrial Mode

Because both the fore and aft pairs of cyclorotors rotate in opposite directions to balance the resultant torque, direction of rotation of the aft pair must be reversed in order to achieve translational motion in terrestrial mode. The direction of rotation is reversed using a programmable electronic speed controller (ESC). The quad-cyclocopter uses the Afro 12A multirotor ESC reprogrammed with the BLHeli firmware. In the terrestrial mode, the front pair of cyclorotors are reversed such that a pure throttle input will cause the vehicle to translate forward (Fig. 15). Providing an additional pitch input, which differentially modulates the fore and aft motor RPMs, while maintaining the throttle enables the vehicle to achieve precise positioning. In addition, the quad-cyclocopter can translate backwards using a combination of pitch input and reversed motor direction. Differentially varying the rotational speeds of the side pairs enables the vehicle to turn with a small radius.

Aerial Mode

Figure 16 provides an overview of the control strategy used in hover. Pitch control is provided by modulating the RPM of the fore and aft pairs of cyclorotor 16(a), roll control is achieved by differentially varying rotational speeds of the side pairs 16(b), and finally yaw control is possible by differentially rotating the thrust vectors of the side pairs 16(c). It is important to note that both pitch and roll can potentially be adjusted using the thrust vectors. However the thrust vectors are only used for yaw control and have been adjusted prior to flight testing in benchtop experiments such that each cyclorotor provides maximum lift at 1600 RPM. One advantage of the quad-cyclocopter configuration over the twin-cyclocopter is since the angular momentum of cyclorotors are balanced, the vehicle is not affected by gyroscopic coupling. Therefore, mixing actuator inputs between the two operational modes only involves disabling the feedback control system and reversing the front pair of cyclorotors in terrestrial mode.

AERIAL AND TERRESTRIAL LOCOMOTION TESTING RESULTS

During the flight tests, the proportional and derivative gains were tuned using the Ziegler-Nichols approach. The gains that offered acceptable stiffness and damping to reject external disturbance with minimal oscillations were selected. Gains and control inputs for trimmed flight varied slightly from the ones obtained through gimbal testing because the gimbal setup provided additional damping in pitch, roll, and yaw. Additional variation was caused by shifts in the moments of inertia because the cyclocopter is equipped with a single 3S 11.1 volt 1300 mAh Lithium-Polymer battery weighing 100 grams.

Flight tests were systematically conducted by first providing a pure throttle command and observing the vehicle response. Any forward translation was counteracted by either reducing the phasing of the cyclic blade pitch or varying the motor RPM. Initial experiments were used to individually test either aerial or terrestrial modes. The vehicle has successfully demonstrated stable hover (Fig. 17). As seen from Fig. 18, a sustained hover is obtained with the pitch and roll angles held close to zero and minimum offsets in yaw rate. Further experiments will be conducted demonstrating both modes during a single operation.

The quad-cyclocopter achieves terrestrial locomotion at a lower operational RPM in hover, resulting in 84% lower power consumption. While throttle input is used to provide maximum forward translational velocity, position can also be further adjusted by rotating a single pair of cyclorotors. The reversible ESCs permit the forward pair of the cyclorotors to rotate in any direction, enabling the vehicle to move backwards without turning. Further tests will be conducted to determine the maximum speed and endurance of the quad-cyclocopter in terrestrial mode.

CONCLUSION

The objective of the present work was to develop a cyclocopter capable of efficient aerial and terrestrial locomotion. Unlike multi-terrain-capable quadcopter platforms developed in the past, the cyclocopter solely relied on its four cyclorotors to achieve both modes of locomotion. During flight, independent control of four rotational speeds and thrust vectors were used to efficiently navigate and maintain stable hover. During operation on land, the quad-cyclocopter maneuvered using differential motor RPM. The main challenge in developing the quad-cyclocopter was overcoming the high empty-weight fraction penalty in order to achieve stable hover. Therefore, emphasis was placed on blade aerodynamic and structural design to achieve maximum rotor performance, integrating wheels used in terrestrial locomotion with cyclorotor endplates, and designing lightweight landing gear system that enabled smooth transition between the operational modes. The

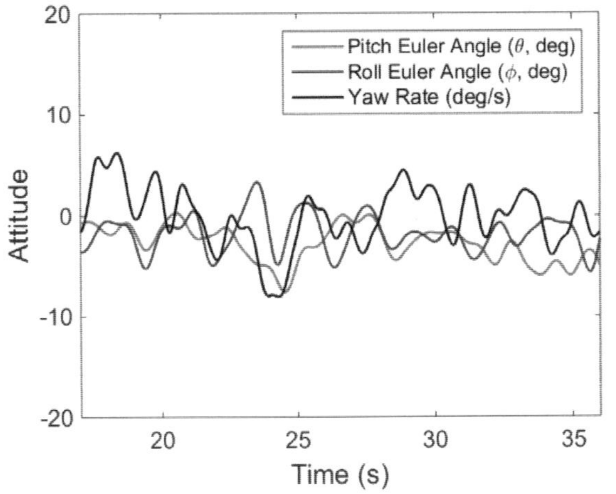

Fig. 18. Onboard attitude data streamed from ELKA-R during hover.

following is the summary and some of the specific conclusions drawn from this study:

1. The 900 gram quad-cyclocopter successfully demonstrated stable, sustained hover and terrestrial locomotion. The vehicle solely relied on its four cyclorotors, using a combination of rotational speed and thrust vectoring to achieve both modes of locomotion. Transition between operational modes are enabled using a retractable landing gear system and carbon fiber wheels that are integrated with the cyclorotor endplates. The 0.23 meters (9.25 inches) diameter wheels are directly controlled by motor RPM through a 7.5:1 single-stage gear transmission.

2. The optimized cyclorotor on the quad-cyclocopter uses four blades, symmetric 45° pitching amplitude, c/R ratio of 0.67, and pitching axis location at 25% of blade chord. Along with other structural components, the optimized cyclorotor weighs 83 grams and has a thrust-to-weight (T/W) ratio of approximately 3.25 at the operating rotational speed of 1600 RPM. Overall, the cyclorotors contribute up to 37% of total gross take-off weight.

3. Terrestrial locomotion requires 44 Watts compared to 280 Watts, which is a 84% reduction in power consumption. Bench-top experiments showed that the wheels operate at about 200 RPM compared to the 1600 RPM required in hover.

4. Because both the fore and aft pairs of cyclorotors rotate in opposite directions to balance the resulting torque, direction of rotation of the aft pair must be reversed in order to achieve translational motion in terrestrial mode. The direction of rotation is reversed using a programmable electronic speed controller (ESC). The quad-cyclocopter uses the Afro 12A multi-rotor ESC reprogrammed with the BLHeli firmware. In the terrestrial mode, the front pair of cyclorotors are reversed such that a pure throttle input will cause the vehicle to translate forward.

5. One advantage of the quad-cyclocopter configuration over the twin-cyclocopter is because the angular momentum of cyclorotors are balanced, the vehicle is not affected by gyroscopic coupling. Therefore, mixing actuator inputs between the two operational modes only involved disabling the feedback control system and reversing the front pair of cyclorotors in terrestrial mode.

ACKNOWLEDGMENT

This research was supported by the Army's MAST CTA Center for Microsystem Mechanics with Dr. Brett Piekarski (ARL) and Mr. Chris Kroninger (ARL-VTD) as Technical Monitors. Fellowship for the main author was provided by the National Science Foundation Graduate Research Fellowships Program (NSF GRFP) under Grant No. DGE 1322106. Authors would also like to acknowledge Brian Davis for his assistance with the design and manufacturing process, and Dr. Derrick Yeo for piloting the vehicle.

REFERENCES

[1]Ramasamy, M., Leishman, J., and Lee, T., "Flowfield of a Rotating-Wing Micro Air Vehicle," *Journal of Aircraft*, Vol. 44, No. 4, 2007, pp. 1236-1244.

[2]Kawasaki K., Zhao M., Okada K., Inaba M., "MUWA: Multi-field universal wheel for air-land vehicle with quad variable-pitch propellers," Proceedings of the 26th IEEE/RSJ International Conference on Intelligent Robots and Systems: New Horizon, 2013, Tokyo, Japan.

[3]Kalantari A. and Spenko M., "Design and Experimental Validation of HyTAQ, a Hybrid Terrestrial and Aerial Quadrotor," Proceedings of the IEEE International Conference on Robotics and Automation, 2013, Karlsruhe, Germany.

[4]Benedict, M., Jarugumilli, T., and Chopra, I., "Effect of Rotor Geometry and Blade Kinematics on Cycloidal Rotor Hover Performance," *Journal of Aircraft*, Vol. 50, No. 5, 2013, pp. 1340–1352.

[5]Benedict, M., Jarugumilli, T., and Chopra, I., "Experimental Optimization of MAV-Scale Cycloidal Rotor Performance," *Journal of the American Helicopter Society*, Vol. 56, No. 2, April 2011, pp. 022005-1–022005-11.

[6]Jarugumilli, T., Benedict, M. and Chopra, I., "Wind Tunnel Studies on a Micro Air Vehicle-Scale Cycloidal Rotor," *Journal of the American Helicopter Society*, Vol. 59, No. 2, April 2014, pp. 02208 (1–10).

[7]Shrestha, E., Hrishikeshavan, V., Benedict, M., Yeo, D., and Chopra, I.,"Development of Control Strategies for a Twin-Cyclocopter in Forward Flight," *Journal of the American Helicopter Society*, Vol. 61, No. 4, October 2016, pp. 1–9.

[8]Shrestha, E., Yeo, D., Hrishikeshavan, V., Benedict, M., and Chopra, I., "Gust Disturbance Rejection Study of a Cyclocopter Micro Air Vehicle," Proceedings of the 72nd Annual National Forum of the American Helicopter Society, West Palm Beach, Florida, May 17–19, 2016.

[9]Benedict, M., "Fundamental Understanding of the Cycloidal-Rotor Concept for Micro Air Vehicle Applications," Ph.D Thesis, Department of Aerospace Engineering, University of Maryland College Park, December 2010.

[10]Benedict, M., Jarugumilli, T., Lakashminaryayan, V. and Chopra, I., "Effect of Flow Curvature on Forward Flight Performance of a Micro-Air-Vehicle-Scale Cycloidal-Rotor," *AIAA Journal*, Vol. 52, No. 6, June 2014, pp. 1159-1169.

[11]Migliore, P.G., Wolfe, W.P., and Fanuccif, J.B., "Flow Curvature Effects on Darrieus Turbine Blade Aerodynamics," *Journal of Energy*, Vol. 4, (2), 1980, pp. 49–55.

[12]Benedict, M., Shrestha, E., Hrishikeshavan, V., and Chopra, I., "Development of Micro Twin-Rotor Cyclocopters Capable of Autonomous Hover" *Journal of Aircraft*, Vol. 51, (2), March–April 2014, pp.672–676.

[13]Benedict, M., Gupta, R., and Chopra, I., "Design, Development, and Open-Loop Flight Testing of a Twin-Rotor Cyclocopter Micro Air Vehicle," *Journal of the American Helicopter Society*, Vol. 58, No. 4, October 2013, pp. 1–10.

[14]Zachary, H., A., Benedict, M., Hrishikeshavan, V., and Chopra, I., "Design, Development, and Flight Test of a Small-Scale Cyclogyro UAV Utilizing a Novel Cam-Based Passive Blade Pitching Mechanism," *International Journal of Micro Air Vehicles*, Vol. 5, No. 2, June 2013, pp. 145–162.

[15]Hrishikeshavan, V., Benedict, M., and Chopra, I., "Identification of Flight Dynamics of a Cyclocopter Micro Air Vehicle in Hover," *Journal of Aircraft*, Vol. 52, No. 1, 2015, pp. 116–129.

[16]Hrishikeshavan, V., and Chopra, I., Refined Lightweight Inertial Navigation System for Micro Air Vehicle Applications, AHS Specialists Meeting on Unmanned Rotorcraft and Network-centric Operations, Chandler, AZ, Jan 20-22, 2015.

[17]Georgy, J., Noureldin, A., Korenberg, M., and Bayoumi, M., "Modeling the Stochastic Drift of a MEMS-Based Gyroscope in Gyro/Odometer/GPS Integrated Navigation," IEEE Transactions on Intelligent Transportation Systems, Vol. 11, (4), Dec 2010, pp. 856–872.

[18]Y. K. Thong, M. S. Woolfson, J. A. Crowe, B. R. Hayes-Gill, and R. E. Challis, "Dependence of inertial measurements of distance on accelerometer noise," Meas. Sci. Technol., Vol. 13 , (8), pp.1163–1172 , 2002.

[19]Hrishikeshavan V., and Chopra I., "Performance, Flight Testing of Shrouded Rotor Micro Air Vehicle in Edgewise Gusts," *Journal of Aircraft*. Vol. 49, pp.193–205

Development of a Cantilevered Rotor-Based Meso-Cyclocopter

Carl Runco
Graduate Research Assistant

David Coleman
Graduate Research Assistant

Moble Benedict
Assistant Professor

Advanced Vertical Flight Laboratory, Department of Aerospace Engineering
Texas A&M University, College Station, TX, 77843

ABSTRACT

This paper describes the design, development and flight testing of a meso-scale cyclocopter. Weighing only 29 grams, the present vehicle is the smallest cycloidal rotor based aircraft ever built. Unlike the previous cyclocopters, the current prototype utilizes a novel, light weight (3 grams) cycloidal rotor design, with cantilevered blades, having semi-elliptical planform shape and no exposed rotor shaft. To minimize bending deflections the blades use a unique, lightweight (0.15 grams each) but high strength-to-weight ratio unidirectional carbon-fiber based structural design and are fabricated using a specialized manufacturing process. The cycloidal rotor design was chosen through systematic performance measurements conducted using a custom-built miniature three-component force balance. Based on experimental parametric studies, a 4-bladed rotor and symmetric blade kinematics with pitch amplitude of 45° provided the highest thrust and power loading (thrust/power) and was used in the final rotor design. The airframe is fabricated using a combination of carbon-fiber and state-of-the-art 3D printing techniques. The attitude control strategy utilizes a combination of rpm-control of the two cycloidal-rotors/tail-rotor and thrust vectoring of the cycloidal rotors. The control strategy is implemented on a custom-built 1.3 gram autopilot, which uses a closed-loop proportional-derivative controller for hover stability. The vehicle has been systematically flight tested by tuning the feedback gains and has demonstrated stable hovering flight.

INTRODUCTION

Over the past decade there has been increased interest in developing different types of micro-air vehicles (MAVs), largely because of the potential mission capabilities associated with small-scale flying devices. Such miniature, lightweight vehicles offer superior portability, rapid deployment capability, and feature a low radar cross section. MAVs can further support tactical military operations or explore environments that would otherwise be too dangerous for a human, such as surveillance and reconnaissance in the battlefield, biochemical sensing, fire and rescue operations, border surveillance, and traffic monitoring. With their reduced noise signatures and lower production costs, MAVs are poised to effectively accomplish any and all of these types of tasks. However, these missions often require flying conditions outside of the flight capabilities of current systems and factors such as gusts, confined spaces, and increased flight times are prohibitive barriers to most MAVs in use today. To overcome these difficulties, flying platforms need to be developed that can navigate through these scenarios and must be designed to exhibit high gust tolerance and the ability to operate in confined spaces. To date, many small, fixed-wing MAVs have been developed, however these types of flying devices are unsuitable for missions which require hovering/low-speed flight or tight turns and other demanding maneuvers. Rotorcraft, on the other hand, are much more suited for these flight regimes, with most vehicles employing single, coaxial, or multi-rotor configurations. The primary aerodynamic limitation of conventional rotor blade designs for use in MAVs is the reduced efficiencies at low Reynold's numbers (10,000–50,000), in particular, low values of maximum figure of merit of around 0.65 (Ref. 1,2). The reduced aerodynamic performance is caused by the large values of profile drag associated with thick boundary layer formations on the blades, large induced losses, and higher rotational and turbulent losses in the downstream wake of the rotating blades (Ref 3, 4). This demands that a more efficient, hover-capable propulsion method be investigated that can harness the potential of low Reynold's number unsteady aerodynamics, while simultaneously providing the maneuverability and gust tolerance required of MAVs.

A promising solution to overcoming the deficiencies associated with standard rotors is the cycloidal rotor, or cyclorotor. The cycloidal rotor is a revolutionary vertical take-off and landing (VTOL) propulsion concept, and consists of several blades that rotate about a horizontal axis with the blade span parallel to the axis of rotation (Fig. 1). The pitch angle of each of the blades is varied periodically as the blade moves around the azimuth of the rotor such that the blade is at a positive geometric angle of attack both at the top and bottom halves of its circular trajectory. The blade kinematics and aerodynamic forces on a functioning cycloidal rotor are shown schematically in Fig. 2. Here, θ is the pitch angle, ϕ represents the direction of resultant thrust vector from vertical, and Ω is the rotational speed of the rotor. By varying both the rotational speed and the cyclic pitch phasing, the magnitude and direction of the net thrust vector of the rotor can be carefully controlled.

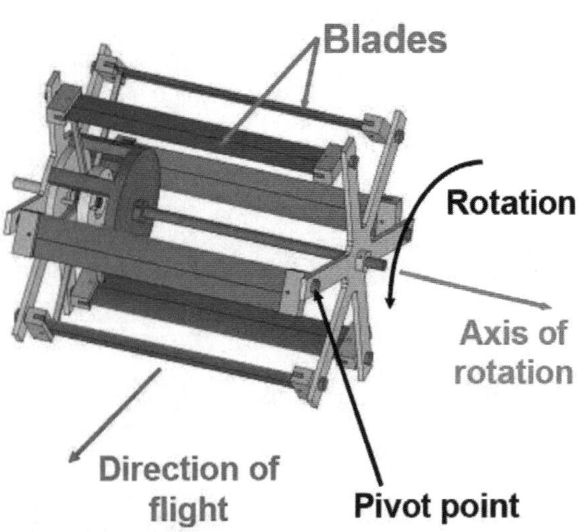

Figure 1. Cycloidal rotor model with key components identified.

There are several advantages of the cycloidal rotor over conventional rotors, one of the most significant of them being the fact that all the spanwise elements of the blade operate at the same aerodynamic conditions, which allows all the blade elements along the blade span to be set at optimum condition. This is in contrast with conventional rotors, in which each blade element along the span experiences a different flow velocity, Reynold's number, and angle of attack. Recent experiments conducted at the University of Maryland by the co-author have also shown that a cycloidal rotor may be more aerodynamically efficient in terms of power loading (thrust per unit power) than a conventional rotor of the same scale and similar actuator area (Ref. 5, 6). Additionally, a cycloidal rotor is able to obtain the required thrust at a significantly lower rotational speed than an equivalent conventional rotor, and therefore has a reduced acoustic signature, making it further

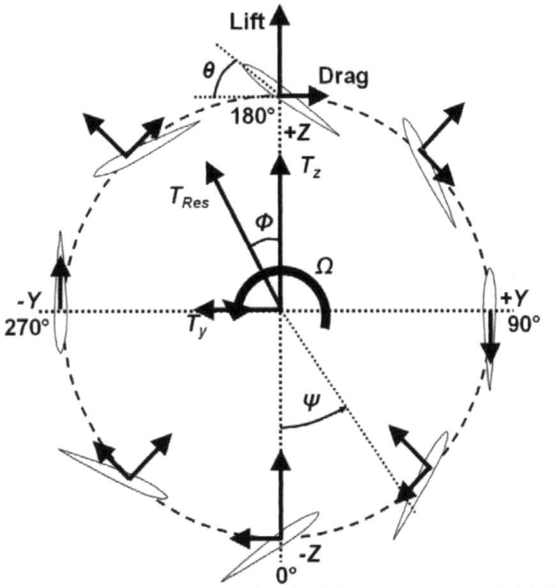

Figure 2. Blade kinematics and forces on a cycloidal rotor in hover.

advantageous for indoor reconnaissance missions. Furthermore, since the blades are cyclically pitched once per revolution (1/rev), unsteady flow mechanisms may delay blade stall onset and in turn may augment the lift produced by the blades. Finally, since the thrust vector of a cycloidal rotor can be instantaneously set to any direction perpendicular to the rotational axis, the concept may also have better maneuverability compared to a conventional rotor based MAV, further demonstrating its usefulness in highly constrained applications and indoor operations.

The concept of the cycloidal rotor has been around for nearly a century, and many attempts, predominantly at full-scales, have been made to develop a cycloidal rotor based aircraft (Ref. 7-9). However, none of these vehicles were able to achieve flight because of the lower rotor performance (due to lack of understanding of the physics of the problem) and the large structural weight of the full-scale cycloidal rotors. Note that the cycloidal rotor might not be suited to full-scale aircraft applications anyway, and instead is more appropriate for MAVs where the design benefits from the inherent increase in strength-to-weight ratio as structures are scaled down. Now, with the advent of lightweight materials and electronics, improved manufacturing methods, and the development of high-efficiency motors, the cycloidal rotor concept has finally flown recently, proving the technique is possible with MAV-scale rotorcraft (Ref. 10-12). These vehicles, a sampling of which is shown in Fig. 3, have been developed and systematically studied by the co-author in order to optimize the performance of an MAV-scale cycloidal rotor (Ref. 11, 12). The experimental and computational studies have included detailed performance measurements, flowfield studies using Particle Image Velocimetry (PIV) (Ref. 10, 13-16), and the development and validation of an aeroelastic model to predict the hover performance of a cyclorotor (Ref. 17). These results have veritably shown that the cyclorotor is a viable propulsion device for hover-capable MAVs.

The previous vehicles developed have been in the 200-800 gram range, which is far outside the envisioned weight range of the initial MAV concept, which was approximately 100 grams gross take-off weight (Ref. 18). The focus of the present work, therefore, is to push the limits of current state-of-the-art technologies to design, build, and flight test the world's smallest cyclocopter, targeted to be an order of magnitude smaller than smallest cyclocopter in existence today. The focus of the present study, therefore, is to develop and experimentally validate a meso-scale cycloidal rotor aircraft capable of greater maneuverability, stealth capability, and portability, beyond that offered by the larger-scale cyclocopters. This meso-scale cyclocopter, envisioned to be a versatile hand-held flying device, is shown conceptually in Fig. 4. Developing such an unconventional vehicle at these small scales is a brand new challenge not only from a vehicle design/construction standpoint, but also because it requires more fundamental research to be conducted to understand the 3-dimensonal unsteady aerodynamics (due to the low aspect ratio of the blades and

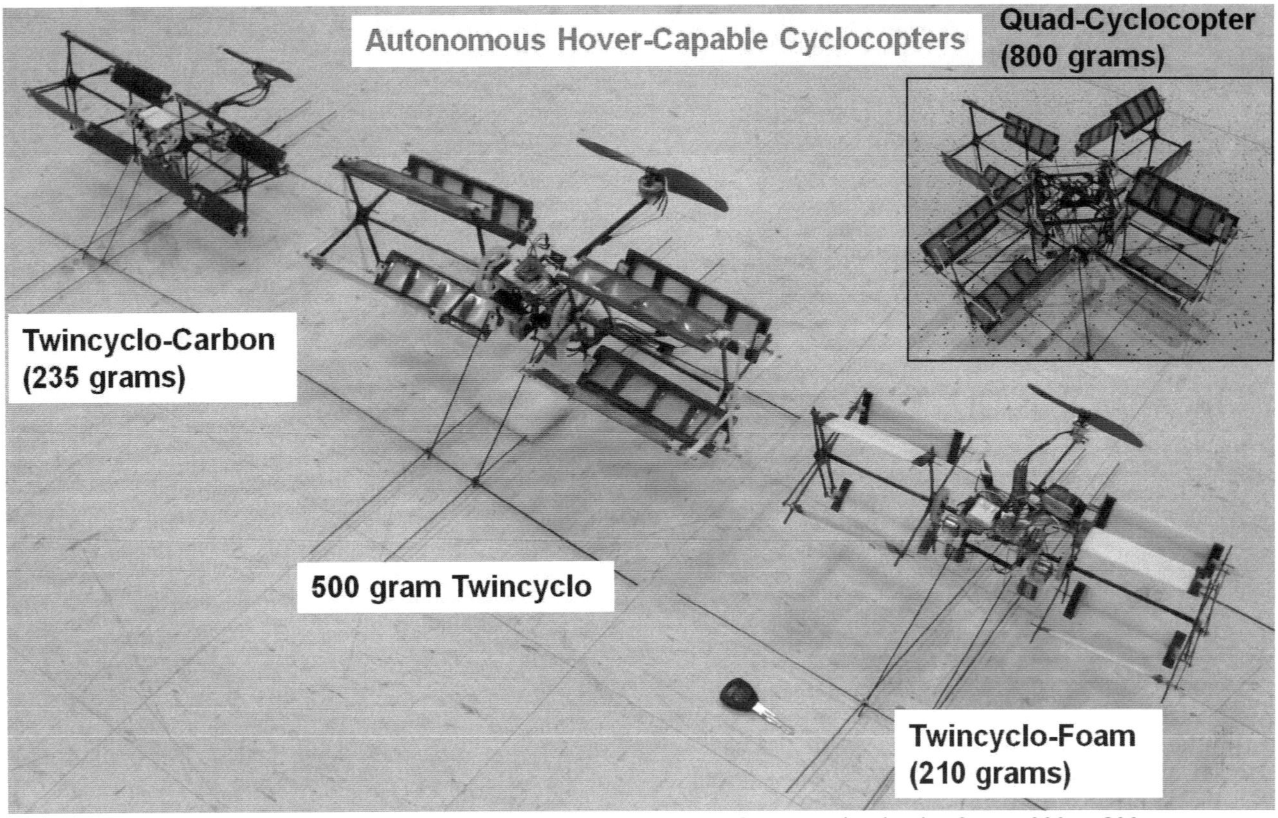

Figure 3. Hover-capable cyclocopters developed by the co-author, ranging in size from ~200 to 800 grams.

high reduced frequencies) at ultralow Reynolds numbers experienced by the blades. Also, at small scales, owing to the lower inertia, vehicle dynamics tends to be much faster requiring high-bandwidth closed-loop control with very minimal actuator delays.

VEHIICLE CONFIGURATION

The current vehicle has a pair of miniature 1-inch radius cycloidal rotors, each weighing 3 grams total and capable of producing more than 10 grams of thrust, shown in Fig. 5. The current vehicle is shown in Fig. 6 with key components labeled. In this configuration, the vehicle is called a twin-cyclocopter. A conventional tail rotor and motor are incorporated in order to counteract the nose-up reaction pitching moment produced by the two cyclorotors rotating in the same direction, as well as provide an additional amount of lift for the vehicle. The rotational speed of the two cycloidal rotors and the tail rotor, combined with the ability to vector the thrust of the cyclorotors, gives the vehicle, in its current configuration, 5 control degrees of freedom (Ref. 11, 12), which results in the cyclocopter having greater actuation potential than a typically under-actuated system such as a quad-rotor. This means that the cyclocopter will be able to command instantaneous accelerations in more directions than a quad-rotor; therefore, the control authority and superior maneuverability have to be systematically investigated in order to fully reap the benefits and superior flying qualities offered by this type of thrust generating system. An airframe was developed using 3D printed parts and carbon fiber to integrate the rotors, tail prop and electronics into a single unit, resulting in a vehicle weighing 29 grams. A weight breakdown of the subsystems of the vehicle is provided in Table 1. The final vehicle is a twin meso-scale cyclocopter capable of both hovering and forward flight.

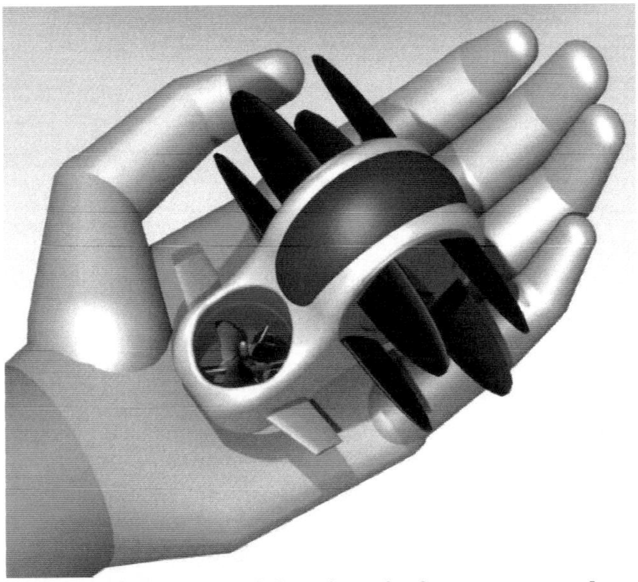

Figure 4. Conceptual drawing of a future meso-scale cyclocopter.

The remainder of this paper outlines the detailed design and experimental testing of the cyclorotors to evaluate the performance and choose an optimal blade and rotor design,

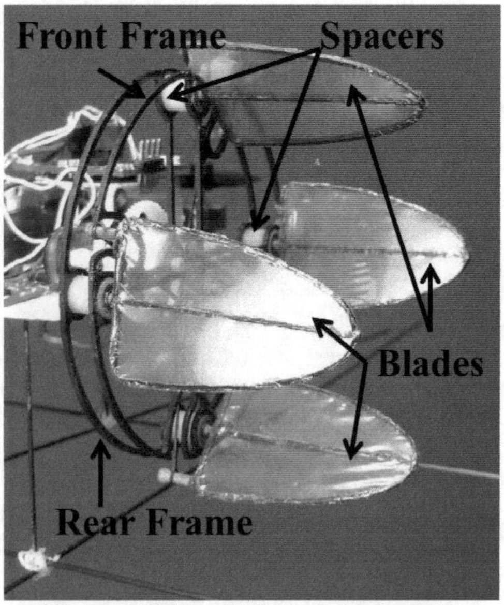

Figure 5. Meso-scale cycloidal rotor.

which can generate the required thrust efficiently. The manufacturing process, developed specifically for creating strong, lightweight blades and rotor is explained in detail. Next, the vehicle control and stabilization techniques are described, as well as the vehicle telemetry and flight data collection. Finally, flight tests were conducted in hover, and the results are presented which show the vehicle in relatively stable hovering flight, demonstrating successful vehicle operation.

ROTOR AND PITCHING MECHANISM DESIGN

As stated previously, designing a light-weight cyclorotor with simplified blade-pitch mechanism is a challenge at small scales. Using knowledge gained from previous studies, a light-weight cyclorotor was designed that is different from any previously built cyclorotor. The rotor (shown in Fig. 5) uses cantilevered blades with a semi-elliptical planform shape to reduce rotor weight (by eliminating the shaft and the additional tip end-plate) and improve aerodynamic efficiency (by reducing induced drag). As a baseline design, a rotor radius of only one inch was chosen in order to keep

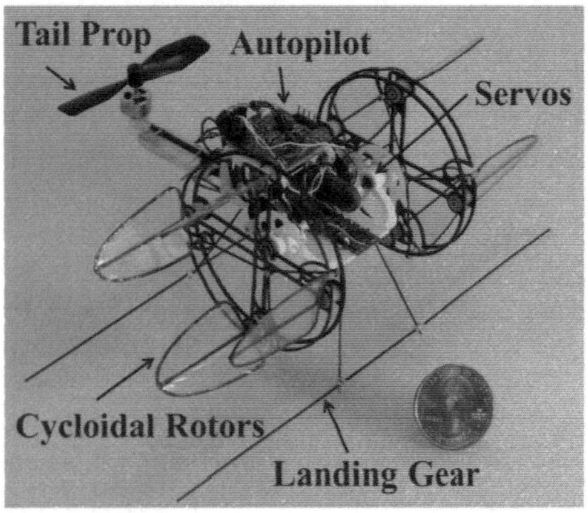

Figure 6. 29 gram meso-scale twin-cyclocopter.

Component	Weight (g)	Total (%)
Motors + transmission	7	24
Cyclorotors (combined)	6	21
Structure + wires	5	18
Li-Po Batteries	4.9	17
Electronics	3.7	13
Tail rotor + motor	2.4	8
Total	29	100

Table 1. Component Weight Breakdown of Cyclocopter

the size and weight of the rotor low, as well as reduce the overall cross-sectional profile of the craft and thus fit well within the targeted vehicle size. The construction of rotors this small was challenging because they had to be lightweight, structurally stiff, and must accommodate the pitching mechanism. Because the blades on this current vehicle are cantilevered, it was not possible to simply scale down the structure of the rotor from previously built cyclocopters (shown in Fig. 3) and adopt it for use in this design. For this reason, a single end-plate-based rotor was developed, which was designed to resist flexing under the centrifugal loads experienced during high-speed rotation. As shown in Fig. 5, the endplate was made of two separated carbon fiber frames connected with spacers made from Delrin® which also house the two root pitch bearings for each blade. This resulted in a lightweight end-plate with high stiffness. The carbon fiber frames are made out of 1/32" thick laminate. It has been shown that deflection of the blades at the root can reduce thrust by up to 40 percent; therefore blade-root stiffness is equally critical and the carbon fiber frames were made thicker at the attachment points to provide the necessary support (Ref. 19).

The cycloidal rotor utilizes a passive blade pitching mechanism which is based on a 4-bar linkage system where the blade pitching is kinematically coupled to the rotation. More details on the operation of the blade pitching mechanism can be found in Ref. 11. The pitching mechanism presented its own set of unique challenges due to its small size. Figure 7 shows an exploded view of the current rotor and pitching mechanism with several of the tiny, intricate parts labeled accordingly. One of the important considerations in any cycloidal rotor configuration is the fact that the pitch linkages must be hinged to the pitch offset beyond the end of the main rotor shaft in order to avoid interfering with it. For this cantilevered rotor design, the blades were mounted on the front carbon fiber frame, resulting in a root offset for each of the blades from the end of the main rotor shaft by the Delrin® spacer (see Fig 5). This allows the pitching linkages to run in the space between the carbon fiber frames directly from the end of the main rotor shaft to the blades. By strategically positioning the pitching links between the frames in this manner, they are protected during vehicle crashes. The pitch links are made of

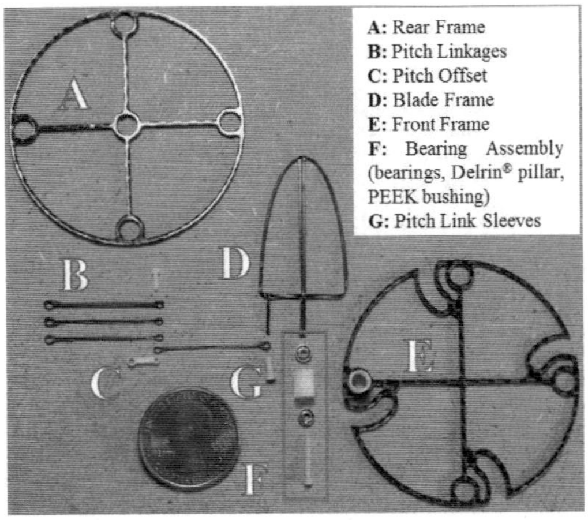

Figure 7. Exploded view of the rotor, pitching mechanism and blades with parts labeled accordingly.

unidirectional carbon fiber prepreg. They are manufactured using a Teflon™ base mold and a silicone mat to compress the fibers into the mold, thus achieving the desired shape. This process, outlined in the subsequent sections on blade design, allows the production of parts of consistent quality that are too delicate to be fabricated with other methods. The resulting linkages weigh only 0.01 grams each, and, since all of the carbon fibers are aligned in the direction of loading, they are very strong in that direction. The rest of the parts in the pitch mechanism, such as the bushings that interface between the blade frames and blade bearings and allow the blades to freely pitch about the pitching axis, are milled from PEEK plastic (for abrasion resistance). The central pitch offset is machined out of Delrin® and is glued onto the shaft with a very strong, plastic-bonding cyanoacrylate adhesive. The pitch link sleeves are fastened to the ends of the blade frames, creating a hinge between the blades and pitch links which allow free and smooth rotation of the pitching mechanism. A pair of carbon fiber rods 0.7 mm in diameter was attached to the root chord of the blade, and serves as the pitching axis and pitch link pegs. The fully assembled rotor is shown in Figure 8.

BLADE DESIGN AND MANUFACTURING

The hallmark of the current cycloidal rotor is the cantilevered blades, a design feature possible only at smaller scales due to the increase in strength-to-weight ratio as the structure is scaled down. The blade design is biologically inspired, utilizing a symmetrically pitched, flat-plate airfoil. This design was chosen because of the low operating Reynolds number of approximately 11,000, and because constructing an airfoil at these scales is too impractical to implement. For the blade geometry, a semi-elliptical planform was selected because it is known that this shape improves aerodynamic efficiency. Manufacturing these blades proved to be one of the most difficult tasks, and required developing a specialized fabrication technique that ensured consistency and reproducibility in creating strong, lightweight blades. The method developed to accomplish this was to employ a layup technique using a Teflon™ mold and silicone compress which resulted in the successful fabrication of stiff, light-weight blades. This was the same process used to manufacture the pitch linkages, and ensured that the unidirectional carbon fiber frame would bond sufficiently and securely to the pitch axis and pitch link pegs. The Teflon™ material was chosen because it will not bond with the carbon fiber when heat treated, is heat resistant, and is strong enough to maintain the mold shape under the clamping pressure. Figure 9 shows the step-by-step process employed to produce the blades using this method.

First, a strip of unidirectional carbon fiber is laid into the Teflon™ mold so that it makes a complete loop (Fig. 9a). Then 0.7 mm diameter carbon fiber rods for the pitch axis and pitch link pegs are laid in their respective slots. The pitching axis of each of the blades is carefully positioned at exactly the chordwise center of gravity in order to reduce the pitching moment on the blades due to centrifugal loads (Fig. 9b). After that, another complete loop around the mold is made with the unidirectional strip so that the rods are sandwiched between layers of it (Fig. 9c). This is then layered, as shown in Fig. 9d, and clamped tightly together. The aluminum plates evenly apply the clamping pressure. The flexible, heat-resistant silicone sheet transfers the pressure to the carbon fiber and pushes it into the mold without developing damaging pressure points. The anti-stick/releasing film prevents the carbon fiber from bonding with the silicone. When the full mold layup is complete (Fig. 9e), it is cured at 350° F for 135 minutes. After the curing is finished and the cured epoxy and frame have cooled to room temperature, a 3 micrometer thick Mylar sheet is added to both sides of the now complete frame, effectively creating a closed surface within the frame shape. This is done by applying contact cement in a very thin layer to the frame and the Mylar sheet is stretched over it. A heating iron is used to remove any imperfections by stretching out the Mylar. The resulting blades weigh less than 0.2 grams, and have a maximum chord of about 0.8 inches and a span of 1.3 inches. These blades are currently produced only two at a time, but the quality, strength, and weight are consistent and therefore this method is used for manufacturing all new blades.

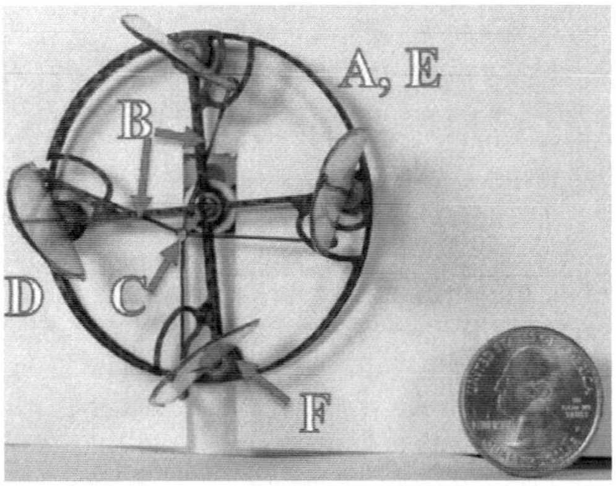

Figure 8. Fully assembled rotor for the twin micro-cyclocopter.

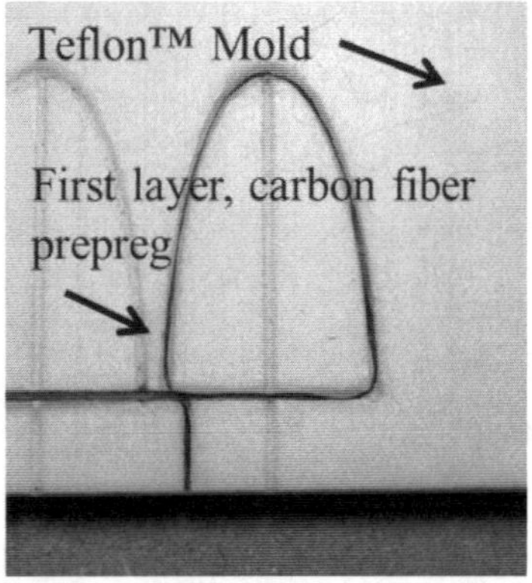

a) Looping the unidirectional carbon fiber around the blade mold

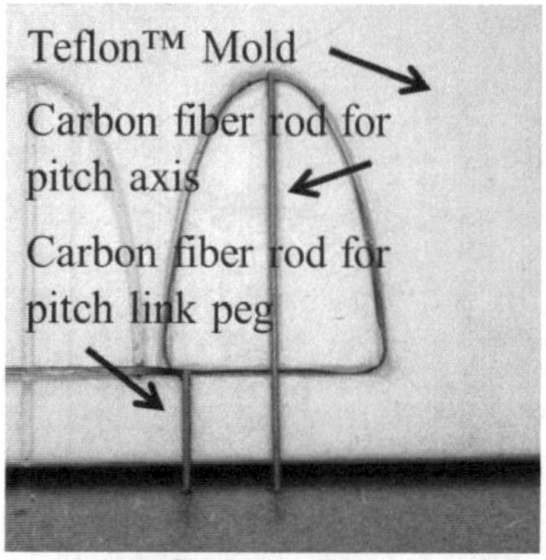

b) Inserting the carbon fiber rods for the pitch axis and pitch link pegs

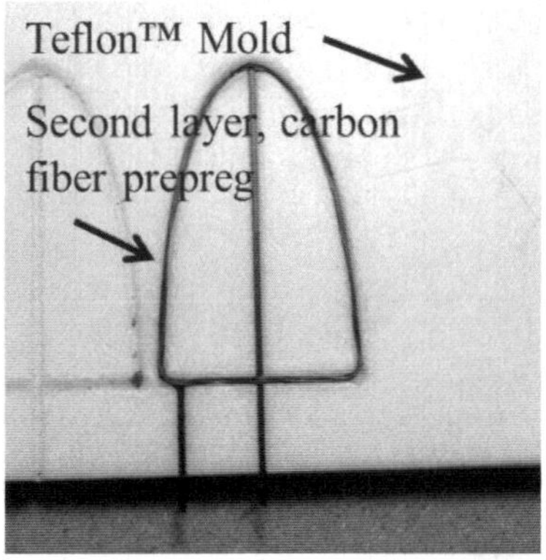

c) Looping the second layer around the mold

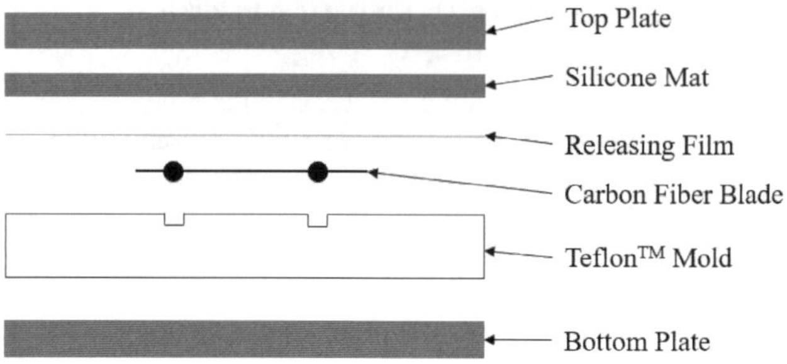

d) Schematic showing constitutive layers of the blade layup

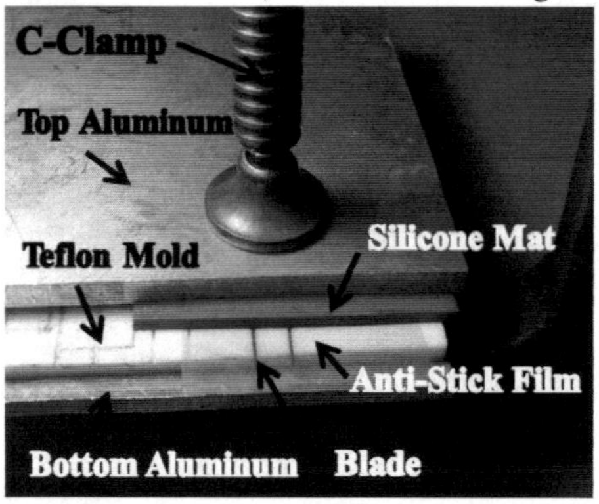

e) Layup assembled in C-clamp and ready for curing

Figure 9. Multi-step process for manufacturing cycloidal rotor blades.

ROTOR PERFORMANCE

In order to maximize the thrust and aerodynamic efficiency of the rotor design, a series of systematic experiments was conducted. These tests were to compare the performance of various rotor configurations. The pitch amplitude of the blade during the rotation of the rotor was varied from ±35° to ±50° in increments of 5° for blades of different size. The goal of this study was to obtain the blade size, pitch amplitude and number of blades that produces the highest thrust at a fixed rotational speed and best thrust/power ratio (or power loading) at the operating thrust. To accurately measure rotor performance (vertical force, sideways force, and power), a 3-component force balance was constructed and systematically calibrated to ensure accurate results. This new miniature 3-component force balance was used to measure the small forces produced by the 3 gram cycloidal rotors developed for the present vehicle. The complete force balance is shown in Fig. 10, with key components labeled. As can be seen, the balance is composed of force load cells that measure the two orthogonal components of thrust (T_Z and T_Y), and a reaction torque sensor that measures the torque generated by the rotor. In addition, a Hall-effect switch is used to measure the RPM of the rotor in real time. Mechanical power is calculated from the torque and rpm measurements.

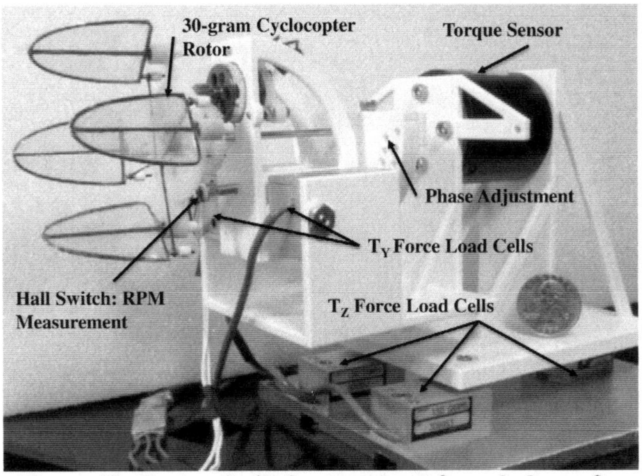

Figure 10. Custom built 3-component force balance for the meso-scale cyclocopter rotor.

The results from the experimental parametric studies using the 3-component balance are presented in the graphs below. The first step was to analyze the effect of blade size on rotor performance. The goal of comparing these different blade sizes was to explore which size would provide the maximum thrust for the lowest weight penalty. To this end, a set of three blades with identical planform, but scaled to different sizes, was built and tested. These blades were called Wing 1, Wing 2, and Wing 3, and are shown in Fig. 11. After testing, it was proven that Wing 3 was too heavy and therefore unable to withstand the inertial loads and it suffered from large deflections at the required RPM, so only limited data was collected at low rotational speeds. Wing 1, on the other hand, was lightweight enough to rotate at 4000 RPM, but was too small to generate sufficient thrust. Wing 2, however, was able to produce the necessary thrust at the targeted RPM, and was subsequently chosen as the final design. Fig. 12 plots the variation in thrust with rotational speed for Wing 1 and Wing 2, demonstrating that the larger wing can produce the required thrust at a reasonable rotational speed.

The rest of the studies were conducted using Wing 2 while changing other rotor parameters. The next tests were conducted to analyze the effects of maximum blade pitch amplitude. A plot showing the variation in thrust with rotational speed for the 4-bladed pitch amplitude sweep of 35° to 50° is shown in Fig. 13. This graph conclusively shows that 45° maximum pitch amplitude generates the highest thrust for the 4-bladed rotor. These results are consistent with the previous findings that have found that

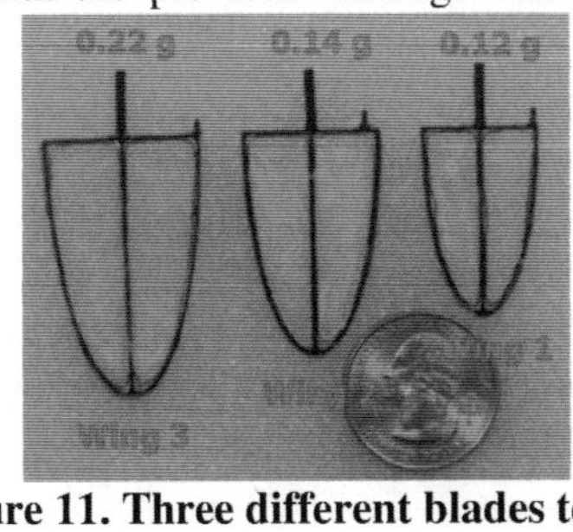

Figure 11. Three different blades tested.

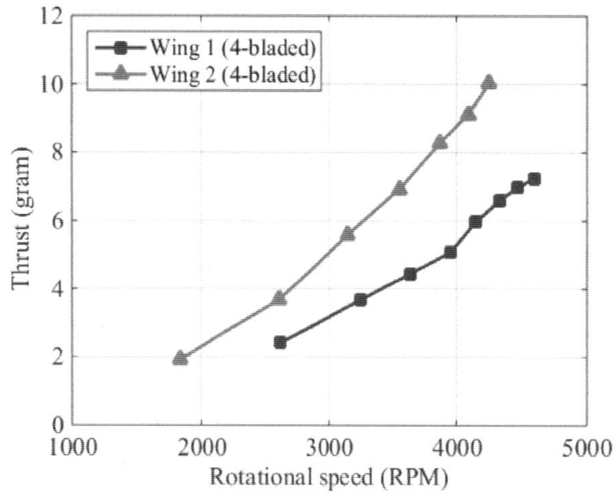

Figure 12. Thrust vs. rpm for two different wing sizes on a 4-bladed cycloidal rotor.

large pitch amplitudes (40°–45°) produce the greatest amount of thrust for 4-bladed cyclorotors (Ref. 19). Of equal importance is the thrust to power ratio (or power loading) plotted in Fig. 14, which is a quantitative metric describing the aerodynamic efficiency of the rotor. Again, the 45° pitch amplitude demonstrates superior performance relative to the other pitch amplitudes. This same pitch sweep was repeated for the 2-bladed case and the results are shown in Figs. 15 and 16. The results of this test show that, for the 2-bladed rotor, the maximum thrust (Fig. 15) and power loading (Fig. 16) are produced between 40° and 45° pitch amplitude. Based on these results, we selected 45° as the optimal pitch angle and then varied the number of blades of the cyclorotor. The next step was to understand the effect of the number of blades on performance. Figures 17 and 18 show thrust and power loading, respectively, for 2, 3, 4, and 6-bladed rotors at 45° pitch amplitude. This data shows that the 6-bladed rotor produces the most thrust. However, the extra blades add weight to the rotor and as seen from Fig. 18, the 4-bladed one is more efficient for an equivalent level of thrust. Therefore, the final cyclorotor configuration chosen for the present vehicle is 4-bladed with a pitch amplitude of 45°.

It is interesting to see that on a 4-bladed rotor the 45° pitch amplitude had significantly higher performance than the 40° case; however, on a 2-bladed rotor both 45° and 40° had comparable performance. This is because the pitch amplitude alone does not fully reflect the aerodynamic conditions experienced by the blade. It is important to bring in the inflow through the rotor in each of the cases. Inflow velocity is proportional to the square root of thrust for a constant rotor area and it reduces the effective angle of attack seen by the blades. Since the 4-bladed configuration has a higher inflow than a 2-bladed one (due to higher thrust) at the same rotational speed, the effective angle of attack seen by the blade on a 4-bladed rotor is lower than the 2-bladed rotor. This means that the 45°, 4-bladed rotor may have an approximately equivalent angle of attack as the 40°, 2-bladed one. This is also the reason why the 4-bladed rotor does not produce twice the thrust of the 2-bladed at the same rotational speed. Based on these results, a blade pitch

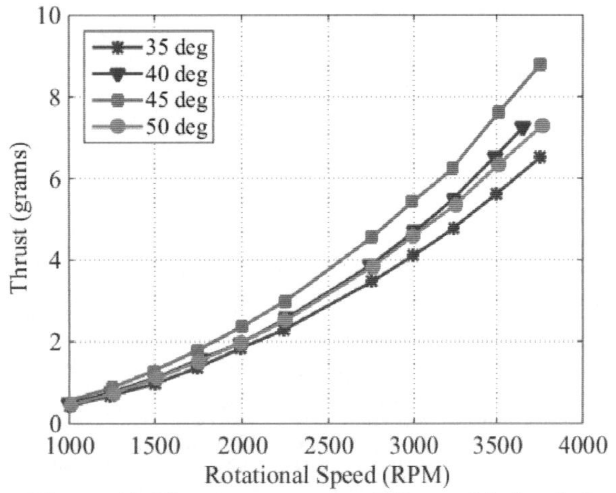

Figure 13. Thrust vs. rpm at different blade pitch amplitudes for 4-bladed cyclorotor.

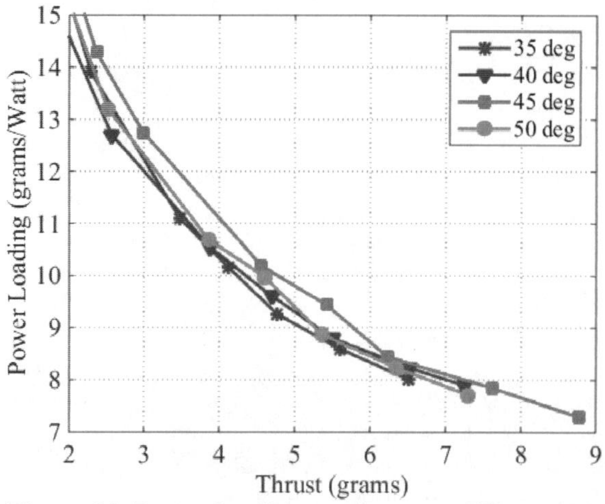

Figure 14. Power loading vs. thrust at different blade pitch amplitudes for 4-bladed cyclorotor.

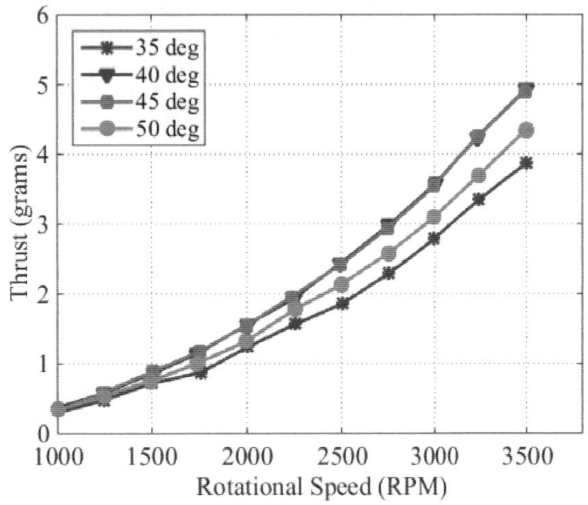

Figure 15. Thrust vs. rpm at different blade pitch amplitudes for 2-bladed cyclorotor.

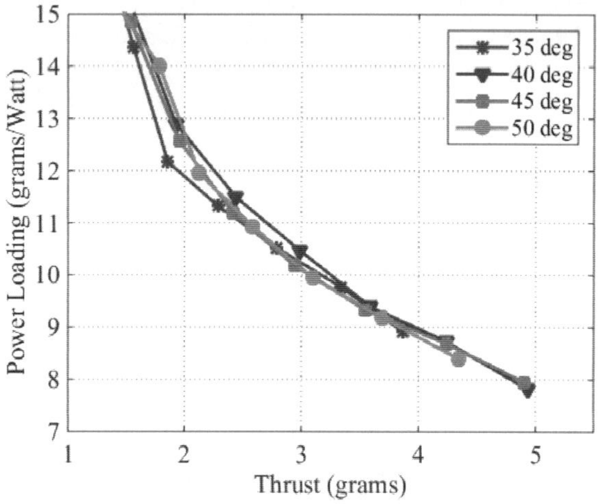

Figure 16. Power loading vs. thrust at different blade pitch amplitudes for 2-bladed cyclorotor.

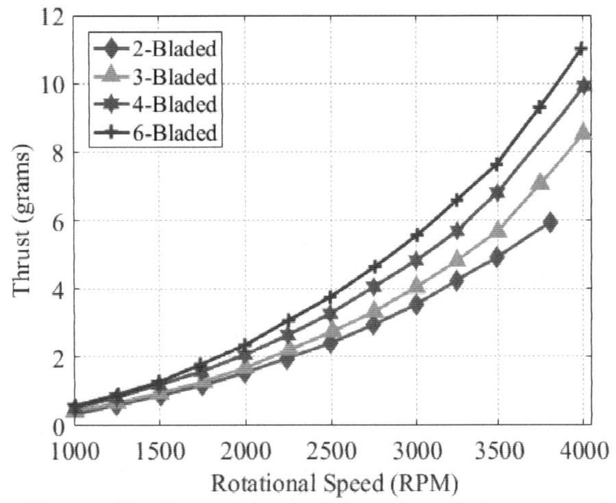

Figure 17. Comparison of thrust at 45 degree pitch between 2, 3, 4, and 6-bladed cyclorotors.

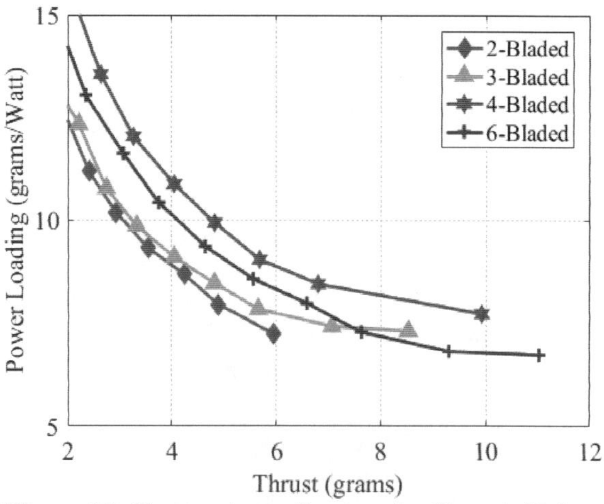

Figure 18. Comparison of power loading at 45 degree pitch between 2, 3, 4, and 6-bladed cyclorotors.

amplitude of 45° and a 4-bladed rotor configuration was chosen as the final design. This rotor outperforms all of the other rotor configurations we tested and it produces the required thrust of 10 grams at our target operating RPM of about 4000, expending 1.28 watts of mechanical power, as shown in Figs. 19 and 20. Because the remaining thrust will be produced by the tail rotor, a lightweight propeller was selected and tested over a range of rotational speeds using the motor that will be on the vehicle. The required tail rotor thrust to counteract the torque of the main rotors is around 8 – 10 grams. A plot of this data is presented in Fig. 21 which indicates that the selected rotor-motor combination is capable of producing the required thrust.

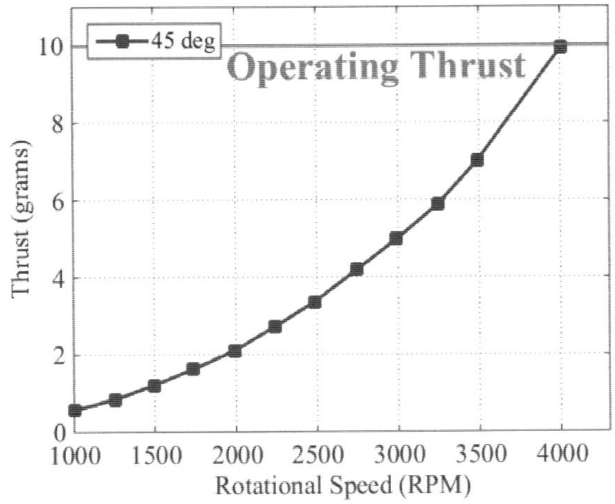

Figure 19. Thrust produced by final rotor design

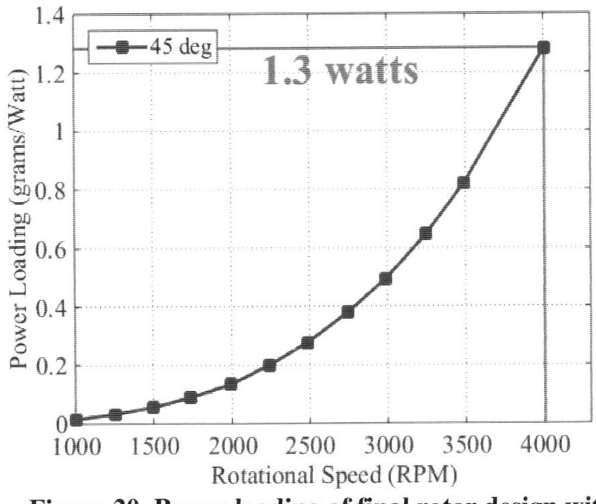

Figure 20. Power loading of final rotor design with mechanical power labeled.

With blade size, geometry and number of blades systematically determined, the final analysis conducted on the blades and rotor setup was an evaluation of the blade deflection, either from a weak root boundary condition or bending along the span of the blade. The method used to qualitatively evaluate blade deflection was a series of experiments conducted using a strobe light, which was flashed in synchronization with the rotational frequency of the rotor. This gives the appearance that the rotor is stopped, and the amount of deflection experienced by the rotor was qualitatively compared with that of the non-rotating rotor. These results are shown in Fig. 22, and it can be clearly seen that when comparing the image with that of Fig. 5, there is only minimal deflection at the operating frequency of 4000 RPM. Based upon these results, it was concluded that, on the present rotor design, bending deflections do not substantially reduce rotor efficiency.

VEHICLE INTEGRATION

Using two rotors in the configuration previously described, a flight-ready vehicle was constructed as previously shown in Fig. 6. Each component on the vehicle was carefully selected or manufactured to be as light as possible without compromising vehicle integrity. The airframe, which functions as an anchor point for all electrical and mechanical subsystems, is designed to hold the components in a compact configuration, thus, minimizing the profile of the craft. The airframe is rapid prototyped from ABS plastic and features carbon fiber rods extending outwards that function as landing gear. The thrust vectoring of each of the

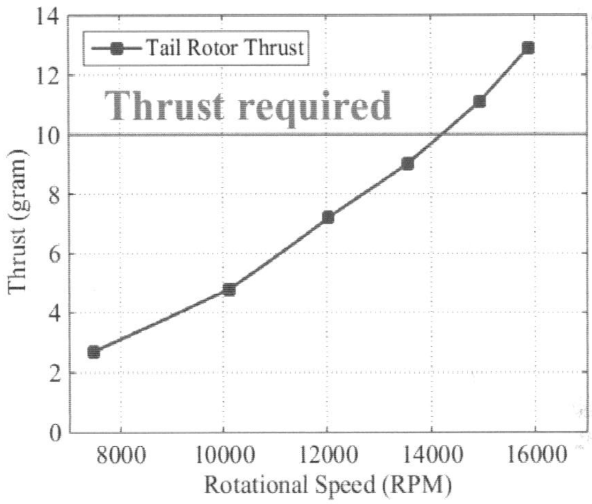

Figure 21. Thrust of tail rotor vs. rpm.

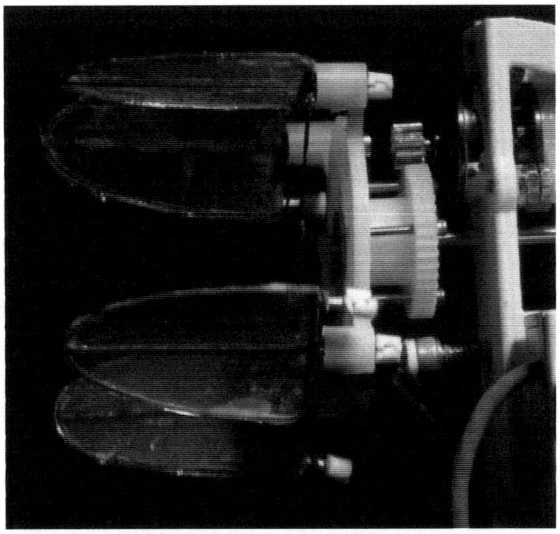

Figure 22. Blade deflection tests at rotor operating conditions visualized using strobe light.

cyclorotors is actuated by means of a shape-memory alloy based servo actuator weighing 1 gram, the lightest possible COTS servo available. The rotors themselves are each driven by a 2.5 gram brushless, out-runner motor. Both the motors and actuator servos are powered by a 160mah single cell Li-Po battery weighing ~3.7 grams. The total weight of the first prototype is approximately 29 grams. Table 1, provided above, gives a breakdown of the system components by weight and Fig. 6 shows the completed vehicle.

ATTITUDE CONTROL AND VEHICLE TELEMETRY

Control of the vehicle is accomplished in two ways: varying the RPM of the drive motors for either the main rotors or the tail prop (changes thrust magnitude) and rotating the offset link in the pitching mechanisms, which changes the blade pitch phasing (thrust vectoring). Modulating the rotor RPMs and cycloidal rotor pitch offset link angle provides complete control over rolling, pitching and yawing motion. Because of this ability to instantaneously change the magnitude and direction of the thrust vectors from the rotors, superior levels of agility and maneuverability can be achieved over traditional rotorcraft. The specifics of the control strategy implemented for attitude control are demonstrated in Fig 23. For roll control, the RPM of each of the main cyclorotors is changed differentially to increase the thrust of one rotor while decreasing that of the other rotor, effectively creating a moment about the longitudinal axis of the vehicle (Fig. 23a.). To control pitch, the RPM of the tail prop will be varied to either increase or decrease the thrust created and generate a moment about the lateral axis due to the offset of the tail prop from the C.G. (Fig. 23b). Finally, yawing moments will be generated by rotating the pitch offset links in unison which tilts the thrust vectors of the main rotors in opposite directions, effectively creating a moment about the vertical axis of the vehicle (Fig 23c).

An important consideration of cyclocopter flight is the gyroscopic coupling present between roll and yaw degrees of freedom when either a roll or yaw control input is given. Because both main cycloidal rotors are rotating in the same direction (clockwise when viewed from the starboard side of the vehicle) about the lateral axis, any force which causes rotation of the vehicle about either the longitudinal or vertical axis will result in a gyroscopic moment 90° out of phase. Thus, if the vehicle is rolled, for example, to the right, a positive yawing motion (clockwise when viewed from above) will be introduced. To compensate for this coupled behavior between roll and yaw, the autopilot is programmed to automatically provide input commands to yaw when roll is activated, and vice versa. Feedback gains are handled similarly and provide coupled commands between both degrees of freedom. The result is a vehicle capable of many different types of flight regimes and maneuvering.

The attitude stabilization is implemented onboard using an autopilot – a custom-built, embedded processor-

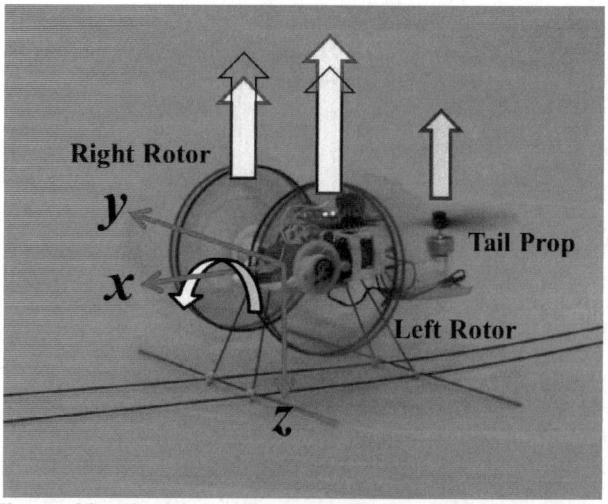

Figure 23a. Positive roll control illustration of increased thrust on left rotor and decreased thrust on right cyclorotor.

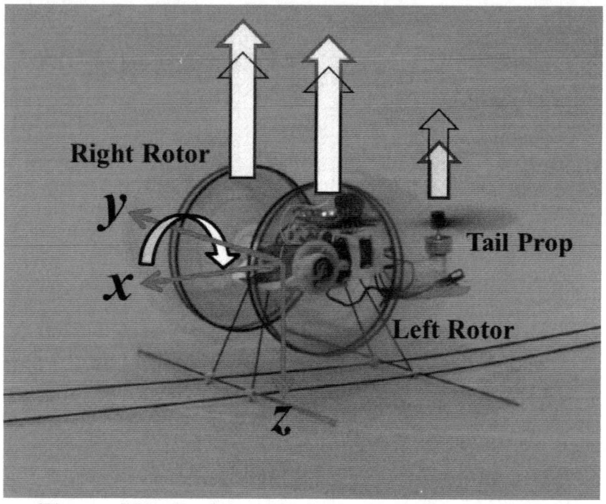

Figure 23b. Positive pitch control illustration of increased thrust on main rotors and decreased thrust on tail prop.

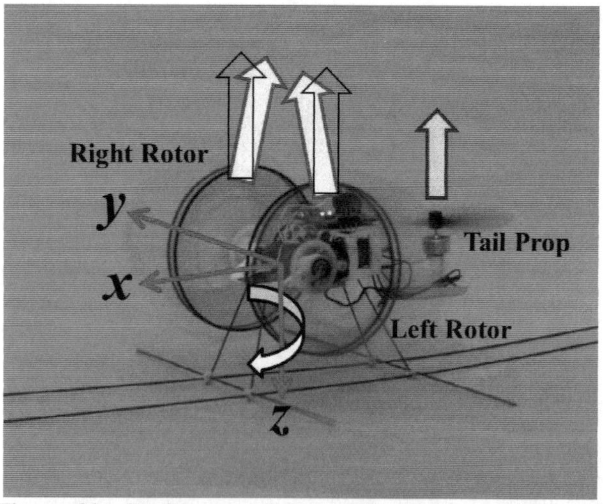

Figure 23c. Positive yaw control illustration of forward tilted thrust on left rotor and backward tilted thrust on right rotor.

sensor board (Fig. 24). It weighs 1.3 grams and is powered by a single 1-cell 3.7 volt 30mAh Li-Po battery. The autopilot houses an STM32 microprocessor with a 32-bit ARM Cortex M3 core for high-end onboard computation tasks. The MPU-9150 IMU integrated on the board includes tri-axial gyroscopes, accelerometers, and magnetometers. Wireless communications are serviced by an on-board nRF24L01 chip, a low-power 2.4 GHz RF transceiver. The autopilot has a sensor update rate of 500 Hz and is capable of streaming vehicle attitude data and actuator controls data to the base station with a short latency (Ref. 20). The autopilot is capable of sensing vehicle attitude and angular rates and sending corrective signals to the servos for stabilization by varying the pulse width input to the motors and servos.

Figure 24. Custom built 1.3-gram kinematic pilot with U.S. quarter for size comparison.

To communicate wirelessly with the onboard controller, the operator uses a LabVIEW interface which includes a wireless IEEE 802.15.4 data link. The base station LabVIEW program allows the operator to control the vehicle, modify the feedback gains, change the sensitivity of pilot inputs, and record attitude data transmitted by the onboard processor. The LabVIEW program receives pilot inputs through the use of a commercially available DX6i Spectrum transmitter which is hardwired to the base station, and features a user interface to collect the feedback gains and pilot sensitivity specifications. The program then connects to the microcontroller through a wireless 2.4 GHz radio link and uses this connection to send the control inputs and receive the vehicle attitude and rates data. The data processing and inner-loop feedback control calculations are performed onboard by the microprocessor.

The on-board gyros measure the pitch (q), roll (p) and yaw (r) angular rates while the accelerometers record the tilt of the gravity vector in the body frame. The body-fixed frame directions and rates are sketched in Fig. 25. The vehicle attitude can then be extracted by integrating the gyro measurements with time. However, it is known that this leads to drift in attitude measurements (Ref. 21). Accelerometers on the other hand offer stable bias, but are sensitive to vibrations and in general offer poor high frequency information (Ref. 22). Therefore a complementary filter was incorporated to extract the pitch and roll Euler angles using a high-pass filter for the gyros (4 Hz cut-off) and a low-pass filter for accelerometers (6 Hz cut-off). The rotational vibrations were filtered out since they are sufficiently higher than the body dynamics. On-board inner loop feedback was implemented using a proportional-derivative (PD) controller as shown in Fig. 26. The feedback states are the pitch and roll Euler angles θ and ϕ, and the

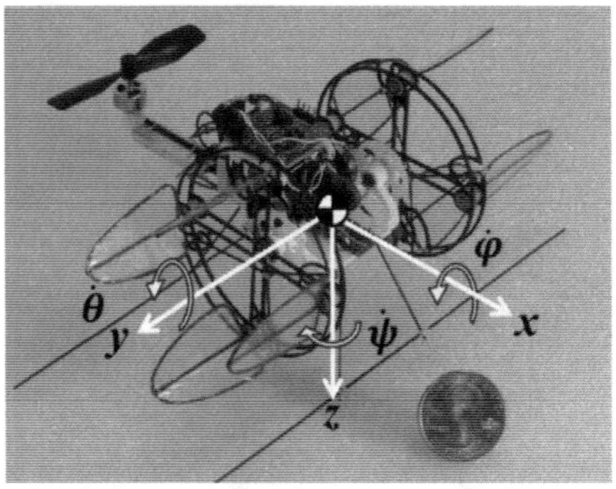

Figure 25. Micro cyclocopter body-fixed frame and with angular rates as measured by the on-board IMU.

attitude rates, p, q and r. The outer loop feedback capability was provided for translational positioning by a human pilot or for a position tracking system.

HOVERING FLIGHT TESTING

Since the flying qualities and behavior of this newly developed vehicle are largely unknown and initially unpredictable, an important flight testing consideration was safety of the vehicle during crashes in order to prevent destruction of the airframe and reduce maintenance requirements. For this reason, an *8ft x 8ft x 8ft* safe enclosure was utilized in which to conduct flight tests (Fig. 27). The sides are covered by a 90 micron thick plastic sheet to retain the vehicle in the enclosure during flying and offer a soft impact should the vehicle collide with the walls. Additionally, a sheet of plastic suspended from the edges of the frame a few inches off the ground provides protection to the vehicle in the case of a crash landing. As expected, flying in this enclosure substantially reduces the number of mechanical breakdowns and improves the efficiency of the flight testing operation. The vehicle takes off from the center of the suspended sheet which provides a stable platform for vehicle lift-off.

The feedback method used to stabilize the vehicle is an experimentally tuned PD controller which stabilizes the vehicle about a set point. Thus, it is important to trim the vehicle about the desired flight condition in order to appropriately utilize the controller feedback. For the initial

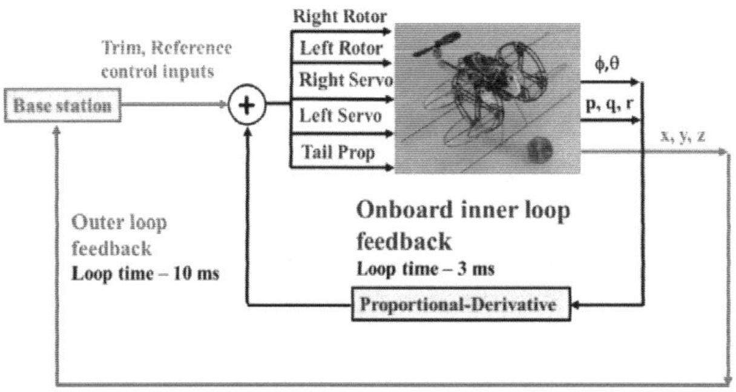

Figure 26. Diagram of closed - loop feedback implemented in control strategy.

Figure 27. Safe-enclosure used for testing vehicle.

flight tests, the desired flight condition was hovering flight; therefore, effort was focused toward modifying the direction of the individual thrust vectors from the main rotors until the thrust force was perfectly vertical. Making sure the thrust vectors are vertical also decouples the yaw and roll trim inputs. To accomplish this, the motor outputs were adjusted so that one rotor would spin much faster than the other. This rotor is then the primary source of lift for the vehicle and any yawing behavior the vehicle experiences is predominantly caused by the faster rotor. By observing the vehicle, we could adjust the trim of the appropriate servo until the vehicle no longer yawed. The servo was set to each extreme to ensure that there was enough control authority. Then the bisection method was used to converge on the point where the thrust vector was vertical. Once each rotor was adjusted in this manner the thrust vectors of the rotors were nearly vertical. The next step was to trim for eliminating roll. This was done simply by introducing a constant offset between rotor RPMs. Additionally, it was necessary to correct the throttle offset for the tail prop so as to ensure it generated enough thrust to balance out the moment generated by the rotating rotors and supplement the lifting force. After these steps were complete, the vehicle was tested by hopping it to make sure it lifted off vertically in a stable fashion. Minor adjustments (e.g. forwards and backwards trim, pitch, and roll) needed to be made to further trim the vehicle.

After successfully trimming the vehicle for hovering flight, the next step was to introduce feedback gains in order to provide the additional level of control necessary beyond a human pilot's capability in order to stabilize the vehicle. For flight testing with on-board stabilization, only roll and pitch damping (derivative) gains were initially introduced which allowed the vehicle to achieve a marginal level of stability. No yaw gains were necessary. The derivative gains had the effect of slowing down the vehicle dynamics and allowing us to further trim the vehicle. However, it was necessary to introduce proportional gains to eliminate drift. These were added in slowly until the vehicle behaved as expected, and then the derivative gains were slightly adjusted to improve hover stability. Shown in Figure 28 is a short, stable flight test conducted which successfully demonstrated the vehicle operation and flight testing procedure. The associated gyro data (Fig. 29) and Euler angles (Fig. 30) for this flight are provided. The onboard feedback control system was able to restrict the angular rates and the vehicle did not deviate from level flight by more than ±10 degrees in roll or pitch. The data begins at 2 seconds because that is when the hover begins after takeoff. The large spike around 7 seconds in the roll gyro and the associated smaller spike in the pitch gyro (Fig. 29) are when the throttle is reduced for landing.

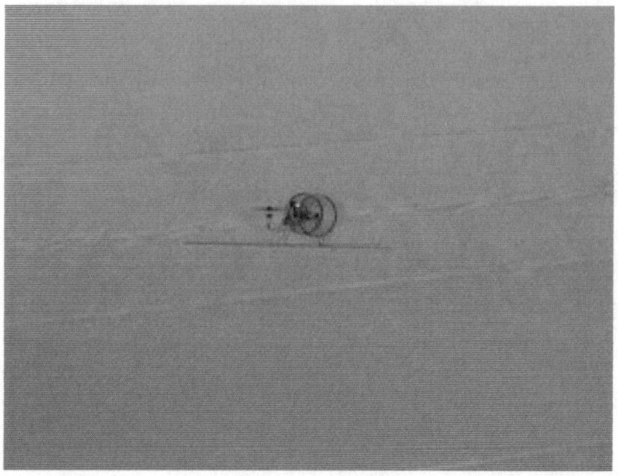

Figure 28. Meso-scale cyclocopter in relatively stable hovering flight.

SUMMARY AND CONCLUSIONS

The cyclocopter concept, while having been conceived many years ago, has only recently seen materialization in the form of a fully functioning and operational micro air vehicle. The advantages of employing a cycloidal thrust generation mechanism instead of a traditional helicopter rotor include the possibility for much greater aerodynamic efficiency, agility, maneuverability, and gust tolerance. Thus, it is of great interest to reduce the size of the vehicle to the smallest scale possible for improved reconnaissance and surveillance for indoor scenarios. This paper has outlined the design, development, and flight testing of the world's smallest cycloidal-rotor based air vehicle. The first iteration of this cyclocopter has demonstrated that the concept can be created and implemented at small scales with careful designing and manufacturing. For this particular design, a cycloidal rotor with a single endplate and cantilevered, elliptical wings was developed. Through a systematic testing procedure, the blade design was optimized and a specialized manufacturing technique was created to reliably produce strong, lightweight blades. Several key conclusions of this study are as follows:

1. The present cyclorotors have a unique design, which takes advantage of the small scale to achieve higher thrust/weight ratios compared to previous cyclorotor designs.

2. The blade manufacturing process developed in the present study allows ultra-lightweight blades to be fabricated with consistent quality and physical properties.

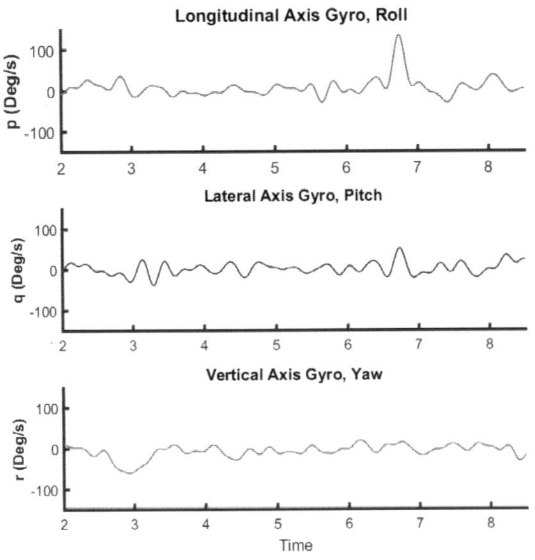

Figure 29. Angular rates for flight test.

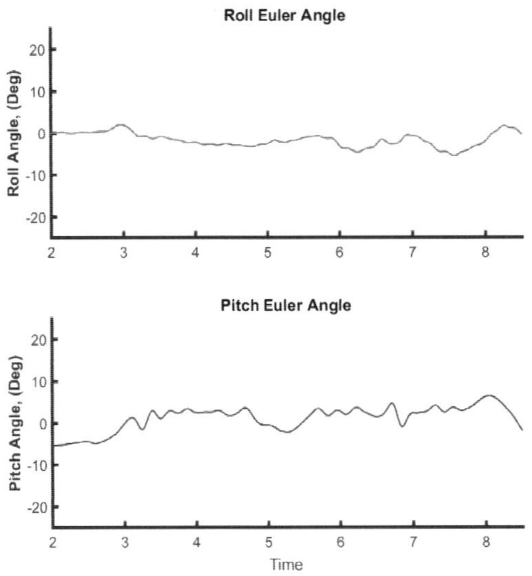

Figure 30. Euler angles of flight test.

3. Using the custom-built miniature 3-component balance developed, experimental parametric studies were conducted to optimize cyclorotor performance.

4. Using an inner stabilization loop implemented on a custom-built onboard autopilot, the vehicle has been able to achieve stable hovering flights. This demonstrated the downwards scalability of this revolutionary flying concept.

5. The cyclocopter at these scales requires low latency between state measurements and control outputs due to the inherently fast vehicle dynamics. As such, the hardware and software must be tailored to achieve this.

Immediate future work involves improving the flight qualities of the vehicle and more parametric studies. Additionally, once the vehicle has achieved a greater level of flying quality with the on-board inner-loop attitude stabilization, a human pilot will be introduced in the outer-loop for position control with the goal of increasing the total flight capabilities of the vehicle. Once the endurance has improved, the next step is to extract a linearized state-space model of the vehicle through experimental system identification studies using a camera-based 3-D motion tracking system (i.e. VICON™). The motion capture system will also be used to replace the pilot in the outer loop with the purpose of flying the cyclocopter completely autonomously for an extended period of time. Thus, the overall goal of the project is to develop an efficient meso-scale cyclocopter MAV (shown in Fig. 4) that could perform a stable, autonomous, 20-minute hover/forward flight indoors/outdoors, and even in the presence of moderate gusts.

Author contact: Carl Runco runc10@tamu.edu, David Coleman dcaero14@tamu.edu, Moble Benedict benedict@tamu.edu

ACKNOWLEDGMENTS

This research was supported by the U.S. Army's Micro Autonomous Systems and Technology–Collaborative Technology Alliance (MAST-CTA) with Chris Kroninger (Army Research Laboratory–Vehicle Technology Directorate) as Technical Monitor. The authors would like to thank Elena Shrestha for her help with setting up the flight testing, Vikram Hrishikeshavan for developing the autopilot, and Brett Himmelberg for conducting rotor performance testing and analysis.

REFERENCES

[1] Pines, D., and Bohorquez, F., "Challenges Facing Future Micro-AirVehicle Development," Journal of Aircraft, Vol. 43, (2), March–April 2006, pp. 290–305.

[2] Chopra, I., "Hovering Micro Air Vehicles: Challenges and Opportunities," Proceedings of American Helicopter Society Specialists' Conference, International Forum on Rotorcraft Multidisciplinary Technology, Seoul, Korea, October 15–17, 2007,

[3] Leishman, J. G., Principles of Helicopter Aerodynamics, Cambridge University Press, New York, NY, 2006, pp. 229, 280, 334–337.

[4] Ramasamy, M., Johnson, B., and Leishman, J. G., "Understanding the Aerodynamic Efficiency of a Hovering Micro-Rotor," Journal of the American Helicopter Society, Vol. 53, (4), October 2008, pp. 412–428.

[5] Benedict, M., Jarugumilli, T., and Chopra, I., "Effect of Rotor Geometry and Blade Kinematics on Cycloidal Rotor Hover Performance," *Journal of Aircraft*, Vol. 50, (5), 2013, pp. 1340–1352.

[6] Benedict, M., Jarugumilli, T., and Chopra, I., "Experimental Optimization of MAV-Scale Cycloidal Rotor Performance," *Journal of the American Helicopter Society*, Vol. 56, No. 2, April 2011, pp. 022005-1 - 022005-11.

[7] Wheatley, J. B. and Windler, R., "Wind-Tunnel Tests of a Cyclogiro Rotor," NACA Technical Notes No. 528, May 1935.

[8] Kirsten, F. K., "Cycloidal Propulsion Applied to Aircraft," *Transactions of the American Society of Mechanical Engineers*, Vol. 50, No. AER-50-12, 1928, pp 25—47.

[9] Benedict, M., "Fundamental Understanding of the Cycloidal-Rotor Concept for Micro Air Vehicle Applications," Ph.D Thesis, Department of Aerospace Engineering, University of Maryland, College Park, MD, December 2010.

[10] Benedict, M., Jarugumilli, T., and Chopra, I., "Design and Development of a Hover-Capable Cyclocopter MAV," Proceedings of the 65th Annual National Forum of the American Helicopter Society, Grapevine, TX, May 27—29, 2009.

[11] Benedict, M., Gupta, R., and Chopra, I., "Design, Development and Open-Loop Flight Testing of a Twin-Rotor Cyclocopter Micro Air Vehicle," *Journal of the American Helicopter Society*, Vol. 58, No. 4, October 2013, pp. 1 – 10.

[12] Benedict, M., Shrestha, E., Hrishikeshavan, V., and Chopra, I., "Development of a Micro Twin-Rotor Cyclocopter Capable of Autonomous Hover," *Journal of Aircraft*, Vol. 51, No. 2, 2014, pp. 672 – 676.

[13] Sirohi, J., Parsons, E., and Chopra, I., "Hover Performance of a Cycloidal Rotor for a Micro Air vehicle," *Journal of the American Helicopter Society*, 52(3), July 2007, pp. 263—279.

[14] Benedict, M., Ramasamy, M., Chopra, I., and Leishman, J. G., "Performance of a Cycloidal Rotor Concept for Micro Air Vehicle Applications," *Journal of the American Helicopter Society*, Vol. 55, No. 2, April 2010, pp. 022002-1 - 022002-14.

[15] Benedict, M., Ramasamy, M., Chopra, I., and Leishman, J. G., "Experiments on the Optimization of the MAV-Scale Cycloidal Rotor Characteristics Towards Improving Their Aerodynamic Performance" Proceedings of the International Specialists' Meeting on Unmanned Rotorcraft, Scottsdale, Arizona, January 20—22, 2009.

[16] Benedict, M., Jarugumilli, T., and Chopra, I., "Experimental Performance Optimization of a MAV-Scale Cycloidal Rotor," Proceedings of the AHS Specialists' Meeting on Aeromechanics, San Francisco, CA, Jan 20—22, 2010.

[17] Benedict, M., Mattaboni, M., Chopra, I., and Masarati, P., "Aeroelastic Analysis of a Micro-Air-Vehicle-Scale Cycloidal Rotor in Hover," *AIAA Journal*, Vol. 49, No. 11, November 2011, pp. 2430 – 2443.

[18] McMichael, J. M., and Francis, USAF (Ret.), C. M. S., "Micro Air Vehicles Toward a New Dimension in Flight," *U.S. Department of Defense Weapons Systems Technology Information Analysis Center (WSTIAC) Newsletter*, Vol. 1, No. 13, JanJul 2000 Originally published online at http://www.darpa.mil/tto/MAV/mav auvsi.html, Aug. 7, 1997.

[19] Benedcit, M. and Chopra, I., "Design and Development of an Unconvnetional VTOL Micro Air Vehicle: The Cyclocopter," *Proceedings of SPIE Micro- and Nanotechnology Sensors, Systems, and Applications IV*, 83731F, International Society for Optics and Photonics, Baltimore, 2012.

[20] Hrishikeshavan, V., Chopra, I., "Refined Lightweight Inertial Navigation System for Micro Air Vehicle Applications," AHS Specialists' Meeting on Unmanned Rotorcraft and Network-centric Operations, Chandler, AZ, Jan 20-22, 2015.

[21] Georgy, J., Noureldin, A., Korenberg, M., and Bayoumi, M., "Modeling the Stochastic Drift of a MEMS-Based Gyroscope in Gyro/Odometer/GPSIntegrated Navigation," *IEEE Transactions on Intelligent Transportation Systems*, Vol. 11, (4), Dec 2010, pp. 856–872.

[22] Thong, Y. K., Woolfson, M. S., Crowe, J. A., Hayes-Gill, B. R., and Challis, R. E., "Dependence of inertial measurements of distance on accelerometer noise", *Measurement Science and Technology*, Vol. 13 , (8), pp.1163–1172 , 2002.

Adaptive Pure Pursuit based Autonomous Landing of Quadrotors

Alvika Gautam IIIT Delhi India

P.B. Sujit IIIT Delhi India

Srikanth Saripalli Texas A & M University College Station, Texas,USA

Abstract

Landing of quadrotors is a complex and challenging maneuver. Developing autonomous landing technologies has been an active area of research over the past decade. This paper presents a novel guidance strategy for vision based autonomous landing process for both stationary and moving landing targets. We develop a modified pursuit guidance strategy, wherein we integrate a novel velocity controller with the pure pursuit guidance law to achieve faster and accurate landing. The proposed controller is validated for a stationary and moving landing pad through simulations. Further, we also evaluate the said controller in real time conditions for a stationary landing pad through outdoor hardware experiments. Both simulation and experiment results show that the proposed guidance strategy performs better than traditional vertical landing approaches.

Introduction

Rotary wing UAVs like quadrotors are ubiquitous in both civilian and defense applications like, surveillance (Ref. 1), agriculture (Ref. 2) (Ref. 3) and search and rescue operations (Ref. 4), because of their maneuvering, hovering capabilities and ease of handling. One of the important endeavors for UAVs is to autonomously complete a full mission, from takeoff to land. Autonomous landing of quadrotor is a critical phase as any uncontrolled deviations can result in crash. Moreover, time critical requirements of certain applications and limited flight time of quadrotors make it imperative to complete the landing process in the shortest time possibles.
Autonomous landing of quadrotors (also referred to as UAV), at a desired location involves two steps: (a) sensing the landing pad or predetermining the landing pad (also referred to as the target) location and (b) generating control commands for the UAV such that, it lands on the desired landing pad. Typically, landing systems use INS, GPS for localization and additional sensors like radar altimeter, ultrasound, barometric pressure sensor for altitude as the GPS based altitude measurement is inaccurate (Ref. 5) (Ref. 6) (Ref. 7). Autonomous landing using GPS is also prone to localization errors of as high as 5 to 6 meters (Ref. 8). Thus, there is a need to augment GPS/INS based systems with other sensing techniques to enhance accuracy and reliability.
Camera vision has come up as a popular sensing techniques for target detection. It is passive, lightweight and easy to integrate with the autopilot system. Once the location of the landing pad is known using a sensing technique, two types of approaches can be considered to maneuver a UAV towards the target. The first approach separates the problem of vehicle guidance and control into two separate loops as shown in figure 1. Second approach integrates both the loops into one and designs them simultaneously. In the first approach, the outer loop, using target information, generates guidance commands which in general refers to the determination of heading from vehicle's current location to a designated target. The inner loop controls the UAV attitude according to the reference guidance commands. Typically, autonomous landing has been attempted using integrated approach techniques, such as nonlinear control techniques (Ref. 9) (Ref. 10), intelligent control techniques (Ref. 11) (Ref. 12) and robust control techniques (Ref. 13). These techniques have been exhaustively tested on simulations but not on hardware. Moreover, segregated loop approach is more suited for low cost vehicles and real time flight applications, since it is more flexible and offers finer control on each module.

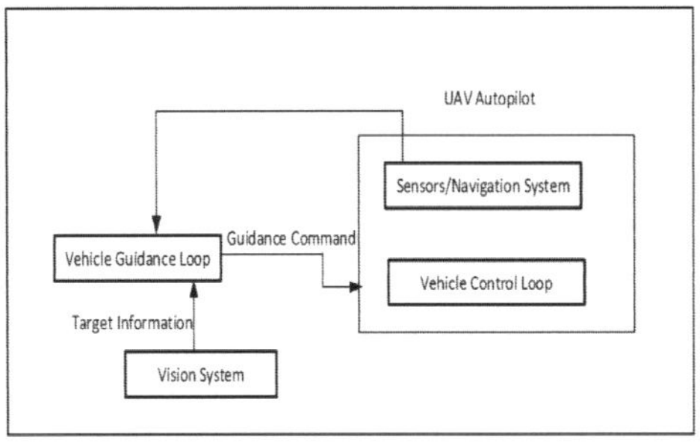

Fig. 1. Block diagram of a guidance based landing system.

Guidance based landing techniques have been studied for fixed wings. Kim et al. (Ref. 14) proposed a fully autonomous vision-based net recovery system for a fixed wing UAV. It consisted of a ground vision system whose aim was to detect a recovery net during landing. Pursuit guidance law was used for longitudinal guidance and a nonlinear pursuit guidance law for lateral guidance in order to land the UAV. In (Ref. 15) the landing phase was divided into spiral descent and final approach. The aircraft aligned along the flight path angle towards the approach direction by the end of the spiral descent phase. The aircraft was then guided from the approaching

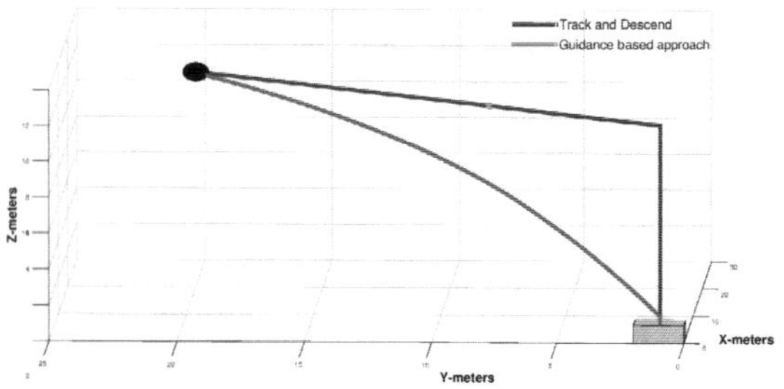

Fig. 2. Trajectory of typical landing of a quadrotor and the proposed landing using pure pursuit guidance law.

waypoint to the recovery net using a pseudo pursuit guidance law. Bang et al. (Ref. 16) proposed a t_{go} guidance law for vision based autonomous landing for both fixed wing and rotary wing UAVs. Vision based guidance techniques for quadrotor landing have been limited to vertical landing approaches only (Ref. 17) (Ref. 18). In such approaches, quadrotor tracks the target while moving along a linear path till certain distance and then descends vertically to land on the target. This can result in increased flight time during the landing process. In this paper, we propose a pure pursuit based guidance law to land on a stationary as well as moving targets. Using the guidance commands, the quadcopter approaches the target while simultaneously descending in altitude, thus following a more efficient trajectory as compared to the vertical landing approaches, as shown in figure 2 to reach the target in a time efficient manner. We test the efficacy of the guidance law extensively on simulations and also in real world conditions through outdoor hardware experiments.

Nomenclature

- xy and z : horizontal plane and vertical axis respectively
- u and t : UAV and target respectively.
- V_u, χ_u and γ : UAV speed (m/s), course angle and flight path angle (radians).
- V_t and χ_t : Target speed (m/s) and course angle (radians).
- V_u^0, V_t^0 : Initial UAV and target speed (m/s).
- R_{xy}, $\dot{R}_{xy}$: LOS distance in xy plane (meters) and rate of change of this distance (m/s).
- R_{xy}^0 : Initial LOS distance in xy plane (meters).
- R_z, $\dot{R}_z$: vertical distance between the UAV and the target and rate of change of this distance.
- R : LOS distance between UAV and Target (meters).
- θ, $\dot{\theta}$: LOS angle (radians) and rate of change of LOS angle .
- ψ, $\dot{\psi}$: Projection of LOS angle (radians) on xy plane and its rate of change.
- a_{xy}, a_z: Guidance commands generated for the UAV in 3D space.
- $x_{t_t}, y_{t_t}, x_{u_t}, y_{u_t}\ z_{u_t}$: Position coordinates of target and UAV at time t.

Basics of Pure Pursuit (PP) guidance

In this section, we provide basics of 2D and 3D pure pursuit guidance laws for completeness and then use these guidance laws for landing in the subsequent sections.

2D Pure Pursuit Guidance

Consider a planar UAV-target engagement geometry as shown in figure 3. The speed of the UAV and the target are V_u and V_t respectively. The notation χ_u, χ_t, ψ, and R_{xy} represent the UAV course angle, target course angle, line-of-sight (LOS) angle, and range separation respectively. The target is assumed to be non maneuvering, that is χ_t is assumed to be constant. Using the engagement geometry in figure 3, $\dot{R}_{xy}$ and $\dot{\psi}$ are calculated as,

$$\dot{R}_{xy} = V_t\cos(\chi_t-\psi) - V_u\cos(\chi_u-\psi), \quad (1)$$
$$\dot{\psi} = (V_t\sin(\chi_t-\psi) - V_u\sin(\chi_u-\psi))/R_{xy}. \quad (2)$$

The guidance command, denoted as a_{xy} is the lateral accel-

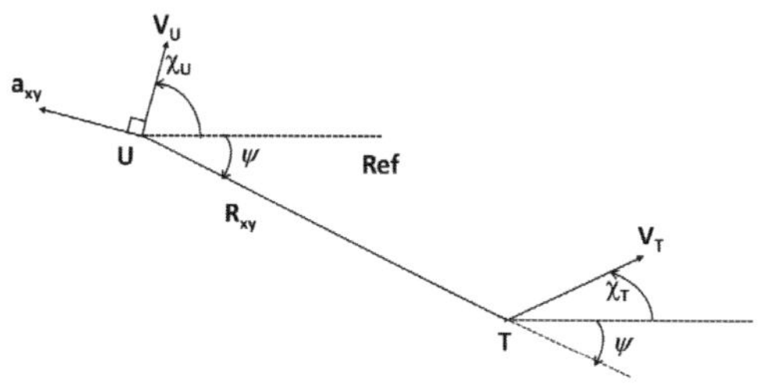

Fig. 3. Planar engagement geometry of UAV-target.

eration applied normal to the velocity vector. The guidance command is expressed as,

$$\dot{\chi}_u = a_{xy}/V_u. \quad (3)$$

Pure pursuit (PP) [(Ref. 19)] works on the principle, that if the tracking vehicle persistently points towards the target then it will ultimately intercept it. In other words, PP aligns χ_u towards ψ. A practical implementation of the pure pursuit uses a feedback law as,

$$a_{xy} = -K_a(\chi_u-\psi), K_a > 0, \quad (4)$$

where K_a is the gain. However, for PP to be effective, it is important to maintain $\dot{\chi}_u = \dot{\psi}$, so that the vehicle can always point towards the target. Combining equation (3) and (4), the guidance command for simple pure pursuit is given as,

$$a_{xy} = V_u\dot{\psi} - K_a(\chi_u-\psi). \quad (5)$$

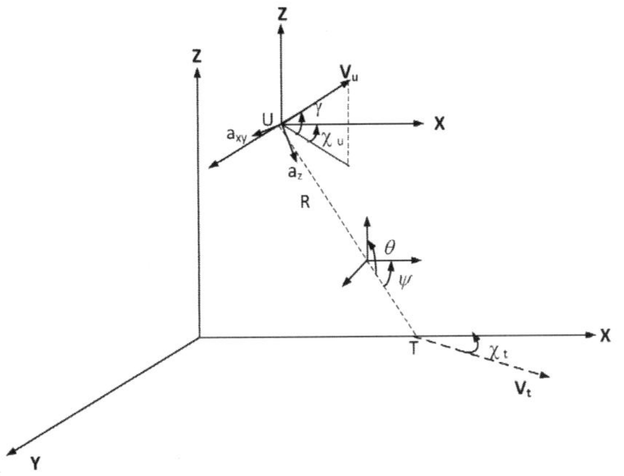

Fig. 4. Engagement geometry for a UAV landing on a moving target, 3D scenario

3D Pure Pursuit guidance

An actual landing process occurs in three dimensional space. We extend the 2D pursuit guidance principle to 3D scenario. We consider the target moving on the ground represented by XY horizontal plane. The UAV starts from some position (x_0, y_0, z_0) assuming that the target is visible from this initial location.
The parameters to be considered are the course and flight path angles of the UAV. Expressions for the distance between UAV and target and these angles are given as,

$$R = \sqrt{(x_t - x_u)^2 + (y_t - y_u)^2 + (z_t - z_u)^2}, \quad (6)$$
$$R_{xy} = \sqrt{(x_t - x_u)^2 + (y_t - y_u)^2}, \quad (7)$$
$$R_z = z_t - z_u, \quad (8)$$
$$\psi = \tan^{-1}(y_t - y_u / x_t - x_u), \quad (9)$$
$$\theta = \tan^{-1}(z_t - z_u / R_{xy}). \quad (10)$$

Figure 4 shows the engagement geometry for a UAV landing on the target in a 3D scenario. Expressions of $\dot{R}_{xy}$, $\dot{R}_z$, $\dot{\psi}$ and $\dot{\theta}$ are derived similar to the 2D case.

$$\dot{R}_{xy} = V_t \cos(\chi_t - \psi) - V_u \cos(\theta)\cos(\chi_u - \psi), \quad (11)$$
$$\dot{R}_z = V_u \sin(\theta), \quad (12)$$
$$\dot{\psi} = (V_t \sin(\chi_t - \psi) - V_u \cos(\gamma)\sin(\chi_u - \psi))/R_{xy}, \quad (13)$$
$$\dot{\theta} = -(V_u \sin(\gamma)\cos(\theta) + \dot{R}_{xy}\sin(\theta))/R. \quad (14)$$

Guidance commands in 3D space, responsible to maneuver the UAV both in xy plane and z axis are given by,

$$a_{xy} = V_u \cos(\chi_u)\dot{\psi} - K_a(\chi_u - \psi), \quad (15)$$
$$a_z = V_u \dot{\theta} - K_a(\gamma - \theta), \quad (16)$$
$$\dot{\chi}_u = a_{xy}/V_u \cos(\gamma), \quad (17)$$
$$\dot{\gamma} = a_z / V_u. \quad (18)$$

The target here is assumed to be a non maneuvering target with constant velocity and heading. The target and UAV model are given as

$$x_{t_{t+\Delta t}} = x_{t_t} + V_t \cos(\chi_t)\Delta t, \quad (19)$$
$$y_{t_{t+\Delta t}} = y_{t_t} + V_t \sin(\chi_t)\Delta t, \quad (20)$$
$$x_{u_{t+\Delta t}} = x_{u_t} + V_u \cos(\gamma)\cos(\chi_u)\Delta t, \quad (21)$$
$$y_{u_{t+\Delta t}} = y_{u_t} + V_u \cos(\gamma)\sin(\chi_u)\Delta t, \quad (22)$$
$$z_{u_{t+\Delta t}} = z_{u_t} + V_u \sin(\gamma)\Delta t. \quad (23)$$

The guidance objective is to land smoothly on the target, while persistently tracking it. Pure pursuit aims at keeping the velocity vector of UAV directed towards the target which is one of the goals for landing and moreover it is simple to implement.
The simple pure pursuit guidance law given in equation (5) does not constrain the UAV-target closing velocity. It is the velocity with which the tracking vehicle closes on to the target and is given as negative of range rate, that is, $-\dot{R}$. As UAV approaches the target, closing velocity between UAV and the target should asymptotically decrease for a safe and precise landing. To achieve this, it must satisfy the condition shown in equation (24),

$$\lim_{R \to 0} -\dot{R} \to 0. \quad (24)$$

In subsequent sections, we first explain the integration of PP guidance with velocity controller, to control UAV's velocity for a 2D engagement scenario. We then extend this to 3D and show its application to an actual landing scenario.

PP with integrated velocity control

Velocity of UAV and target must become equal at touchdown to achieve desired closing velocity. Additionally, velocity of UAV must be relatively high for significant part of the flight to approach the target in time efficient manner. To achieve both these properties, we propose a log of polynomial function to control the velocity of UAV. It is given as,

$$f(x) = N^{log_n(1+ax+bx^2+cx^3)}, \quad (25)$$
$$x = R_{xy}/R^0_{xy}. \quad (26)$$

subject to,

$$1 + a + b + c = n. \quad (27)$$

Theorem 0.1. *Let f(x) be a log of polynomial function of the ratio of current to initial line of sight distance (R_{xy}/R^0_{xy}). Then, the closing velocity ($-\dot{R}_{xy}$) exponentially decreases to zero as the distance goes to zero if $V_u = f(x)V_t$.*

Proof. In PP, the aim is to continuously point towards the target and the guidance command is aimed at maintaining χ_u=ψ. Therefore equation (1) reduces to

$$\dot{R}_{xy} = V_t \cos(\chi_t - \psi) - V_u. \quad (28)$$

As, the UAV approaches the target, $R_{xy} \to 0$ and $\chi_t \to \psi$, reducing equation (28) to

$$\lim_{R_{xy} \to 0} \dot{R}_{xy} = V_t - V_u. \quad (29)$$

The variation in velocity is given as,

$$V_u = V_t f(x), \quad (30)$$

where,

$$N = V_u^0 / V_t^0, \quad (31)$$
$$x = R_{xy} / R_{xy}^0. \quad (32)$$

Initially, R_{xy}=R_{xy}^0, x=1. Therefore, from equations (26) and (30):

$$\log_n(1+ax+bx^2+cx^3) = 1, \quad (33)$$
$$V_u = NV_t \ if \ R_{xy} = R_{xy}^0. \quad (34)$$

As UAV approaches the target and $R_{xy} \to 0$, $x \to 0$ and $log_n(1+ax+bx^2+cx^3) \to 0$. As a result,

$$\lim_{R_{xy}\to 0} V_u \to V_t, \quad (35)$$

$$\lim_{R_{xy}\to 0} \dot{R}_{xy} \to 0. \quad (36)$$

□

The velocity of UAV is varied as product of target velocity and log polynomial function of distance. The variation depends on N i.e ratio of magnitude of initial UAV to target velocity and x i.e ratio of current line of sight to initial line of sight distance. The variation of UAV velocity based on distance allows us to vary it appropriately during different phases of the flight. Besides achieving the desired closing velocity conditions, log of polynomial function has other desirable properties of UAV flight,

1. It has a slower decay for most of the flight which makes sure that velocity of the UAV is significantly higher than the target.
2. Faster decay towards the end which quickly drives the closing velocity to minimum as UAV approaches the target.

In 3D, $-\dot{R}_{xy}$, ψ reflect the velocity with which the UAV approaches the target and its alignment with the target while doing so. $-\dot{R}_z$ reflects the velocity with which the UAV descends. The extension of theorem 0.1 to a 3D scenario to drive both $\dot{R}_{xy}$ and $\dot{R}_z$ to minimum is described next.

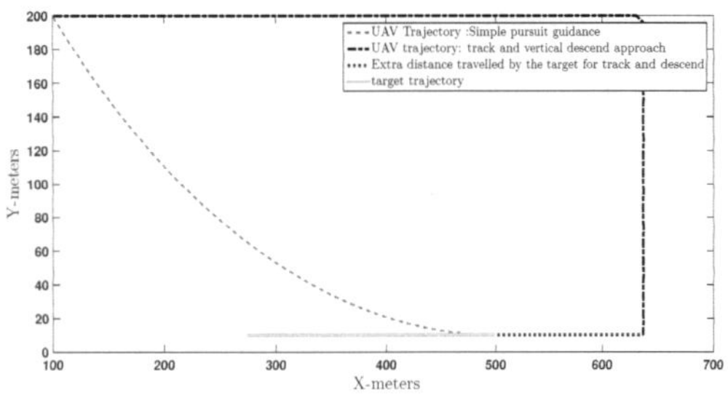

Fig. 5. Landing trajectories comparison between simple pure pursuit and track descend approach.

Lemma 0.2. *Let f(x) be a log of polynomial function of the ratio of current to initial line of sight distance (R^0/R). Then, the closing velocity in xy plane and z direction ($-\dot{R}_{xy}$ and $-\dot{R}_z$) exponentially decreases to zero as the distance goes to zero if $V_u = f(x)V_t$.*

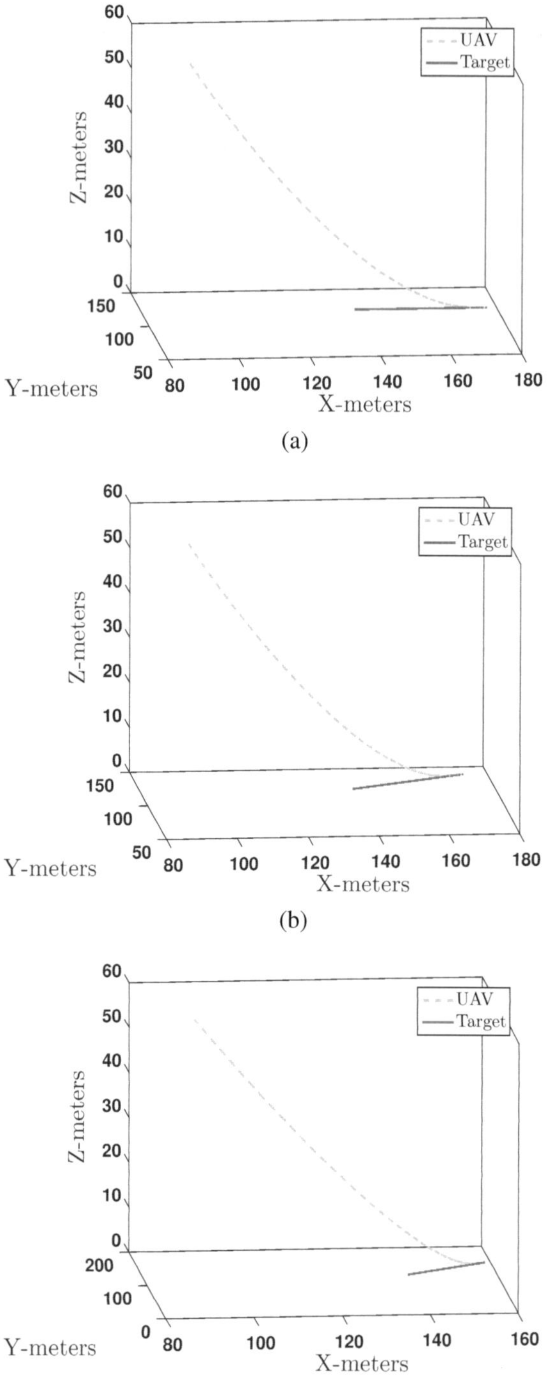

Fig. 6. MATLAB results showing the trajectory followed by the UAV to land, for target heading of (a) 0 deg **(b)** 30 deg **(c)** 60 deg

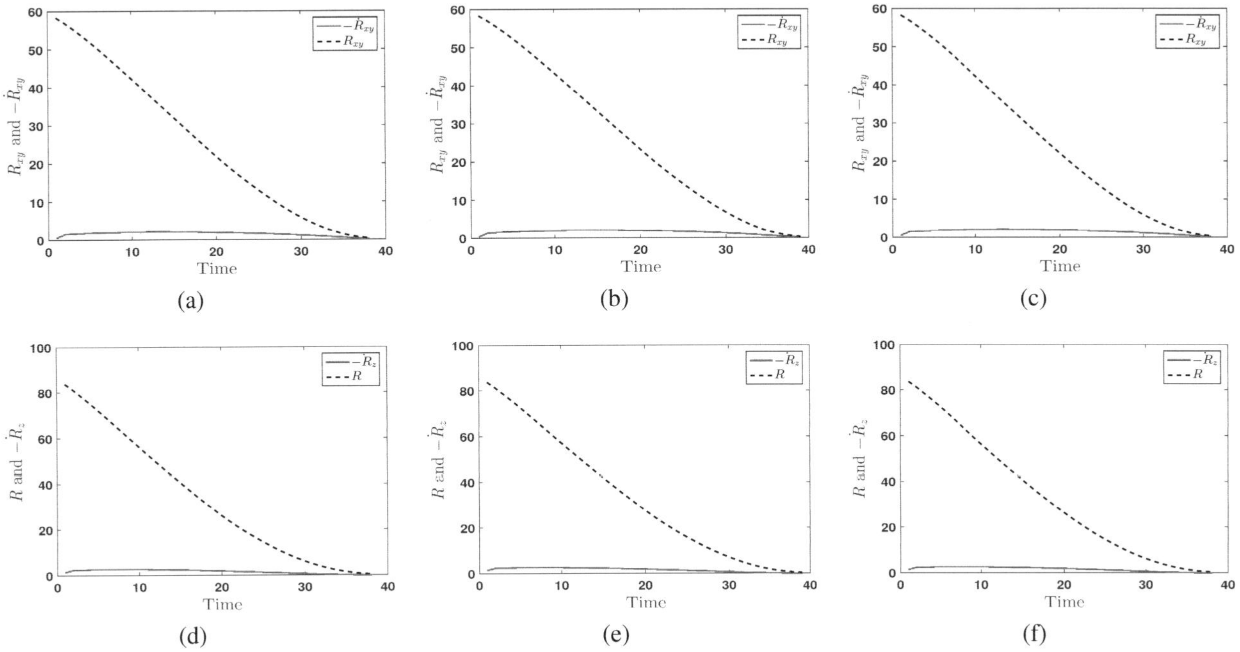

Fig. 7. Closing velocity profile with time in MATLAB, for a target heading of (a) 0 deg **in** xy **plane (b)** 30 deg **in** xy **plane (c)** 60 deg **in** xy **plane (d)** 0 deg **in** z **direction (e)** 30 deg **in** z **direction (f)** 60 deg **in** z **direction.**

Proof. At the time of landing, as $R \to 0$, θ becomes very small. Taking a small angle approximation, equation (10) reduces to,

$$\dot{R}_{xy} = V_t \cos(\chi_t - \psi) - V_u \cos(\chi_u - \psi). \tag{37}$$

Also, as UAV approaches the target $\chi_t \to \psi$ and $\sin(\theta) \to 0$. Therefore, as $R \to 0$,

$$\dot{R}_z \to 0. \tag{38}$$

$$\dot{R}_{xy} \to V_t - V_u. \tag{39}$$

As $R_{xy} \to 0$, $\dot{R}_{xy}$ given by equation (39) approaches to 0 as explained in theorem 0.1. □

Simulation Results

Consider a scenario as shown in figure 5, where the UAV lands on a moving target. For the first case, we use a simple pure pursuit guidance, while in the second case the UAV tracks the target and then vertically descends towards the target. Pure pursuit guidance method took 45 seconds to reach the target where as the *track and descend* method took 72 seconds. This shows better performance of guidance based landing in terms of time to land by 37.5 percent in a very simple scenario. The parameters for the simulation were $V_u^0 = 2$ m/s, $V_t = 1$ m/s and $R_{xy}^0 = 40$ meters. The termination criteria was $R_{xy} \leq 0.1$ meters in respective cases.
We further present the results of guidance based landing method for a 3D scenario in MATLAB, followed by the results obtained in ROS using a quadrotor simulator which takes into account quadrotor dynamics.

Landing using 3D guidance in MATLAB

As a sample scenario, a UAV was considered with $R_{xy}^0 = 40$ meters, at an altitude of 60 meters above the ground. χ_t was 0 deg which was constant and χ_u, γ were initialized to be equal to ψ and θ respectively. V_u^0 and V_t were 2m/s and 1m/s respectively. The target speed and the heading was constant throughout the simulation. UAV and target trajectories are plotted for a target heading of 0 deg in figure 6(a). Trajectories for different target headings of 30 deg and 60 deg, with same initial positions of UAV and target are plotted in figures 6(b) and 6(c). The termination criteria was $R \leq 0.1$ meters with $|\dot{R}_{xy}| \leq 0.02$ m/s. The closing velocity profile for 0 deg target heading is shown in figures 7(a) and 7(d), where both $\dot{R}_{xy}$ and $\dot{R}_z$ go to 0 as $R \to 0$ thereby achieving the desired terminal conditions. Similar trends were observed in landing for same initial positions and different target headings of 30 deg and 60 deg as shown in figures 7(b), 7(c), 7(e) and 7(f). The acceleration profiles for the three heading angles are shown in figure 8.

Sinusoidal target motion Consider a target moving in a sinusoidal manner. The target motion equations are given by,

$$x_{t_{t+\Delta t}} = x_{t_t} + \frac{(V_t * \Delta t)}{\sqrt{1 + (2\cos(x_t)^2}}, \tag{40}$$

$$y_{t_{t+\Delta t}} = \sin(x_{t_{t+\Delta t}}). \tag{41}$$

Target course angle is given by,

$$\chi_t = \tan^{-1}(\cos(x_t)). \tag{42}$$

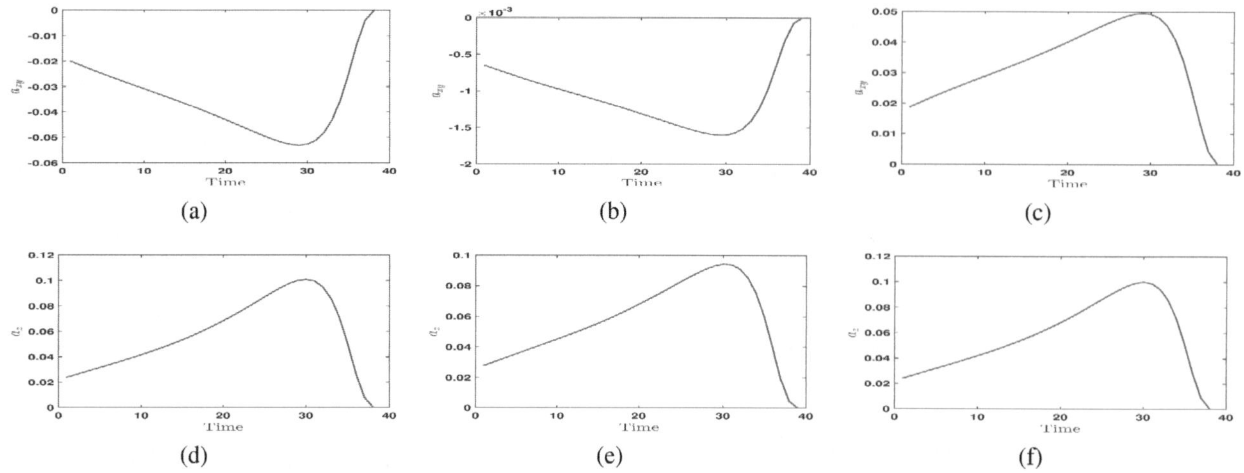

Fig. 8. Lateral acceleration profile with time in MATLAB, for a target heading of (a) 0 deg **in** xy **plane (b)** 30 deg **in** xy **plane (c)** 60 deg **in** xy **plane (d)** 0 deg **in** z **direction (e)** 30 deg **in** z **direction (f)** 60 deg **in** z **direction.**

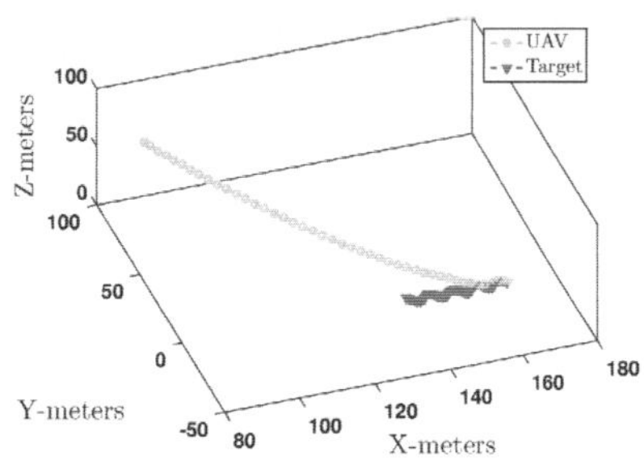

Fig. 9. Landing trajectory of a UAV for a sinusoidal target

Trajectory for the sinusoidally moving target is shown in figure 9, where the UAV is able to land on the landing target successfully with minimum closing velocity. Closing velocity and lateral acceleration profiles are shown in figures 10 and 11.

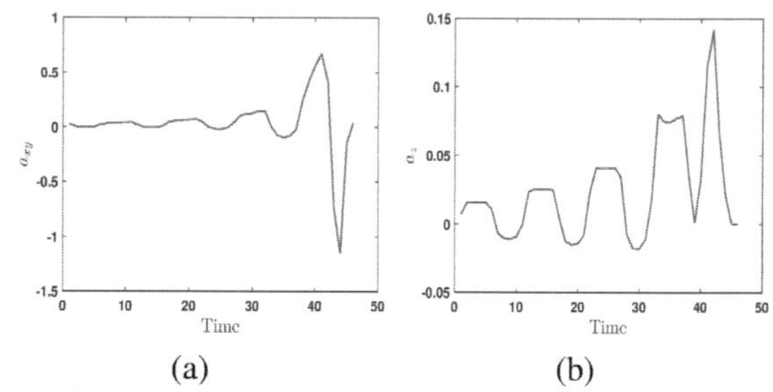

Fig. 11. Lateral acceleration profiles for a sinusoidal target(a) in xy **plane with time (b) in** z **direction with time.**

Hardware setup

We use an off the shelf quadcopter (IRIS+) for hardware validation of our landing algorithm. The landing algorithm runs on a linux based embedded computer: Odroid U3. This is mounted on top of the quadrotor and connected to a downward facing 720P USB camera. The hardware setup along with the landing target is shown in figure 12. The target is chosen in the form of a red colored circle of known diameter. The landing algorithm is implemented using ROS and target detection is performed with opencv, using color based blob detection approach. Guidance commands are sent to the quadrotor over a USB link using ROS package mavros which converts ROS messages to mavlink format and vice versa, to aid communication with the autopilot.

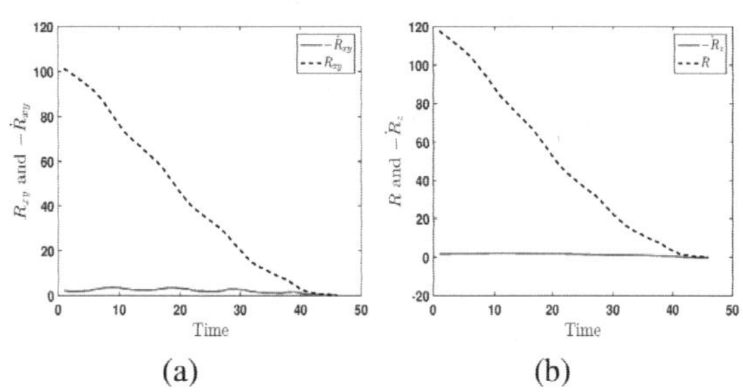

Fig. 10. Closing velocity profiles for sinusoidal target (a) in xy **plane with time (b) in** z **direction with time.**

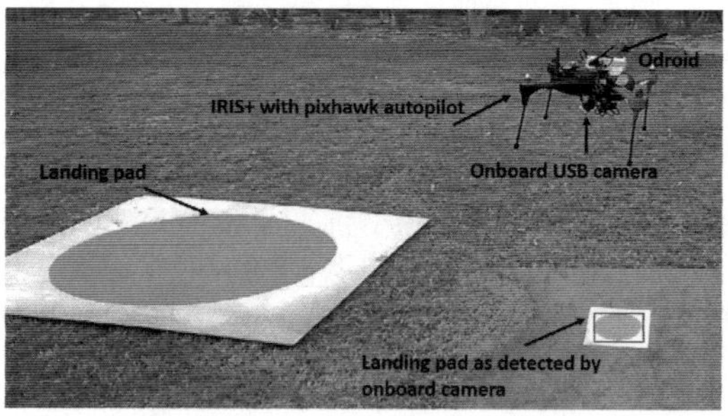

Fig. 12. Quadcopter with the whole hardware interface and the landing pad as seen by the onboard camera (inset).

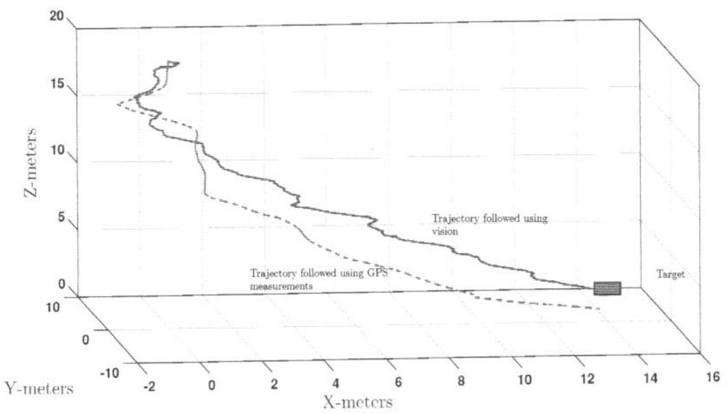

Fig. 13. Comparison of the trajectories followed by the quadcopter to land on the target using vision and GPS measurements.

Hardware results

In order to validate our algorithm, we have performed outdoor experiments with the hardware setup described in section .
We have performed two sets of experiments. The first experiment tests the validity of the proposed guidance algorithm without vision in the controller feedback loop. This is done by generating guidance commands based on GPS measurements. In this case, the GPS coordinates of the target are assumed to be known. In the second experiment, the target is detected using vision and guidance commands are issued to quadrotor based on the extracted target information. To initiate the landing process, the quadrotor was commanded to takeoff to an altitude of 20 meters. The target was stationary and the image data was processed online.
The inset in figure 12 shows the target as detected by the onboard camera (bounding box in the figure). At each step of the landing process, the guidance system generates (a) velocity commands in xy plane with respect to image frame (b) velocity commands in z direction based on the altitude. These commands are converted to velocity north, east and down commands based on the heading of the quadrotor. Figure 13 shows the trajectories followed by quadrotor with GPS measurements and vision based input. With GPS measurements, the quadrotor follows an efficient trajectory towards the target. The error in landing location is due to error in the GPS measurements. By including the vision in loop, the quadrotor is able to land precisely on the target following a similar trajectory. Figures 14 and 15 show the performance of vision based controller. The quadrotor velocity in both lateral and longitudinal directions is proportional to the error in target distance (measured in pixels) in the respective direction.
Figure 16 shows the performance of the closing velocity controller with GPS input. It can be observed that the variation in closing velocity is same as that obtained through simulations and the closing velocity can be seen to be approaching zero towards the end. As observed from the hardware results with stationary target, the guidance based approach enables the quadrotor to land on the landing pad accurately. The integrated closing velocity controller provides velocity commands in reference to quadrotor position from target and also achieves zero closing velocity, necessary for event-less landing.

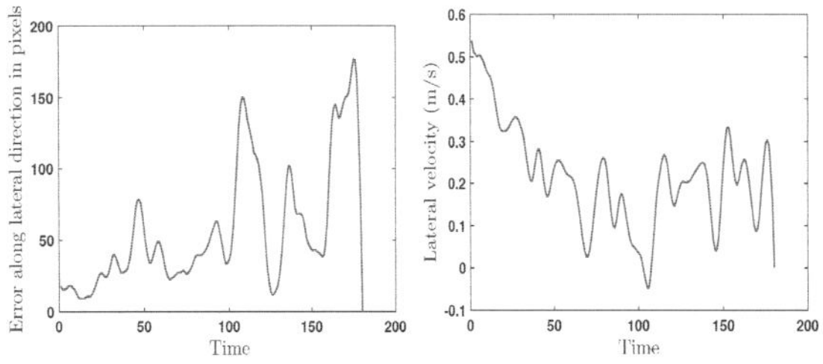

(a) (b)

Fig. 14. (a) Error output by controller in lateral direction (b) Lateral velocity profile.

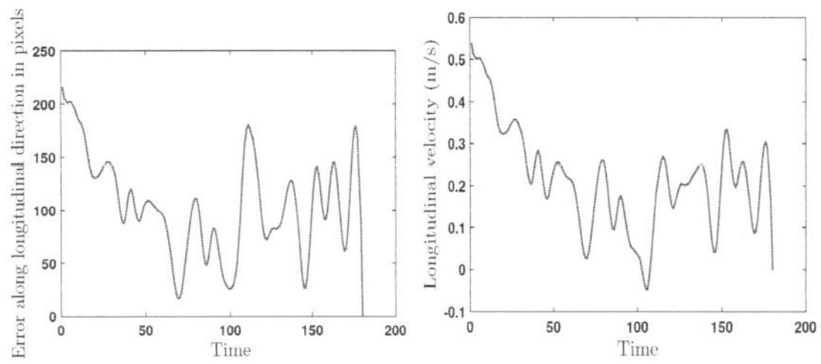

(a) (b)

Fig. 15. (a) Error output by controller in longitudinal direction (b) Longitudinal velocity profile.

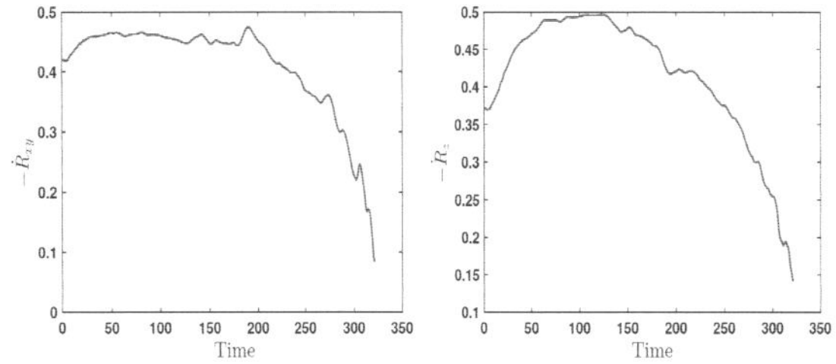

(a) (b)

Fig. 16. Closing velocity profile with time from GPS measurements (a) in xy plane (b) in z direction.

Conclusion

In this paper, the challenges of the autonomous landing problem for quadrotors have been discussed. A pure pursuit guidance based landing approach for quadrotors is proposed for a faster and accurate autonomous landing. Guidance based landing approach is shown to be 37.5 percent faster as compared to the vertical landing approach. A closing velocity controller has been integrated with the PP guidance based landing approach to satisfy velocity constraints and achieve faster landing with minimum closing velocity. The efficacy of the proposed approach is validated exhaustively through simulations. We have also performed validation of the proposed approach for a stationary landing pad through extensive outdoor hardware experiments. The results show that the proposed approach follows an efficient trajectory and lands on the target precisely with vision in feedback. The future scope of this work involves evaluating the proposed controller for a moving landing target through outdoor hardware experiments.

References

[1]Geng, L., Zhang, Y. F., Wang, J. J., Fuh, J. Y. H., and Teo, S. H., "Mission planning of autonomous UAVs for urban surveillance with evolutionary algorithms," 2013 10th IEEE International Conference on Control and Automation (ICCA), June 2013.
doi: 10.1109/ICCA.2013.6564992

[2]Tokekar, P., Hook, J. V., Mulla, D., and Isler, V., "Sensor planning for a symbiotic UAV and UGV system for precision agriculture," 2013 IEEE/RSJ International Conference on Intelligent Robots and Systems, Nov 2013.
doi: 10.1109/IROS.2013.6697126

[3]Herwitz, S., Johnson, L., Dunagan, S., Higgins, R., Sullivan, D., Zheng, J., Lobitz, B., Leung, J., Gallmeyer, B., Aoyagi, M., Slye, R., and Brass, J., "Imaging from an unmanned aerial vehicle: agricultural surveillance and decision support," *Computers and Electronics in Agriculture*, Vol. 44, (1), 2004, pp. 49 – 61.
doi: http://dx.doi.org/10.1016/j.compag.2004.02.006

[4]Waharte, S. and Trigoni, N., "Supporting Search and Rescue Operations with UAVs," Emerging Security Technologies (EST), 2010 International Conference on, Sept 2010.
doi: 10.1109/EST.2010.31

[5]Zhang, J. and Yuan, H., "Analysis of unmanned aerial vehicle navigation and height control system based on GPS," *Journal of Systems Engineering and Electronics*, Vol. 21, (4), Aug 2010, pp. 643–649.
doi: 10.3969/j.issn.1004-4132.2010.04.018

[6]Cho, A., Kim, J., Lee, S., Choi, S., Lee, B., Kim, B., Park, N., Kim, D., and Kee, C., "Fully automatic taxiing, takeoff and landing of a UAV using a single-antenna GPS receiver only," Control, Automation and Systems, 2007. ICCAS '07. International Conference on, Oct 2007.
doi: 10.1109/ICCAS.2007.4407014

[7]Hsiao, F.-B., Huang, S.-H., and Lee, M.-T., "The study of real-timed GPS navigation accuracy during approach and landing of an ultralight vehicle," Recent Advances in Space Technologies, 2003. RAST '03. International Conference on. Proceedings of, Nov 2003.
doi: 10.1109/RAST.2003.1303946

[8]Cesetti, A., Frontoni, E., Mancini, A., Zingaretti, P., and Longhi, S., *A Vision-Based Guidance System for UAV Navigation and Safe Landing using Natural Landmarks*, Springer Netherlands, Dordrecht, 2010, pp. 233–257.

[9]Voos, H., "Nonlinear control of a quadrotor micro-UAV using feedback-linearization," Mechatronics, 2009. ICM 2009. IEEE International Conference on, April 2009.
doi: 10.1109/ICMECH.2009.4957154

[10]Ahmed, B. and Pota, H. R., "Backstepping-based landing control of a RUAV using tether incorporating flapping correction dynamics," 2008 American Control Conference, June 2008.
doi: 10.1109/ACC.2008.4586905

[11]Malaek, S. M. B., Sadati, N., Izadi, H., and Pakmehr, M., "Intelligent autolanding controller design using neural networks and fuzzy logic," Control Conference, 2004. 5th Asian, Vol. 1, July 2004.

[12]Mori, R., Suzuki, S., Sakamoto, Y., and Takahara, H., "Analysis of visual cues during landing phase by using neural network modeling," *Journal of aircraft*, Vol. 44, (6), 2007, pp. 2006–2011.

[13]Shue, S.-P. and Agarwal, R. K., "Design of Automatic Landing Systems Using Mixed H_2/H_∞ Control," *Journal of Guidance, Control, and Dynamics*, Vol. 22, (1), 1999, pp. 103–114.

[14]Kim, H. J., Kim, M., Lim, H., Park, C., Yoon, S., Lee, D., Choi, H., Oh, G., Park, J., and Kim, Y., "Fully Autonomous Vision-Based Net-Recovery Landing System for a Fixed-Wing UAV," *IEEE/ASME Transactions on Mechatronics*, Vol. 18, (4), Aug 2013, pp. 1320–1333.

[15]Yoon, S., Kim, H. J., and Kim, Y., "Spiral Landing Trajectory and Pursuit Guidance Law Design for Vision-Based Net-Recovery UAV," , (AIAA 2009-5682), Chicagi, Illinios, 2009.

[16]Byoung-Mun Min, D. H. S. H. B., Min-Jea Tahk, "Guidance law for vision-based automatic landing of UAV," *International Journal of Aeronautical and Space Sciences*, Vol. 8, 2007, pp. 46–53.

[17]Barber, D. B., Griffiths, S. R., McLain, T. W., and Beard, R. W., "Autonomous landing of miniature aerial vehicles," *Journal of Aerospace Computing, Information, and Communication*, Vol. 4, (5), 2007, pp. 770–784.

[18]Huh, S. and Shim, D. H., *A Vision-Based Automatic Landing Method for Fixed-Wing UAVs*, Springer Netherlands, Dordrecht, 2010, pp. 217–231.

[19]Valavanis, K. P. and Vachtsevanos, G. J., *Handbook of Unmanned Aerial Vehicles*, Springer Publishing Company, Incorporated, 2014.

[20]Robotics, D.

[21]odroid, "odroid," .

Obstacle Detection Algorithms in Thermal Infrared Images for Collision Avoidance

Adrian Carrio · Srikanth Saripalli · Pascual Campoy

Abstract Obstacle detection is an essential component in any Sense and Avoid system. Thermal Infrared (TIR) imaging technology can provide enhanced capabilities to current vision-based SAA systems, allowing for operation under extreme illumination conditions, such as direct sun exposure, and during the night.

This paper presents a comparison of image processing algorithms for the detection of flying obstacles in TIR images, including both methods based in morphological operations and methods based in the Discrete Haar Wavelet Transform (DHWT). To the author's knowledge, the use of these methods for obstacle detection in TIR images is completely novel.

These algorithms have been evaluated on a dataset of more than 5K image pairs, including both ground-based and airborne TIR images. This dataset includes images captured under very different illumination conditions (normal exposure, direct sun exposure and at night) and with various sky conditions (clear and cluttered skies).

Conclusions validate these algorithms by presenting measurements in terms of precision and recall for the detection of flying obstacles in several video sequences and a study of the computational cost of the evaluated algorithms.

Adrian Carrio · Pascual Campoy
Computer Vision & Aerial Robotics Group,
Centro de Automática y Robótica(CAR), UPM-CSIC,
Universidad Politécnica de Madrid,
C/ José Gutiérrez Abascal 2, 28006 Madrid, Spain
Tel.: +34-913363061
Fax: +34-913363010
E-mail: adrian.carrio@upm.es
URL: www.vision4uav.com

Srikanth Saripalli
School of Earth and Space Exploration
Arizona State University
Tempe, AZ 85287, USA

1 Introduction and related works

A lot of effort is currently being put in the development of systems which are able to detect, localize and avoid manned aircraft, as these systems will enable safer operability for Unmanned Aerial Vehicles (UAVs) in non-segregated aerospace. Several countries are holding limitations on the autonomous operability beyond the operator's line of sight, restricting the use of UAV platforms for civilian tasks such as: disaster rescue [12], wildlife surveillance [13] and building inspection [2].

So far, no standard requirements have been proposed by aviation agencies in USA or Europe on the technical specifications of SAA systems, while several systems are being developed and tested. Some rules of thumb have been proposed by organizations such as, ASTM which stated a 500 ft miss distance in resolution maneuvers [1] or the Highly Capable UAV (HCU) Payloads Planning Document, which set the minimum detection distance in 3 statute miles (4.83Km).

Active sensing technologies such as radar and laser-based systems have been proposed for the long-range detections required in order to avoid fast-moving aircraft. Given the power limitations onboard micro and mini UAVs, the use of these technologies for the development of Sense-And-Avoid systems is not feasible.

RGB cameras have been widely considered for See-and-Avoid [3, 6, 11, 14, 17] for their advantageous features: low cost, lightness and low energy consumption. However, several limitations still prevent the development of solutions based only in RGB image sensors. An example is the need for both high angular resolution (125 μrad for the detection of general aviation [5]), wide field of view in order to guarantee visual contact and enough magnification to ensure aircraft in the image has features enabling its detectability. Another important limitation of RGB cameras is their dependency on

Figure 1 Prototype system with both RGB and TIR cameras mounted on a static tripod.

illumination, which limits their usability under extreme lighting conditions and at night.

In this paper, we present the integration and evaluation of a system capable of detecting flying obstacles using TIR image sensors using two different approaches: morphological operations and the Discrete Haar Wavelet Transform (DHWT). Both methods have been proposed for the detection of aircraft in RGB images and only DHWT has been proposed for the detection of objects other than aircraft in TIR images, such as pedestrians [8]. We also show that TIR cameras are a good alternative to RGB cameras for obstacle detection under extreme lighting conditions. To the best of the author's knowledge, no previous evaluation of SAA detection algorithms based on thermal image sensors has been published.

This paper is organized as follows: Section 2 presents the proposed system; Section 3 introduces the vision-based obstacle detection algorithms proposed; Section 4 describes the collected data; Section 5 discusses the experimental validation; and Section 6 presents the conclusions and future work.

2 System description

The proposed system uses a TIR camera for the detection of flying aircraft. An RGB camera has also been integrated in the system for visualization purposes. The system was initially integrated on a static tripod, shown in Fig.1 which logged and stored both RGB and TIR images with the help of a desktop computer. Afterwards the system was integrated onboard a quadrotor UAV.

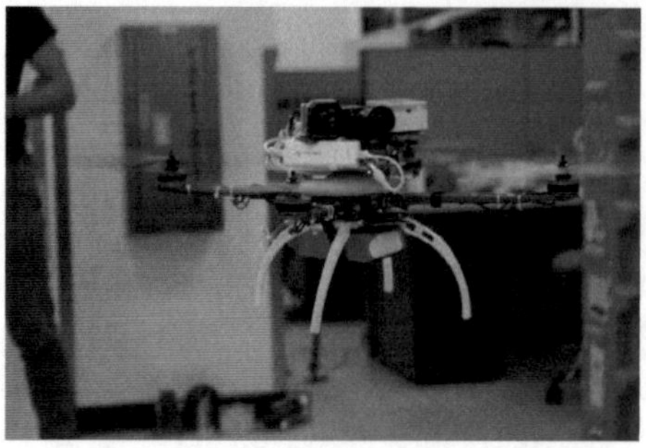

Figure 2 UAV prototype. Both RGB and TIR cameras where integrated on a small quadrotor.

2.1 Image sensors

TIR images were captured using a Flir Tau 2 LWIR camera featuring an image resolution of 640×512, a weight of 90 grams and a field-of-view of 69°×56°. Thermal images were acquired at 10Hz and digitalized using a standard A/D video converter. No radiometric measurements were done.

RGB images were captured with a AVT Prosilica GT 1920C camera with 1936×1456 resolution and 83°×67° wide field of view. The camera weights about 310 grams, including the lens.

2.2 UAV system integration

Cameras were integrated onboard a custom-built quadrotor UAV, shown in Fig.2. Images from both cameras are simultaneously acquired at 10 fps and stored using an onboard Intel NUC D53427RKE Computer, featuring an Intel i5 processor. The UAV with the integrated SAA system has a takeoff weight of 3 Kg and can fly for approximately 15 minutes using a 10000 mAh Li-Po battery.

3 Vision-based obstacle detection algorithms

Flying aircraft will appear as small-sized (subpixel to a few pixels) contrast features in both RGB and TIR images. While in RGB images features may appear brighter or darker than the background, it can be expected that flying aircraft will appear brighter in TIR images, as there is more thermal radiation coming from them into the sensor. A problem which is common to both RGB and TIR images is the existence of spurious contrast points, especially in the cluttered ground regions.

A simplification for the contrast features detection problem has been proposed in many vision-based SAA prototype systems which consists on segmenting the sky and detecting obstacles in sky pixels only [4, 10]. By only focusing on sky pixels, the computational cost of the algorithms can be greatly reduced while avoiding the detection of false positives in the ground regions.

As in other vision-based SAA protoype systems, such as [4, 10], we assume that aircraft will appear only above the horizon line for simplicity. Our algorithm for horizon line detection divides the image into two regions: sky and ground and we apply two contrast feature detection algorithms in the segmented sky region, one based on the top-hat morphological operation and the other one based on the Discrete Haar Wavelet Transform. Both algorithms run independently, each of them providing the obstacles present in the image. A mean-shift strategy can be used over the outputs of the feature detection algorithms in order to track the detected targets along time.

3.1 Horizon line detection

Previous to the detection of the horizon line the image histogram is equalized for normalization purposes. Then, the position of the horizon line in an image of width w and height h is estimated by finding the positions in the image of P_l and P_r. The resulting horizon line will be the line joining these two points, which can be computed using Algorithm 1. Please note that $I(i, j)$ is the image intensity function, and n and t are constants which have been empirically set to 1/8 of the image height and 3/20, respectively.

Algorithm 1 Horizon line detection

for each row i in $[1, h]$ **do**
 $f_l(i) = \frac{\sum_{k=1}^{w/2} I(i,k)}{w/2}$
 $f_r(i) = \frac{\sum_{k=w/2}^{w} I(i,k)}{w/2}$
end for
for each row i in $[1, h-n]$ **do**
 $g_l(i) = \frac{\sum_{k=i}^{i+n} f_l(k)}{n}$
 $g_r(i) = \frac{\sum_{k=i}^{i+n} f_r(k)}{n}$
end for
$i = 1$
while $g_l(i+1) - g_l(i) > t \wedge i < h-1$ **do**
 $P_l = (i+1, 1)$
 $i = i + 1$
end while
$i = 1$
while $g_r(i+1) - g_r(i) > t \wedge i < h-1$ **do**
 $P_r = (i+1, 1)$
 $i = i + 1$
end while

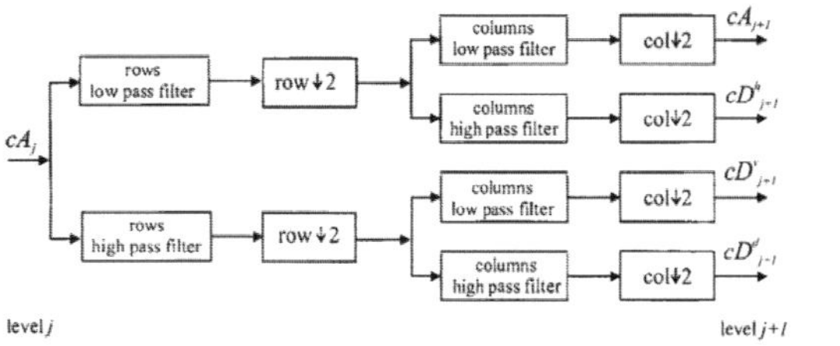

Figure 3 2D Discrete Wavelet Transform. A decomposition step.

3.2 Morphological image processing

Several contrast feature detection methods have been proposed for the detection of flying obstacles in TIR images [7,9,16]. As mentioned earlier, thermal radiation received from the flying obstacles can be expected to be higher than the local image background, thus causing its corresponding pixels to have higher intensity than those corresponding to the background for TIR images. A good approach for capturing this positive-contrast features is the top-hat transformation. Let $Y \oplus S$ and $Y \ominus S$ represent respectively the dilation and erosion morphological operations of a grayscale image Y by a structuring element S, which acts as a cut-off parameter for filtering out features that are too large to be of interest, the top-hat transformation is defined by Eq. 1.

$$TH(Y, S) = Y - [(Y \ominus S) \oplus S] \quad (1)$$

After applying the morphological filtering operation to the TIR image, the image is thresholded to remove spurious low-contrast points, which usually appear in cluttered skies or as a result of image noise. Non-zero pixels in the thresholded image will be considered detections if their intensity in the original TIR image is above a certain constant threshold, which helps minimizing false detections.

3.3 2D Discrete Haar Wavelet Transformation

A 2D discrete wavelet transformation decomposes a single approximation coefficient at level j into four components at level $j+1$: the approximation coefficient cA_{j+1} and detail coefficients at three orientations (horizontal cD^h_{j+1}, vertical cD^v_{j+1} and diagonal cD^d_{j+1}). The process is illustrated in Fig. 3. The symbol col $\downarrow 2$ stands for down sampling columns by keeping only the even indexed columns. Similarly, row $\downarrow 2$ means down sampling rows by keeping only evenly indexed rows [15].

For N even, the Discrete Haar Wavelet Transform is defined as:

$$W_N = \begin{bmatrix} \sqrt{2}/2 & \sqrt{2}/2 & 0 & 0 & \cdots & 0 & 0 \\ 0 & 0 & \sqrt{2}/2 & \sqrt{2}/2 & & 0 & 0 \\ \vdots & & & & \ddots & & \vdots \\ 0 & 0 & 0 & \cdots & 0 & \sqrt{2}/2 & \sqrt{2}/2 \\ -\sqrt{2}/2 & \sqrt{2}/2 & 0 & 0 & \cdots & 0 & 0 \\ 0 & 0 & -\sqrt{2}/2 & \sqrt{2}/2 & & 0 & 0 \\ \vdots & & & & \ddots & & \vdots \\ 0 & 0 & 0 & \cdots & 0 & -\sqrt{2}/2 & \sqrt{2}/2 \end{bmatrix} \quad (2)$$

The upper half of W_N acts as a low-pass filter as it produces a weighted average while the lower half of W_N acts as a high-pass filter since it produces a weighted difference.

Given an input image A of MxN dimensions the 2D Discrete Haar Wavelet Transform can be computed as:

$$B = W_M A W_N^T \quad (3)$$

In the case of a A being a square matrix this can be expressed as:

$$B = W_M A W_N^T = \begin{bmatrix} H \\ G \end{bmatrix} A \begin{bmatrix} H \\ G \end{bmatrix}^T = \begin{bmatrix} HAH^T & HAG^T \\ GAH^T & GAG^T \end{bmatrix} \quad (4)$$

The lower left-hand block is GAH^T. GA differences columns of A and the rows of this product are averaged by multiplication with H^T. As a result, the lower left-hand corner holds information about horizontal intensity changes in the image. Large values indicate a large horizontal change as we move down the image while small values indicate little horizontal change. GAH^T therefore provides a downscaled image where high-frequency horizontal edges are enhanced. A thresholding operation on the resulting image will enhance objects with high thermal contrast, which can be marked as detections if their intensity in the original TIR image is above a certain constant threshold, as in the case of the morphological operations approach.

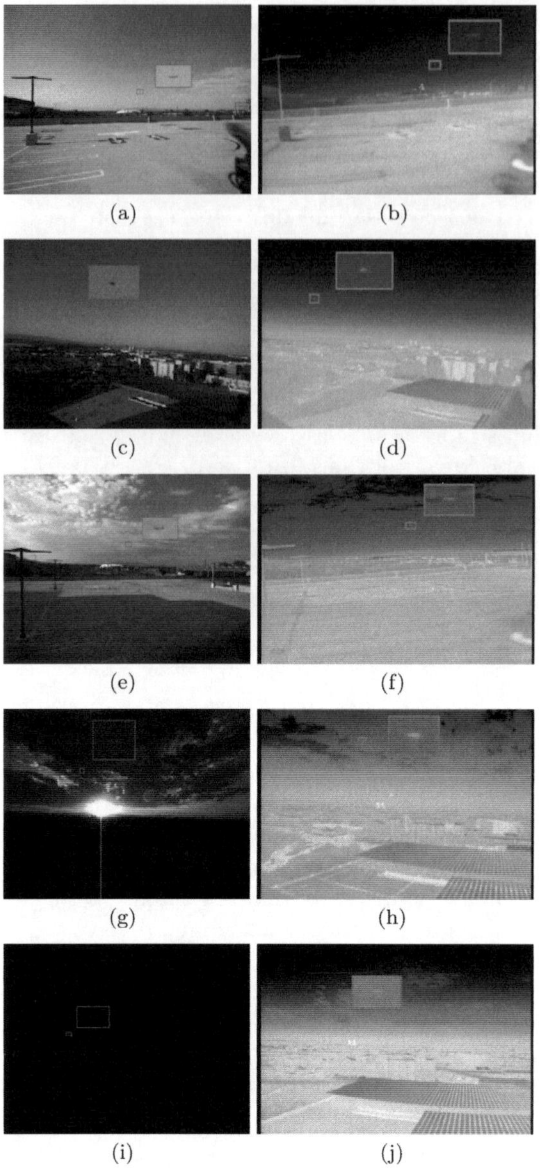

Figure 4 Samples of RGB (left) and TIR (right) frames, captured at different times of the day and under different cloud conditions. All images except c and d show highlighted flying aircraft. A bird is highlighted in images c and d. Smaller rectangles indicate the detected objects, while larger rectangles have been superimposed and magnify the detections for visualization purposes.

4 DATA COLLECTION

Images of fly-by aircraft scenes from both a static tripod and from cameras on-board a UAV were collected. The data was acquired in Tempe, AZ, during the months of April to June 2015. More than 5000 synchronized image pairs of RGB and TIR images were acquired, containing various lighting and cloud conditions, shown in Fig. 4. Flying obstacles captured on camera not only include aircraft but also birds, shown in Fig. 4 c and d.

Static-camera images were collected from a building rooftop at a height of approximately 35m from the ground. Four manual flights of the UAV prototype were performed from the top of a parking facility, flying up to 20m above the ground. Both cameras were configured in full auto mode.

As it was expected, Fig. 4 shows how thermal images are not so greatly affected by the different cloud and

Figure 5 Automatic horizon line detections under different challenging scenarios. Above, a frame captured from the ground showing a cluttered sky and solar panels with low thermal radiation. Below, images captured during a flight over concrete being heated by the sun.

lighting conditions. Detections seem more clearly visible to the human eye in the TIR images, especially under cluttered skies (Fig. 4e), with front sunlight (Fig. 4g) or at night (Fig. 4i).

5 Experimental evaluation

In this section, results of the proposed algorithms running on the captured data are presented. First, some results showing the robustness of the horizon line detection are shown. Then, a comparison in terms of precision and recall for the top-hat morphological operations approach and the discrete Haar Wavelet Transform approach are presented. Computational cost of both approaches is also analyzed.

5.1 Horizon line detection

Our proposed algorithm showed robust and accurate results for horizon line detection. Occasionally, the position of the horizon line may shift by a few pixels under extreme thermal radiation conditions. Examples of the detection of the horizon line with our approach are shown in Fig. 5.

5.2 Detection algorithms comparison

Examples of intra-frame detections under different lighting and thermal radiation conditions are shown in Fig. 8. The detection capabilities of the system in both airborne and ground video sequences are shown in Fig. 6 and Fig. 7, respectively.

The thermal camera being configured in full auto mode causes pixel intensity to be frequently adjusted for different thermal radiation scenarios, making contrast-based detections difficult. For instance, the concrete surfaces shown in Fig. 6 are a challenging problem for automatic detection. The camera is continuously adjusting its dynamic range to the huge thermal radiation received by the heated concrete, thus decreasing the contrast of objects whose thermal radiation signal is lower, such as distant aircraft. Despite this effect, our implementation of both algorithms with a single set of parameters was used to perform all the automatic detections shown in this paper. Parameters were set to minimize false positive rate at the cost of false negative detections. TH outperforms DHWT in all cases. Moving camera does not affect detection accuracy. Night images cause a drop in performance, probably because of the lower thermal radiation reaching the sensor at night, which might be causing lower contrast and lower number of detections triggered in the presence of an aircraft. Results are shown in Table 5.2.

The average processing time for the proposed detection algorithm was 17ms per frame in the case of the method based on morphological operations and 22ms per frame in the case of the Discrete Haar Wavelet Transform method. These execution times include the computation of the horizon line detection. Algorithms were running on an Intel i7 Quad-core desktop computer running at 3.4 Ghz without any parallelization optimizations.

The proposed system can work day/night and under challenging lighting conditions, outperforming other vision-based SAA systems in the literature.

6 Conclusions

A lightweight, low-power aircraft detection system for small UAVs integrating RGB and TIR cameras has been developed and evaluated, demonstrating flying obstacle detection capabilities on TIR images.

An evaluation dataset was collected, including more than 5K images of synchronized RGB and TIR images of fly-by aircraft scenes. Algorithms for horizon line detection and contrast features detection using both morphological operations and the Discrete Haar Wavelet Transform were discussed and tested using the logged data.

TIR imaging has been proven as an effective technology for vision-based SAA systems, extending the cap-

Video sequence #	Method	Camera	Cloud condition	Precision	Recall
1	TH	Static	Cloudy	1.0000	0.8734
1	DHWT	Static	Cloudy	1.0000	0.5949
2	TH	Static	Dusk	1.0000	0.6716
2	DHWT	Static	Dusk	1.0000	0.5970
3	TH	Static	Night	1.0000	0.3296
3	DHWT	Static	Night	1.0000	0.2234
4	TH	Moving	Clear	1.0000	0.7630
4	DHWT	Moving	Clear	1.0000	0.1481

Table 1 Evaluation of both detection algorithms proposed, TH (Top Hat) and DHWT (Discrete Haar Wavelet Transform), in four of the images sequences acquired. Parameters on both algorithms have been tuned in order to minimize false positive rate at the cost of false negative detections. TH outperforms DHWT in all cases. Moving camera does not affect detection accuracy. Night images cause a drop in performance, probably because of the lower thermal radiation reaching the sensor at night, which might be causing lower contrast and lower number of detections triggered in the presence of an aircraft.

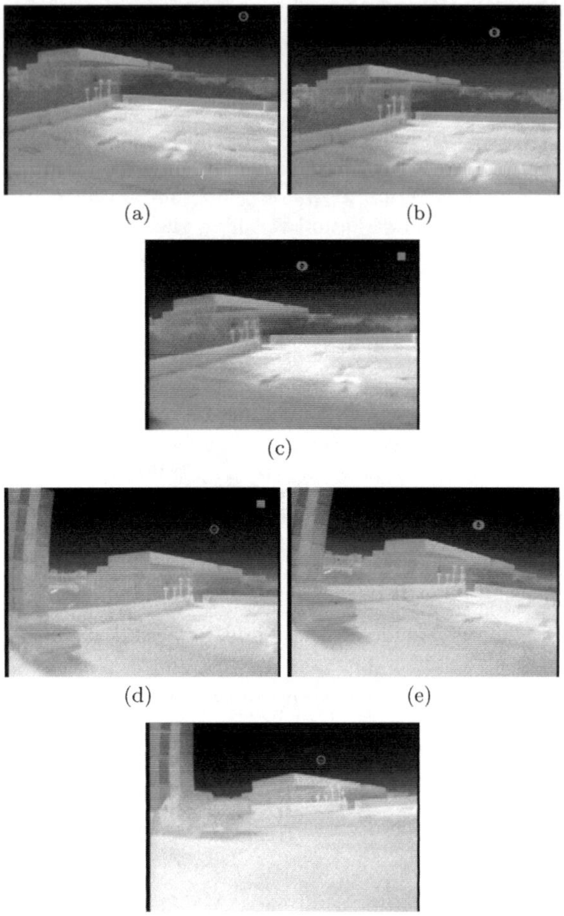

Figure 6 Automatic aircraft detections in a video sequence captured onboard using the method based on morphological operations. The hot concrete surfaces around become a challenging problem for the detection, as the camera is continuously adjusting its range to the received radiation, therefore decreasing the contrast of objects whose thermal radiation signal is low, such as far aircraft.

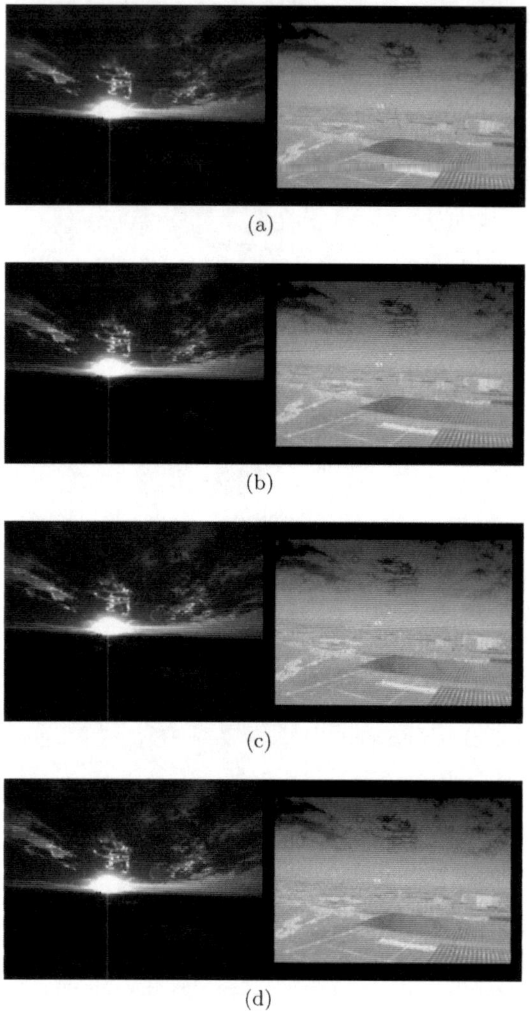

Figure 7 RGB images (left) show a sequence of four subsequent frames corresponding to a challenging dawn scene with front sunlight captured from the ground. Automatic aircraft detections using the Discrete Haar Wavelet Transform method on TIR images are shown on the right.

Figure 8 Automatic aircraft detections under different lighting and thermal radiation conditions using the method based on morphological operations.

abilities of RGB cameras for the operation under extreme illumination conditions such as during the night or with direct exposure to the sun. Experimental results demonstrate the feasibility of using TIR imagery for aircraft detection. In contrast to other RGB camera-based SAA systems, our prototype system can be used during the night and under challenging lighting conditions.

Besides the higher cost of thermal sensors as compared to RGB image sensors, an expected limitation would be the operation under foggy or dense cloud meteorological conditions, where TIR cameras show a weak performance.

The system provides good detection performance results at real-time frame rates with 100% accuracy in all cases and recall rates above 67% during the day. In night conditions recall rate drops to 33%, maintaining 100% accuracy.

Acknowledgements This work has received financial support from the Spanish Ministry of Science in the project DPI2014-60139-R and from the European Commision Programm IRSES in the project UECIMUAVS. The authors would like to thank Benjamin Stinnett for his valuable help carrying out the test flights.

References

1. ASTM F2411-07, Standard Specification for Design and Performance of an Airborne Sense-and-Avoid System (Withdrawn 2014), ASTM International, West Conshohocken, PA, 2007, www.astm.org
2. Carrio, A., Pestana, J., Sanchez-Lopez, J.L., Suarez-Fernandez, R., Campoy, P., et al.: UBRISTES: UAV-Based Building Rehabilitation with Visible and Thermal Infrared Remote Sensing. In: Robot 2015: Second Iberian Robotics Conference, pp. 245–256. Springer International Publishing (2016)
3. Dey, D., Geyer, C., Singh, S., Digioia, M.: A cascaded method to detect aircraft in video imagery. The International Journal of Robotics Research p. 0278364911412807 (2011)
4. Geyer, C., Dey, D., Singh, S.: Prototype sense-and-avoid system for UAVs. Robotics Institute, Carnegie Mellon University, Tech. Rep. CMU-RI-TR-09-09 (2009)
5. Geyer, C., Singh, S., Chamberlain, L.J.: Avoiding collisions between aircraft: State of the art and requirements for uavs operating in civilian airspace (2008)
6. Hajri, R.: UAV to UAV Target Detection and Pose Estimation. Ph.D. thesis, Monterey, California. Naval Postgraduate School (2012)
7. JiCheng, L., ZhengKang, S., Tao, L.: Detection of spot target in infrared clutter with morphological filter. In: Aerospace and Electronics Conference, 1996. NAECON 1996., Proceedings of the IEEE 1996 National, vol. 1, pp. 168–172. IEEE (1996)
8. Li, J., Gong, W., Li, W., Liu, X.: Robust pedestrian detection in thermal infrared imagery using the wavelet transform. Infrared Physics & Technology **53**(4), 267–273 (2010)
9. Liu, T., Li, X.: Infrared small targets detection and tracking based on soft morphology top-hat and sprt-pmht. In: Image and Signal Processing (CISP), 2010 3rd International Congress on, vol. 2, pp. 968–972 (2010). DOI 10.1109/CISP.2010.5646926
10. McGee, T.G., Sengupta, R., Hedrick, K.: Obstacle detection for small autonomous aircraft using sky segmentation. In: Robotics and Automation, 2005. ICRA 2005. Proceedings of the 2005 IEEE International Conference on, pp. 4679–4684. IEEE (2005)
11. Minwalla, C., Thomas, P., Ellis, K., Hornsey, R., Jennings, S.: Flight test evaluation of a prototype optical instrument for airborne sense-and-avoid applications. In: SPIE Defense, Security, and Sensing, pp. 83,870R–83,870R. International Society for Optics and Photonics (2012)
12. Naidoo, Y., Stopforth, R., Bright, G.: Development of an UAV for search & rescue applications. In: AFRICON, 2011, pp. 1–6. IEEE (2011)
13. Olivares-Mendez, M.A., Fu, C., Ludivig, P., Bissyandé, T.F., Kannan, S., Zurad, M., Voos, H., Campoy, P.: Towards an Autonomous Vision-Based Unmanned Aerial System against Wildlife Poachers. Sensors **15**(12), 31,362–31,391 (2015)
14. Rozantsev, A., Lepetit, V., Fua, P.: Detecting flying objects using a single moving camera. IEEE transactions on pattern analysis and machine intelligence (2016)
15. Sonka, M., Hlavac, V., Boyle, R.: Image processing, analysis, and machine vision. Cengage Learning (2014)
16. Zeng, M., Li, J., Peng, Z.: The design of top-hat morphological filter and application to infrared target detection. Infrared Physics & Technology **48**(1), 67–76 (2006)
17. Zsedrovits, T., Zarandy, A., Vanek, B., Peni, T., Bokor, J., Roska, T.: Visual detection and implementation aspects of a UAV see and avoid system. In: Circuit Theory and Design (ECCTD), 2011 20th European Conference on, pp. 472–475. IEEE (2011)

Turbulence Modeling of a Small Quadrotor UAS Using System Identification from Flight Data

Ondrej Juhasz **Mark J.S. Lopez**
Research Associate Research Associate
San Jose State University
Ames Research Center
Moffett Field, CA

Marcos G. Berrios **Tom Berger** **Mark B. Tischler**
Aerospace Engineer Aerospace Engineer Senior Scientist
U. S. Army Aviation Development Directorate (AMRDEC)
Ames Research Center
Moffett Field, CA

ABSTRACT

The extraction of a turbulence model for a small quadrotor UAS using flight test data is presented and validated using simulation. The turbulence model is a control equivalent turbulence input (CETI) model, which drives the controls of the aircraft to reproduce realistic motion caused by turbulence. Since the quadrotor UAS is unstable, the extraction methodology must be done with a control system engaged. Effects of signal length, noise in the feedback path, and model uncertainty on the extraction methodology are quantified in simulation and recommendations are given to obtain high quality flight test data for accurate extraction of turbulence models. The extracted turbulence model is shown to have the same low-order form as existing turbulence models.

NOTATION

Symbols

G_c	Control system
G_{δ_t}	CETI Transfer function
G_{xx}	Signal autospectrum, plotted in power dB; $10\log 10\|G_{xx}\|$
J	Model Cost: Error of fit between model and truth data
q_a	Acutal aircraft pitch rate (rad/sec)
q_m	Measured pitch rate (rad/sec)
δ_a	Bare-airframe input (mixer units: +/- 4500)
δ_c	Commanded input (mixer units)
δ_l	Commanded rate response (rad/sec)
δ_t	Turbulence input (mixer units)
η	Noise (units of bare-airframe output)
$\bar{\sigma}_{q_m}$	Signal-to-noise ratio
τ	Time delay (seconds)

Acronyms

ADD	Aviation Development Directorate
GNC	Guidance Navigation and Control
CETI	Control Equivalent Turbulence Input
PSD	Power Spectral Density

INTRODUCTION

The popularity of small unmanned aerial systems (UAS) has continued to increase in both the military and commercial markets. For the military, small VTOL UAS offer the ability to dramatically increase the warfighter's situational awareness on the battlefield, while commercially they may deliver packages or medicine to a company's customers. Quadrotors in particular have become popular due to their mechanical simplicity over traditional single main rotor helicopters that have swashplates.

Due to the low cost of the aircraft themselves, small UAS are generally developed with a "fly-crash-fix" cycle, where engineering design and analysis methods may be overlooked in favor of quickly getting in the air. However,

Presented at the Seventh AHS Technical Meeting on VTOL Unmanned Aircraft Systems, Mesa, AZ, January 24-26, 2017. Approved for public release; distribution unlimited

as the technology contained in a UAS becomes more advanced and the acquisition process becomes more rigorous, properly vetted engineering methods will need to be utilized. For example, more emphasis will need to be placed on the control system design and validation effort.

System identification has long been used to develop accurate bare-airframe models of fixed- and rotary-wing aircraft (Refs. 1, 2). These identified models are often used as the basis for control law design and development, especially when physics-based models are not available or reliable. Recently, frequency-domain system identification methods have successfully been demonstrated on quadrotors in hover/low-speed flights (Refs. 3, 4). Low-order state-space models are fit to the frequency responses to get an accurate representation of aircraft dynamics over a broad frequency range. The identification results show that small VTOL UAS have unstable modes that are at much higher frequency than manned size VTOL and require high-bandwidth control systems to stabilize them.

Control systems for these UAS must provide good stability characteristics, closed-loop tracking performance, and disturbances rejection capabilities. Realistic turbulence models are important for successfully designing and validating controller disturbance rejection performance in a simulation environment. The Dryden turbulence model represents turbulence based on a "frozen-field" through which the helicopter flies through. This type of model is very common in fixed wing aircraft, but is not applicable to hovering flight. Therefore, a turbulence model based on the control equivalent turbulence inputs (CETI) is developed (Refs. 5, 6). The CETI model uses flight data taken in turbulence to reproduce control inputs needed to generate the same levels of motion as were seen during the flight in turbulence. The extraction process requires an accurate model of the bare-airframe. While applied differently, both the CETI and Dryden turbulence models capture low-order representations of turbulence. Within control design, a similar approach to the CETI model has been successfully used to generate the required inputs to negate the effects of disturbances based on estimated control equivalent disturbance inputs (Ref. 7).

This paper will outline a frequency-domain based CETI model extraction methodology and apply it to data extracted from time histories taken in flight in turbulence. The extraction methodology will first be validated in simulation by investigating the effects of time history length, sensor signal-to-noise ratio, and quadrotor bare-airframe uncertainty. It will be shown that an accurate CETI model can be extracted from the given time-history data, and that the extracted CETI model in simulation exactly reproduces the model derived from flight.

The work in this paper falls under the Aviation Development Directorate (ADD) Quadrotor Guidance, Navigation, and Controls (GNC) Project (Ref. 8). This project aims to adapt the state-of-the-art modeling, guidance, and controls technologies and methodologies used at ADD for large-scale vehicles to small-scale UAS.

QUADROTOR AIRCRAFT DESCRIPTION

The aircraft used is the 3DR Iris+ (herein Iris) quadrotor shown in Fig. 1. The Iris weighs about 2.6 lbs, has four twin bladed 10 inch propellers, and uses a Pixhawk flight control processor, which runs the opensource Ardupilot codebase. This software provides all sensor processing and flight control, and runs at 400 Hz. The base code was modified to include additional logging to track inputs going into the mixer. All logging data is generated at 200 Hz. The ability to add inputs directly to the mixer was also incorporated. Direct mixer inputs were used during the bare-airframe identification process to excite the bare-airframe response of the quadrotor without the filtering of the control system.

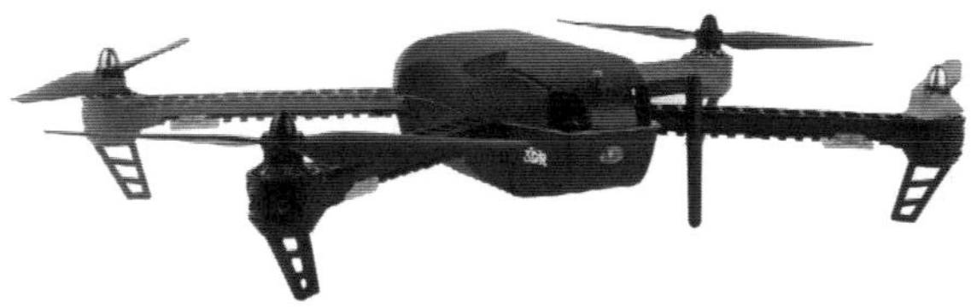

Fig. 1. 3DR Iris+ quadcopter

Bare-Airframe System Identification

As part of the ADD Quadrotor GNC project, the frequency-domain based system identification tool, CIFER® (Ref. 1), was used for both the state-space model identification, as well as the generation of autospectra of CETI time histories presented here.

The Iris longitudinal dynamics are dominated by an unstable mode between $\omega = 3-4$ rad/sec (Fig. 2). In order to stabilize this mode, the crossover frequency in this axis must be significantly higher, and it is near $\omega = 30$ rad/sec in the stock control system. This need for a high crossover frequency drives the need to have an accurate model up to at least 60-90 rad/sec (Ref. 1) for control system design.

The bare-airframe identification model structure and results align well with those previously published (Refs. 3,4). A total of 15 states were required to model the full aircraft, including the 9 rigid-body states as well as high-frequency "lags". The lags represent high-frequency roll off in magnitude and phase that can be attributed to motor dynamics, structural dynamics, and on-board sensor filtering. A time-delay was also used in each axis that accounts for sensor and processing delays.

The identified longitudinal dynamics are overlayed with flight data in Fig. 2. The state-space longitudinal dynamics

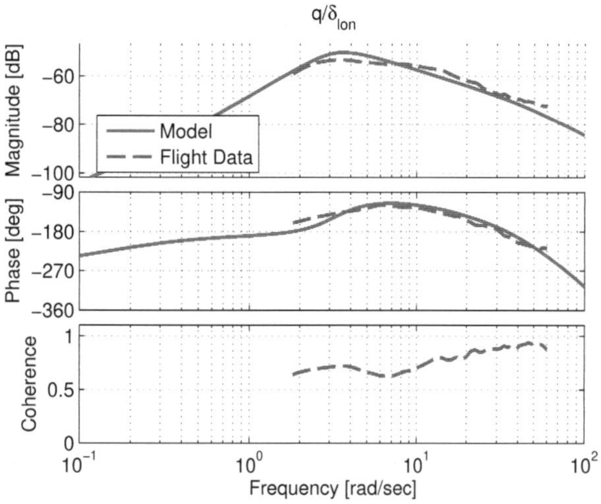

Fig. 2. Hover pitch rate response to longitudinal mixer inputs

equations in hover are shown in Eqn. (1) and the parameters are given in Table 1, along with their Cramer-Rao bounds and insensitivities (Ref. 1). The implementation of the lag puts the control derivative ($M_{\delta_{lon}}$) in the A matrix, which is now driven by the lagged control input, δ'_{lon}. All parameters are well identified with low Cramer-Rao bounds and insensitivities, giving a high confidence in the physical parameters of the model. The pitch rate damping derivative (M_q) was found to have a negligible influence on the model and was eliminated from the identification. Excellent fits to flight data were also obtained for the lateral, direction, and heave axes. This paper will demonstrate the CETI extraction method on the longitudinal axis, so figures are included for that axis only. The average cost of the identification for all axes, $J_{ave} = 43$, is is well below the $J_{ave} \leq 100$ threshold (Ref. 1), meaning an excellent fit has been obtained.

$$\begin{Bmatrix} \dot{u} \\ \dot{q} \\ \dot{\theta} \\ \dot{\delta}'_{lon} \end{Bmatrix} = \begin{bmatrix} X_u & 0 & -32.174 & 0 \\ M_u & M_q & 0 & M_{\delta_{lon}} \\ 0 & 1 & 0 & 0 \\ 0 & 0 & 0 & -lag \end{bmatrix} \begin{Bmatrix} u \\ q \\ \theta \\ \delta'_{lon} \end{Bmatrix} + \begin{bmatrix} 0 \\ 0 \\ 0 \\ lag \end{bmatrix} \{\delta_{lon}(t-\tau)\} \quad (1)$$

CETI EXTRACTION

Extraction Methodology

Since the Iris bare-airframe is highly unstable, all turbulence data must be collected with the control system engaged. The baseline Iris control system called "Stabilize" mode provides attitude stabilization. Once in a hover, the aircraft was kept in roughly the same location but was allowed to drift to minimize additional inputs into the control system. Long time histories were used to ensure the aircraft was adequately perturbed and to capture low frequency dynamics. The use of closed-loop responses is in contrast to the results in (Ref. 5), where only time histories in which aircraft motion was uncorrelated with pilot inputs were used. This meant the pilot input (δ_c) was uncorrelated with the reconstructed bare-airframe input (δ_{lon}) and that the aircraft motion was coming purely from excitation by turbulence.

Table 1. Iris Longitudinal Identification Results

Engineering Symbol	Value	CR (%)	Insens. (%)
$M_{\delta_{lon}}$	0.01362	3.62	1.36
X_u	-0.1428	7.32	2.53
M_u	1.102	6.90	2.14
M_q	0.000	-	-
lag[a]	50.0	-	-
τ	0.02811	6.47	3.22

[a] Fixed parameter
[b] Eliminated parameter

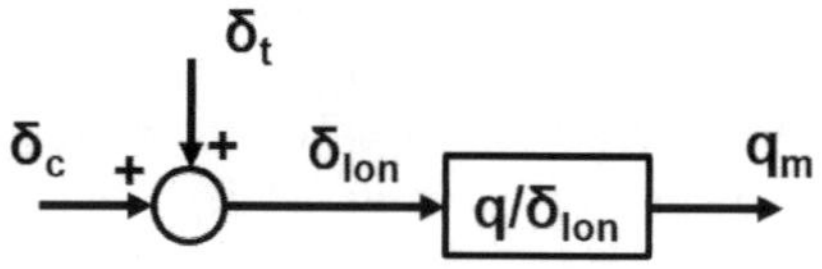

Fig. 3. Simplified block diagram of aircraft pitch rate response including turbulence inputs

The turbulence extraction and identification methodology aims to identify an equivalent control input (δ_t) which excites the aircraft similarly to real turbulence. This is the same approach used in Ref. 5. The concept is shown in Fig. 3 for the longitudinal axis, where the measured aircraft response (q_m) is driven by the commanded inputs (δ_c) as well as turbulence inputs (δ_t) . In equation form the total measured pitch rate is:

$$q_m = \frac{q}{\delta_{lon}}(\delta_c + \delta_t) \quad (2)$$

Solving for the turbulence input gives:

$$\delta_t = \frac{\delta_{lon}}{q} q_m - \delta_c \quad (3)$$

Measured aircraft rates are passed through the inverse of the identified aircraft model, reconstructing the control inputs needed to generate the rates. Measured commanded inputs are then subtracted from these inputs, leaving control equivalent turbulence inputs. The results generated here are solely in the frequency domain, so a "stabilized inverse" is not required, as was in Ref. 5, which used a time-domain approach to reconstruct the bare-airframe inputs. In the frequency domain, the arithmetic of Eqn. (3) (using a linear, not logarithmic scale) is (Ref. 9):

$$G_{\delta_t\delta_t} = \left|\frac{\delta_{lon}}{q}\right|^2 G_{q_mq_m} - G_{\delta_c\delta_c} \quad (4)$$

VALIDATION OF THE CETI EXTRACTION METHOD USING SIMULATION

The block diagram in Fig. 3 is valid for an open-loop system where there is no effect of feedback. Since the CETI extraction process was done on a closed-loop system, it is important to understand the effects of sensor noise and plant modeling uncertainty on the resulting model. It is known that for good signal-to-noise ratios, the bare-airframe model extracted from closed-loop simulation is unbiased for output measurement noise (Ref. 1), but the effect of measurement noise on the CETI model extraction is unclear. It is also important to investigate the required signal lengths. In order to get adequate low frequency data, long time histories may be needed, and the exact length is unknown. An analysis was conducted on a closed-loop simulation of an Iris in order to clarify the effects of:

- Signal length
- Measurement noise
- Bare-airframe model uncertainty

Simulation Setup

The simulation model is shown in Fig. 4 and contains the closed loop longitudinal axis attitude-hold dynamics similar to the stock Iris. In order to provide a realistic model of the Iris, the simulation uses the longitudinal dynamics from the identified state-space Iris model ($H = q/\delta_{lon}$) from Fig. 2 and a control system (G_c) which has a similar crossover frequency to the stock "Stabilize" mode used in the flight test. The CETI model used is a low-order turbulence model (Eqn. 5) where the gain (K) and break frequency (a) were set to generate meaningful levels of turbulence in the simulation and to be consistent with results in Ref. 5. The CETI model is driven by a white noise (w_n) signal and the resulting turbulence input (δ_t) enters the system at the mixer and sums with the commanded inputs from the control system (δ_c) to form the aircraft bare airframe input (δ_{lon}). The aircraft pitch rate (q) is summed with sensor noise (η) to form the total measured pitch rate (q_m) which is used in the feedback. Pilot and outer loop control inputs (δ_l) are assumed to be small and uncorrelated with the turbulence and noise and so should not have a large effect on the extracted CETI model.

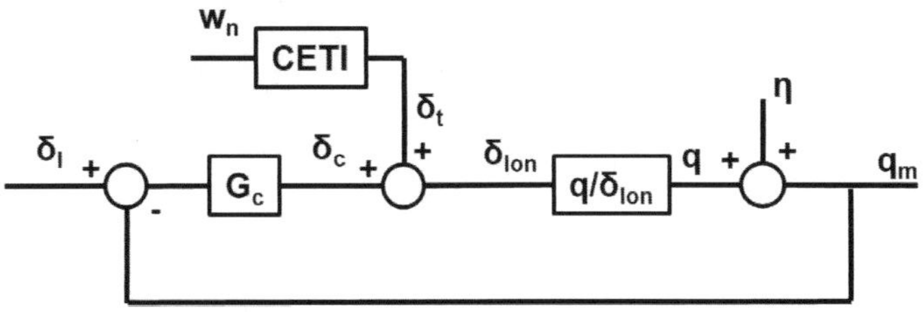

Fig. 4. Block diagram of simulation dynamics

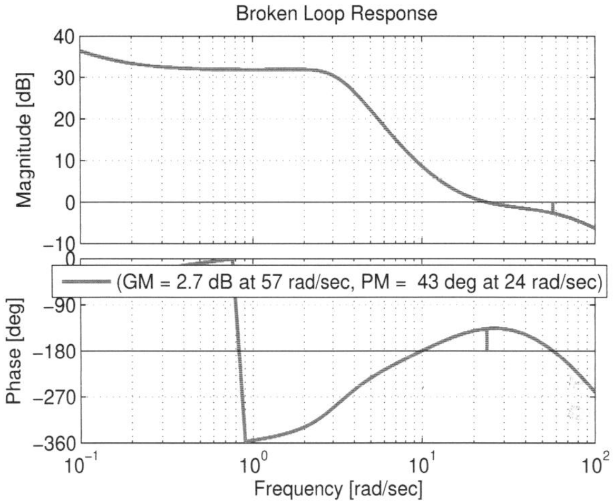

Fig. 5. Simulation broken loop response

$$G_{\delta_t}(s) = \frac{K}{s+a} \quad (5)$$

The broken loop response for the loop broken at the actuators (G_cH) is shown in Fig. 5. Also shown are the control system crossover frequency and stability margins. This control system has been tuned to have a high disturbance rejection bandwidth and a crossover frequency similar to the stock Iris. Details of the controller optimization process are given in Ref. 10.

Baseline CETI Model Extraction

The baseline model extraction uses the arithmetic in Eqn. (4) to re-identify a known CETI model using an exactly known plant and CETI model with no sensor noise and so represents a best case scenario. Time signals of 300 seconds were used to align the extraction with signal lengths obtained in flight. The autospectra of the truth CETI model and PSD as well as the extracted CETI model are shown in Fig. 6. The truth CETI PSD curve represents the PSD of the truth CETI model (Eqn. 5) using the 300 second generated white noise signal. The extracted CETI model (using Eqn. (4)) aligns nearly exactly with the truth CETI PSD at low freqeuncies up to $\omega = 7$ rad/sec, validating the extraction in perfect conditions. The extraction process then oscil-

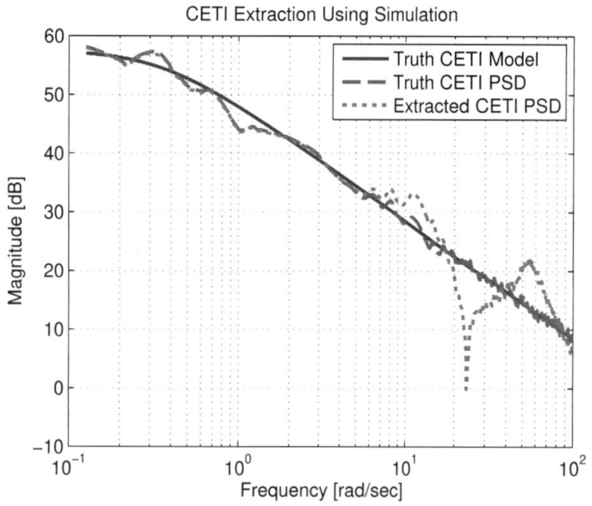

Fig. 6. Extraction of CETI model using simulation

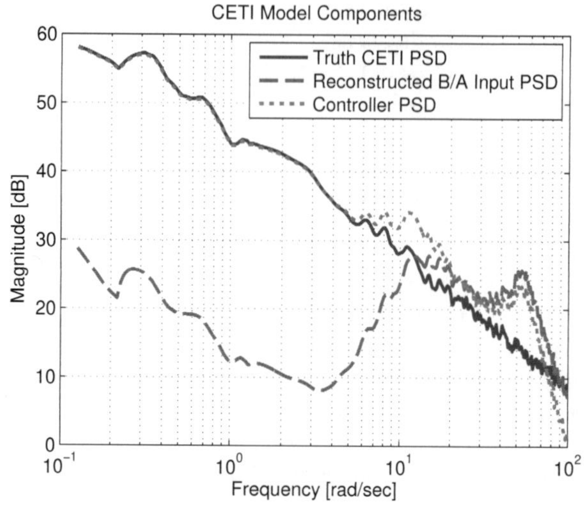

Fig. 7. Component signal PSDs used to generate CETI response

lates around the true response at frequencies near the control system crossover. Well above crossover (not shown), the extracted CETI response aligns again with the truth data.

The CETI turbulence is obtained by subtracting the mixer input (δ_c) from the reconstructed bare-airframe input (δ_{lon}), both of which are plotted in Fig. 7. The controller input dominates at low frequency up to about $\omega = 5$ rad/sec, and the reconstructed bare-airframe input is nearly two orders of magnitude (30 dB) lower. At these frequencies, the aircraft does not move and the control system takes out all turbulence through the feedback of pitch rate. The mixer input is exactly opposite of the CETI turbulence input. At $\omega > 5$ rad/sec, the broken loop response (Fig. 5), shows a dramatic drop; the control system is becoming less effective at removing turbulence and the aircraft begins to move. Signal autospectra contain only magnitude information, so the relative phasing of the control system input and bare-airframe input signals are not known, and the extraction process breaks down. At frequencies much higher than crossover (not shown), the control system is ineffective and the turbulence input directly drives aircraft motion, and the CETI model is again cleanly extracted from the data.

This simple analysis shows that if the crossover frequency is high enough, the CETI model is able to be cleanly identified using the extraction process.

Length of Required Flight Data

White noise has a PSD of unity across all frequencies and for infinite time scales. However, the PSD at low frequency for short signal lengths might not equal exactly one. This section identifies the signal durations necessary to accurately identify low frequency dynamics when driven in the time domain by white noise. White noise is driven through the CETI model from Eqn. (5) and the corresponding PSDs ($G_{\delta_t \delta_t}$) are shown in Fig. 8 for a range of time histories between 100 to 1200 seconds. For short time histories, the

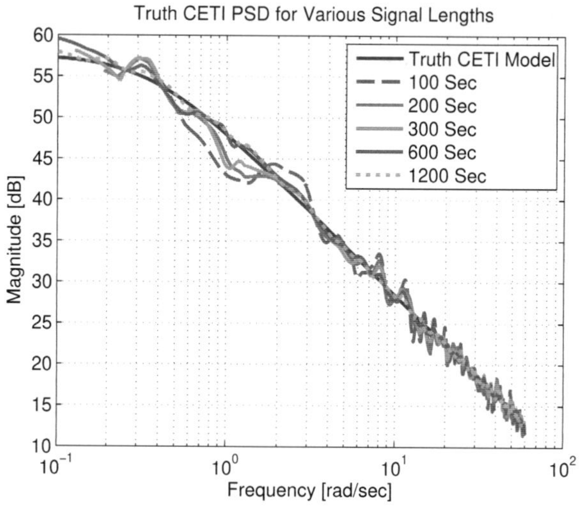

Fig. 8. PSD of white noise of different lengths being passed through CETI model

PSD at low frequency shows considerable variability from the truth data, and the PSD converges on top of the model for very long time histories.

The cost based on the error between the autospectrum of the time signals and the model (see Ref. 1) is shown in Fig. 9. As the signal length increases, enough averages of the white noise are achieved and the cost approaches 0, an exact fit to the model. Guidelines from Ref. 1 indicate a cost of less than 50 gives an excellent match to truth data. For the turbulence to achieve a good fit to the model, more than 200 seconds of data are required. This indicates that the Iris must be flown for at least 200 seconds in a turbulence field to achieve adequate spectral content to extract an accurate CETI model. The time histories obtained in turbulence in flight meet this criteria.

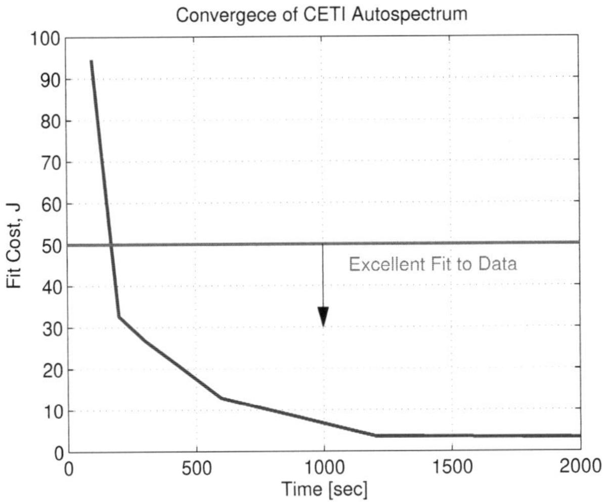

Fig. 9. Cost of fit between white noise of different lengths being passed through CETI model and the exact CETI model

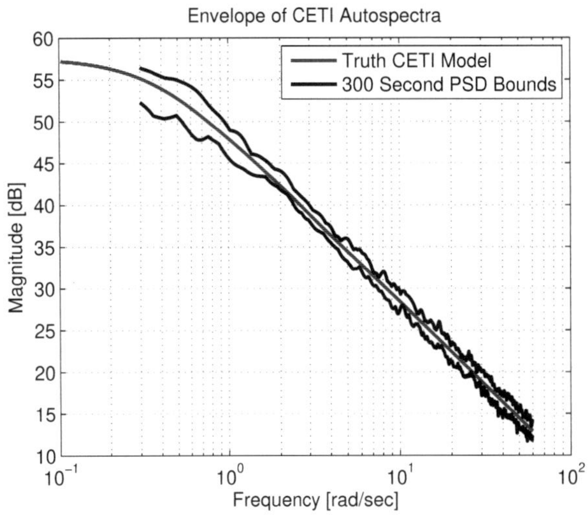

Fig. 10. Envelope of upper and lower bounds of CETI autospectra that may be obtained from different time histories

Convergence of Autospectra

The simulations are driven by random white-noise processes. If time histories are generated with different noise seeds, they will form slightly different autospectra. This is true for flight data as well, as different time histories taken right after one another will have different spectral content. Figure 10 shows the band within which the real turbulence model will fall into when driven by 300 seconds of white noise with different noise seeds.

CIFER® can average multiple time histories to produce a single autospectrum (Ref. 1). As the number of time histories used increases, they will average to produce the exact CETI turbulence model. The cost decreases from $J = 20$ to 6 just by using two, 300 second time histories. Two time histories were obtained for all flights in turbulence, so an excellent representation of the CETI dynamics should be extracted from the flight data.

Influence of Measurement Noise

In a closed-loop system, measurement noise is correlated around the loop and may be amplified, degrading the accuracy of the extracted CETI model. To better understand the affects of output noise, various levels of measurement noise (η) are added to the system in Fig. 4 and the resulting models are overlayed with each other. In transfer function form, the effect of measurement noise on the measured bare-airframe output signal as shown in Fig. 4 is:

$$q_m = -\frac{1}{1+G_cH}\eta + \frac{H}{1+G_cH}\delta_t \tag{6}$$

Solving for the turbulence input:

$$\delta_t = \frac{1+G_cH}{H}q_m + \frac{1}{H}\eta \tag{7}$$

The two components that characterize the predicted turbulence are the measured rate (q_m) as well as noise (η), both easily measurable in this simulation case. Figure 11 shows the magnitude component of the two transfer functions from Eqn. (7) that multiply the measured pitch rate and noise input, respectively. At frequencies well below crossover ($\omega = 24$ rad/sec), the noise signal is attenuated by nearly 40 dB when compared to the pitch rate signal. At these frequencies, noise will have a small impact on the extraction of the CETI model. Near and above crossover, the noise signal is the same as, or amplified more than, the measured rate signal, meaning an accurate representation of turbulence will not be obtained at these frequencies if there is high signal to noise. This simple analysis demonstrates that for a closed loop CETI extraction, a high bandwidth control system is needed as reasonable results cannot be expected above crossover frequency if there is any noise in the measurements.

To provide further insight, the simulation in Fig. 4 was run for 300 seconds with varying levels of sensor noise. Based on the results from Fig. 9, this signal length should allow for excellent identification results close to the known CETI model. Noise levels were chosen to give a broad range of signal-to-noise ratio, defined as the the signal root-mean-square (RMS) value divided by the noise RMS value. Large signal-to-noise ratios signify high quality, low noise data.

Figure 12 shows PSDs of the varying simulated measurement noise levels plotted along with the simulated measured aircraft rates, q_m. At low frequencies below crossover, all aircraft rates align well with each other and the no noise case. The noise is well attenuated by the control system when compared to turbulence inputs, as predicted by Fig. 11, and the simulated measured rate does not change. At around $\omega = 10$ rad/sec for the highest noise

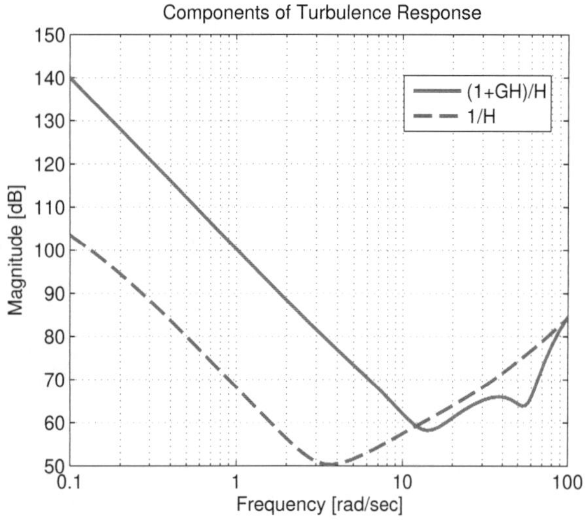

Fig. 11. Frequency responses comparing the scaling of measured output and measurement noise on CETI extraction

case, the measured response begins to be fully dominated by the noise and the control system is becoming ineffective at removing it. Well above crossover, the noise is directly fed into the measured rate for all cases.

Figure 13 shows the extracted CETI model for the varying measurement noise cases as well as the truth CETI model and PSD signal. It is clear that as the signal-to-noise ratio increases, the CETI extraction procedure breaks down. This figure also clearly shows where the identification of the CETI model from flight data would need to take place and that a max frequency of about 10 rad/sec should be used in the identification.

The CETI costs are determined by using the autospectra of the extracted CETI model and the truth CETI model and are plotted in Fig. 14. The costs is determined between $\omega = 0.4$ and 10 rad/sec (beyond which CETI extraction breaks down). Minimum signal-to-noise ratios of $\bar{\sigma}_{q_m} = 2.5$ are needed in order to accurately extract a model from turbulence. This is consistent with recommended signal-to-noise ratio $\bar{\sigma}_{q_m} > 3$ that is needed to accurately ($< 10\%$ bias error) extract a bare-airframe model from a closed loop response (Ref. 1).

To quantify the signal-to-noise ratio from flight data, two time histories taken during flight in turbulence were analyzed. Two additional time histories taken from indoor flights were also analyzed, where the turbulence was characterized as minimum. The measured pitch rate autospectra are shown in Fig. 15. Ref. 11 defines the signal-to-noise ratio from flight data ($\bar{\sigma}_{q_m}$) as a ratio of the RMS of two portions of the autospectrum. The autospectrum is assumed to be noise once it plateaus at high frequency. This plateau occurs at $\omega = 30$ rad/sec for the data in Fig. 15 and an additional on-board filter around $\omega = 50$ rad/sec filters the noise

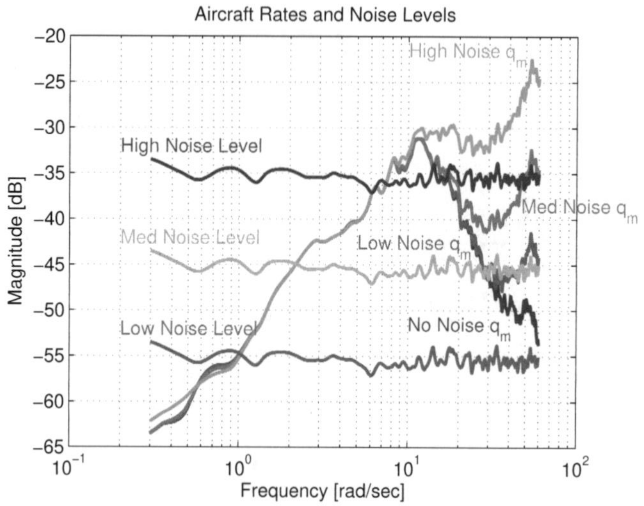

Fig. 12. Measured aircraft rates with increasing measurement noise levels

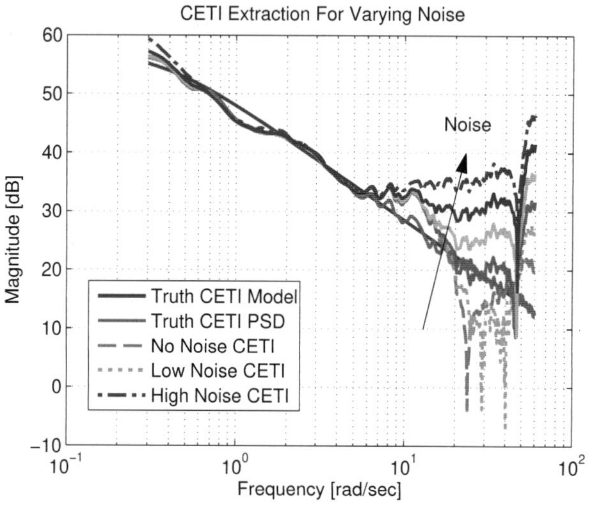

Fig. 13. Extracted CETI models with varying measurement noise

out completely. The autospectrum is considered to be signal at frequencies below the noise plateau.

Following this analysis, the flight data for the Iris has a signal to noise ratio of $\bar{\sigma}_{q_m} = 2.6$ for the outdoor data and a better $\bar{\sigma}_{q_m} = 3.9$ for the indoor data. Based on Fig. 14, the Iris data is adequate for accurate extraction of the CETI model.

Bare-Airframe Uncertainty

Variability of the bare-airframe model could also impact the CETI extraction. The exact Iris dynamics are not known. The identified state-space model parameters have Cramer-Rao bounds, which give a measure of 1σ uncertainty of the parameters. In the proceeding analysis, each parameter is varied by $\pm 2\sigma$ to capture a 95% probability of capturing

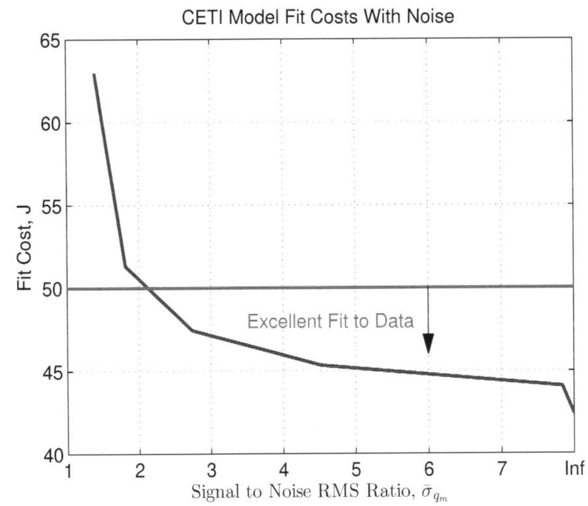

Fig. 14. Extracted CETI model cost changes with varying signal-to-noise ratio, $\bar{\sigma}_{q_m}$

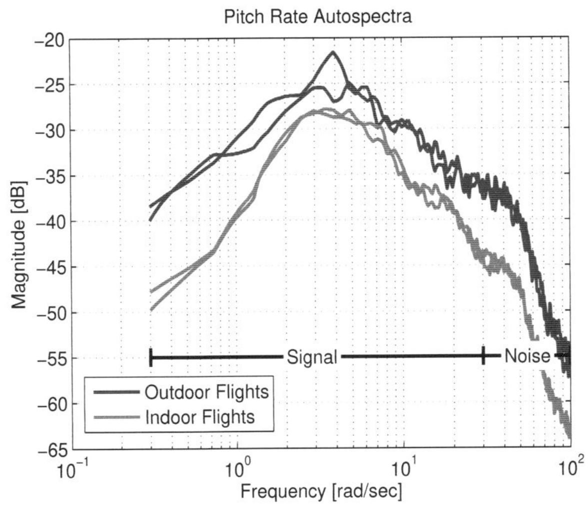

Fig. 15. Pitch rate autospectra from flight used to measure signal-to-noise ratio

the correct value of the parameter. This is done for all possible combinations of parameters from Table 1. Time histories of 300 seconds are used, as is a signal-to-noise ratio of $\bar{\sigma}_{q_m} = 2.6$, to closely match the Iris flight data. The CETI extraction method remains the same for each case. The nominal bare-airframe model is used for the inverse in Eqn. (4). The resulting CETI models are plotted along side the truth data in Fig. 16. The parameters from Table 1 are relatively well identified, with Cramer-Rao percentages generally below 7%, so large changes to the extracted CETI model were not anticipated between the original and perturbed results. The cost of fitting the perturbed models did not vary by more than $\Delta J = 5$ from the cost of fitting the original extracted CETI model.

Since the bare-airframe model is well identified, it is not anticipated that uncertainty will have an impact the CETI extraction using flight data.

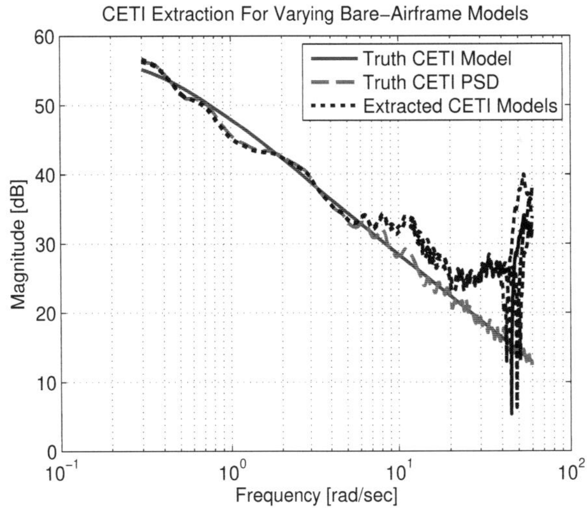

Fig. 16. Effect of bare-airframe 2σ uncertainty on the extracted CETI model

Guidelines

The simulation model of the Iris was used to validate the extraction methodology and develop requirements for accurate CETI extraction from flight. These guidelines are as follows:

1. Around crossover, the CETI extraction process breaks down. It is important that the control system crossover either be sufficiently high (around $\omega = 30$ rad/sec), or the flights in turbulence are done with the controller off and the pilot stabilizes the vehicle with small inputs.

2. Time histories of at least 200 seconds are required to provide adequate spectral content at low frequency.

3. Signal-to-noise ratios of $\bar{\sigma}_{q_m} > 2.5$ are needed to provide clean data with low CETI model extraction costs.

4. Bare-airframe uncertainty was not shown to have a large impact. However, all parameters in the state-space model were accurately identified and the resulting model matched well with the non-parametric frequency response. If an accurate model is not know there will certainly be effects on the CETI extraction.

CETI MODEL EXTRACTION USING FLIGHT DATA

Time histories taken in the same turbulence level were processed to determine the equivalent turbulence input using Eqn. (4). An anemometer was not available during the testing, so the exact variation in wind-speeds is not know. The resulting extracted CETI autospectrum is shown in Fig. 17, along with a corresponding first-order transfer function fit of the response.

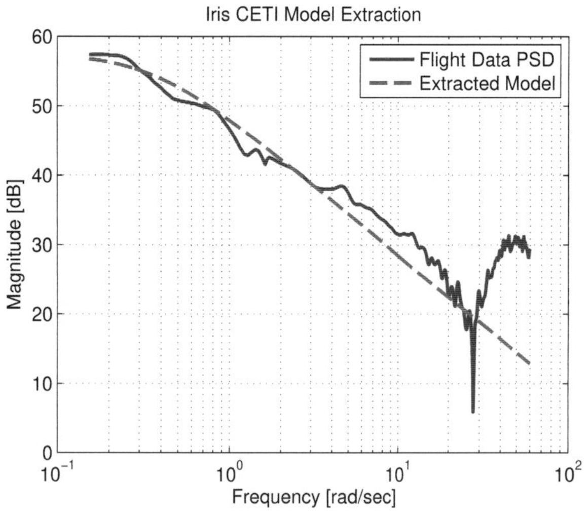

Fig. 17. Extraction of a CETI model in the longitudinal axis from flight data

$$G_{\delta_t}(s) = \frac{264}{s+0.351} \tag{8}$$

This transfer function represents the control equivalent turbulence input for the turbulence level from flight. The fit tracks well up to around $\omega = 4$ rad/sec and is consistent with the first-order of the CETI model given in Refs. 5, 6.

The extracted CETI autospectrum is nearly identical to the ones generated during the simulation analysis. Impacts from the control system and noise are seen near the crossover frequency just as they were in simulation. This figure also shows robustness to variations in the control laws since the extracted CETI model from flight data has the same shape as in simulation even though the control systems were different.

SAMPLE APPLICATION OF CETI MODEL

The CETI model developed has two primary uses.

- Direct control system design in the frequency domain. There is no need to run time marching simulations to determine the effects of flying in turbulence. Quantities like aircraft attitude or position RMS due to varying levels of turbulence can be directly calculated in the frequency domain.
- Real-time simulation, piloted or not, to assess the aircraft response in turbulence. For a piloted aircraft, handling quality ratings could be taken in varying levels of turbulence.

In this sample application, two control systems will be simulated and compared to each other and flight data taken in the stock onboard "Stabilize" mode in both the frequency and time domains. The first control system is the same as

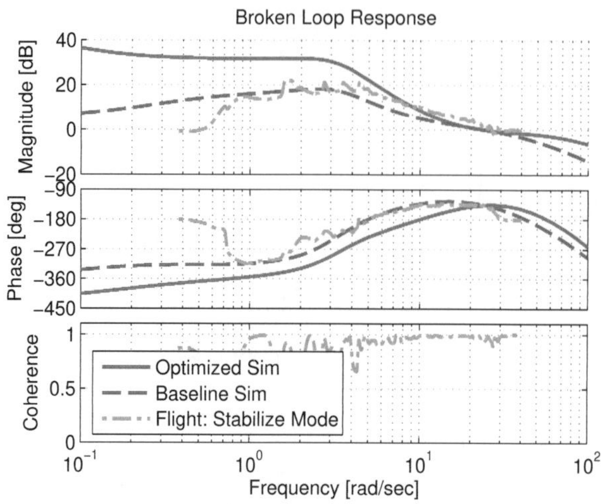

Fig. 18. Comparisons of the broken-loop response of two different Iris control systems and flight data

was used in the simulation analysis. This *optimized* control system has been tuned to have a high crossover frequency and high disturbance rejection bandwidth. An additional *baseline* control system was designed that more closely resembles the stock Iris "Stabilize" control system, and should perform worse than the more optimized one. The turbulence levels will be forced to be the same in simulation as in flight by using the identified CETI model of Eqn. (8). An output noise level giving a signal-to-noise ratio of $\bar{\sigma}_{q_m} = 2.6$ was used in simulation to match the flight data, and 300 second time simulations were generated.

Figure 18 shows the comparison of the broken loop response for the two control system in simulation as well as flight data using the stock "Stabilize" mode. While both have similar crossover frequencies, the optimized control system has much higher magnitude at low frequency, meaning the control system is more active at attenuating turbulence than the baseline control system at those frequencies. Also, the flight "Stabilize" mode data and baseline control system are very similar and so should behave similarly in the presence of turbulence.

Figure 19 shows the disturbance rejection plot for disturbances to measured aircraft pitch attitude. The baseline control system and the stock "Stablize" mode do not attenuate disturbances to as high a degree as the optimized one below the bandwidth frequency. The baseline control system has the lowest disturbance rejection bandwidth (DRB), while the optimized and "Stabilize" mode have high DRBs.

Figure 20 shows the pitch rate autospectra of the flight data and the two control systems in the example. The flight data has a slightly higher pitch rate response to turbulence than the baseline control system. The similarity between the flight test data and baseline control system was anticipated by the similarity of the broken-loop response. As was seen in the disturbance rejection figure, the optimized control system does an excellent job at rejecting the turbulence

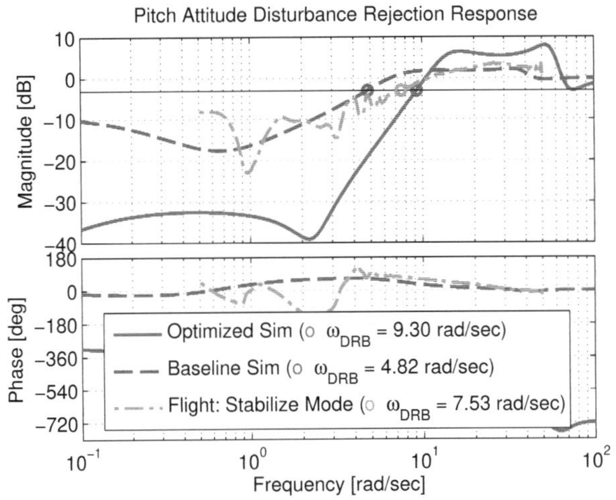

Fig. 19. Comparison of pitch attitude disturbance rejection response of two different Iris control systems

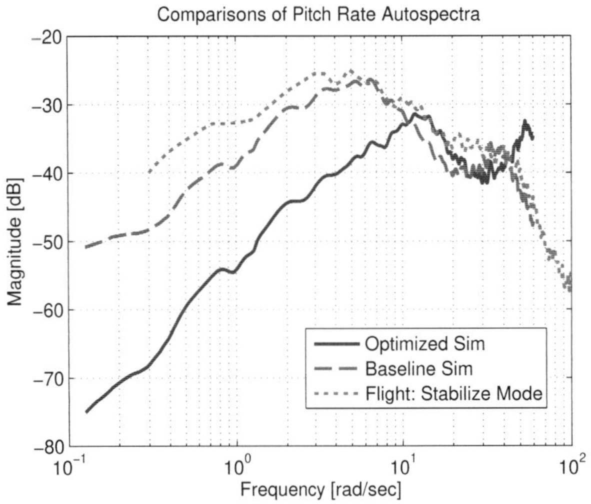

Fig. 20. Comparisons of pitch rate PSD in CETI turbulence between two simulated control systems and flight data

Table 2. RMS values of pitch rate due to turbulence from simulation and flight data between 0.3 and 30 rad/sec

	Pitch Rate RMS [deg/sec]
Optimized Simulation	2.72
Baseline Simulation	3.90
Flight in "Stabilize" mode	5.15

levels, however there is a peak above 50 rad/sec. RMS values are taken between 0.3 and 30 rad/sec and are given in Table 2. This table quantifies the trends seen in the plot of the pitch rate autospectra as well as the disturbance rejection and broken loop figures. The optimized control system produces about half the RMS as the stock "Stabilize" mode.

To get a better sense for the oscillation that can be seen in flight, the pitch attitude is plotted for 20 seconds in Fig. 21. The flight data shows larger low frequency oscillations which are similar to the baseline controller. The optimized controller has small, high frequency excursions from the trim pitch attitude.

This sample application showed that the CETI model is clearly able to reproduce an accurate turbulence level for the Iris. Additionally, frequency domain metrics such as signal RMS are able to capture the predicted trends in aircraft response due to turbulence.

TOWARDS A GENERALIZABLE TURBULENCE MODEL

In addition to the CETI model carried through the sample application, an additional outdoor set of time histories were taken on a less windy day. A set of data was also taken indoors to provide a baseline. Since actual wind data was not available, the resulting autospectra can only be compared to each other, and cannot be associated with specific turbulence levels. The autospectra for the two outdoor days and the indoor data set are shown in Fig. 22. All three autospectra have the same characteristic first order form at low frequency and are separated only by a gain, K.

$$G_{\delta_t}(s) = \frac{K}{s+a} \tag{9}$$

In future work the gain and break frequency will be a function of turbulence RMS, mean velocity, or a length scale as in Ref. 5. Multiple quadrotors of varying sizes will be tested to develop length scales suitable for this configuration and make the CETI model generic. The turbulence levels will also be directly measured using an anemometer, and a larger array of turbulence data will be collected.

CONCLUSIONS

This work described and validated a methodology to extract a control equivalent turbulence input (CETI) model from closed-loop flight of a quadrotor UAS in turbulence. The key conclusions from this paper are:

1. A CETI model may be extracted from closed-loop flight test. The extraction process is effective for a control system crossover frequency that is well above the CETI dynamics.

2. Time histories greater than 200 seconds and signal-to-noise ratios $\bar{\sigma}_{q_m} > 2.5$ are needed to extract an accurate CETI model. This required signal-to-noise ratio was obtained even for the low-cost sensors that are used in the Iris.

3. The methodology is robust to plant modeling uncertainty in the Iris. A cost difference of $\Delta J < 5$ was obtained for all variations of 2σ perturbations to the bare-airframe parameters.

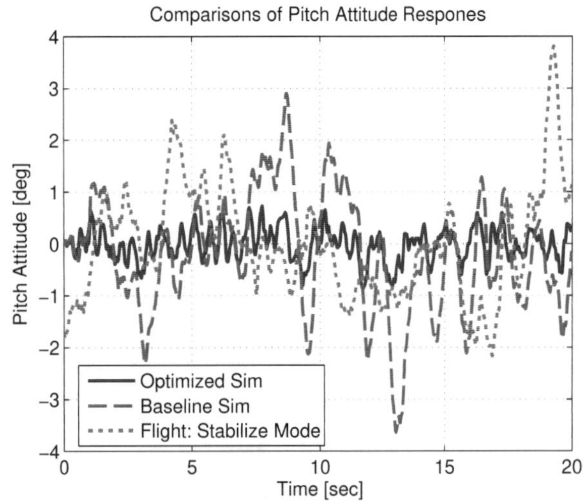

Fig. 21. Comparisons of pitch attitude time histories between two simulated control systems and flight data

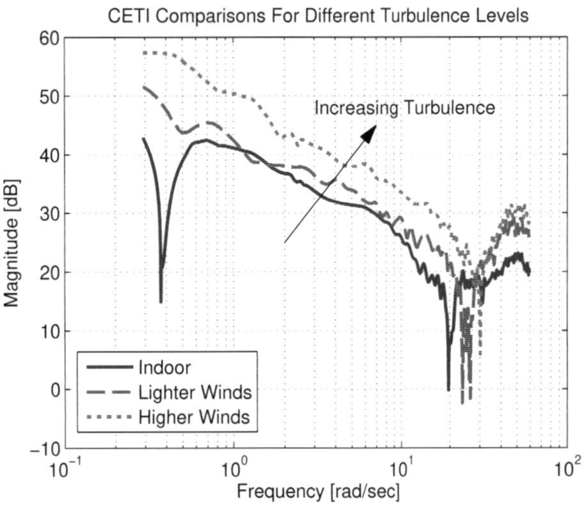

Fig. 22. Comparisons of CETI autospectra in different outdoor wind conditions as well as indoors

4. Using flight extracted noise levels and similar length time histories, a simulation was able to produce the shape of the extracted CETI autospectrum as seen in flight test.

5. A sample application verified that the CETI model produces realistic levels of turbulence in the time domain. A control system similar to the one on-board the Iris generated nearly the same oscillations as were seen in flight.

REFERENCES

[1]Tischler, M. B. and Remple, R. K., *Aircraft and Rotorcraft System Identification: Engineering Methods with Flight Test Examples*, AIAA, 2nd ed., 2012, Reston, VA.

[2]Klein, V. and Morelli, E. A., *System Identification: Theory and Practice*, AIAA, 2006, Reston, VA.

[3]Wei, W., Cohen, K., and Tischler, M. B., "System Identification and Controller Optimization of a Quadrotor UAV," American Helicopter Society 71st Annual Forum Proceedings, May 2015, Virginia Beach, VA.

[4]Niermeyer, P., Raffler, T., and Holzapfel, F., "Open-Loop Quadrotor Flight Dynamics Identification in Frequency Domain via Closed-Loop Flight Testing," Paper AIAA 2015-1539, AIAA SciTech Proceedings, January 2015, Kissimmee, FL.

[5]Lusardi, J. A., Tischler, M. B., Blanken, C. L., and Labows, S. J., "Empirically Derived Helicopter Response Model and Control System Requirements for Flight in Turbulence," *Journal of the American Helicopter Society*, July 2004, pp. 340–349.

[6]Lusardi, J. A., vonGruenhagen, W., and Seher-Weiss, S., "Parametric Turbulence Modeling for Rotorcraft Applications: Approach, Flight Tests and Verification," Rotorcraft Handling Qualities Conference, November 2008, Liverpool, UK.

[7]She, J. H., Fan, M., Ohyama, Y., Hashimoto, H., and Wu, M., "Improving Disturbance-Rejection Performance Based on Equivalent-Input-Disturbance Approach," *IEEE Transactions on Industrial Electronics*, Vol. 55, No. 1, January 2008, pp. 380–389.

[8]Cheung, K. K. et al., "An Overview of the US Army Aviation Development Directorate Quadrotor Guidance, Navigation, and Control Project," To be presented at American Helicopter Society 73rd Annual Forum, May 2017, Fort Worth, TX.

[9]Bendat, J. S. and Piersol, A. G., *Random Data: Analysis and Measurement Procedures*, John Wiley and Sons, Inc., 2nd ed., 1986, New York, NY.

[10]Berrios, M. G. et al., "Performance-Based Disturbance Rejection Criteria for Small-Scale UAS/Quadrotor with Flight Test Results," To be presented at American Helicopter Society 73rd Annual Forum, May 2017, Fort Worth, TX.

[11]Ivler, C. M., Mansur, H. M., Morford, Z. G., Kalinowski, K., and Knapp, M. E., "Flight Test of Explicit and Implicit Rotor-State Feedback Fly-By-Wire Control Laws," American Helicopter Society 72nd Annual Forum Proceedings, May 2016, West Palm Beach, FL.

Landing Path Generation Algorithm for Autonomous Shipboard Helicopter Recovery

Junfeng Yang
jqy5262@psu.edu
Graduate Research Assistant

Joseph F. Horn
joehorn@psu.edu
Professor

Department of Aerospace Engineering
The Pennsylvania State University
University Part, PA

ABSTRACT

A landing path generation algorithm is developed to enable autonomous shipboard helicopter recovery, and the algorithm is demonstrated in high fidelity simulations. The algorithm uses a forecasting algorithm (developed in previous work) that predicts future motion of the deck. Deck forecasts are used to set terminal conditions for the path. This approach enables less dynamic motion during descent and smaller relative velocity at touchdown. The algorithm generates inertial paths with certain kinematic constraints. The approach includes a polynomial representation of the path geometry with unknown parameters, and an optimization algorithm determines the parameters that satisfy the path constraints while minimizing accelerations. Landing quality was assessed by extensive simulations performed using a high fidelity FLIGHTLAB simulation of a utility helicopter with the SCONE ship motion data. Results show that the algorithm is able to generate a reasonable trajectory to support the autonomous landing task. Some ad hoc control logic was needed to ensure desired performance with landing gear touchdown, by matching the deck attitude just before deck contact.

NOMENCLATURE

V_{Zh}	Helicopter vertical velocity
H_h	Helicopter altitude
V_{Zdeck}	Deck vertical velocity during touchdown
H_{deck}	Deck altitude during touchdown
H_1, H_2	Altitude of intermediate control points on landing path
A_Z, V_Z, H	Acceleration, velocity, and position profile of landing path
t_f	Final time of landing
t_{fcast}, H_{fcast}	Time array and altitude array of deck motion prediction
V_{lim}, A_{lim}	Helicopter operational limits on velocity and acceleration
a, b, c, d, e, f	Path polynomial coefficients
H_{LG}	Height over deck of landing gears
L_B	Wheel base of landing gear
L_T	Wheel tread of landing gear
A-V-P	Acceleration, velocity and position
CG	Center of gravity
LG	Landing gear

INTRODUCTION

Landing is one of the most challenging tasks in helicopter shipboard operations due to the ship motions in high sea states, gusty aerodynamic environment, and the confined landing area. These adverse factors cause heavy workload for human pilot, affect the safety of a ship-based landing, and in extreme cases can push the vehicle to its operational limits. Many promising research programs have been performed towards enhancing the ease of pilot operation and improving the precision of the landing task. Ref.1 suggested an algorithm to identify the quiescent ship motion periods, an effective approach but can extend the landing process by waiting for a quiescent condition. In Ref. 2, a ship referenced Translation Rate Command (TRC) with relative position hold and snap to grid control mode has been shown to have potential to reduce workload. Another solution to helicopter recovery on a moving ship is through autonomous control. In early work (Ref. 3) the authors used tracking control laws in longitudinal and lateral coordinates to follow the ship motion while for the vertical axis a flare control law of altitude was applied.

Through the use of Fly-By-Wire controls systems, high-precision differential GPS, and ship state sensors, a fully autonomous ship landing system is now technically feasible. In fact, the FireScout MQ-8B and MQ-8C unmanned rotorcraft have performed autonomous landings at sea (Ref. 4), although there is not a lot of public domain information on the control laws or the operating limits of these autonomous modes. In any case, both autonomous and piloted landings in high sea states and high winds continue to be problematic for the U.S. Navy. Technologies that provide novel path planning for autonomous landing with large deck motions are certainly of interest in order to enhance performance and expand operational envelopes of autonomous landing systems.

In previous work, a Dynamic Inversion (DI) control law was developed to achieve advanced response types for piloted shipboard landing (Ref. 5), and was extended to provide a fully autonomous control mode (Ref. 6). The autonomous DI controller enabled simultaneous regulation of acceleration, velocity and position; therefore the key to a successful trajectory following task is the generation of feasible and kinematically consistent acceleration, velocity and position commands. This same control law architecture and simulation environment from Ref. 6 is used in this research, but the current effort focuses on enhancing the path generation algorithm.

In Ref. 6 two types of landing strategies were evaluated, as shown in Fig. 1. The first approach used deck tracking; i.e. the helicopter tracks flight deck motion in all three axes while closing the vertical gap with a steady relative rate. Simulation tests in Ref. 6 showed good landing quality, but significant maneuvering was observed in following the lateral and vertical periodic ship motion. This maneuvering is unnecessary and likely to push the aircraft to its operational limits. The other approach takes the advantage of deck motion prediction, as provided by a Minor Components Analysis (MCA) prediction algorithm developed by Advanced Rotorcraft Technologies (ART) and discussed in detail in Ref. 6. By estimating the future deck state at the touchdown time, an inertial landing trajectory was obtained by solving the Optimal Rendezvous control problem which then served the commanding path for the acceleration-velocity-position (A-V-P) controller. However, the rendezvous solution was compromised by the possibility of premature deck contact and violations to kinematic constraints, which were not strongly imposed in the Optimal Rendezvous solution. Those kinematic conditions have to be checked after the path was generated, and the solution was then discarded and landing sequence was aborted if the constraints were not satisfied. For the present work, the MCA deck forecasting algorithm included multipoint prediction, which estimates a sampled trajectory of future deck motion over a certain period instead of at only a single instant. The new path planning algorithm presented in this research takes advantage of estimates of deck state at five 1-second intervals into the future over (a five second forecasting horizon).

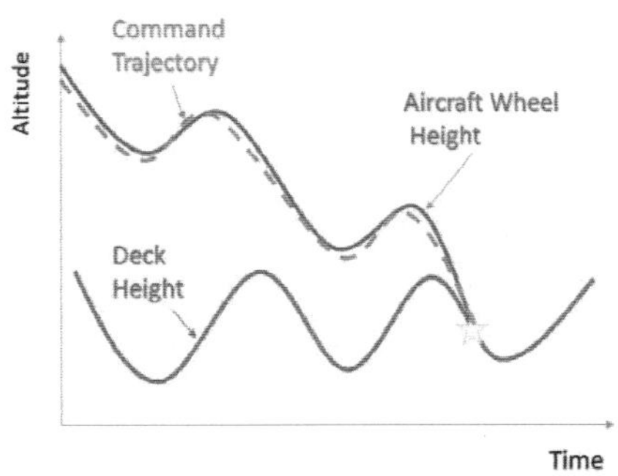

Figure 1a. Landing with Deck Tracking

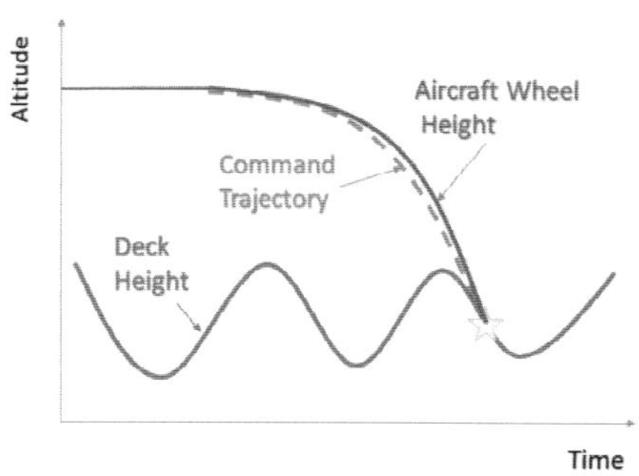

Figure 1b. Landing with Deck Prediction

METHODOLOGY

The path planning algorithm in this paper is designed to integrate with the controller developed in Ref. 6, where the path-following control law manages to regulate the three inertial coordinates in a decoupled manner, therefore, the spatial path planning problem breaks into three 1-D time trajectories. Due to the dynamic nature of ship motion, the lateral and vertical axes control can benefit the most from predictive landing. Without losing generality, only the vertical axis is discussed here. The landing path starts at the station-keeping position and ends at the touchdown position. A reasonable landing trajectory must be a smooth curve connecting the two initial and terminal conditions with proper velocity and acceleration properties. The relative positions of the landing path and the deck trajectory in the vertical plane are demonstrated in Fig. 2.

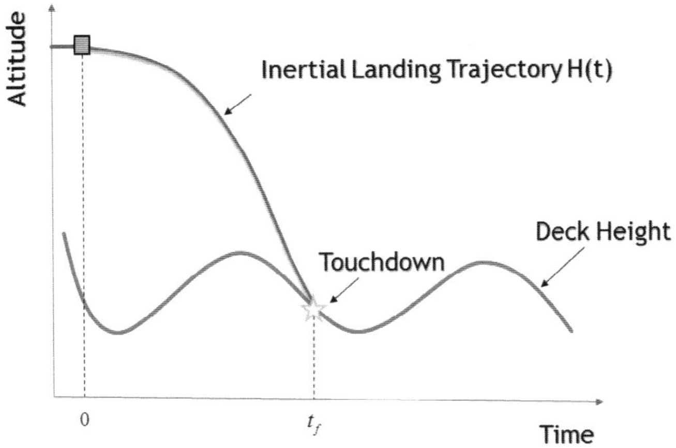

Figure 2. Relative Positions of Landing Trajectory and Deck Trajectory

The inertial descent time-trajectory is denoted by $H(t)$; the descent is initiated at $t=0$ and terminated at $t=t_f$ by a touchdown. The trajectory can be approximated as a fixed-end polynomial in Eq. 1 with six coefficients to be determined.

$$H(t) = at^5 + bt^4 + ct^3 + dt^2 + et + f \tag{1}$$

The consistent velocity and acceleration can be obtained by differentiating Eq. 1

$$V_Z(t) = 5at^4 + 4bt^3 + 3ct^2 + 2dt + e \tag{2}$$

$$A_Z(t) = 20at^3 + 12bt^2 + 6ct + 2d \tag{3}$$

At the two ends, the trajectory and its derivative are specified to match the current helicopter state and deck state at touchdown respectively. Quantitatively this can be expressed as the following conditions:

I. At the t=0, the position $H(0)$ must be the current helicopter height H_h, the velocity $V_Z(0)$ must be zero or current helicopter velocity V_{Zh} whichever is smaller.

II. At $t=t_f$, the position $H(t_f)$ must be equal to the predicted deck height H_{deck}, the velocity $V_Z(t_f)$ must be equal to the deck velocity corrected by a desired relative sink rate e.g. 1 ft/sec

In the meantime, the landing path should not have an intersection with deck trajectory until touchdown, also the velocity and acceleration should not exceed the operational limits V_{lim} and A_{lim}. From I and II four mathematical constraints can be constructed, but there are six unknown coefficients, and thus the path properties are under-specified. The two additional degrees of freedom should affect the internal shape of the position profile and thus the velocity profile and acceleration profile. The parameters can then be optimized to meet some objective. The approach in this paper is to specify two parameters that define $H(t)$ at $t=\frac{1}{3}t_f$ and $t=\frac{2}{3}t_f$. This condition can be quantified as in Eq. (6) and Eq. (7)

$$H_1 = a(\frac{1}{3}t_f)^5 + b(\frac{1}{3}t_f)^4 + c(\frac{1}{3}t_f)^3 + d(\frac{1}{3}t_f)^2 + e(\frac{1}{3}t_f) + f \tag{6}$$

$$H_2 = a(\frac{2}{3}t_f)^5 + b(\frac{2}{3}t_f)^4 + c(\frac{2}{3}t_f)^3 + d(\frac{2}{3}t_f)^2 + e(\frac{2}{3}t_f) + f \quad (7)$$

The equations for determining polynomial coefficients together can be expressed in matrix form Eq. (8)

$$\begin{bmatrix} 0 & 0 & 0 & 0 & 0 & 1 \\ 0 & 0 & 0 & 0 & 1 & 0 \\ t_f^5 & t_f^4 & t_f^3 & t_f^2 & t_f & 1 \\ 5t_f^4 & 4t_f^3 & 3t_f^2 & 2t_f & 1 & 0 \\ \left(\frac{1}{3}\right)^5 & \left(\frac{1}{3}\right)^4 & \left(\frac{1}{3}\right)^3 & \left(\frac{1}{3}\right)^2 & \frac{1}{3} & 1 \\ \left(\frac{2}{3}\right)^5 & \left(\frac{2}{3}\right)^4 & \left(\frac{2}{3}\right)^3 & \left(\frac{2}{3}\right)^2 & \frac{2}{3} & 1 \end{bmatrix} \begin{bmatrix} a \\ b \\ c \\ d \\ e \\ f \end{bmatrix} = \begin{bmatrix} H_h \\ V_{Zh} \\ H_{deck} \\ V_{Zdeck} \\ H_1 \\ H_2 \end{bmatrix} \quad (8)$$

So far, H_1 and H_2 are unknown. An optimization task is set to find out those two variables in pursuit of the minimum value of the objective function with the following constraints:

1. $H(t) > H_{fcast}(t)$ over $t \in [0, t_f)$ to avoid premature contacts
2. $\max|V_Z(t)| < V_{\lim}$
3. $\max|A_Z(t)| < A_{\lim}$

These are inequality constraints. One approach to imposing such constraints is to implement barrier functions in the objective function. The barrier functions become increasingly large when the constraints are not satisfied, but will be negligible when the constraint is not violated. Exponential functions are used to impose constraint barriers for this algorithm. For example, the velocity constraint can be converted into Eq. (9)

$$J = e^{-\lambda(V_{\lim} - \max|V_Z(t)| - 0.1)} \quad (9)$$

The tunable parameter λ can be adjusted to achieve different steepness levels. The objective function in (9) increases rapidly as the velocity constraint is violated, but asymptotically approaches zero when velocities move away from the constraint. If the desired property is not strong enough, a larger value of λ is suggested. The overall objective function has the following form in Eq. 10 where w_1, w_2, w_3 are the weighting factors.

$$J = w_1 \cdot e^{-\lambda_1 \max(H(t) - H_{fcast}(t) - 0.1)} + w_2 \cdot e^{-\lambda_2 (V_{lim} - \max|V_Z(t)| - 0.1)} + w_3 \cdot e^{-\lambda_3 (A_{lim} - \max|A_Z(t)| - 0.1)} \quad (10)$$

A classical gradient descent method was used to solve the optimization problem. Having the optimized inertial trajectory, its consistent velocity and acceleration are obtained through Eq. 2 and 3. The obtained inertial kinematics may be applied to shape the inertial A-V-P command for the helicopter; however this sort of implementation renders the landing quality solely dependent on forecasting accuracy. If the actual deck trajectory differs significantly from the prediction, a hard touchdown or missed contact can occur. In this study, the concept of relative motion control was adopted where the difference of A-V-P between optimized inertial trajectory and that of deck prediction was extracted by subtraction. The relative A-V-P values are then corrected by the actual deck state to reform the inertial A-V-P command. The advantage of the method is in that: if the prediction is accurate, the superposition of relative command and actual deck state perfectly replicates the optimized inertial trajectory, in case of inaccurate predictions, the reformed inertial command may be distorted from the optimized version, but the safety and quality of landing can still be guaranteed due to the nature of relative motion control. Note that the current algorithm designs a trajectory for the touchdown spot on helicopter, while the autonomous control law generates commands for helicopter CG, thus a conversion from touchdown spot to CG must be applied.

SIMULATION REULTS

The path generation algorithm is designed to enable landing of a 17,000 lb utility-class helicopter (similar to the H-60 class of rotorcraft) on a destroyer in high sea states and then tested in simulations. The isolated path generation was first tested and refined using specified deck motion data, before coupling to the vehicle dynamics and controller. The algorithm is then integrated with a high fidelity simulation model using the FLIGHTLAB software. The software includes a model of a moving ship based on the Systematic Characterization of the Naval Environment (SCONE) deck motion cases for a generic small deck ship (Ref. 7). The SCONE deck motion cases are provided by the U.S. Navy to provide realistic generic ship motion cases for use in fundamental research on ship landing. The database includes extended time series data for ship motion, so that numerous randomized cases we can be evaluated in simulations. The particular SCONE case used here is the heave-dominated "SCONE 2" set, which results in deck motions as high as ± 13 ft of heave displacement and ± 11 ft/sec heave velocity. These are substantial deck motions that would be outside current helicopter flight envelopes.

First, the isolated landing path generation was tested for numerous cases using prescribed deck motion based on the SCONE 2 ship motion cases. In the example below, the following settings are used:

$$t_f = 5 \text{ sec}, H_h = 45 \text{ ft}, V_{Zh} = 0 \text{ ft/sec}, H_{deck} = 16 \text{ ft}, V_{Zdeck} = -3 \text{ ft/sec}, V_{lim} = 6.7 \text{ ft/sec}, A_{lim} = 11.3 \text{ ft/sec}^2$$

The forecasting time array is: $T_{fcast} = \begin{bmatrix} 0 & 1 & 2 & 3 & 4 & 5 \end{bmatrix}$ sec

The forecasted deck position array is: $H_{fcast} = \begin{bmatrix} 15 & 20 & 22.12 & 21.95 & 19.46 & 16 \end{bmatrix}$ ft

The optimized position, velocity and acceleration profiles versus time are plotted in Figure 3. It is shown that the initial guess of landing path violates the velocity constraint, while the optimization algorithm successfully addresses this problem by yielding a feasible trajectory.

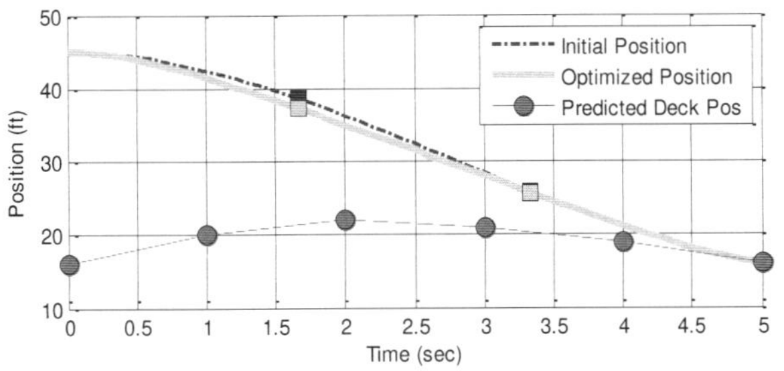

Figure 3-a. Position Profile of Optimized Landing Path

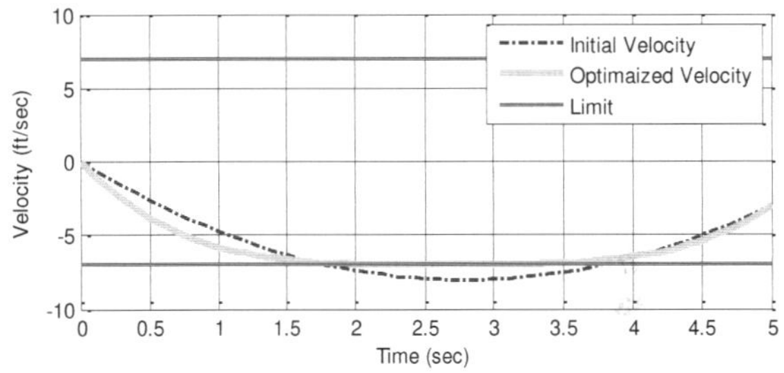

Figure 3-b. Velocity Profile of Optimized Landing Path

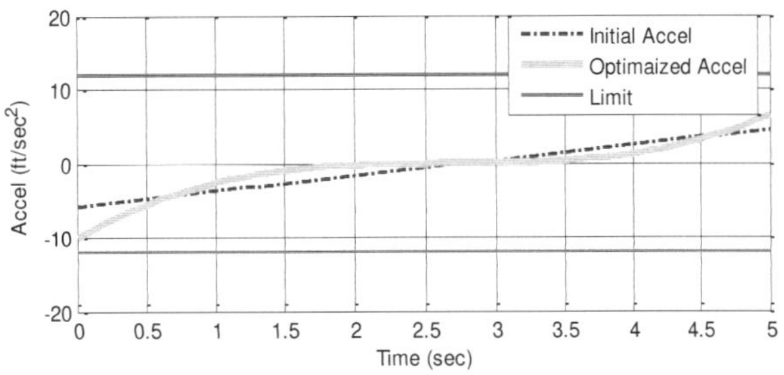

Figure 3-c. Acceleration Profile of Optimized Landing Path

The landing path generation algorithm was then validated in high fidelity simulations that integrate: 1) The path generation algorithm, 2) a high fidelity FLIGHTLAB simulation of a utility helicopter, 3) a model of the ship deck motion and airwake, 4) a dynamic inversion control law that follows the path, 5) the deck motion prediction algorithm. Items 2 to 5 were described in more detail in Ref. 6, while only the path generation algorithm is significantly updated here. The FLIGHTLAB model includes high order dynamics of the blade flapping and lagging degrees of freedom and finite state dynamic inflow. One additional model fidelity improvement was added after the work in Ref. 6. An engine derive train model was used to model rotor RPM variations, and thus expected to provide higher fidelity representation of the heave dynamics. The ship airwake model uses a stochastic turbulence model derived from CFD solutions of flow over a generic ship structure (Ref. 8). The DI control law, described extensively in Ref. 6, tracks the desired A-V-P commands while rejecting disturbances due to airwake, while the MCA deck motion prediction algorithm developed by ART provides reasonably accurate and robust predictions of deck motion out to 5 seconds (although the prediction is by no means exact).

The approach and station keeping control law are the same as in Ref. 6. In the landing phase, since the ship undergoes a relatively steady motion in the longitudinal axis, a conventional tracking control strategy was used for longitudinal control – the commanded longitudinal acceleration, velocity, and position simply match that of the deck. The predictive landing control law with path optimization was implemented only in lateral and vertical axes. Figs. 4 -7 demonstrate a typical case with flight states recorded in approach, station keeping and landing with the quantitative landing metrics shown in Table 1. The plots and metrics imply that the path optimization algorithm together with the hybrid control strategy resulted in a satisfactory landing for this particular case. However, the vertical velocity of the front landing gear at touchdown was larger than expected or desired.

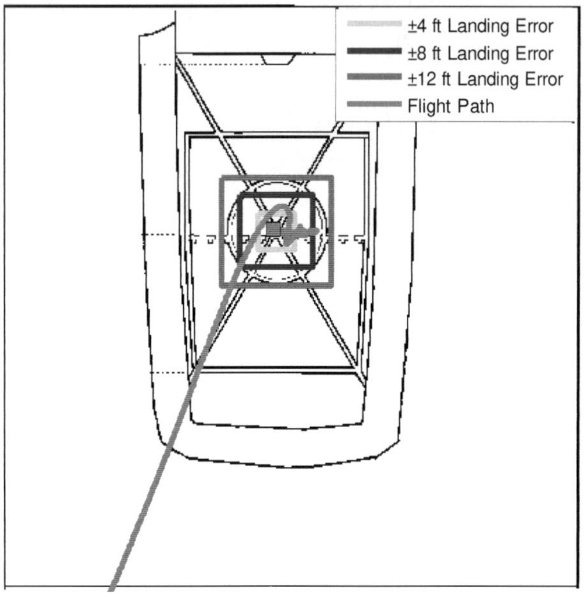

Figure 4. Top View of Trajectory of Approach, Station Keeping and Landing

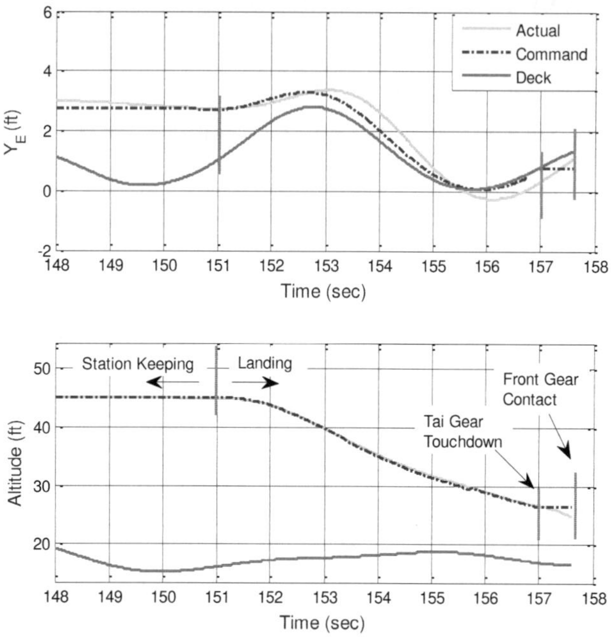

Figure 5. Lateral and Vertical Position of C.G.

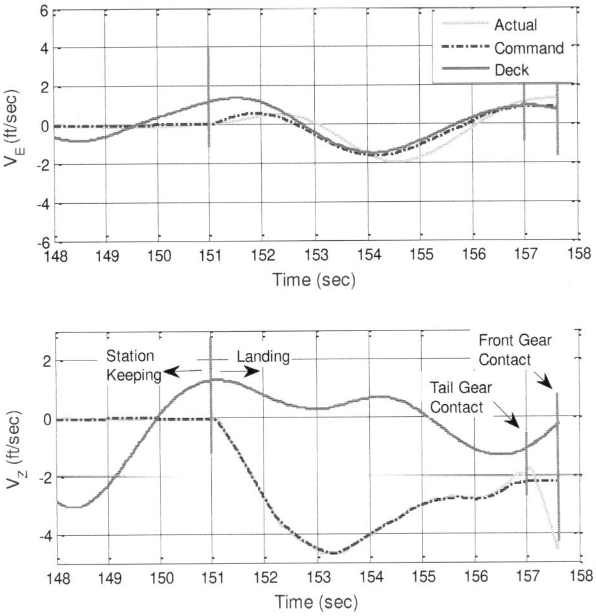

Figure 6. Lateral and Vertical Velocities of C.G

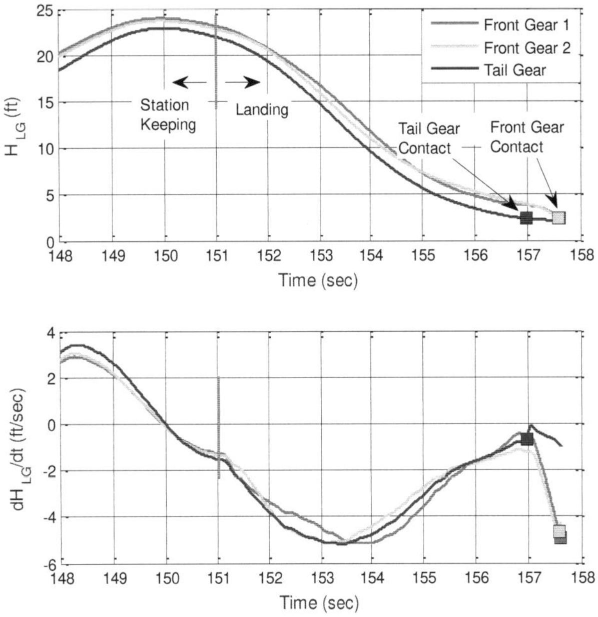

Figure 7. Height and Vertical Velocity of Landing Gear Relative to Deck

Longitudinal Position Error (ft)	Lateral Position Error (ft)	Lateral Velocity Error (ft/sec)			Vertical Velocity Error (ft/sec)		
		Tail Gear	Front Gear 1	Front Gear 2	Tail Gear	Front Gear 1	Front Gear 2
0.433	0.499	0.226	0.679	0.679	0.6651	4.648	4.919

Table 1. Landing Quality Metrics

Additional flight cases were tested with randomized initial conditions in order to evaluate the robustness and reliability of the path optimization algorithm and landing control law. The statistics for 30 cases are summarized in Figures 8-9 which show the landing scatter of the aircraft CG, and the velocities of each landing gear relative to the ship deck point of contact at touchdown. These results also show that the vertical impact on the front landing gear was higher than desired. On the positive side, the final position tolerance for the aircraft CG at touchdown was excellent, and the tail gear touched down softly in all cases.

The issue of the hard landing of the front landing gear was investigated in more detail. In initial developments, the control law has special logic when landing gear contact is first detected. Once there is deck contact, the aircraft dynamics change significantly due to friction forces acting on the gear. The strategy was to simply reduce collective at first contact to allow the aircraft weight to immediately push the aircraft on the deck. We call this strategy "*lift-dumping*". Since the aircraft hovers with a nose up attitude, the tail gear almost always contacts the deck first, at which point the lift-dumping control law is activated, and the aircraft develops significant sink and nose down pitch rate causing the front gear to hit the deck faster than desired.

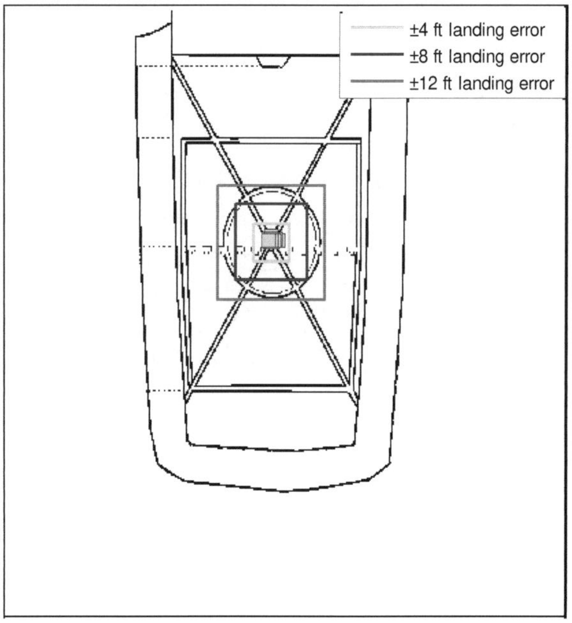

Figure 8. Scatters of Landing Location

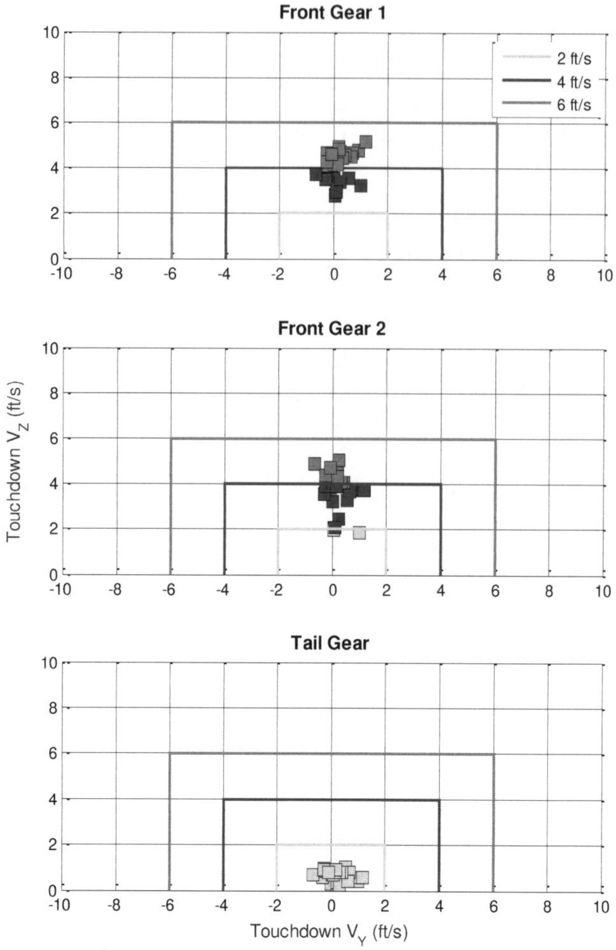

Figure 9. Scatters of Touchdown Velocity Error

Various modifications to the lift-dumping control were investigated, such as reducing the collective rate during the lift dumping or phasing in the lift –dumping control over a longer period of time. In many case these strategies resulted in desired performance. However, there were outlier cases with very poor performance when the lift-dumping was executed too slowly. Notably, if the deck accelerated downward soon after the tail gear contact, the tail gear would lift off and then land very hard.

This problem is a consequence of attitude mismatch at touchdown; where the aircraft attitude does not comply with the deck orientation. Attitude mismatch is a common for helicopter; even a land-based helicopter face the problem during landing since the helicopter attitude is not level in hover trim and they sometime need to land on a sloped surface. However, the attitude mismatch is much more complex with a moving landing deck. In high sea states, a quick sealing of touchdown is required to avoid the complex deck-aircraft interactions with landing gear contact dynamics.

A possible solution to this problem is to place the aircraft parallel to deck plane just before touchdown so that all three landing gear contact simultaneously, and then quickly dump the rotor lift to press aircraft tightly on flight deck. This can be achieved by a quick level-out operation at the moment before anticipated contact. The so called "*level-out*" maneuver cannot occur too early, as the helicopter begins to translate as soon the attitude changes away from the trim. The start time of level-out depends on the quickness of response to attitude command. Since the control law switches from trajectory tracking to attitude command, the aircraft during level-out is expected to drift from the landing center. The level-out must be initiated at a proper time, which cannot be too early to avoid notable drifting, nor to be too late to avoid premature contact. Based on the characteristics of the UH-60 model being investigated, the level-out is initiated 1.0 second before contact.

The level-out control law commands attitude corrections quantitated for removing the difference among heights of three landing gears over deck. The schematic of the geometric parameters of the landing gear is shown in Figure 10:

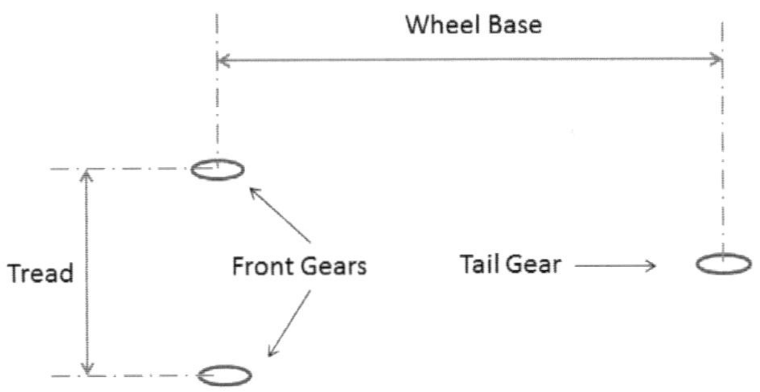

Figure 10. Geometric Parameters of Landing Gears

The amount of pitch and bank angle corrections required to level out the front and tail gear is based on a small angle assumption quantified in Eq. (11) and (12) respectively:

$$\Delta\theta = \frac{(H_{LG1} + H_{LG2})/2 - H_{LG3}}{L_B} \tag{11}$$

$$\Delta\phi = \frac{H_{LG1} - H_{LG2}}{L_T} \tag{12}$$

The real-time value of the above corrective attitude angles are fed into longitudinal and lateral ACAH controller to promptly drive the helicopter to the desired attitude. The maximum response quickness allowed by the command filter in the ACAH controller is used. In the meantime, the vertical axis continues descent along the trajectory planned by algorithm presented before. Since three landing gear are designed to hit deck at the same time, the lift-dumping takes place at a fast rate – 20% per second to seal the touchdown as soon as one gear strikes deck. Figures 11 to 16 represent a typical landing with the level-out strategy.

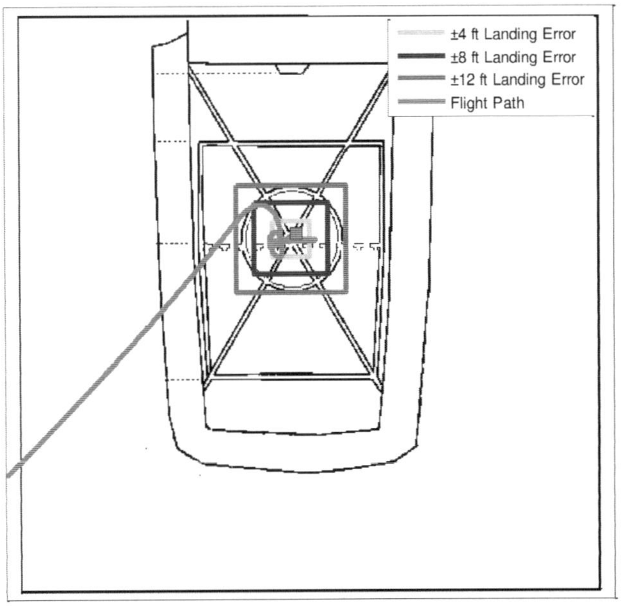

Figure 11. Top View of Approach and Landing Trajectory

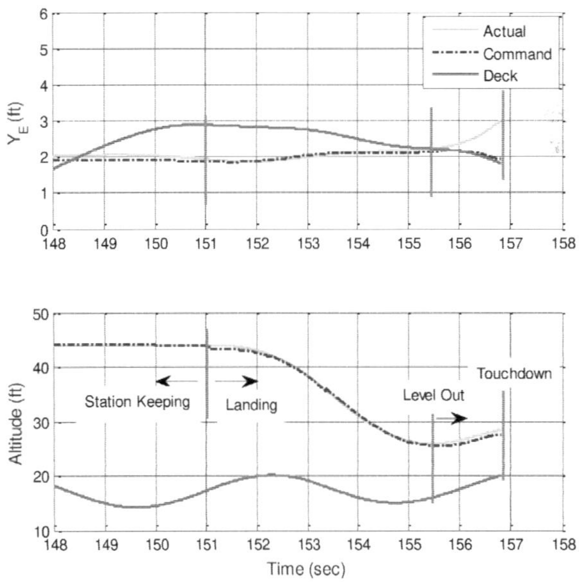

Figure 12. Lateral and Vertical Position vs. Time

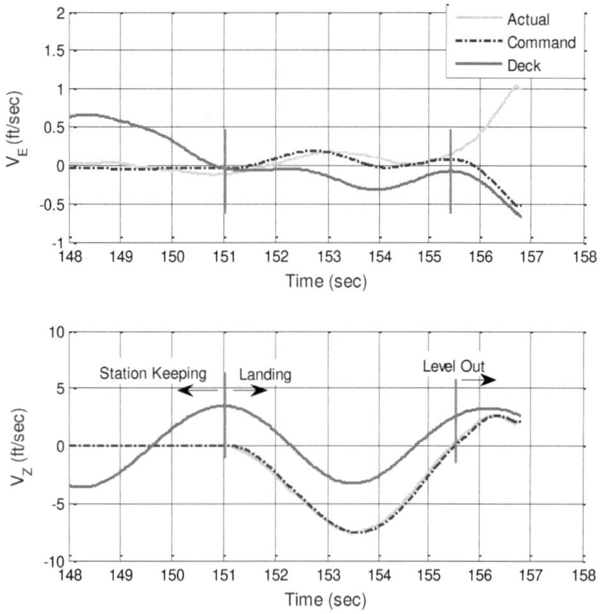

Figure 13. Lateral and Vertical Velocities vs. Time

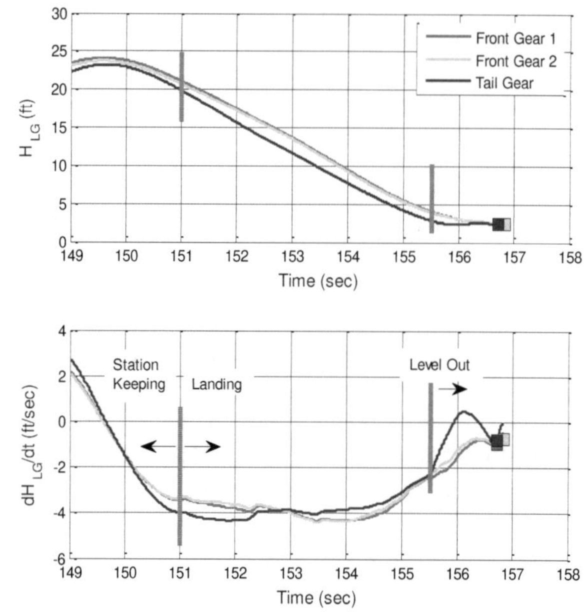

Figure 14. Height and Vertical Velocity of Landing Gear Relative to Deck

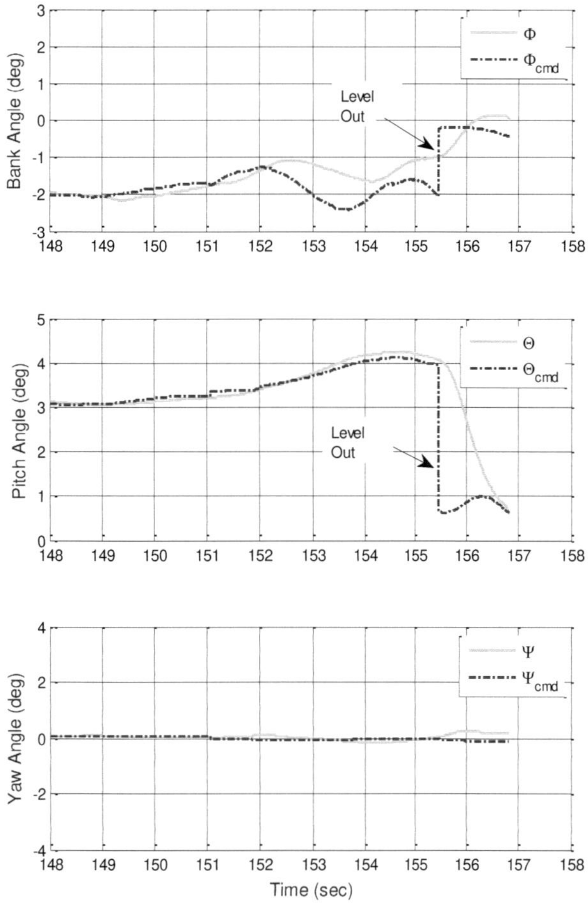

Figure 15. Helicopter Attitude Angles vs. Time

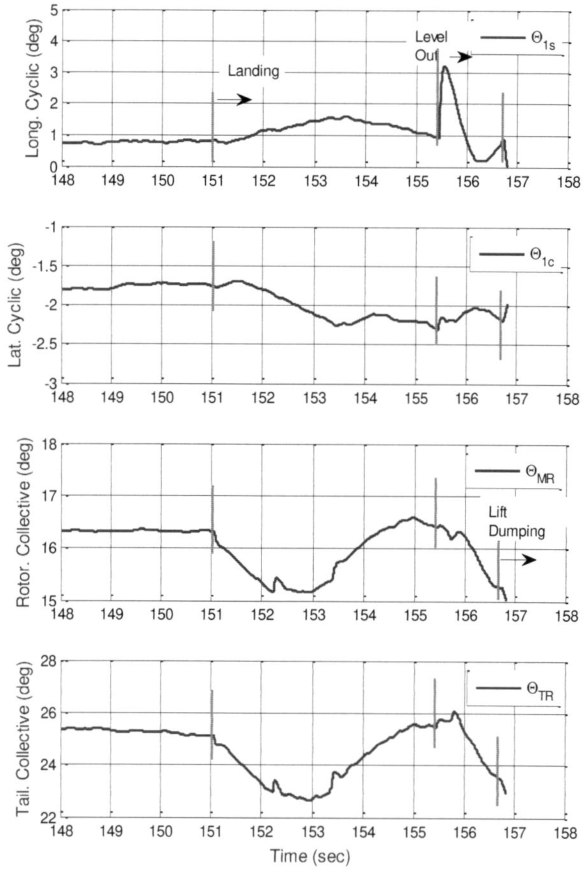

Figure 16. Blade Pitch Angles vs. Time

Figure 14 clearly shows the elevation of tail gear during level out, as it is expected. In Figure 15, the vehicle attitude experienced a step-command in the beginning of level-out action, namely - pitch down to elevate tail gear and bank right to elevate left gear. The above control action was also reflected in the control activity in Figure 16. However, the undesired drifting can also be observed in the lateral position plot of Figure 12-13, although the amount was acceptable. Quantitative metrics of landing quality are summarized in Table 2.

Longitudinal Position Error (ft)	Lateral Position Error (ft)	Lateral Velocity Error (ft/sec)			Vertical Velocity Error (ft/sec)		
		Tail Gear	Front Gear 1	Front Gear 2	Tail Gear	Front Gear 1	Front Gear 2
2.1	1.1	1.635	1.610	1.625	0.553	0.672	0.732

Table 2. Landing Quality Metrics

To verify the control strategy and reasonability of parameters, extensive test case with randomized flight conditions were performed, with the results shown in Figures 17-18. The level-out control law successfully addressed the impact on front gears due to sequence of touchdown at a tolerable cost in terms of landing location drifting and lateral speed with respect to the deck.

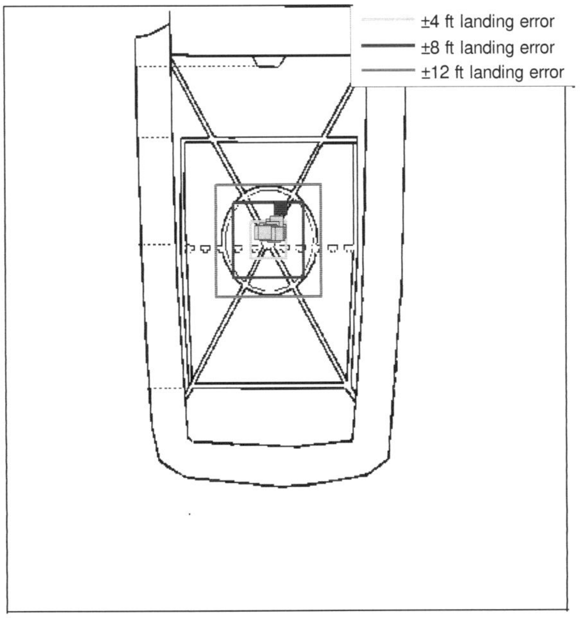

Figure 17. Scatters of The Landing Location

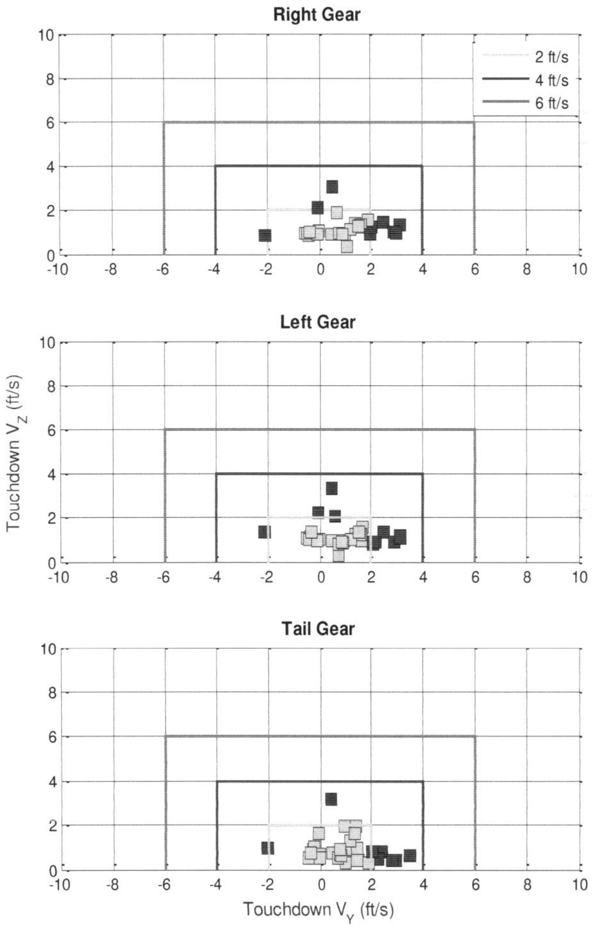

Figure 18. Scatters of Touchdown Velocity Error

SUMMARY AND CONCLUSIONS

This paper presented an algorithm for landing path generation based using forecasted deck state. The work builds off of the control laws and simulation environment developed in previous work (Ref. 6) by improving the path generation algorithm. The new path generation results in less dynamic descent maneuvering and softer touchdown. The algorithm is based on a polynomial representation of trajectory with unknown coefficients which are determined by optimization with respect to objective functions that impose desirable kinematic properties. Isolated tests of the algorithm verified that the optimization scheme could meet design objectives. A hybrid implementation of the generated inertial trajectory has been proposed to combine the inertial trajectory commands with the well-established relative motion control. To address the attitude mismatching problem, an additional level-out control law was developed. Extensive simulation tests were performed with a wide range of flight conditions using the FLIGHTLAB simulation model of a medium utility helicopter operating to a destroyer class ship. The following conclusions can be made regarding the techniques developed in this paper.

1. The polynomial parameterization provides a simple but flexible enough representation of trajectory for guidance of the landing on a moving flight deck. The method is simple in that polynomial coefficients can be calculated immediately given the boundary condition and height of internal control points. It is flexible because different heights of intermediate control points lead to paths with various position, velocity and acceleration profiles. The proper shape of the trajectory can then found by optimization of the two internal control points.

2. The hybrid implementation of inertial trajectory is critical in reducing the sensitivity of predictive landing to the accuracy of the deck motion forecasting. The extraction of relative trajectory and modification of the inertial command alleviates the dependency on accurate prediction of terminal deck state.

3. The rolling and pitching flight deck, along with the asymmetric trim of the helicopter raises the issue of the aircraft-deck attitude mismatching problem. The approach of level-out and lift-dumping presented in this paper uses terminal attitude adjustment to accommodate close to simultaneous contact on all landing gear resulting in softer impact. The level-out operation results in some drifting which undermines the positional precision of the landing, and therefore careful timing of the level-out maneuver is required. This drifting can hopefully to be solved by incorporating an attitude expectation into the trajectory planning, and this should be investigated in future work.

ACHNOWLEDGMENTS

This work was sponsored by the Office of Naval Research, ONR, under contract number N00014-14-C-0004. The views and conclusions contained herein are those of the authors and should not be interpreted as necessarily representing the official policies or endorsements, either expressed or implied, of the Office of Naval Research, or the U.S. government.

REFERENCES

[1]Ferrier, B. and Manning, T., “Simulation and Testing of the Landing Period Designator(LPD) Helicopter Recovery Aid,” Naval Engineers Journal, January 1998

[2]Perrins, J.A., and Howitt, J., “Development of a Pilot Assisted Landing system for Helicopter/Ship Recovery,” American Helicopter Society 57th Annual Forum Proceedings, Washington, DC, May 2001

[3]Gavert, G. and Schulze, E., “Shipboard Launch and Recovery of RPV Helicopters in High Sea States,” AIAA Guidance and Control Conference, Palo Alto, CA,1978

[4]Freedberg, S.J., “Fire Scout Drone’s First At-Sea Takeoff,” breakingdefense.com, December 23, 2014.

[5]Soneson, G.L., and Horn, J.F., “Simulation Testing of Advanced Response Types for Ship-Based Rotorcraft,” American Helicopter Society 70th Annual Forum Proceedings, Montreal, Quebec, Canada, May 2014

[6]Horn, J.F., Yang, J.F., He, C.J., Lee, D., Tritschler, J.K., “Autonomous Ship Approach and Landing Using Dynamic Inversion Control with Deck Motion Prediction”, 41st European Rotorcraft Forum, Munich, Germany, September 2015.

[7]Schwartz, A., “ (SCONE) – Standard Deck Motion Data for a Generic Surface Combatant,” Memorandum from Office of Naval Research and Naval Surface Warfare Center - Carderock Division, May 2015.

[8]Xin, H. and He, C., “A Statistical Turbulence Model for Shipboard Rotorcraft Simulations," Proceedings of the American Helicopter Society 63rd Annual Forum, Virginia Beach, VA, May 2007.

Wind Gust Estimation on a Small VTOL UAV

Venkatasubramani S. R. Pappu
vms5250@psu.edu
PhD Student

Yande Liu
yxl5197@psu.edu
MS Candidate

Joseph F. Horn
joehorn@psu.edu
Professor

Jared Cooper
cooper@barron-associates.com
Research Scientist

Department of Aerospace Engineering
The Pennsylvania State University
University Park, PA, United States

Barron Associates, Inc.
Charlottesville, VA,
United States

ABSTRACT

This paper describes a Kalman filter based gust identification technique for estimating wind gusts on a small rotary-wing unmanned aerial vehicle (UAV) during flight. The algorithm is designed to estimate wind gusts using only inertial, position, and control actuation sensors (no air data measurements). The technique is demonstrated using flight tests of an off-the-shelf unmanned quad-rotor UAV. An accurate dynamic model of the aircraft was developed using system identification techniques, and the model is used in a Kalman filter to estimate the external wind disturbances. Testing was conducted in an indoor flying lab using a high speed fan to simulate gusts and a motion tracking system to provide accurate state measurements. The motion capture system was used for both closed loop control of the aircraft and in the gust estimation algorithms. The technique can be used to enhance gust rejection on small UAVs or to survey wind fields around objects to understand turbulent airwakes and to validate physics models of these flow fields.

NOMENCLATURE

A,B,C,D,G,H	= linear model system, control, output, feedforward, and disturbance matrices
D	= specific drag force, m/s^2
D_u, D_{θ^2}, D_w	= non-linear dynamic model drag parameters
F_k, H_k	= discrete-time linearized system and measurement matrices
g	= acceleration due to gravity, 9.81 m/s^2
K	= Kalman gain and command filter sensitivity gain
K_P, K_I, K_D	= proportional, integral and derivative gain
M_u, $M_q, M_{\delta_{long}}$	= pitching moment stability and control derivatives
P_k, Q_k, R_k	= estimation, process, and measurement noise covariance matrices
q	= pitch rate, deg/sec or rad/sec
S_g, ω_g	= gust model scaling and frequency parameters
T	= specific thrust force, m/s^2
$T_0, T_{u^2}, T_w, T_{\delta_{th}}$	= non-linear dynamic model thrust parameters
U,V,W	= forward, lateral, and vertical velocity, m/s
$\hat{w}$	= white noise signal
X_u, X_θ	= longitudinal force stability derivative
x, y, u, w, v	= state, output, control input, disturbance, and measurement noise vectors
x, y, z	= aircraft position, m
Y_v	= lateral force stability derivative
Z_w, Z_w, $Z_{\delta_{th}}$	= vertical force stability and control derivatives
Δ	= difference between commanded and measured value (tracking error)
Δt	= sampling time step, sec
$\delta_{long}, \delta_{lat}, \delta_{th}$	= longitudinal, lateral and throttle control input, PWM signals in μsec

τ	= time constant, sec
θ, ϕ, ψ	= roll, pitch, and yaw Euler angles, deg or rad
ω_n, ζ, p	= natural frequency, damping ratio, and real-valued pole of error dynamics

Subscripts:

a	= relative to airmass
$bias$	= bias value
cf	= command filter value
cmd	= commanded value
g	= wind gust
I	= relative to inertial frame

Superscripts:

hf	= in aircraft heading frame coordinate system
I	= in inertial frame coordinate system

INTRODUCTION

Small rotary-wing UAVs are known to be sensitive to wind gusts. In addition they are capable of operating near structures that might generate unsteady airwakes in windy conditions. If small UAV autopilots had capability to measure or estimate wind gust velocities, this information might be used to enhance disturbance rejection in the control laws. Such information might also be used to map the wind field (both time averaged and unsteady flow properties) in a region of interest. For example, the wind gust estimates could be used to better understand flows in urban canyons, or to gather data in the airwake behind a ship with a flight deck. The U.S. Navy is particularly interested in this latter application, as they spend considerable resources in testing all combinations of helicopters and ship in various wind conditions to establish safe operating envelopes. Measurements of ship airwakes are of significant interest for use in validation of computational models that might be used to aid in qualification, design, and training in support of helicopter shipboard operations. Snyder et al. have demonstrated the use of small rotary-wing UAVs to survey and understand airwake disturbances behind ships (Ref. 1). Paley et al. have demonstrated the use of flow measurements on small UAVs to estimate winds and augment flight controllers (Ref. 2), and Langelaan et al. developed a method for estimating wind fields for small fixed-wing UAVs without requiring accurate vehicle models (Ref. 3).

While airspeed sensors are standard equipment on fixed-wing aircraft, accurate low speed airspeed measurement is typically not available on rotary-wing VTOL aircraft. Flow measurements on rotary-wing UAVs are problematic as the measurements are usually affected by the rotor downwash. Thus, accurate estimation of gusts based solely on inertial measurements and measurements of the control compensation would be useful for surveying flow fields and developing and validating models of these flow fields. Such a method clearly needs to be model-based, since without airspeed measurement, a relationship between gust disturbances and the aircraft response must be known. Model based approaches were used by Lusardi et al who developed a gust turbulence model (the CETI model) through flight test experimentation, where the turbulence properties are characterized by equivalent control inputs (Ref. 4). The method was derived using measurements of control activity and inertial sensors from flight test of a UH-60 helicopter in the airwake behind a hangar. A method similar to CETI was developed to derive models based on gust velocities instead of control inputs (Ref. 5). These models require only measurement of aircraft inertial response and an accurate dynamic model of the aircraft and do not require direct measurement of flow. The method developed in Ref. 4 was also incorporated in an adaptive control augmentation and shown to improve disturbance rejection in flight simulations (Ref. 6). The estimation algorithm is known as the Gust Identification Algorithm and will be leveraged in this research.

This paper expands upon previous research on the use of small rotary-wing UAVs for estimating wind gusts (Ref. 7). Whereas in previous research, a conventional RC helicopter was used to estimate wind gusts, this paper focuses on the use of small quad-rotor UAV. The quad-rotor UAV has simplified dynamics compared to the conventional

RC helicopter which uses fly bar stabilization and a tail rotor. The quad-rotor dynamics are lower order and there is less cross-coupling. In addition, the simplicity and high performance of the on-board flight controller makes testing more reliable. In previous research, accurate linear dynamic models were derived for the Iris+ quad-rotor UAV using both frequency domain and time domain system identification methods (Ref. 8) for the aircraft operating near hover. This paper presents further developments in the dynamic models and the gust identification algorithm for the quad-rotor UAV, with the use of non-linear dynamic models and a non-linear estimator. These algorithms are demonstrated using indoor flight tests.

Figure 1: Iris+ in Indoor Flight Tests.

TEST FACILITIES

The UAV used in this study is an Iris+, a commercially available quad-rotor UAV from 3D Robotics equipped with sensors, Inertial Navigation System (INS), and autopilot. The Iris+ can be flown in an attitude stabilized mode without GPS, which is done for this work as the tests were conducted indoors. The aircraft was flown in an indoor flight lab equipped with a Vicon motion capture system. This system has the ability to measure position and attitude of the aircraft to millimeter level accuracy. The Vicon is capable of sampling the position and attitude at 100 Hz, but due to latencies in the network and computation load, a 20 Hz sampling rate was found to be consistently achievable and suitable for this work. The Vicon data were incorporated into a data management system operated within the Robot Operating System (ROS) framework with distributed nodes. A node is conceptually a distinct process in the framework, e.g., a global clock, Iris+ vehicle, controller, and Vicon system. Each node runs independently and can communicate with other nodes as needed.

A controller node was developed for the Iris+ for use in the indoor tests. This node, operating on the ground station computer, gathers Vicon state data and sends control signals to the Iris+ over a 3DR radio link to control and regulate the aircraft position (i.e. this node provides translation rate and position hold control). The system was designed to control the Iris+ while it flies in "Stabilize" mode (STAB MODE), so the onboard autopilot on the IRIS+ provides inner loop attitude stabilization, while the ground based controller node regulates velocity and position. In addition the controller node records flight data for the Iris+.

The flight lab is also equipped with a 30 inch high speed fan used to generate simulated wind gusts. A layer of honeycomb was placed on the outflow of the fan to reduce swirl effects. In this configuration, the fan generates a core flow of about 7 m/s. A rearward facing step structure was also used to generate airwake turbulence behind the fan for some tests, while in other tests the Iris+ was subjected to straight wind gusts from the fan. An ultrasonic anemometer was used to measure flow velocities. Velocity measurements were taken upstream of the quad-rotor during flight, and the anemometer was also used to survey the spatial variations in fan flow. The same fan, anemometer, simulated ship structure, and lab space were used in earlier studies and described in more detail in Ref. 7. The indoor flight lab facility with fan, anemometer, ship structure, and Iris+ in flight is shown inFigure 1.

GUST IDENTIFICATION ALGORITHM

The concept of the gust identification algorithm is described in detail in Ref. 7 and related papers. The idea is that using a good dynamic model of the aircraft (from validated physics models and/or system identification) and measurement of the control inputs, external disturbances on the aircraft can be estimated. The gust identification concept outlined here and in Refs. 5, 6, and 7, assumes that if the aircraft is subjected to wind disturbances, the discrepancies between measured response and the response predicted by the model can be attributed to these wind gusts. This same general concept was also considered in the work of Refs. 1 and 4 which correlates the disturbance with control inputs or state discrepancies. However, unlike that work, the gust identification algorithm seeks to directly estimate the gust velocities based on this discrepancy.

Consider a linear dynamic model of the aircraft in state space form:

$$\dot{x} = Ax + Bu + Gw$$
$$y = Cx + Du + Hw + v \tag{1}$$

where w is the wind gust disturbance vector, and G is the corresponding disturbance matrix. Similarly the measurements might be affected by the disturbance through H. We seek to estimate wind gusts using a Kalman filter, a commonly used algorithm to estimate system states given noisy and sometimes incomplete measurements. The standard Kalman filter assumes the plant is disturbed by Gaussian white noise of known covariance. However, wind gusts are not represented by white noise (they have time correlation) and are not system states in the current formulation. This is addressed by assuming gusts are represented by filtered white noise, with an assumed scaling matrix, S_g, and bandwidth parameter, ω_g:

$$\dot{x}_g = -\omega_g I x_g + S_g \hat{w} \tag{2}$$

where $\hat{w}$ is unity white noise and x_g is a vector of gust disturbances on the aircraft, typically expressed as velocities and/or angular rates. When coupled with the full state equations we have the system:

$$\begin{bmatrix} \dot{x} \\ \dot{x}_g \end{bmatrix} = \begin{bmatrix} A & G \\ 0 & -\omega_g I \end{bmatrix} \begin{bmatrix} x \\ x_g \end{bmatrix} + \begin{bmatrix} B \\ 0 \end{bmatrix} u + \begin{bmatrix} 0 \\ S_g \end{bmatrix} \hat{w}$$
$$y = \begin{bmatrix} C & H \end{bmatrix} \begin{bmatrix} x \\ x_g \end{bmatrix} + Du \tag{3}$$

A Kalman filter can be developed to estimate the system states, which now includes the gust disturbance, x_g.

$$\dot{\tilde{x}} = \begin{bmatrix} A & G \\ 0 & -\omega_g I \end{bmatrix} \tilde{x} + \begin{bmatrix} B \\ 0 \end{bmatrix} u + K\left(y - \begin{bmatrix} C & H \end{bmatrix} \tilde{x} - Du\right) \tag{4}$$

$$\text{where } \tilde{x} = \text{Estimate of} \begin{bmatrix} x \\ x_g \end{bmatrix} \tag{5}$$

The matrix K is the Kalman gain matrix determined through solution of a Riccati equation. The dimension of the gust vector can be tailored for the specific application. For example, it can include only velocities, assuming a uniform gust velocity over the body of the vehicle. This assumption, which might be suitable for a small aircraft subject to a wind field with large length scales, is used for the quad-rotor used in this study. Previously, simulation studies with a full-scale helicopter model also needed to include angular rate disturbances, which represent linear spatial distribution of the flow field over the aircraft body (Refs. 5, 6, 7).

As the project progressed, it became clear that the linear Kalman filter approach had limitations, and that a non-linear model and gust estimation algorithm would have better performance (this will be discussed in more detail below). Therefore, a non-linear Extended Kalman Filter (EKF) approach was developed. The EKF uses a non-linear model of the form (in discrete time):

$$x_k = f\left(x_{k-1}, u_{k-1}\right) + w_{k-1} \tag{6}$$

$$y_k = h\left(x_k\right) + v_k \tag{7}$$

where, $f(\ )$ and $h(\ )$ are the functions to compute the predicted state and predicted measurements from previous estimates, respectively. The vector w_{k-1} is the process noise and v_k is the observation noise.

The state estimates are then achieved via the EKF state estimation algorithm, which is well documented and widely used. This formulation is based on the textbook of Brown and Hwang (Ref. 9). The EKF for a sampled system includes a prediction step for the state and covariance estimates as in Eqs. (8) and (9) respectively.

$$\hat{x}_{k|k-1} = f\left(\hat{x}_{k-1|k-1}, u_{k-1}\right) \tag{8}$$

$$P_{k|k-1} = F_{k-1}P_{k-1|k-1}F_{k-1}^{T} + Q_{k-1} \tag{9}$$

where, the state transition matrix is a linearization of the discretized plant model, $F_{k-1} = \frac{\partial f}{\partial x}|_{\hat{x}_{k-1|k-1}, u_{k-1}}$, and Q_{k-1} is the discretized process noise covariance matrix. Note that this step simply propagates the current state estimate and measured control through the model of the system dynamics and does not include measurements. The EKF then uses an update equation when a measurement is available, and performs updates per the following:

$$\text{Measurement residual, } \tilde{y}_k = y_k - h\left(\hat{x}_{k|k-1}\right) \tag{10}$$

$$\text{Residual covariance, } S_k = H_k P_{k|k-1} H_k^T + R_k \tag{11}$$

$$\text{Kalman Gain, } K_k = P_{k|k-1} H_k^T S_k^{-1} \tag{12}$$

$$\text{State estimate update, } \tilde{x}_{k|k} = \tilde{x}_{k|k-1} + K_k \tilde{y}_k \tag{13}$$

$$\text{Covariance estimate update, } P_{k|k} = \left(I - K_k H_k\right) P_{k|k-1} \tag{14}$$

where, the observation matrix, $H_k = \frac{\partial h}{\partial x}|_{\hat{x}_{k|k-1}}$, is a linearization of the measurement model and R_k is the sensor noise covariance matrix.

A closed-form linearization of this model is used to calculate continuous time linear models which are then converted to discrete time via state transition model. The aircraft dynamic model is initially formulated in continuous time (as documented below), but then a discrete time approximation must be used for implementation of the EKF. This discrete time model for the prediction step in Eq. 8 is derived via Adams-Bashforth integration.

DYNAMIC MODEL IDENTIFICATION AND CONTROL DESIGN

Linearized Dynamics

The conventional Kalman filter requires a linearized model of the aircraft dynamics (A, B, G matrices) around the operating point of interest. Furthermore, the linearized models are also used in the design of the outer loop controller (as described in the following section). When extending to non-linear estimation methods, the linearized stability derivatives provided an initial estimate of the non-linear model parameters. Thus, the first step was to perform linear system identification on the Iris+ dynamics. Initial development of the identified model of the Iris+ was performed in Ref. 8 and is now further refined in this study.

Simplified decoupled models of the Iris+ lateral and longitudinal dynamics were developed. Computer generated frequency sweeps in the longitudinal, lateral, vertical, and yaw axes, were given to the Iris+ to gather the flight response data for frequency domain system identification (Ref. 10). Doublet maneuvers were conducted to collect data for verification cases. Due to fast and unstable open-loop dynamics of the Iris+, the vehicle must be operated with a closed-loop controller active. Thus, the identified model represents the closed-loop dynamics with attitude stabilization using the STAB MODE. The Iris+ uses an attitude command controller in the pitch and roll channels and a PID controller, which results in a third-order model structure for attitude dynamics in both roll and pitch axes. It should be noted that the attitude command to the Iris+ are filtered through a dead zone, which allows the pilot to more easily hold level attitude trim. The control signals from the flight data were processed to reconstruct the control signal after the dead zone, to allow identification of a linear dynamic response. The aircraft position and attitude information were collected using the motion capture system. These data were used along with recorded pilot input signals to do system identification of the aircraft.

The longitudinal dynamics of the IRIS+ were modeled using the equations of motion:

$$\dot{U} = X_u U - g\theta \quad (15)$$

$$\dot{q} = M_u U + M_q q + M_{\delta_{long}} \delta_{long} \quad (16)$$

$$\dot{\theta} = q \quad (17)$$

where

$$\delta_{long} = K_P\left(\theta_{cmd} - \theta\right) + K_I \int \left(\theta_{cmd} - \theta\right) + K_D\left(\dot{\theta}_{cmd} - \dot{\theta}\right) \quad (18)$$

The state space model shown in Eq. (19) was setup in the CIFER® tool to identify the longitudinal state and control parameters of IRIS+. The vertical axis is open-loop in the stabilize mode of the IRIS+, so a simple heave axis dynamic model was identified as shown in Eq. (20).

$$\begin{bmatrix} \dot{U} \\ \dot{X}_2 \\ \dot{X}_3 \\ \dot{\theta} \end{bmatrix} = \begin{bmatrix} X_u & 0 & 0 & X_\theta \\ M_u & 0 & 1 & -K_P \\ 0 & 0 & 0 & -K_I \\ 0 & 1 & 0 & (M_q - K_D) \end{bmatrix} \begin{bmatrix} U \\ X_2 \\ X_3 \\ \theta \end{bmatrix} + \begin{bmatrix} 0 \\ K_{Pc} \\ K_{Ic} \\ K_{Dc} \end{bmatrix} \theta_{cmd} \quad (19)$$

$$\dot{W} = Z_w W + Z_{\delta_{th}} \delta_{th} \quad (20)$$

In addition, a pure time delay resulting from communication latencies was also estimated and included in the models. These latencies lead to a 0.1 to 0.2 sec delay from ground station input to output measurement on the Vicon system. Figure 2 illustrates the frequency domain system identification of the longitudinal dynamics using CIFER.

Vicon Controller Setup

An outer-loop controller using motion capture state feedback was designed and implemented for this study. The controller, operating on the ground station, uses the Vicon motion capture system to provide position and velocity feedback. It also uses yaw attitude from the Vicon for coordinate transformations. The controller sends PWM signals to the IRIS+ while it operates in STAB MODE. The controller was able to precisely hold the aircraft in a fixed position for the fan tests. The fan gusts generate large disturbances for this class of vehicle, and in order to gather repeatable data the controller needed to hold a desired position. The controller must also safely control the aircraft in high speed flight tests, which were used for development of a non-linear model. Thus, the performance of this controller turned out to be an important factor for this study.

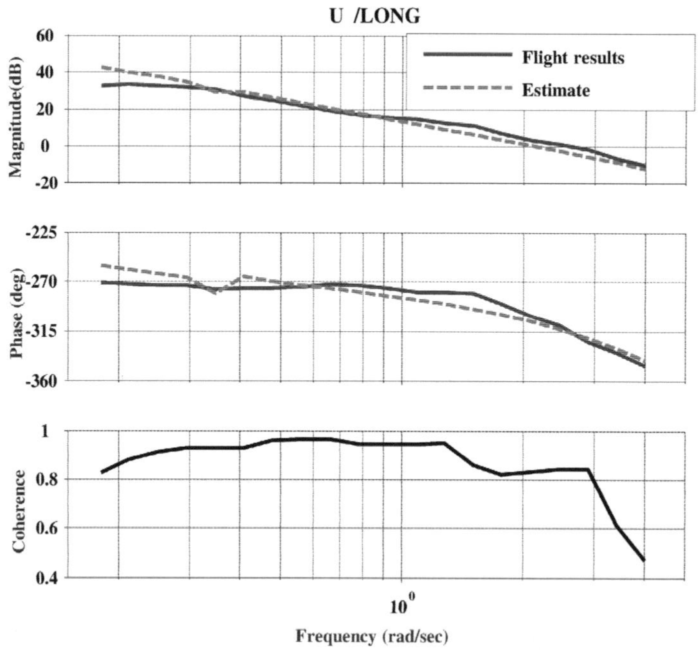

Figure 2: Frequency response of velocity to longitudinal input.

The inner-loop controller on board the IRIS+ autopilot was not modified. The PWM signals sent to the autopilot in STAB MODE are converted to commanded roll and pitch attitude in the autopilot software. Thus, command roll and pitch attitude are effectively the control inputs to the closed loop dynamics of the IRIS+ in this mode. The inner loop control laws for roll and pitch achieve an Attitude Command Attitude Hold response type. One detail that needs to be accounted for is the deadband between the roll and pitch PWM signals and the commanded attitude. While useful for a human pilot, the deadband degrades performance of the outer loop controller. An inverse deadband is used in the outer loop control law to negate the deadband of the inner loop. This allows fine adjustment of roll and pitch attitude for precise position hold. In STAB MODE, the thrust axis uses direct control over collective motor speed with no feedback, which is difficult to fly for a human pilot, but was easily regulated when tied into the outer loop controller. The yaw axis achieves a rate command / heading hold response, which is easy to fly for a human pilot and holds heading very precisely. Thus, no outer loop Vicon compensation was used for yaw, as the heading could be locked in place without pilot or outer loop compensation.

The outer loop control law was designed to achieve a Translational Rate Command / Position Hold response type. A joystick inceptor was used (in place of standard RC radio controller) and allows a human pilot to manually control velocity and position of the aircraft within the flying space. The system was designed such that simple inputs allow a novice pilot to easily maneuver the aircraft to any point in the Vicon lab. Once the stick is released the controller holds the aircraft at an exact position. In addition, a centering mode was developed, where a press of a button commands (0,0) position in the lab, corresponding to the main hovering location used for the fan gust testing. The controller is capable of fully autonomous control given a prescribed trajectory. Use of a human operator was simply found to be a convenient mode for testing, as different locations can be surveyed without developing new trajectory files or a waypoint interface.

The Vicon system provides direct measurement of x, y, z position, which are passed through a filtered differentiator to estimate velocities in the inertial frame. The velocities are then used in the control laws and eventually the gust identification algorithm. Pilot commands are given in the aircraft heading frame, i.e. forward motion of the stick causes forward motion based on the current heading of the aircraft, and lateral stick displacement causes sideward flight. Likewise, the control compensation in roll and pitch attitude commands must be given in the aircraft heading frame, such that nose down pitch commands result in forward motion and roll right commands result in right sideward motion. However, Vicon measurements are in the inertial frame (fixed relative to the room). Thus, transformations between the inertial frame and the aircraft heading frame are required. This is achieved through a simple transformation involving the aircraft heading. For example, transformation of velocities from inertial to heading frame is given by:

$$\begin{bmatrix} u^{hf} \\ v^{hf} \\ w^{hf} \end{bmatrix} = \begin{bmatrix} \cos\psi & \sin\psi & 0 \\ -\sin\psi & \cos\psi & 0 \\ 0 & 0 & 1 \end{bmatrix} \begin{bmatrix} u^{I} \\ v^{I} \\ w^{I} \end{bmatrix} \tag{21}$$

The inverse transformation simply uses the transpose of the matrix. The right superscripts hf and I indicate heading frame and inertial frame respectively. The vertical component is identical between the two frames, thus the superscript will be dropped for the vertical components, and the transformation is reduced to two-dimensional.

$$T_{hf/I} = \begin{bmatrix} \cos\psi & \sin\psi \\ -\sin\psi & \cos\psi \end{bmatrix} \qquad T_{I/hf} = \begin{bmatrix} \cos\psi & -\sin\psi \\ \sin\psi & \cos\psi \end{bmatrix} \tag{22}$$

Figure 3 shows an overall schematic of the outer loop controller setup with the IRIS+, ground station computer, and Vicon motion capture system. Note, that data were recorded in flight so that the gust identification algorithm can be assessed in post-processing. More details on the TRC/PH Outer Loop CLAW block will be discussed in the following section.

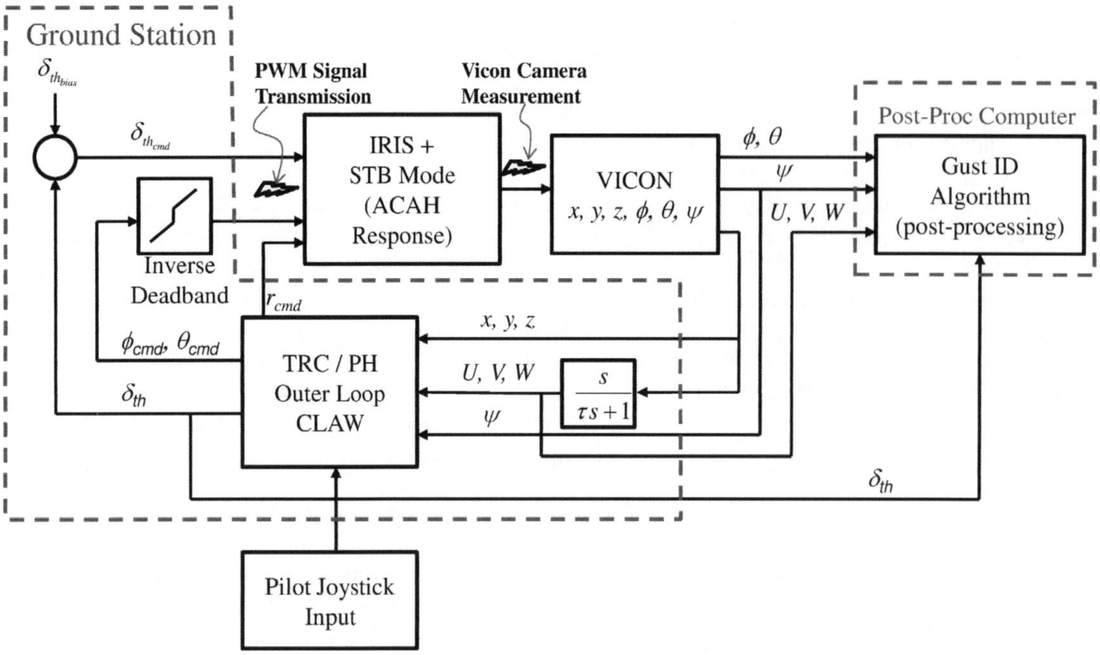

Figure 3: Block diagram of the IRIS+ outer loop controller setup.

Outer Loop Control Law Design

The control law is based on dynamic inversion using a simplified model of the translation dynamics of the aircraft:

$$\begin{aligned} \dot{U}^{hf} &= X_u U^{hf} - g\theta \\ \dot{V}^{hf} &= Y_v V^{hf} + g\phi \\ \dot{W} &= Z_w W + Z_{\delta_{th}}\left(\delta_{th} - \delta_{th_{bias}}\right) \end{aligned} \tag{23}$$

Applying dynamic inversion of these decoupled system yields the control laws:

$$\begin{aligned} \theta_{cmd} &= -\frac{1}{g}\left(\dot{U}_{cf}^{hf} + K_D \Delta U^{hf} + K_P \Delta x^{hf} + K_I \int \Delta x^{hf} dt - X_u U^{hf}\right) \\ \phi_{cmd} &= \frac{1}{g}\left(\dot{V}_{cf}^{hf} + K_D \Delta V^{hf} + K_P \Delta y^{hf} + K_I \int \Delta y^{hf} dt - Y_v V^{hf}\right) \\ \delta_{th} &= \delta_{th_{bias}} + \frac{1}{Z_{\delta_{th}}}\left(\dot{W}_{cf} + K_D \Delta W + K_P \Delta z + K_I \int \Delta z dt - Z_w W\right) \end{aligned} \tag{24}$$

where the *cf* subscripts indicate commanded accelerations from the command filters. The command filters are first order filters that translate pilot stick inputs to commanded velocities, and are defined by a stick sensitivity gain, K, and a time constant, τ. The commanded accelerations and position are also generated through differentiation and integration of the velocity command. The delta terms in Eq. (24) represent the tracking errors, the difference between the filtered command and the measured value for both velocity and position. The errors are converted into the aircraft heading frame, so that compensation occurs with the appropriate attitude change given the current heading of the aircraft. Again, care must be taken to properly convert values between the two reference frames to develop appropriate error signals and appropriate control compensation. Figure 4 shows the Translation Rate Command / Position Hold control law. Note, that subtraction of the velocity damping terms ($-X_u U$ and $-Y_v V$) are left out in the diagram, and in practice, these terms can often be ignored, meaning the outer loop x-axis and y-axis controller requires no model specific parameters. The vertical axis control law is very similar in structure as the x and y axis control, but it does not require the coordinate system transformations. However, correct identification of the model specific control sensitivity and damping derivatives ($Z_{\delta th}$ and Z_w) are important for vertical axis controller performance.

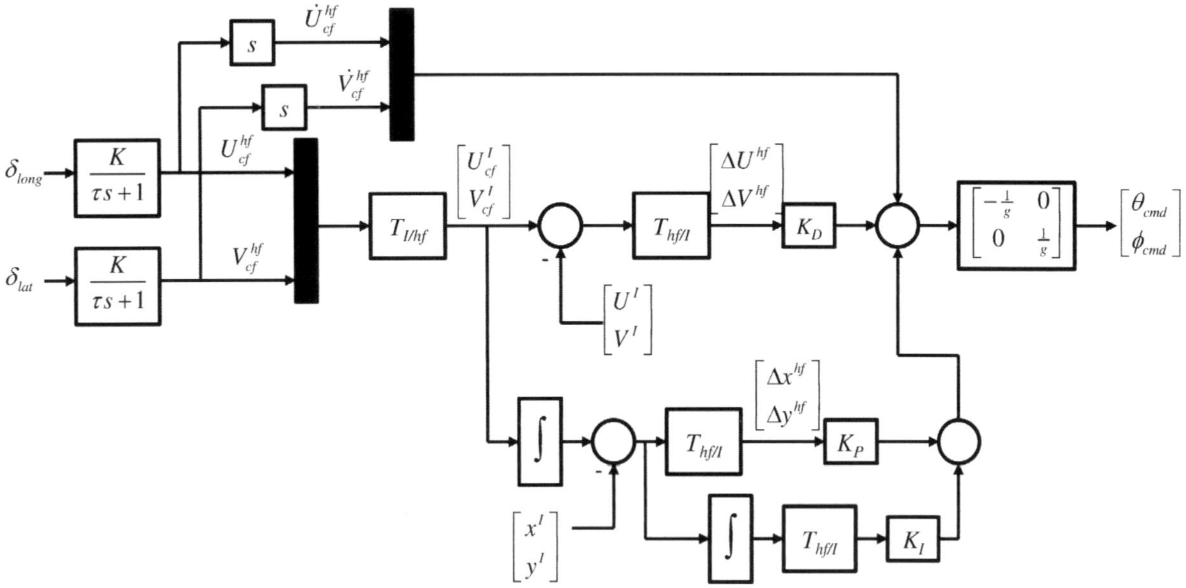

Figure 4: Block diagram of Translational Rate Command/Position Hold control law.

The gain selection for the outer loop controller is based on specification of the error dynamics. We can assume that the inner loop tracks attitude commands with some perturbation due to external disturbance and modelling error, $\delta(t)$. For example, in the longitudinal axis:

$$\theta = \theta_{cmd} + \delta(t) \tag{25}$$

Substituting this expression back into the equations of motion:

$$\dot{U}^{hf} = X_u U^{hf} - g\left(\theta_{cmd} + \delta(t)\right) \tag{26}$$

One derives the longitudinal error dynamics:

$$\begin{aligned} &\Delta\dot{U}_{cf}^{hf} + K_D \Delta U^{hf} + K_P \Delta x^{hf} + K_I \int \Delta x^{hf} dt = g\delta(t) \\ &\Delta\ddot{x}_{cf}^{hf} + K_D \Delta\dot{x}^{hf} + K_P \Delta x^{hf} + K_I \int \Delta x^{hf} dt = g\delta(t) \end{aligned} \tag{27}$$

Differentiating these expressions, then taking Laplace transforms, one can derive the transfer function from perturbation to position tracking error:

$$\begin{aligned} &\Delta\dddot{x}_{cf}^{hf} + K_D \Delta\ddot{x}^{hf} + K_P \Delta\dot{x}^{hf} + K_I \Delta x^{hf} = g\dot{\delta}(t) \\ &\frac{\Delta x}{\delta}(s) = \frac{gs}{s^3 + K_D s^2 + K_P s + K_I} \end{aligned} \tag{28}$$

The same expressions can be derived for the lateral and vertical axes. Note the initial condition terms were dropped, but these would simply result in additional terms in the numerator of this expression and their effects would quickly decay with selection of stabilizing gains. The analysis shows that selection of the gains specifies the poles of the error dynamics. By factoring the polynomial, the gains can be mapped to a frequency and damping ratio of a pair of complex poles, and to a single real valued pole.

$$\frac{\Delta x}{\delta}(s) = \frac{gs}{s^3 + K_D s^2 + K_P s + K_I} = \frac{gs}{\left(s^2 + 2\zeta\omega_n s + \omega_n^2\right)(s+p)} \tag{29}$$

$$K_D = 2\zeta\omega_n + p \qquad K_P = \omega_n^2 + 2\zeta\omega_n p \qquad K_I = \omega_n^2 p \tag{30}$$

The natural frequency, ω_n, parameter effectively defines the bandwidth of disturbance rejection of the controller, the damping ratio, ζ, defines damping, and the real-valued pole, p, primarily drives the integral action. Based on past experience with the controller, selecting $\zeta = 1.0$ generally results in desired damping, and the real-valued pole should have frequency separation with the natural frequency parameters, as a rule of thumb $p = 0.2\omega_n$ is a good selection. The natural frequency, ω_n , can then be selected to achieve the best possible disturbance rejection without inducing closed loop instability or control saturation. The former typically occurs in the lateral and longitudinal axes when the frequency begins to approach the closed-loop bandwidth of inner loop, i.e. the outer loop controller should have frequency separation from the inner loop. Thus, controller tuning is done primarily through a single frequency parameter in each axis. Through testing, the parameters in Table 1 were selected. Note that the vertical axis uses much higher gains, since rotor thrust provides direct actuation in the vertical axis, and gains are only limited by motor response bandwidth and authority. In the lateral and longitudinal axes, the controller provides indirect control through attitude changes and is limited by the attitude response bandwidth. Examples of the controller tracking performance due to both pilot commands and external disturbances will be shown in the following sections.

Table 1. Controller parameters and gains used in different axes.

Axis	Frequency Parameters			Gains		
	ω_n (rad/sec)	ζ (-)	p (rad/sec)	K_P (1/sec^2)	K_D (1/sec)	K_I (1/sec^3)
Lateral and longitudinal	1.0	1.0	0.2	1.4	2.2	0.2
Vertical	2.0	1.0	0.4	5.6	4.4	1.6

Non-linear Dynamics

In the initial development of the Gust Identification algorithms, linear dynamic models were identified and then implemented in a linear Kalman Filter (Ref. 7). Linear models are useful due to the wide range of system identification tools and robust estimation algorithms that can be used to define the model and then estimate system states. However, use of linear systems was found to have two major drawbacks for this application. First, linear models are not necessarily accurate over a wide range of operating conditions. Initial development used simulations of a full size utility helicopter operating in a turbulent airwake. The perturbations in airspeed experienced by the full-sized helicopter (±15 knots) were not large relative to the flight envelope (with flight envelope airspeeds ranging from 40 knots rearward to 150 knots forward). In these tests, we subject RC scale aircraft to gusts on order of 6-8 m/s, which *do* represent large perturbations, as their maximum operating speeds are on the order of 10 m/s. Secondly, it was found that certain important aerodynamic perturbations on the aircraft are not well modeled using a linear perturbation term. A good example is the influence of forward airspeed on lift, normally modeled through the Z_u in a linear perturbation model. While this term is accurate for perturbations around a high speed operating point, near hover this effect is better modeled though a U^2 perturbation.

A non-linear model was developed by assuming a simplified model structure, and then fitting a subset of model parameters to time domain flight data. Fitting was performed using the non-linear least squares fitting algorithms in MATLAB. Parameters already identified via linear system identification were retained wherever this was found to be accurate, and in other cases the linear model parameters were used for initial guesses of certain model parameters used in the non-linear model.

To develop identification data for the non-linear model, flight tests needed to span the range of potential airspeeds experienced by the aircraft during the fan gust tests. The high speed fan produces air velocities as high as 7 m/s, so tests needed to be performed that accelerated the aircraft up to this speed in still air. This presented a challenge due to the confined indoor flying space, and was one of the motivating factors in developing a high performance outer loop controller. Figure 5 shows the aircraft response in a high speed flight test with velocity above 6 m/s. As seen in the trajectory (Fig. 6), the flight was conducted along the diagonal of the indoor flying space to allow sufficient space for the test.

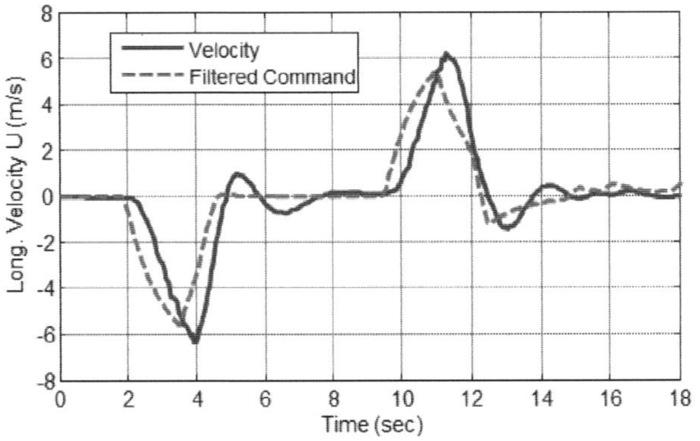

Figure 5: Velocity response to high speed commands.

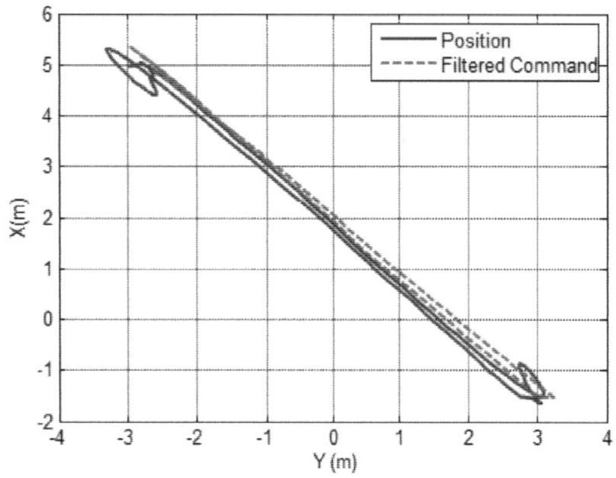

Figure 6: Flight trajectory for high speed commands.

The current investigation focused on a model of the longitudinal / vertical translational dynamics of the aircraft. This model is suitable for estimating longitudinal and vertical wind gusts when the aircraft is nose into the primary flow direction. Investigation of lateral dynamics has also been conducted but is not presented here for the sake of brevity.

A number of non-linear model structures were investigated during the study. Equations (31) summarize the final non-linear model of the aircraft dynamics that was used by the Extended Kalman Filter for the Gust Identification tests:

$$\begin{aligned}
\dot{U}^I &= T\sin\theta - D \\
\dot{W}^I &= T\cos\theta - g \\
D &= D_u U^I\left(1 + D_{\theta^2}\theta^2\right) + D_w W^I \\
T &= T_0 + T_{\delta_{th}}\delta_{th} + T_w\left(W^I\cos\theta - U^I\sin\theta\right) + T_{u^2}\left(U^I\right)^2 \\
T_0 &= \frac{g}{\cos\theta} \qquad \delta_{th} = \frac{10s}{10s+1}PWM_{thrust} \qquad \theta = \frac{10s}{10s+1}\theta_{Vicon} \qquad g = 9.81\ \text{m/s}^2
\end{aligned} \tag{31}$$

The states of the dynamic model are the velocity of the aircraft relative to the airmass with components expressed in the inertial coordinate system. Thus, U^I is forward velocity relative to the air (positive with aircraft moving forward or in a head wind), and W^I is upward velocity relative to the air (positive with the aircraft climbing or in a downdraft). The terms D and T represent the specific drag and thrust forces on the aircraft (force normalized by aircraft weight with units of acceleration). Note that the drag force is in the inertial frame while the thrust force is defined in the body frame of the aircraft. The measured pitch attitude, θ, and the thrust axis command, δ_{th}, are treated as inputs to the dynamic model. The thrust command is the change in the PWM signal for thrust command relative to the trim value sent by the TRC controller to the IRIS+. Likewise the pitch attitude is the pitch angle measured by the Vicon system relative to the trim pitch attitude. The trim PWM thrust command changes slightly with every flight due to battery drain on the IRIS+. The trim pitch attitude in hover is nominally 0, but in practice there is always a slight bias in the trim pitch angle due to the initialization of the Vicon system. Since the trim biases for both thrust command and pitch vary from flight to flight, the perturbations from trim are estimated by passing the signals though a washout filter with time constant 10 sec. The trim thrust value used in the model, T_0, is simply the weight of the aircraft (normalized to g = 9.81 m/s^2), which enforces trim in hover. Note the 1/cosθ term in the trim thrust accounts for an automatic compensation in thrust that occurs within the IRIS+

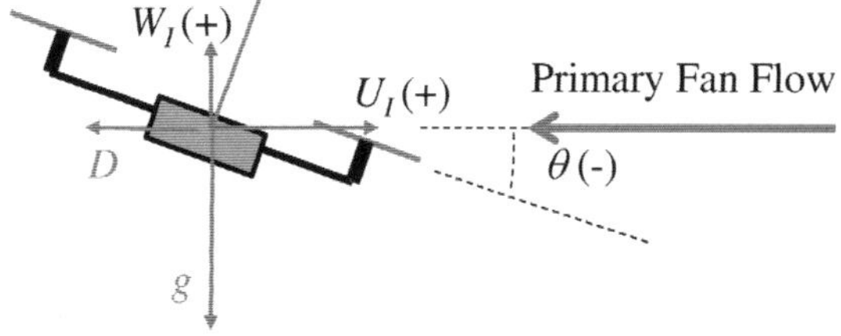

Figure 7: Force and velocity diagram used in the non-linear model.

autopilot. The autopilot increases the rotor RPM to account for the increased thrust required with attitude change. This thrust compensation is not visible to the ground station computer, so the term is assumed to be part of the plant model. Figure 7 shows a schematic of the vehicle with the various specific force terms and inertial velocities due to the wind from the fan.

The parameters in the non-linear model were identified through a variety of methods.

1. Linear system identification was conducted with the CIFER® tool as described before. In the final non-linear model, the T_w and $T_{\delta th}$ terms were fixed to correspond with the Z derivatives identified with CIFER®. The initial guess of the D_u term was set via the identified X_u derivative.
2. The AUSPEX system ID tool (Ref. 11) was used to estimate a series of non-linear extensions of the linear model to identify sensitivity and uncertainty in various parameters. For example, using AUSPEX we were able to determine that the change in vertical force with forward velocity squared (T_{u2}) helped improve the vertical / longitudinal cross-coupling. Initial models assumed that the drag term could be modeled by a velocity squared effect (as expected for aerodynamic drag as discussed below). However AUSPEX analysis showed that linear drag result in better matching of the flight data.
3. The final non-linear model was defined via a least squares fitting of the time domain response for a set of doublet inputs in the longitudinal and vertical axes. This was used to identify the terms: D_u, D_w, $D_{\theta 2}$, T_{u2}.

It was expected that the drag term would be best modelled using a U^2 term of the form $D = D_{u^2} U|U|$ as would be expected for aerodynamic drag with constant drag coefficient. However, when analyzing flight data, this was found to be a surprisingly poor representation of the drag force. This was revealed through large uncertainties in the AUSPEX identification results and in poor fitting of the flight data. The fact that drag is dominated by a linear term indicates that the drag is predominantly from flapping of the rotors. While the rotor blades are not articulated, there is sufficient flexibility in the blades for the rotors to flap aft when subjected to forward speed as induced by the asymmetric lift and a 90° phase lag in flapping response. Similar models have been suggested for quad-rotor UAVs, for example in Ref. 12 a flapping model was used and resulted in a linear drag force with change in velocity. The $D_{\theta 2}$ term accounts for a reduction of this effect as the aircraft is more nose down or nose up relative to the flow, and was found to provide an improved fit of the data for larger amplitude attitude changes in high speed cases.

Figure 8 shows a sample of the time domain model fit for a series of longitudinal and vertical pulse inputs. Results indicate a very good fit of the flight data, and these were consistent across several different flight tests with longitudinal and vertical commands of different magnitudes.

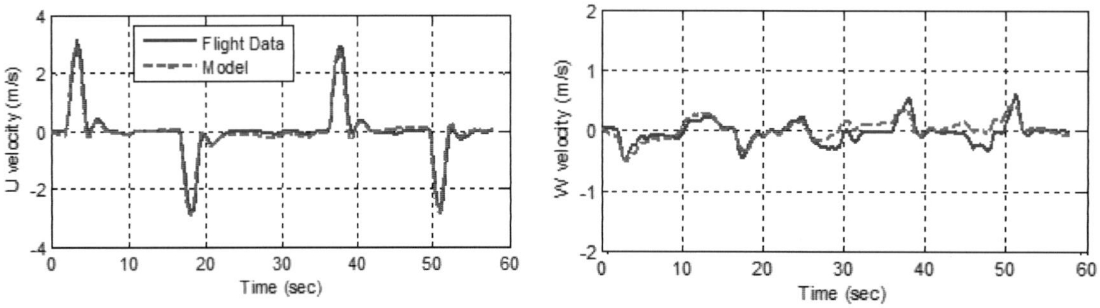

Figure 8: Comparison of flight test and non-linear model responses to longitudinal and vertical pulse inputs.

FAN FLIGHT TESTS

Test Setup

The aircraft was flown in the indoor flight lab and subjected to wind gusts using the high speed fan, while the ultrasonic anemometer was used to simultaneously measure the fan flow half way between the fan and the aircraft during the flight tests. The anemometer was also used to survey the fan flow around the hovering location without the aircraft (when the aircraft was not flying) to measure the distribution of the mean flow as well as unsteady fluctuations (as measured by standard deviation). These tests were conducted both with pure fan flow (no obstructions) and with the rearward facing step structure intended to simulate the structure of a ship. The tests included subjecting the aircraft to steady fan flow and to a series of pulse gusts, where the fan is turned on for 10-20 seconds and then turned off. Note that the fan, anemometer, and simulated ship structure used for these tests are identical to those used in Ref. 7, and more details on this equipment can be found in that paper.

Figure 9 shows a plan view of the test arrangement roughly to scale, while Fig. 10 shows a side view. A sample flight trajectory extracted from recorded flight data is also shown roughly to scale. When the fan gusts hits the aircraft, it tends to blow the aircraft backward, as seen in the x displacement in the flight data. This displacement was as much as 1.3 m for the case with no ship structure and 0.8 m for the case with the simulated ship structure. However, the controller holds lateral position within ±0.2 m and altitude within ±0.3 m of the desired hover location. Based on anemometer surveys of the fan flow, this was found to keep the aircraft within the core fan flow. Surveys of the flow field can be found in Ref. 7, which shows that the mean flow does not change significantly for small x displacement downstream of the nominal hover location, and thus the results are not considered to be significantly affected by these displacements. Flow velocities are sensitive to vertical and lateral location, which was one of the motivating factors for optimizing the position and altitude control laws for the IRIS+. The relatively small displacements seen in the test are considered to be tolerable for this study.

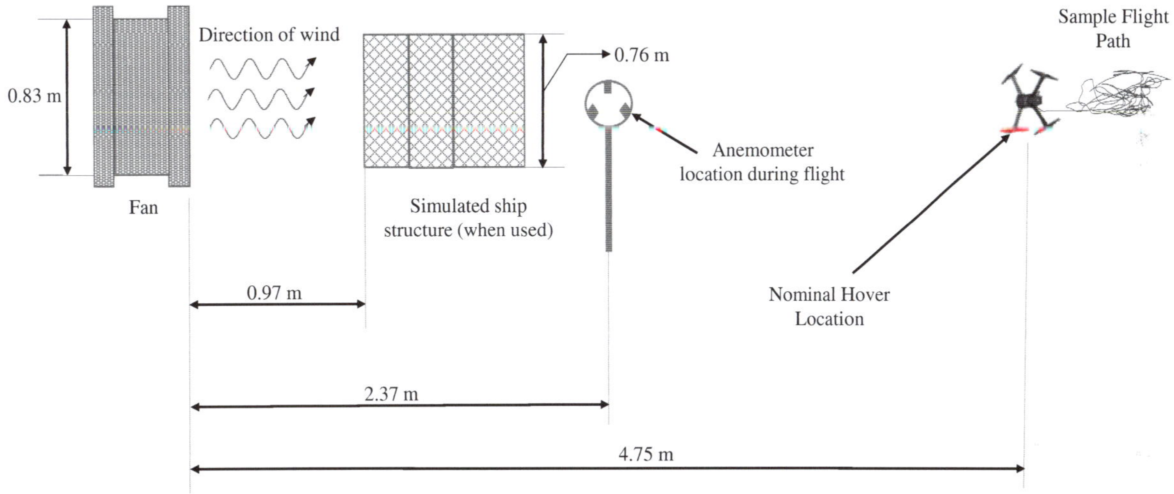

Figure 9: Fan test setup (view from top).

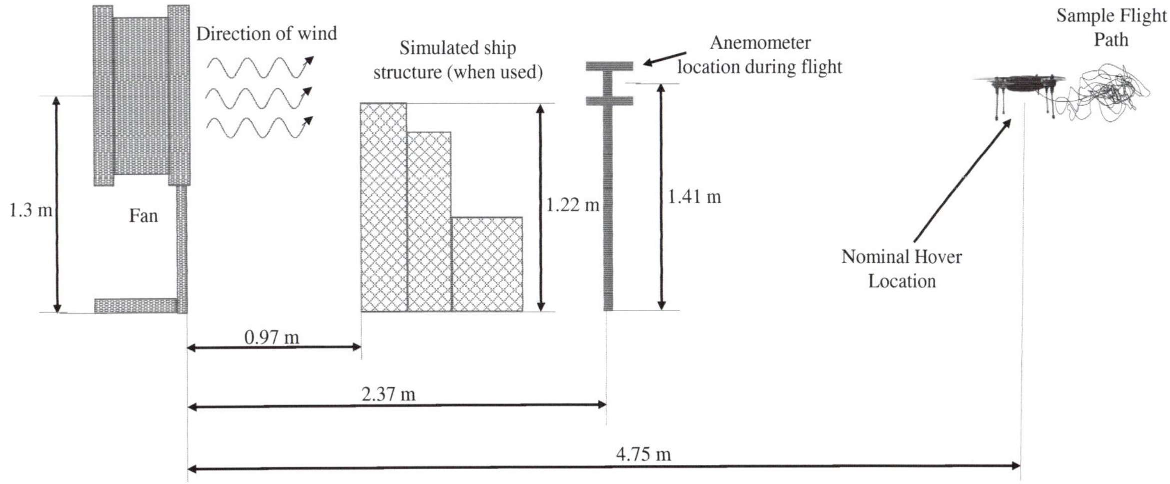

Figure 10: Fan test setup (view from side).

Gust Identification Algorithm for Fan Flight Tests

Initial tests used a very simple linear model for identification of longitudinal and lateral gust velocities, involving decoupled models of longitudinal and lateral translation dynamics. The inertial velocities and attitudes are measured through the Vicon system, and a first order state space system based on Eq. 15 was used. In the case of longitudinal gusts the linear model was of the form:

$$\dot{U} = X_u U + X_\theta \theta + X_u U_g \tag{32}$$

However, test results indicated significant coupling between the vertical axis response and the forward airspeed of the aircraft. For example, the aircraft was observed to initially climb and then dive when subjected to a longitudinal gusts. Other coupling effects and non-linear aerodynamic effects were also hypothesized. This led to the non-linear model development discussed in the previous section. For implementation in the EKF, the non-linear augmented plant model was set up as follows.

The aircraft states are the velocity of the vehicle relative to the airmass:

$$\begin{aligned} U_a &= U_I + U_g \\ W_a &= W_I + W_g \end{aligned} \tag{33}$$

where subscript a denotes velocities relative to the air, I denotes velocities relative to the inertial frame, and g denotes wind gust velocity (velocity of air relative to the inertial frame). Differentiating Eq. (33) yields the system of equations (34):

$$\begin{aligned} \begin{bmatrix} \dot{U}_a \\ \dot{W}_a \end{bmatrix} &= f\left(U_a, W_a, \theta, \delta_{th}\right) + \begin{bmatrix} \dot{U}_g \\ \dot{W}_g \end{bmatrix} \\ \begin{bmatrix} \dot{U}_g \\ \dot{W}_g \end{bmatrix} &= \begin{bmatrix} -\omega_g & 0 \\ 0 & -\omega_g \end{bmatrix} \begin{bmatrix} U_g \\ W_g \end{bmatrix} + \begin{bmatrix} S_{ug} & 0 \\ 0 & S_{wg} \end{bmatrix} \begin{bmatrix} \hat{w}_u \\ \hat{w}_w \end{bmatrix} \end{aligned} \tag{34}$$

The function $f(\,)$ represents the aircraft equations of motion, which yield acceleration relative to the inertial frame, which need to be summed with the derivative of the gust velocities. The assumed gust filtering is then defined by the second set of equations. The forces on the aircraft are a function of the velocity relative to the air and the effective control inputs (aircraft pitch attitude and thrust command). The model is converted to discrete time for implementation in the EKF using Adams-Bashforth integration. Note that the white noise disturbances are not used in the EKF implementation (their covariance properties are only used in the formulation). The measurement is the inertial velocity (as measured by the Vicon system), so the measurement model is linear in the states:

$$y = \begin{bmatrix} U_I \\ W_I \end{bmatrix} = \begin{bmatrix} 1 & 0 & -1 & 0 \\ 0 & 1 & 0 & -1 \end{bmatrix} \begin{bmatrix} U_a \\ W_a \\ U_g \\ W_g \end{bmatrix} + \begin{bmatrix} \hat{v}_u \\ \hat{v}_w \end{bmatrix} \tag{35}$$

The noise covariance for the measurements was assumed to be a 20 cm/s RMS on each velocity measurement, which can then be converted into discrete time by dividing by the time step:

$$R = E\left(vv^T\right) = \begin{bmatrix} 0.04 & 0 \\ 0 & 0.04 \end{bmatrix} \text{ and in discrete time } R_k = \frac{R}{\Delta t}$$

This EKF formulation uses additive noise, whereas the model in Eq. 34 has scaling terms on the disturbance. The white noise in that formulation has covariance of 1. To transform to the EKF system with additive process noise, the covariance matrix is derived as:

$$Q = E\left[(Gw)(Gw^T)\right] = \begin{bmatrix} S_{ug} & 0 \\ 0 & S_{wg} \\ S_{ug} & 0 \\ 0 & S_{wg} \end{bmatrix} \begin{bmatrix} 1 & 0 \\ 0 & 1 \end{bmatrix} \begin{bmatrix} S_{ug} & 0 & S_{ug} & 0 \\ 0 & S_{wg} & 0 & S_{wg} \end{bmatrix} \tag{36}$$

The process noise covariance is then converted to discrete time using the upper right hand corner of the matrix exponential as in Eq. (37), where A is the linearized system matrix of the continuous time model and Q the process noise covariance of the continuous time system. The discretized system matrix is found in the lower right hand corner of this matrix, while the discretized covariance is in the upper right hand side.

$$\begin{bmatrix} X & Q_k \\ X & F_k \end{bmatrix} = e^{\begin{bmatrix} -A & Q \\ 0 & A^T \end{bmatrix} \Delta t} \tag{37}$$

For the current application, a sampling time of $\Delta t = 0.05$ seconds was used. The scaling and bandwidth parameters for the gust model were defined as: $\omega_g = 1$ rad/sec, $S_{ug} = 10.0$ m/s, $S_{wg} = 1.0$ m/s.

Results

The gust estimation using the EKF is compared with anemometer measurements for three different test cases. In all cases, the Iris+ is flown with nose towards the fan flow over the nominal hover location (although the aircraft is typically blown back to some degree as discussed before). The anemometer measurements were recorded on a separate data collection computer, and as a result these measurements are not exactly time synchronized with the flight data. There is also expected to be a phase shift between the measured and estimated gust pulse, as the aircraft is over 2 m downstream of the anemometer. For the purposes of data comparisons, a time offset was applied to the anemometer data to line up the initial fan pulse which was clearly seen in both the anemometer and flight data.

In the first case, the Iris+ is subjected to pulsed fan gusts with no ship structure. The fan is turned on intermittently during the test, and the pulsed fan flow can readily be seen in the anemometer data. In Fig. 11, the results show the EKF estimator captures the basic wave form of the pulsed fan in the estimated longitudinal velocity component, U_g. The average velocity when the flow is fully developed is approximately 7 m/s and the EKF estimates the same value, although with more fluctuations than seen in the anemometer measurement. The fan was found to have slight upward component due to its alignment on the mounting stand. This effect can be a small negative (upward) component in the anemometer measurements of the vertical gust velocity, W_g. The estimated values have a similar trend in time, but predict larger amplitudes and larger fluctuations than seen in the anemometer data. The EKF seems to over-estimate the updraft due to the fan. It is not clear at this point if this discrepancy is due to error in the algorithm or due to distortion of the flow as it moves downstream of the anemometer.

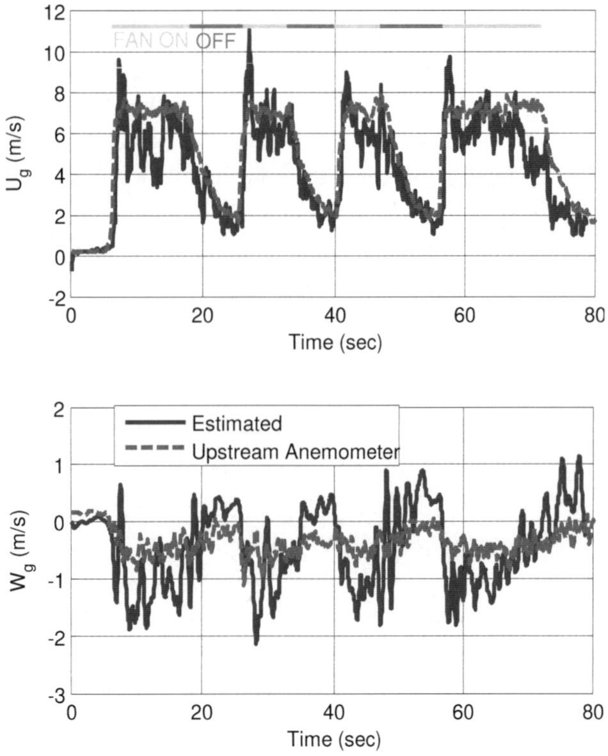

Figure 11: Comparison of estimated gust velocities with anemometer measurements (no ship structure).

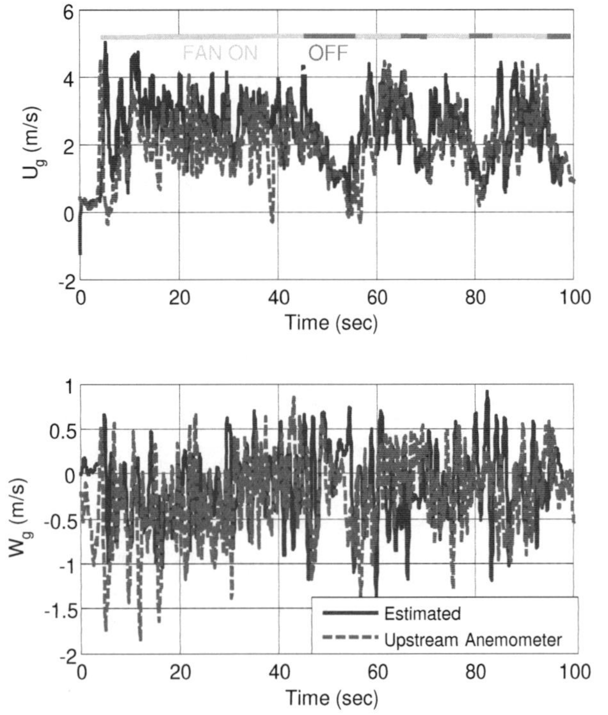

Figure 12: Comparison of estimated gust velocities with anemometer measurements (with ship structure).

Similar tests were carried out with the simulated ship structure between the fan and the Iris+. Due to blockage from the rearward facing step, the mean velocity is lower when compared to the pure fan flow. As seen in the results of Fig. 12, the EKF estimates a similar pulsed longitudinal gust, U_g, when compared to the measured velocities. In addition the estimated unsteady fluctuations appear to be of the same magnitude as measured by the anemometer. Due to distortion of the flow downstream of the anemometer we cannot expect the estimated fluctuations to be exactly in phase with the measurements. The vertical gust components, W_g, appear to have better correlation between the Gust ID estimation and the measurements.

In the next set of tests, the Iris+ was flown for an extended period over the nominal hover location, and the aircraft height was slowly varied between 0.4m and 2.2m above the floor. The tests ran for approximately 100 to 140 seconds of flight without and with the simulated ship structure. The purpose of these tests was to allow the Iris+ to survey the vertical profile of the flow field. After these tests, the anemometer was used to measure flow properties at various heights over the hover location with and without the ship structure. Two minutes of flow data was collected to analyze the mean and standard deviations in flow at these locations.

The EKF estimator was run on the flight data, and samples of the estimated gusts U_g and W_g were collected when the aircraft height was within various 20 cm height bins (0.3 m < h < 0.5 m, 0.5 m < h < 0.7 m, … , 2.1 m < h < 2.3 m). The mean and standard deviations of these were then evaluated and can be compared to the anemometer measurements. Comparison of the estimated and measured mean U_g velocity profile with height is shown in Fig. 13 both with and without the ship structure. The results show the EKF estimates the longitudinal velocity profile very accurately without the ship structure. With the ship structure, there appears to be a consistent bias in the estimation but at least a similar trend with height. Figures 14 and 15 shows comparisons of the mean vertical velocity profile with variations in height and comparisons of the vertical velocity standard deviation (a measure of the unsteady fluctuations in the vertical velocities) respectively. The EKF does not perform as well in estimating the vertical component, but it does show similar trends in the variation of mean and standard deviation with change in height.

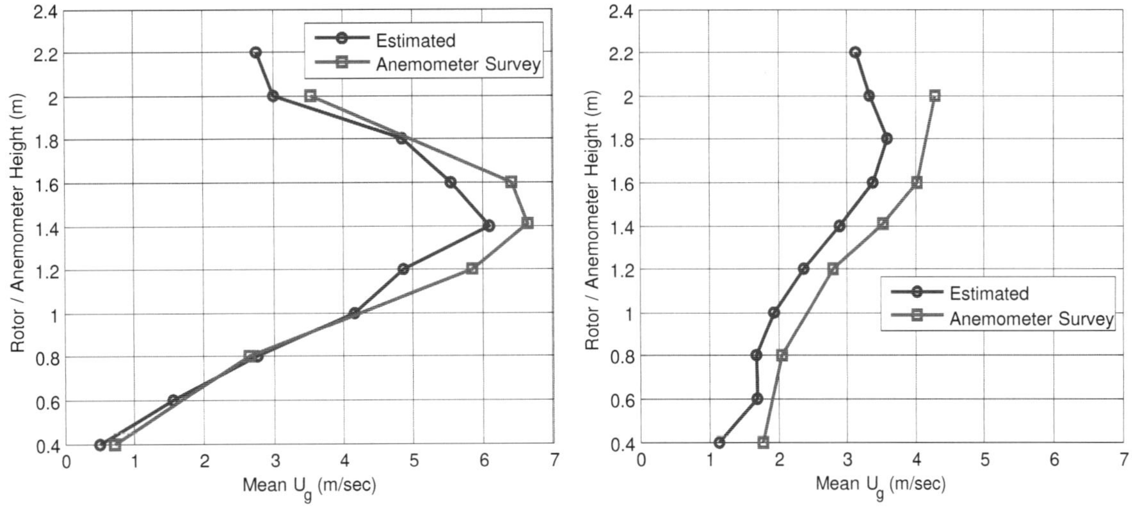

Without Simulated Ship Structure **With Simulated Ship Structure**

Figure 13: Variation of mean longitudinal gust velocity with height.

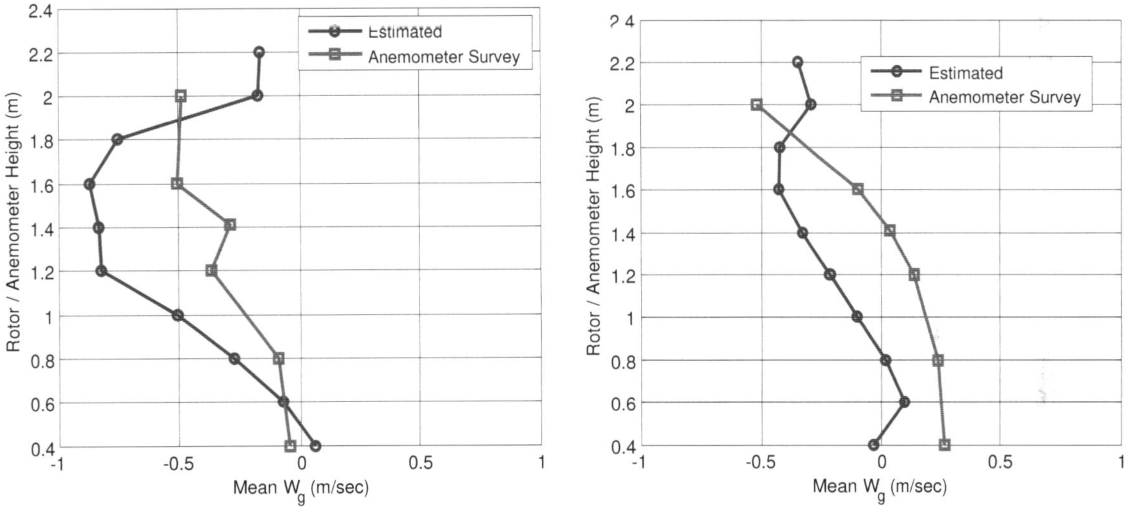

Without Simulated Ship Structure **With Simulated Ship Structure**

Figure 14: Variation of mean vertical gust velocity with height.

CONCLUSIONS

A Kalman filter based gust identification method was refined from earlier research in order to use a non-linear dynamic model within an Extended Kalman Filter framework. The algorithm allows estimation of external wind gusts using a small VTOL UAV without any air data measurements. The algorithms are tested using a small quad-rotor UAV flown in an indoor lab with a Vicon motion capture system, where wind gusts and airwake effects are simulated using a high speed fan and rearward facing step structure. Initial work focused on developing an outer loop control law to hold the aircraft at a fixed location or to follow prescribed velocity commands. Next, an accurate non-linear dynamic model of the UAV dynamics was developed, since earlier research showed that simple linear dynamic models were found to not provide adequate gust estimation results. Fan tests were then performed to evaluate the performance of the new non-linear EKF gust identification estimator. Simultaneous anemometer measurements were performed and compared to estimated gusts to verify the algorithm's ability to estimate time variations in wind. Off-line anemometer measurements were also performed at the same locations where the UAV flew to measure the profile of the wind properties with height. These measurements were compared to estimated

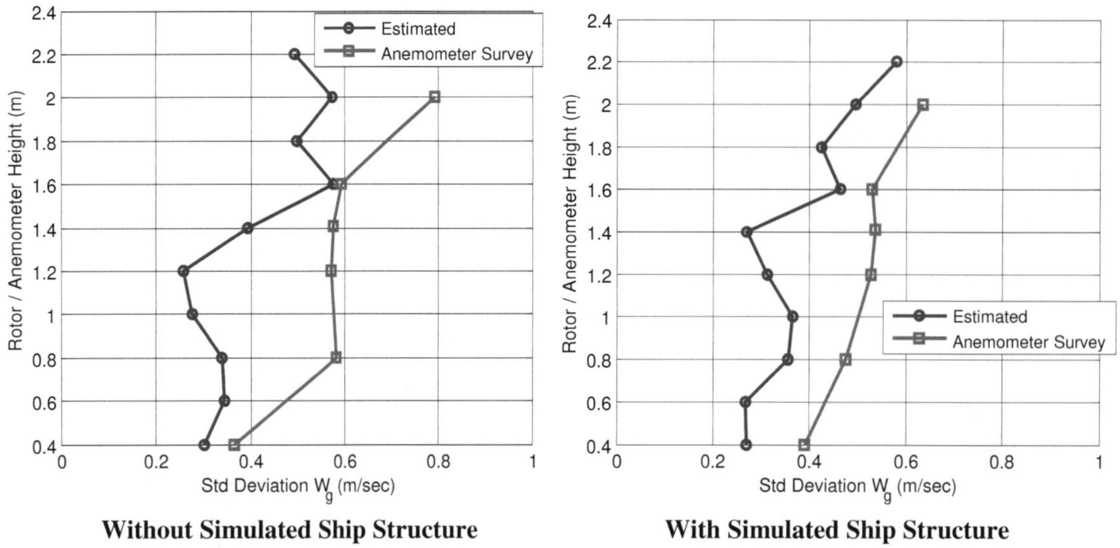

Figure 15: Variation of standard deviation of the vertical velocity with height.

gusts to check the ability of the algorithm to estimate spatial variations in flow properties. Some specific results of the study:

1. A ground-station-based outer-loop controller using Vicon motion measurements and a dynamic inversion formulation was integrated with the onboard attitude stabilizing controller of the Iris+. This proved to be an effective means for indoor flight control, allowing relatively high speed maneuvers within the confined indoor flying space and holding position and altitude when subjected to large fan flow disturbances. This development proved to be a key component of the indoor tests and was crucial for gathering the required flight data. When conducting tests or flow surveys outdoors, it is expected that a differential GPS or high performance INS/GPS solution would be required.
2. A combination of linear and non-linear model development led to an accurate dynamic model of the quadrotor in generalized maneuvering flight over a wide envelope (±7 m/s longitudinal flight speeds and ±3 m/s vertical flight speed). This was also a key development for implementation of the non-linear EKF gust estimation scheme.
3. The EKF algorithm provided accurate estimates of the longitudinal gusts, showing strong correlation with the time and spatial variations in U_g when compared to those values measured by the anemometer. This component was the primary axis of the flow disturbance. On the other hand, there tended to be some bias and scale error in the vertical axis estimates when compared to the measurement. Further refinement of the algorithm is merited to improve vertical velocity estimation.

ACKNOWLEDGMENTS

This work was sponsored by the Office of Naval Research, ONR, under grant number N000141410008. The views and conclusions contained herein are those of the authors and should not be interpreted as necessarily representing the official policies or endorsements, either expressed or implied, of the Office of Naval Research, or the U.S. government.

REFERENCES

1. Snyder, M.R., Kang, H.S., Brownell, C.J. and Burks, J.S., "Validation of Ship Air Wake Simulations and Investigation of Ship Air Wake Impact on Rotary Wing Aircraft," *Naval Engineers Journal*, 125-1, pp. 49-50 (2013).
2. Yeo, D.W., Sydney, N., Paley, D.A., "Onboard Flow Sensing for Rotary-Wing UAV Pitch Control in Wind," AIAA Guidance, Navigation, and Control Conference, San Diego, CA, Jan 4-8, 2016.
3. Jack W. Langelaan, Nicholas Alley, and James Neidhoefer. "Wind Field Estimation for Small Unmanned Aerial Vehicles," *Journal of Guidance, Control, and Dynamics,* Vol. 34, No. 4 (2011), pp. 1016-1030.

4. Lusardi, J. A., Tischler, M. B., Blanken, C. L., and Labows, S. J., "Empirically Derived Helicopter Response Model and Control System Requirements for Flight in Turbulence," *Journal of the American Helicopter Society*, Vol. 49, No.3, July 2004, pp.340-349.
5. Lee, D. and Horn, J.F., "Simulation of Pilot Workload for a Helicopter Operating in a Turbulent Ship Airwake," *Journal of Aerospace Engineering – Proceedings of the Institute of Mechanical Engineers Part G, Special issue on shipborne aviation*, Vol. 219, (G5), October 2005, pp. 445-458.
6. Cooper, J., Schierman, J., and Horn, J.F., "Robust Adaptive Disturbance Compensation for Ship-based Rotorcraft," AIAA Guidance, Navigation, and Controls Conference, Aug. 2010.
7. Schafer, S., Horn, J.F., and Cooper, J., "Flight Test Measurement of Ship Airwake Disturbances using Small-Scale Rotorcraft," American Helicopter Society 71st Annual Forum, Virginia Beach, Virginia, May 5-7, 2015.
8. Cooper, J., Pappu, V.S.R., Liu, Y., and Horn, J.F., "Vehicle Modeling for Flight Test Measurements of Ship Airwake Disturbances," American Helicopter Society 72nd Annual Forum, West Palm Beach, Florida, May 17-19, 2016.
9. Brown, R.G., and Hwang, P.Y.C., *Introduction to Random Signals and Applied Kalman Filtering* (3rd edition), John Wiley & Sons. pp. 214–219.
10. Tischler, Mark B. and Remple, Robert K., *Aircraft and Rotorcraft System Identification*, American Institute of Aeronautics and Astronautics (Education Series), 2012.
11. Cooper, J., Reilly, J., Schlapkohl, T., Gist, J., and Howe, G., "AUSPEX: User's Manual," Tech. Rep., Barron Associates, 2012.
12. Sydney, N., Smyth, B. and Paley, D.A., "Dynamic Control of Autonomous Quadrotor Flight in an Estimated Wind Field," Proceedings of the 52nd IEEE Conference on Decision and Control, Florence, Italy, December 2013.

Fire Scout Operations in Parametric – Like Roll Ship Conditions

Dr. Bernard Ferrier	Mr. Andrew Silver	Mr. Robert Ernst
bferrier@hoffmanengineering.com	Andrew.silver@navy.mil	robert.ernst@navy.mil
SETA (NAVAIR PMA 266)	Naval Architect	Chief Engineer
Hoffman Engineering Corp	SEA 851	PMA 266
Arlington, VA	Bethesda, MD	Patuxent River, MD

ABSTRACT

The MQ-8 (Fire Scout) unmanned aircraft system (UAS) is the U. S. Navy's primary ship-based helicopter UAS. The current Concept of Operations (CONOPS) calls for the MQ-8 to be based aboard any suitably equipped air capable ship, i.e. any ship installed with Mission Control Station (MCS), Tactical Common Data Link (TCDL), Unmanned Aerial Vehicle Common Automatic Recovery System (UCARS), associated radios and other helicopter handling systems. In particular, the Littoral Combat Ship (LCS) is the main platform for MQ-8 operations. LCS has two variant classes of ships, the Freedom and Independence class which MQ-8 will be deployed on. As part of the Navy deployment process, a ship suitability assessment is conducted in the form of Dynamic Interface (DI) testing. DI testing evaluates all aspects of shipboard helicopter suitability and compatibility, including the adequacy, effectiveness, and safety of MQ-8 shipboard Flying Qualities and Performance (FQ&P), aviation support facilities and procedures for all ship-based helicopter types. A significant component parameter centers on the ship's motion characteristics.

In the course of MQ-8B and LCS envelope expansion DI trials, unexpected ship motions have been measured and recorded. These motions were almost exclusively observed to be large roll. These roll motions may have been induced by a seaway, or may have been induced by non-linear interaction with pitch and heave motions. Since seaway measurements were not available, there was no way of knowing with certainty the cause of these motions. On further study it was shown that all of the axial traces were in phase. The ship speed with notable exception of angular roll displacement, all the other degrees of freedom, their rates and their accelerations, were low. Since the ship speed was low, and the pitch and heave motion was low, a possible cause of these motions was a low period swell coming on the beam of the ship. With shallow draft and streamlined hull without traditional rudder or stabilizing surfaces, the LCS can be made to roll without much wave energy.

The implications for Fire Scout operations encountering parametric-like roll events are discussed. The need for careful definition of the ship's motion characteristics are being driven by the combination of limited manpower availability and a requirement to reduce the time needed to execute aircraft launch and recovery (L&R) or to maintain deck tempo. To facilitate the upcoming ship-suitability assessment, a parallel effort was initiated to determine the feasibility of applying simulations and plug-and-play mission module options in support of unmanned DI at-sea testing and experimentation. Initial analysis indicates traditional linear ship motion models do not predict the presence of parametric roll. Real time measurements appear to offer one of the few choices available to detect the on-set of these roll cycles. At-sea preliminary study results are discussed and compared with simulated scenarios.

LIST OF ACRONYMS

AS	Airborne Subsystem
AIS	Automatic Identification System
AV	Air Vehicle
BSBII	BRITE Star Block II
CONOPS	Concept of Operations
DI	Dynamic Interface Study
EI	Energy Index
EO/IR	Electro-Optical / Infrared
FFG	Guided Missile Frigate
FQ&P	Flying Qualities and Performance
HLA	High Level Architecture
IFF	Identification Friend or Foe
ISR	Intelligence, Surveillance and Reconnaissance
LCS	Littoral Combat Ship
LD	Laser Designator
LPD	Landing Period Designator
LPDLOOP	Landing Period Designator sofware
LDR	Laser Designator Rangefinder
LSO	Landing Signal Officer
L&R	Launch and Recovery
MCS	Mission Control Station

MRU	Motion Reference Unit
NATO	North Atlantic Treaty Organization
NOAA	National Oceanic and Atmospheric Administration
NATOPS	Naval Air Training and Operating Procedures Standardization
NAVAIR	Naval Air Systems Command (also NAVAIRSYSCOM)
NAWCAD	Naval Air Warfare Center Aircraft Division
ONR	Office of Naval Research
PMA	Program Manager AIR
PEO	Program Executive Officer
RF	Range Finder
TCDL	Tactical Common Data Link
TCS	Tactical Control Station
TS	Tracking System
UCARS	Unmanned Aerial Vehicle Common Automatic Recovery System
UCSD	University of California, San Diego
UAS	Unmanned Aircraft System
UAV	Unmanned Aerial Vehicle

INTRODUCTION

The MQ-8 Fire Scout System is the culmination of a quarter century of Navy development and evolution. The latest version of the Fire Scout is the MQ – 8C based on the Bell Model 407 helicopter with additional fuel capacity to give the MQ-8C near twice the endurance of the smaller MQ-8B. The basic mission of the Fire Scout remains unchanged; both the MQ-8B and MQ-8C are designed to provide Intelligence, Surveillance and Reconnaissance (ISR) information back to the host platform for dissemination to other users as required [1].

As required of new aircraft on new vessels, MQ-8B and MQ-8C were evaluated for shipboard suitability using envelope expansion techniques during the DI period. In the course of these trials, unexpected ship motions were noticed and recorded. This response appears as an response mechanism similar to natural frequency vibration. In figure 1, the vessel enters into cycle of roll; not static heel. From the photo, it appears the seaway excitation is small. The ship is moving at slow speeds.

Figure 1 – Roll cycle deck motions

..."reported that a vessel, whose frequency of oscillation in heave/pitch is twice its natural frequency in roll, shows undesirable seakeeping characteristics that can lead to the possibility of exciting large roll oscillations". William Froude: Transactions of the Roy Instit of Naval Architects: 2:180-227 (1861)

As indicated by the 150 year old description above [3], parametric roll is caused by the proper phasing of roll, pitch, and heave. The development of these conditions is of considerable interest to the Navy as it may affect the L&R of MQ-8 Fire Scout.

BACKGROUND

MQ-8C is composed of a larger air vehicle (AV) increasing the range of operation on its earlier variant and capable of carrying a larger payload, able to complete missions more rapidly and efficiently than its predecessor (figure 2).

Figure 2 – MQ – 8C on Sea Trials

MQ – 8C Component Description

The Fire Scout system includes an unmanned AV (converted Bell 407), a BRITE Star Block II (BSBII) payload which encompasses electro-optical (EO)/ infrared (IR)/ laser designator and rangefinder (LDR) functions (figure 3), a ship-integrated MCS featuring Tactical Control Station (TCS) software (figure 4), TCDL equipment, and UCARS (figure 5). Additional deployed payloads include the Automatic Identification System (AIS), and other specialty payloads such as Identification Friend or Foe (IFF). The Fire Scout BSBII sensor suite has the necessary capabilities for the detection and classification of targets of interest for counter insurgency and anti-piracy operations [4].

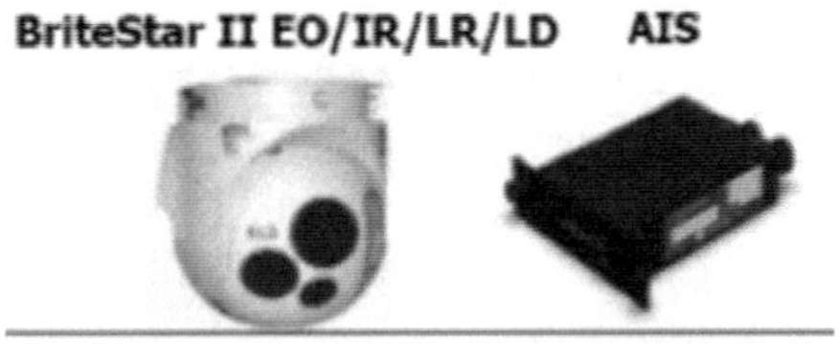

Figure 3 – EO/IR & AIS Payloads

Figure 4 – Mission Control System

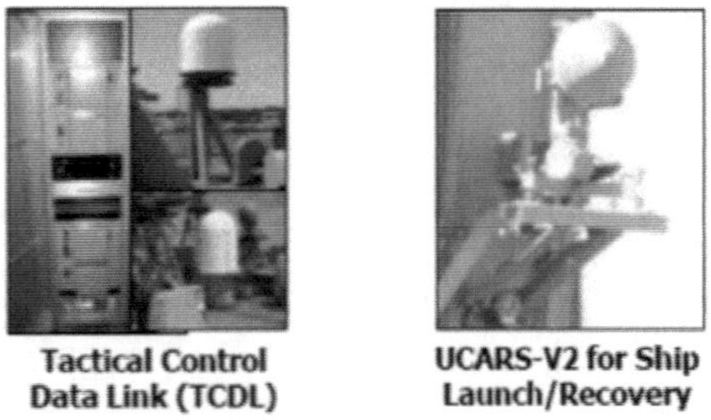

Figure 5 – Digital Data Links

Given the capabilities of a helicopter, MQ-8 Fire Scout (MQ-8B v. MQ-8C comparison in figure 6 below) can operate in confined operational areas from various US Navy air-capable ships, including the new LCS. It is these confined operational spaces located in the littoral maritime environment which presents MQ-8 Fire Scout with one of its greatest challenges. For the shipboard landing task, it is incumbent on the system to accurately sense the AV's position and orientation in space relative to the host ship's deck and to also be able to adjust to the ship's motions. The Program Manager Air (PMA) must use existing systems to achieve these goals without incurring additional delays in the MQ-8 deployment schedule. Many of these systems employ simulation techniques to estimate what modifications, if any, are required to the existing system design. A typical deck interface simulation technique is developed in the annex.

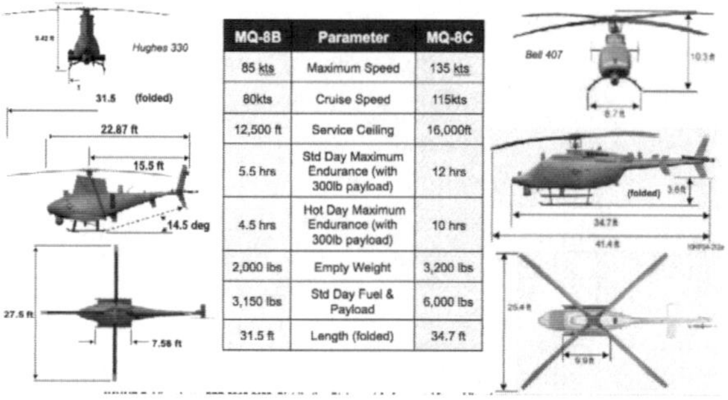

MQ-8B	Parameter	MQ-8C
85 kts	Maximum Speed	135 kts
80kts	Cruise Speed	115kts
12,500 ft	Service Ceiling	16,000ft
5.5 hrs	Std Day Maximum Endurance (with 300lb payload)	12 hrs
4.5 hrs	Hot Day Maximum Endurance (with 300lb payload)	10 hrs
2,000 lbs	Empty Weight	3,200 lbs
3,150 lbs	Std Day Fuel & Payload	6,000 lbs
31.5 ft	Length (folded)	34.7 ft

Figure 6 – MQ-8B vs. MQ-8C Characteristics

Dynamic Interface UAV Envelope Expansion

DI is performed to reduce risks and maximize operational flexibility. Globally, DI is concerned with the effects that one free body has in respect to another. DI investigates L&R of all rotorcraft, and vehicle characteristics, such as, engaging and disengaging of rotors, vertical replenishment and helicopter in-flight refueling envelopes. Shipboard suitability testing assesses the adequacy, effectiveness, and safety of shipboard aviation. While the vast majority of pilot attention is focused on the bubble surrounding a ship test platform, significant cross-referenced ship deck motion data are recorded throughout the L&R flight evolution. Deck motion impacts on aircraft stability at the moment of recovery have been documented both using the Naval Air Systems Command (NAVAIR) Manned Flight Simulator and at-sea. It is instructive to view the extent to which the ship's motion can impact the air vehicle dynamic and mechanical responses. Table 1 displays the primary components affecting the DI rating scales. While the other DI functions may be directly or indirectly measured, visual cues assessments are entirely empirical and can be subjective. For UAS evaluations, visual cues are replaced by instrumental analysis.

Table 1 – Dynamic Interface Measures

o **Relative Wind Speed and Direction**
o **Ship Airwake and Turbulence**
o **Ship Motion**
o **Visual Cues or Instrumented (UAS)**

All four factors listed in Table 1 interact dynamically on the air vehicle and pilot or operator affecting the recovery condition. The unexpected deck motions of the LCS classes (Freedom and Independence) were not identified by simulation. They were recognized during the post-processing of the recorded data.

PARAMETRIC ROLL

Parametric roll resonance is a phenomena caused by the passage of longitudinally propagating waves over a hull form with large bow and stern flares. According to Levandou and Gaillarde (2003), uneven wave surface together with the pitch-heave motions of the ship results in a time varying underwater hull geometry [5]. This varying geometry, in turn, changes the metacentric height or static roll stability. The variation of static roll stability can lead to large changes in roll angle when a small perturbation, such as a rudder movement, is introduced. This small amount of roll motion, under the right circumstances, can lead to significant roll angles. Parametric roll appears to occur under the following conditions (Galeazzi et al, 2012) [6]:

- The natural roll period is equal to twice the wave encounter period
- The wavelength is on the order of the ship length causing heave and pitch frequency to be twice the natural roll frequency of the ship
- The wave height exceeds a critical level
- There is little roll damping

The LCS can be subject to larger than expected rolling because there are very few roll damping appendages such as bilge keels and rudders. If the LCS is sensitive to parametric roll, then the L&R of the MQ-8C can be much more difficult.

DI ENVELOPE EXPANSION

Test Equipment Installed

The purpose of embarking the Landing Period Designator (LPD) unit is to establish the onboard working environment of the LPD system and to acquire ship motion data recorded as a function of the MQ-8. The primary interest is to document deck motion characteristics as it may affect MQ-8 Fire Scout's shipboard L&R capabilities. The LPD is designed to provide ship motion cues to assist in anticipating ship deck quiescent periods that result in acceptable conditions for a shipboard landing and aircraft securing on deck after recovery. The LPD computes the Energy Index (EI) which is a real time measurement of the ship's seaway induced inertial content compared to the mechanical and dynamic limits of the air vehicle. The LPD is designed to fill a technological gap related to all-weather operations. MQ-8 Fire Scout's deck motion responses are calculated using ship motion as a function of the UAS deck motion limit criteria. When touching the deck, the air vehicle is considered an extension of the ship. The vehicle experiences forces and moments from the ship which create rectilinear and angular accelerations on the vehicle. The accelerations can be numerically integrated to determine the forces and attitude of the captured vehicle relative to the ship as function of time, for various ship motions [7].

The system used to measure the ship's motions is the Kongsberg (Seatex) Motion Reference Unit (MRU). Owing to the temporary nature of the installation, the MRU (figure 7) was typically zip-tied fast to an upright beam behind and below an electrical cabinet on the port side of the hangar. The MRU was further isolated in bubble wrap. The EI was calculated based on the MRU's evaluation of the deck's motion.

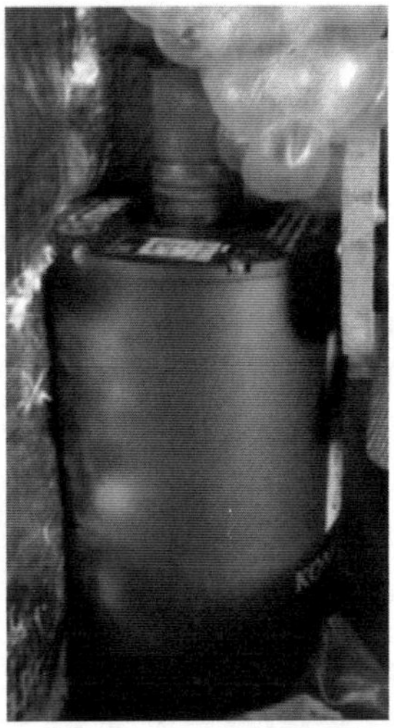

Figure 7 – MRU

Aboard LCS-2 class ships, the entire test kit was assembled and secured on the port side of the hangar near the hangar door on the flight deck level. The MRU was attached to an upright steel beam which itself was welded fast to the deck protecting the instrument from accidental disturbances. It was levelled and calibrated to the ship's integrated AGI Moriah ship motion and wind measurement system located in the helicopter control tower (Figure 8). Moriah provides digital wind speed and direction as well as ship's speed, heading, roll and pitch conditions.

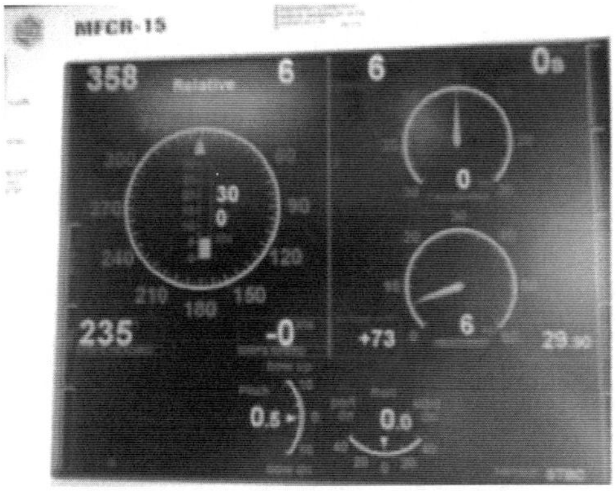

Figure 8 – AGI MORIAH Unit

Figure 9 displays the LPD Processing Unit monitor in secured configuration for easy access and viewing.

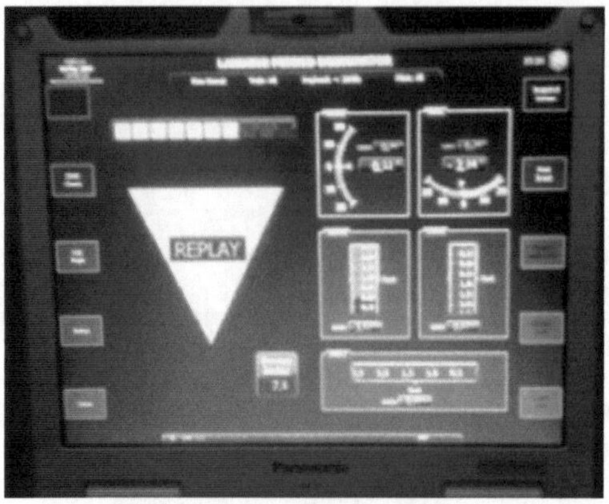

Figure 9 – LPD Monitor

The equipment was marshalled by NAVAIR DI Department. The mounting of the equipment and tapping into the ship's power source was the extent of the system's interaction with the ship. As observed above, the LPD equipment is composed of devices which are designed to measure, calculate and record various ship motion and aircraft deck response parameters.

During the trial periods, LPD data were continuously recorded in files of one-hour segments (table 2). Each file contained a time stamp, the calculated EI value and the engineering units representing the various six degrees of freedom, and their rates and accelerations. In addition, channels were reserved for ship true heading and GPS position, but not used on this trial.

Table 2 – Typical LPD Data File

```
[METRIC DATA]   Roll   RollVel   Pitch   PitchVel   Yaw   YawVel   Surge   SurgeVel
SurgeAcl   Sway   SwayVel   SwayAcl   Heave   HeaveVel   HeaveAcl   Heading Gyro

TIME   EI   (PHI)   (PHIVEL)   (THE)   (THEVEL)   (PSI)   (PSIVEL)   (LG)   (LGVEL)
(LGACEL)   (LT)   (LTVEL)   (LTACEL)   (VT)   (VTVEL)   (VTACEL)   (HDG)   Status

deg   deg/sec   deg   deg/sec   deg   deg/sec   metres   m/sec
m/s/sec   metres   m/sec   m/s/sec   metres   m/sec   m/s/sec   deg

0000.29,  0.1,  0.2083,  0.0140,  -0.0295,  -0.0049,  -39.1524,  -0.0002,  -0.0036,  -
0.0035,  -0.0034,  0.0047,  -0.0008,  -0.0013,  0.0137,  0.0003,  -0.0368,  320.8477,

0000.49,  0.1,  0.2081,  -0.0235,  -0.0299,  -0.0200,  -39.1545,  0.0003,  -0.0033,  -
0.0022,  -0.0030,  0.0047,  -0.0005,  -0.0013,  0.0138,  0.0010,  -0.0361,  320.8456,
```

The following examples are from an earlier trial of the recorded test results with corresponding study graphics. Three basic graphics are shown. The first graphic (figure 10) displays a one-hour trace containing roll and pitch along with the corresponding EI. The second graphic (figure 11) displays a rise and fall time event (time from rest to out of limit) along with a third graphic (figure 12) displaying the corresponding rate displacement for the period.

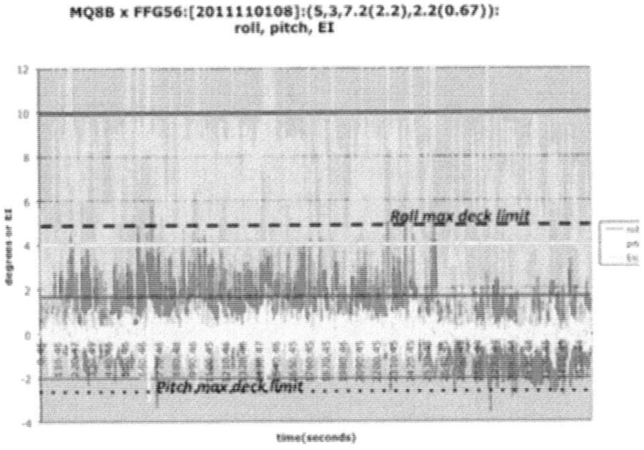

Figure 10 – Trial recorded 01 Nov @ 08:00

Figure 10 time history displays the full one-hour recorded roll and pitch traces along with the corresponding energy index trace. The blue trace is the EI calculation. Any trace above the red EI = 10 line indicates out-of-limit or unsatisfactory conditions (unacceptable for L&R). Anything between the EI 4.0 (yellow) and 10.0 (red) lines refer to high energy but still within limit deck conditions. Traces within the green and yellow bounded area refer to safe deck conditions while the area between zero (0) and the green line is the zone called green or quiescent deck conditions.

Figures 11 and 12 display one of the recorded rise and fall time events contained within this file. In each case, note the response of the EI trace to changing angular or translational axial conditions. Here, red deck was caused by excessive roll and Y' (lateral rate) motions.

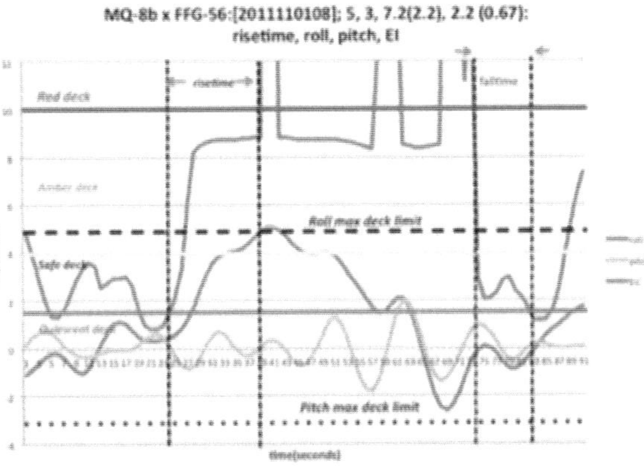

Figure 11 – Rise and Fall time events

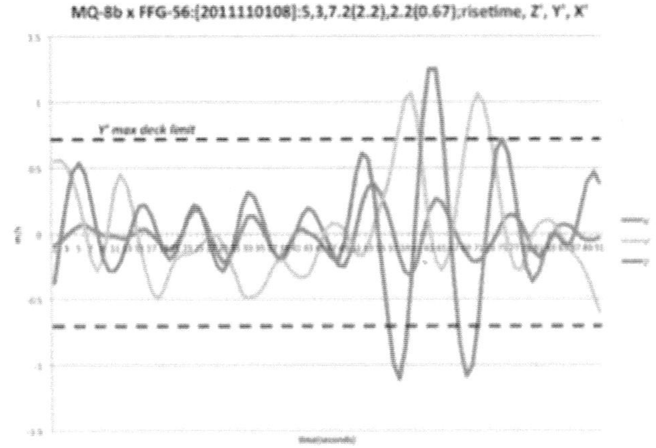

Figure 12 – Corresponding rate motions

Table 3 displays an edited file summary of rise time events, time length by rise time, and the distribution of LPD deck status.

Table 3 – Related Analysis of time history

****NAV11C EVALUATION FILE****
DIAGNOSTICS on data file: 2011110108.csv
This file is type: 2.00 where 0 2 3=recorded 1=simulated
Ship= 'FFG-56' Aircraft= 'MQ-8B' ROLLMAX= '0.00' PITCHMAX= '0.00' ZVEL= '0.0' ft/s LATVEL= '0.0 ft/s'
Data Playback time step= '0.10' seconds'

RISETIME EVENT BEGINNING @ TIME '2270.95' EIA '1.46' ENDING TIME '2278.95' EIA '53.91'
Risetime event = '8.00' seconds
ROLL= '5.07' PITCH= '0.31' ROLLVEL= '0.23' PITCHVEL= '0.40' ZVEL= '-0.06' m/s YVEL= '-0.28 m/s'

RISETIME EVENT BEGINNING @ TIME '2301.95' EIA '1.59' ENDING TIME '2330.96' EIA '14.08'
Risetime event = '29.01' seconds
ROLL= '0.25' PITCH= '-0.40' ROLLVEL= '0.99' PITCHVEL= '0.38' ZVEL= '-0.51' m/s YVEL= '0.70 m/s'

RISETIME EVENT BEGINNING @ TIME '2412.45' EIA '1.52' ENDING TIME '2426.45' EIA '11.41'
Risetime event = '14.00' seconds
ROLL= '-0.70' PITCH= '-0.65' ROLLVEL= '1.48' PITCHVEL= '1.59' ZVEL= '-0.79' m/s YVEL= '0.73 m/s'

RISETIME EVENT BEGINNING @ TIME '2516.95' EIA '1.58' ENDING TIME '2526.44' EIA '25.35'
Risetime event = '9.49' seconds
ROLL= '2.89' PITCH= '0.80' ROLLVEL= '-1.38' PITCHVEL= '-1.09' ZVEL= '0.39' m/s YVEL= '-0.70 m/s'

RISETIME EVENT BEGINNING @ TIME '2554.44' EIA '1.72' ENDING TIME '2561.44' EIA '19.14'
Risetime event = '7.00' seconds
ROLL= '-1.19' PITCH= '-1.01' ROLLVEL= '0.91' PITCHVEL= '0.41' ZVEL= '-0.50' m/s YVEL= '0.69 m/s'

There are '7200.00' points in run
Green Deck = '81.30' seconds = '11.29' percent of run
Green-Amber Deck = '161.00' seconds = '22.36' % of run
Amber Deck = '442.40' seconds = '61.44' percent of run
Red Deck = '35.30' seconds which is= '4.90' percent of run
Total deck availability in run = '684.70' seconds which is = '95.10' percent of run

Good risetimes in run + 5.0 seconds = '44.00'
OK risetimes in run + 4.0 - 4.9 seconds= '0.00'
Error risetimes in run < 4.0 seconds= '0.00'
Total number of risetimes in run= '44.00'

Interpreting the graphics above is straight forward. Figure 10 is composed of deck condition zones: green or quiescent deck (zone between 0 and 1.74), green-amber or safe deck (region between 1.74 and 4.0), amber or caution deck (4.0 – 9.99) and red or danger deck (>=10.0). MQ-8 can operate in any zone except red. Figure 11 displays a close-up of a rise time event (ship rising from rest to out-of-limit). Here again, ship's roll and pitch are indicated along with the corresponding EI calculation. The diagram indicates that both roll and pitch were nearly in trim displacement. Departure from zero indicates normally a ship manoeuver or a list. In Figure 12 the corresponding vertical, lateral and longitudinal rate motions are displayed. For this period of time, Z' an Y' are in opposite phases occurring at the moment in which the ship experiences excessive lateral motion rates pushing the EI out greater than 10 (red deck). Table 3 contains a synopsis of the rise time events and deck conditions. The data file is a recording composed of 3600 seconds (60 minutes or 7200 points at 0.5 seconds rate). Referring to the example, the ship experienced 813.00 seconds of quiescent deck or 11.3% of the run. Safe decks exists for 1610.0 seconds of safe deck or 22.4 % of the run. Amber deck measured 4424.0 seconds or 61.4% and 353.0 seconds were measured as danger or red deck which was a bit greater than 4.9% of this run. MQ-8B could have operated through 95.1% of the run. There were 44 rise time events measuring equal or greater than 5.0 seconds.

Dynamic Interface– establish requirements

In the course of analyzing the ship's motion, other characteristics which may impact the launch, recovery and deck handling, were identified. The time necessary to raise the deck from a stable to an unavailable condition can be derived experimentally from the calculation of the maximum EImax. As described earlier, the time required to raise the deck from minimal motion to unacceptable motion is called the risetime (written as one or two words). The risetime is a thumb print characteristic of the ship's response and rarely changes. In terms of the EI scale, it is defined as the period of time that is measured from the end of a green signal to the positive side of the red line. The risetime is mirrored by a droptime, which is the time period measured from the negative side of the red line to the negative side of the green line [8]. A typical trace as recently recorded, is displayed in Figure 13.

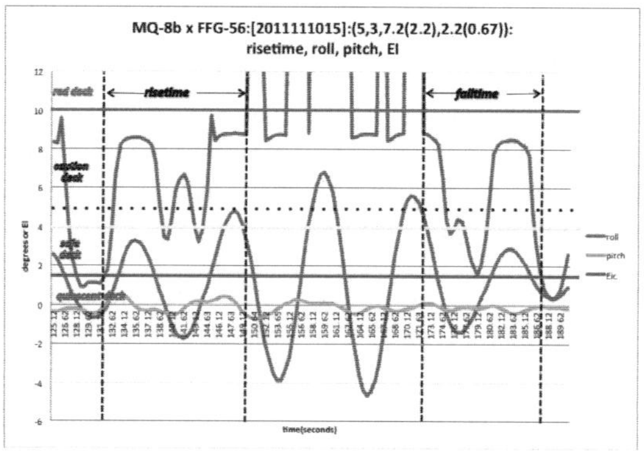

Figure 13 - Rise time events in file "2011111015"

The example shows a roll axis divergence eventually exceeding the MQ-8B deck roll limit of 5.0 degrees. In this case the time it took for the deck to rise from quiescence was 18 seconds. There appears to be a 1° roll list to port whilst pitch appears to be essentially in trim. The impulse converges in a fall time measuring 15 seconds. Interestingly, the index exceeded 10.0 prior to the roll axis exceeding the maximum deck motion limit. Figure 14 displays the deck motion rates for the same time period.

The Y' axis diverged like roll, but is 1/4 cycle out of phase with roll. There was sufficient energy on the second limit crossing to cause a roll strong enough to exceed deck limits. This is another example of the coupling between Y' and roll. During this test, the ship managed to record at various points 18.9° roll, 4.7° pitch, 2.3 m/s Z' vertical rate, and 2.7 m/s lateral rate maximum deck displacement.

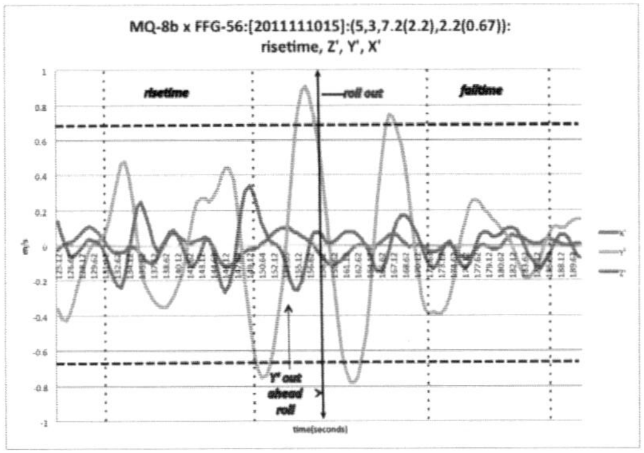

Figure 14 - Risetime events (rates) in 2011111015

Dynamic Interface – All Weather Operations

The majority of the motions recorded alongside the encountered deck accelerations or forces, appeared to equate to "textbook" motion traces for given study platforms. Recorded shipboard data are often compared with related information from other sources such as the National Data Buoy Center run by National Oceanic and Atmospheric Administration (NOAA).

Figure 15 displays a NOAA wave height forecast chart similar to those received during DI Fire Scout trials. It is often compared with anchored NOAA wave buoys mainly as a reference for the shipboard data recordings.

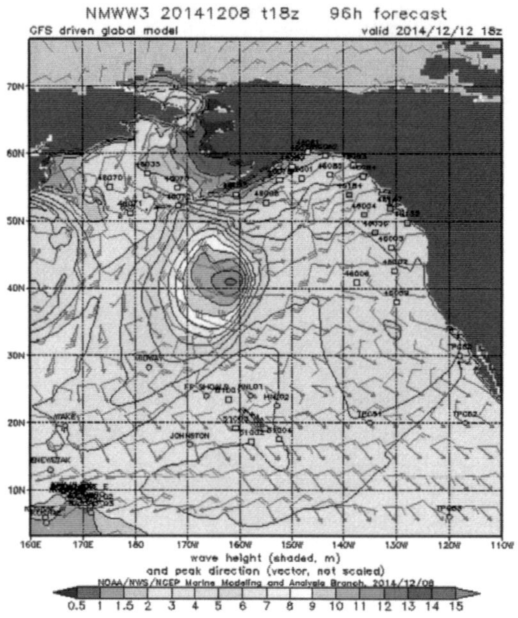

Figure 15 – NOAA Wave Ht Chart

Events during the LCS class DI gave reason for pause. On what appeared to have been a low sea state day, unexpected motion events occurred exceeding air vehicle deck limits uniquely in roll. Figure 16 displays the hour file recorded in which the excessive motions were recorded. Of particular note (green) were the extent of roll cycles. Figure 17 displays a typical Fire Scout (MQ-8B) recovery [9].

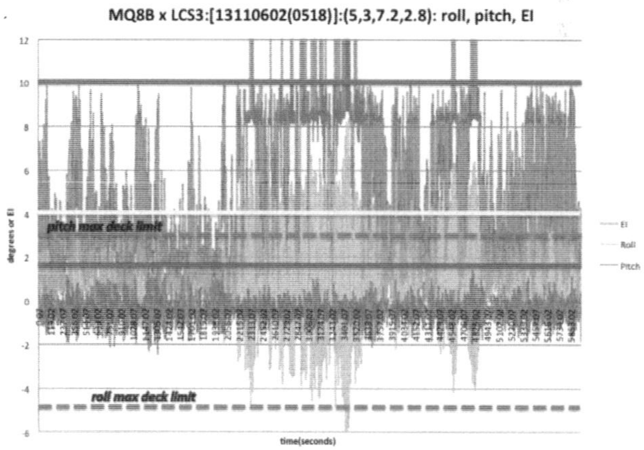

Figure 16 – Typical Time History Trace

Figure 17 –MQ-8B on Final

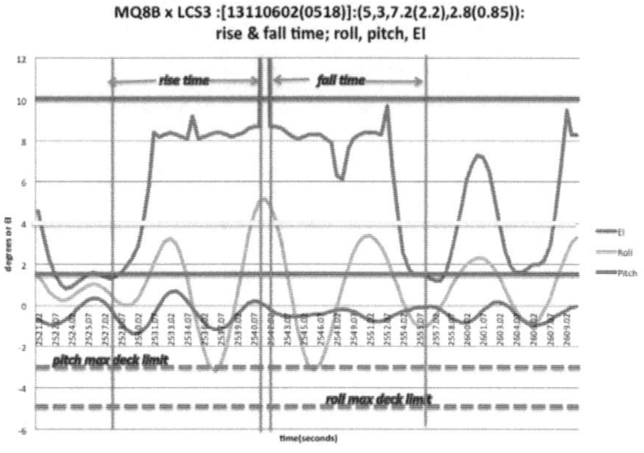

Figure 18 – Rise / Fall time events recorded

Examining the related traces (figures 16, 18, 19), and lining up the graphs (figure 20), it would appear a rare event occurred in which all the degrees of freedom appear to be in phase, but only roll contains sufficient energy to move the ship beyond the MQ-8 landing limits.

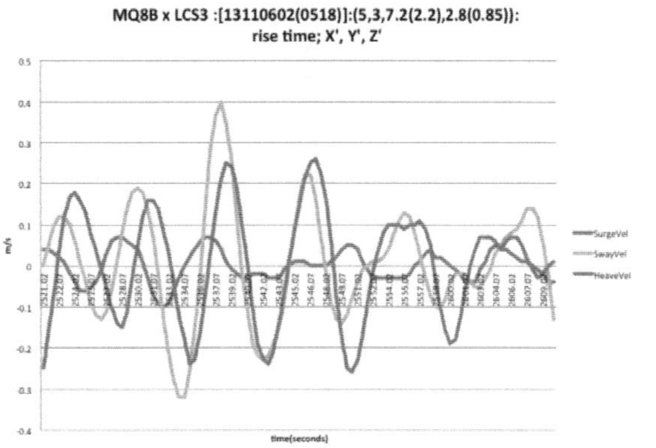

Figure 19 – Corresponding Rise time rates

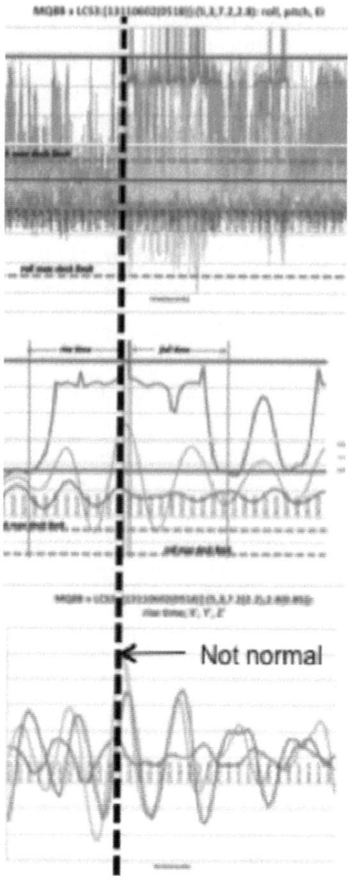

Figure 20 – All DOF axes in-phase in Rise time Event

On closer inspection, it was determined this event was not an isolated case. Figure 21 displays other examples of rise time events composed of in-phase axial motions even those that are diametrically opposed.

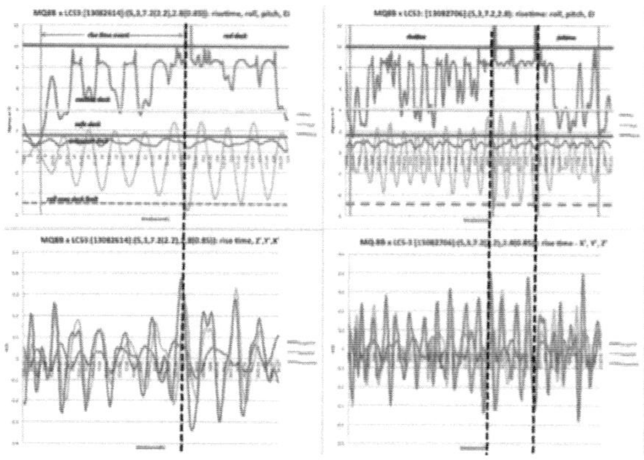

Figure 21 – Parametric-like roll events

Table 4 lists the motion statistics for one of the LCS DI trials.

Table 4 – DI Trial Motion Statistics

- **Total hours recorded at sea: 240**
  - ***Sea State: 0 – 3*** Ship handling: 0 – 5 Ship Speeds: 2 – 15 knots
  - Max motions: roll – 9.9° pitch – 2.5° Z' – 0.7 m/s Z" – 1.0 m/s²
    Y' – 0.6 m/s Y" – 0.6 m/s² X'- .28 m/s X" - .29 m/s²
  - MQ-8 deck limits: roll – 5°, pitch – 3°, Z'– 2.2 m/s(7.2'/s), Y'– 0.86m/s(2.8'/s)
- **The number of rise time events: 252 (non slam)** *This went unnoticed until mugs fell over; raised eyebrows. These events did not interfere generally with the DI.*
  - Average minimum rise time recorded: 10 seconds
  - Computational minimum rise time: 6.3 seconds
  - No of Slam events: 0 (non aviation);
  - ***Excessive ship motion axis causing rise time events***:
    - ***Roll*** : ***100.0%***
    - Y, Z' : 0.0%
    - Pitch 0.0%
- **The number of launch and recovery events recorded = 177 (86 launch; 86 recovery; 5 wave-off): 64 piloted serials; 22 autonomous serials;**
  - Launch: 55 green, 20 gr-amber, 25 amber, 0 red events
  - Recovery: 60 green, 15 gr-amber, 25 amber, 0 red events
- ***Other: ship's power failure events (some occurring during flight ops). System robustness was a requirement for the success of this test.***

Red flags

According to the NOAA wave charts based on local wave buoy data, the test box off of California was experiencing sea states 0 – low sea state 3. Yet the LCS platform recorded a 10° roll. There was motion along the other degrees of freedom but nothing to support a 10 degree roll. Lateral and vertical rates never exceeded the respective deck limits of the MQ-8. Based on the statistics table, only roll exceeded deck limits of the MQ-8. This is very unusual. Referring to figure 22, a video was taken of the deck whilst it was in a roll cycle.

Figure 22 – FireScout on Platform in roll cycle

These resonance cycles of ship motions appears to have occurred while the ship was moving relatively slowly through the sea to an overnight track.

DEFINING MOTION CHARACTERISTICS

The impacts on motion sensitive tasks by ship platform traits as those described in this article, could be significant. Throughout this article, the apparent motions described have been classified as parametric-like roll. While the ship's 6 degrees-of-freedom motions are all in phase, only elevated roll motions are evident. However, there are many related functional variables that have not been verified. Careful analysis of the seaway to include currents, swells, direction, speeds along with significant wave heights and their periods, is required to gain the complete picture or definition. To this point, much of the analysis has been using motions in the ship coordinate system without attentive documentation of the ship's heading and speed relative to any existing seaway.

Logbook climatic observations indicated most of the motion events occurred in calm sea conditions with the ship navigating at under 10 knots of ship's speed. These conditions are contrary to the conditions reported in much of the literature concerning parametric roll. The vast majority of the publications on the subject describes an elevated sea condition equating to storm waves [10].

Future work

Envelope expansion dynamic interface trials are expected to be conducted with an MQ-8C on board one of the LCS class vessels later this year. A key factor related to increased operational capability in landing UASs onboard ship is the ability to repeatedly L&R safely from a ship moving in response to the seaway. The operational benefits of an unmanned air vehicle are increased if it can be safely captured autonomously, i.e. without the aid of an experienced pilot and Landing Signal Officer (LSO) to a quiescent ship. The test instrumentation for this task during DI trails includes ship motion recording devices. The recordings from these trials will join the library of ship motions data already in existence. Coupled with these recordings, careful attention will be paid to maintaining an eye on the ships speeds and headings. This is combined with external seaway recordings both from "nearby" anchored wave buoys (figure 23) managed by NOAA and close up wave buoy information recorded by University of California San Diego's (UCSD) Scripps Institution of Oceanography (figure 24).

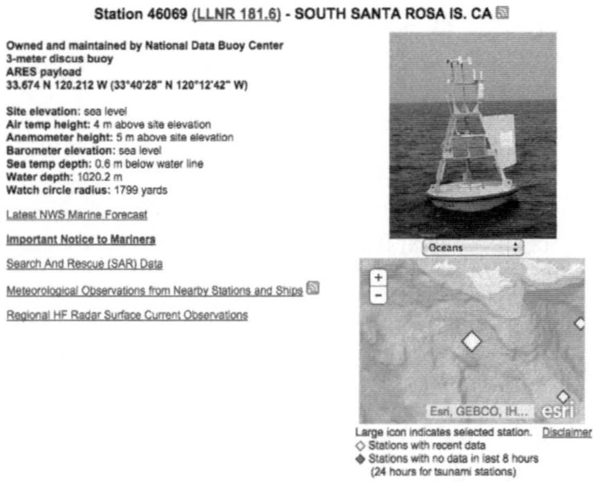

Figure 23 – NOAA Wave Buoy Station

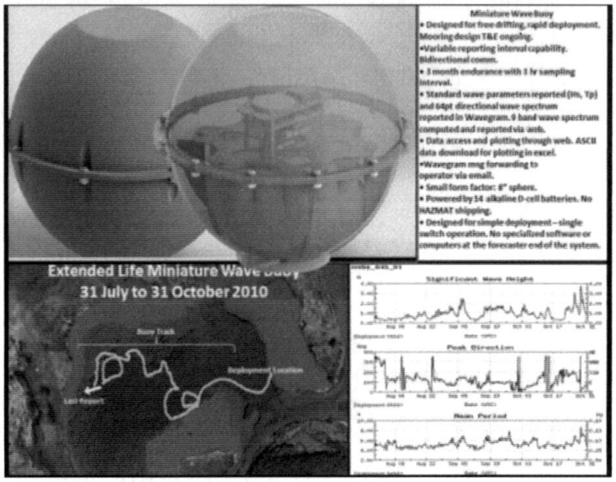

Figure 24 – Scripps Wave Buoy System

In the case of the MQ-8C, some of the possible impact on the aircraft will be mitigated by the system's LPD unit identifying the quiescent deck period.

SUMMARY AND CONCLUSIONS

Parts of the MQ-8 future improvements involve the enhancement of the associated system components, such as that being done for the L&R segment. The primary goal for conducting DI analysis is to expand existing operating envelopes and increase AV's availability thereby improving overall naval effectiveness. The objective of DI is to determine the maximum safe air vehicle/ship platform operational limitations. Given an air/ship system and inherent operational limitations, DI strives to increase tactical flexibility for any set of environmental conditions. Modeling and Simulation complements DI testing to delineate system limitations. The calculated system limitations provide experimental DI with the necessary data to more effectively set testing strategy to probe the limiting conditions.

This paper provides a discussion of the improvements associated with the MQ-8 system focusing on the L&R process and equipment. These improvements are driven by a need for L&R systems requiring few or no people to operate; systems requiring reduced time to execute high tempo L&R on the deck; systems that provides improved reliability whilst being simple to maintain. One potential problem that could seriously affect MQ-8's L&R on both of the LCS class ships is the large magnitude of roll experienced in seemingly benign sea conditions. This phenomenon should be further investigated to determine its cause, whether by a swell or parametric conditions, so that it possibly can be avoided.

There has been a great deal of interest in this project from its inception. The team looks forward to providing a status update at the conclusion of the development and start of deployments.

ACKNOWLEDGEMENTS

This paper is dedicated to the loving memory of Dr. Peter J. F. O'Reilly, a founding father of the DI discipline, and one of the first people to propose the use of the EI to automatically recover Unmanned Air Vehicles. The authors grately appreciate the significant contributions made by former Dept Heads Dean Carico (DI) and Terrence Applebee (Seakeeping). The authors gratefully acknowledge the support afforded by Captain J. Dodge, USN, PMA266, SEA 851 Branch Head Tim Smith and the contributions made by Isaiah Kim NAVAIR 4.3.1 Assigned PMA266, CDR David Eccles, USN (ret), Assistant Program Manager for Engineering PMA266, Wyle Program Manager CDR Steven "Rocky" Rauch, USN (ret), and Wyle Technical Editor, Monica Rauch.

REFERENCES

1) Didonato, Lt M(2015). "The Future of Fire Scout, A Bold Push for Integration". **Rotor Review**, Fall, *number 130*. San Diego

2) *Froude, Dr. W (1861). "On the Rolling of Ships". Republished the Papers of William Froude (1955). Royal Institution of Naval Architects (RINA). London.*

3) *Ferrier, Dr. B., Ernst, R., Sehgal, A., Nelson, J. (2014). "MQ-8C (Fire Scout) Launch and Recovery Program Improvements". Proceedings of the American Society of Naval Engineers Launch & Recovery Symposium. Alexandria*

4) *Levandou, M., & Gaillarde, G. "Operational Guidance to Avoid Parametric Roll," RINA, 2003.*

5) *Galeazzi, R. Blanke, M., and Poulsen, N., "Chapter 2. Detection of Parametric Roll for Ships," Parametric Resonance in Dynamical Systems, T.I. Fossen and H Nijmeijer (eds.) Springer Science +Business Media LLC 2012.*

6) *Ferrier, Dr. B (2014). MQ-8B/C x LCS-4 Deck Characterization using NAVAIR LPD. QR140530. Arlington.*

7) *Ferrier, Dr. B., Ernst, R., Sehgal, A. (2015). "Instrumented Deck Landing Cueing in Unmanned Aircraft Systems". Proceeding of the American Helicopter Society Forum. Virginia Beach.*

8) *Nelson, J. & Danielson, J.(2013). MQ-8B Fire Scout Initial Ship Suitability Tests Aboard USS McInerney (FFG 8). NAWCADPAX/RTR-2009/118. Lexington Park.*

9) *France, W. N., et al. (2001). "An Investigation of Head-Sea Parametric Rolling and its Influence on Container Lashing Systems". Proceedings of the SNAME. New York.*

ANNEX

SIMULATED ENCOUNTERED DECK FORCE - SHIP

The fundamental tools used early in the Landing Period Designator (LPD) development were the Ship Motion Program (SMP) series [1] coupled with the Ship Motion Simulation (SMS) program [2]. The program methodology uses spectral probabilities in order to produce deterministic synthetic time histories. The SMP series is a world standard and well documented.

The SMP is divided into two basic themes, spectral analysis and the calculation motion histories in the time domain. The SMS fundamental relationship is:

$$S_r = S_\omega(\omega) \bullet RAO \bullet f(V, \mu) \qquad (1)$$

where:

S_r: Ship response spectrum

$S_\omega(\omega)$: Seaway spectrum

RAO: Ship transfer functions

$f(V,\mu)$: Frequency mapping

V: Velocity

μ: Relative wave angle

The seaway is most defined by a forcing function called the Bretschneider, which is defined by the following expression:

$$S_\omega(\omega) = H_S^2 \, e^{\left(\frac{-1994.5}{w^4 T_\emptyset^4}\right)} \qquad (2)$$

where:

$T_\emptyset$: period (sec)

ω : wave frequency (rad/sec)

$S_\omega(\omega)$ seaway spectrum (m^2-sec)

H_S : significant wave height (m)

The spectral characteristic of a vessel is defined by experimental or computational developed transfer functions termed Response Amplitude Operators (RAO). The RAO define the dynamic ship responses for a specified load/operating condition (figure 1).

The forcing function or Wave Spectrum is defined as a collection of a great number of simple, regular sinusoidal waves in summation. Relating this to the physical world, the seaway is composed of a great number of waves varying slightly in frequency.

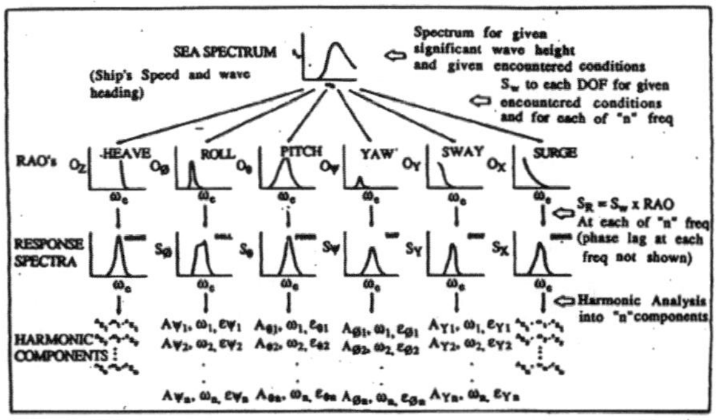

Figure 1- Calculation Methodology

The ship response spectrum is created as the product of the RAO and the driving sea spectrum over the entire range of frequencies. The response spectrum is reduced to sets of harmonic components for each degree-of freedom. Synthetic time histories are created stochastically by summing the harmonic components over a given time period. A typical time history equation is given by:

$$A_Z = \sum_{n=1}^{k} \left(A_{Z_n} \cos\left(w_n - e_{Z_n} \right) \right) \quad (3)$$

where

A_Z:	DOF amplitude
w:	a circular frequency
e:	phase angle

Time histories are produced by the sum of 48 synthetic functions (k=48). In summary, the SMS creates deterministic measures of ship motion from a probabilistic spectrum. These validated measures were used to drive the ship visuals in the Manned Flight Simulator [3].

SIMULATED ENCOUNTERED DECK FORCE – AV

Motion of an aircraft on the flight deck is calculated in terms of ship motion as a function of the aircraft model. The aircraft model is considered an extension of the ship. The model is defined by: aircraft landing gear footprint; on-deck location and orientation; aircraft weight, inertia, center of gravity (CG), lateral drag area and center of pressure. The aircraft experiences ship transferred forces and moments that create rectilinear and angular accelerations on the AV. The accelerations can be numerically integrated to determine the position and attitude of the helicopter relative to the ship as function of time, for various ship motions [4]. In essence, the aircraft is displaced as the sum of all forces to which it is exposed. Figure 2 displays the moments acting on a skid-based aircraft at sea.

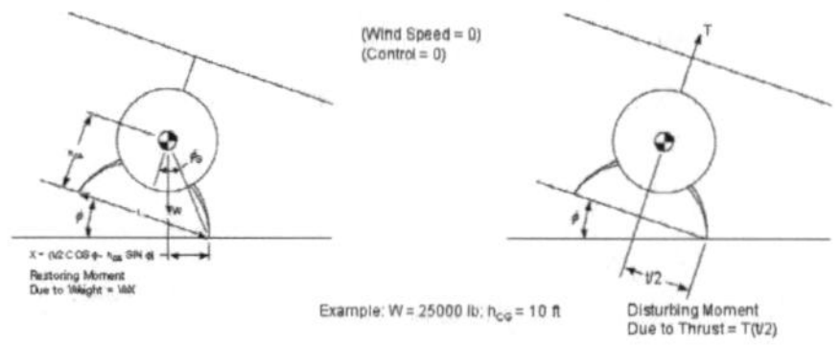

Figure 2 - Moments Acting on a Skid-based Helicopter at Sea

The inertial loads at the helicopter CG induced by ship motion are given by;

$$\begin{aligned} F_{i_x} &= W * AXcg \\ F_{i_y} &= W * AYcg \\ F_{i_z} &= W * AZcg \end{aligned} \quad (4)$$

where,

$\begin{pmatrix} Fix \\ Fiy \\ Fiz \end{pmatrix}$ = inertial forces due to ship motion

W = aircraft weight

$\begin{pmatrix} Axcg \\ Aycg \\ Azcg \end{pmatrix}$ = acceleration

In the longitudinal, lateral and vertical directions, these inertial loads become:

$$\begin{pmatrix} X \\ Y \\ Z \end{pmatrix} = \begin{pmatrix} T_{11} T_{12} T_{13} \\ T_{21} T_{22} T_{23} \\ T_{31} T_{32} T_{33} \end{pmatrix} \begin{pmatrix} Fix \\ Fiy \\ Fiz+W \end{pmatrix} \quad (5)$$

where:

$T_{ij} = T(\varphi, \theta, \psi)$ (transformation matrix from ship's axis system to horizontal level/vertical axis system).

and

φ = roll
θ = pitch
ψ = yaw

A wind force is added to the ship motion induced forces. In SMS, a unidirectional continuous wind model (simplistic model), with its vector in the same direction as the seaway, is applied. Figure 3 displays the inertial components associated with the introduction of wind on the computation of the sum of forces at the CG of the AV. The wind vector is defined by its magnitude (V_{wod}) and its direction (y_{wod}). To compute the lateral force applied at the aircraft Center of Pressure due to the wind, the V_{wod} is resolved along the normal to the aircraft center line (Vw_{long} and Vw_{lat}). The lateral component is used to compute the lateral force, as follows:

$$Fw_y = 3.5\ Ay \left(\frac{Vw_{lat}}{100}\right)^2 \tag{6}$$

where: Ay =Aircraft projected area normal to the Vw_{lat} component

Fw_y =Lateral force applied at the aircraft center of pressure due to wind

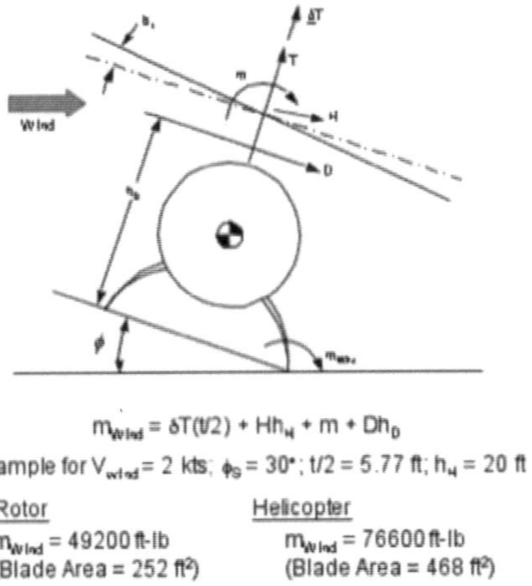

Figure 3- Wind Acting on a Skid-based AVs on the Deck

The standard tricycle coordinate system used to develop the stability of AV on the deck of a moving vessel is displayed in Figure 4. In this study, the helicopter handling system probe replaces the nose wheel and the skids are reduced to point references replacing the main gear.

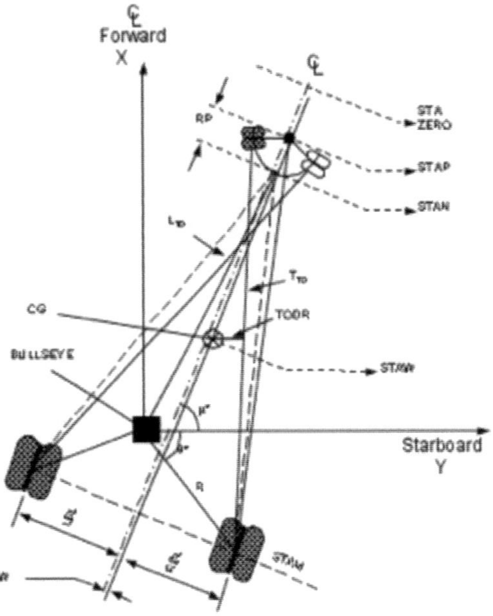

Figure 4- Typical AV Freebody Diagram

The axial forces on the main landing gear due to the wind force Fw_y is given by:

$$FRMG_{WIND} = (Fw_y)\frac{(WL_{CP} - WL_G)}{(L_{BL} - R_{BL})} \tag{7}$$

where:

$FRMG_{WIND}$ = Main Gear (right) axial force
Fw_y = Wind lateral force component
WL_{CP} = Center of pressure waterline
WL_G = Ground waterline
L_{BL} = Left wheel buttline
R_{BL} = Right wheel buttline

The incremental aircraft roll due to the wind is given by:

$$D\phi_{(WIND)} = \tan^{-1}\left(\frac{FRMG_{WIND}}{(K)(L_{BL})}\right) \tag{8}$$

where: K = spring constant

Axial forces on the main landing gear due to aircraft inertial forces in the plane of the main gear, is given by:

$$F_{RMG(Inertia)} = Y\left(\frac{WL_{CG} - WL_G}{L_{BL} + R_{BL}}\right) \quad (9)$$

where:

WL_G = CG waterline

$F_{RMG(Inertia)}$ =Right main gear axial force due to the lateral inertial force Y defined in equation (5).

Assuming perfect rocking, the axial force on the left main gear is vectorally opposite to the force acting on the right main gear:

$$F_{LMG(Inertia)} = -F_{RMG(Inertia)} \quad (10)$$

Where:

$F_{LMG(Inertia)}$ = Left main gear axial force

The incremental aircraft roll due to inertial loads is determined by:

$$Df_{(Inertia)} = \tan^{-1}\left(\frac{F_{RMG(Inertie)}}{(K)(L_{BL})}\right) \quad (11)$$

The simulation model assumes a constant wind; therefore, $Df_{(WIND)}$ is constant throughout the simulation run. However, $Df_{(Inertia)}$ is continuously changing with ship motion.

The distance from the aircraft CG to each line is computed as Turnover to Occur Right (TODR) and Turnover to Occur Left (TODL). They define the distance that the CG should move for a turnover to occur (right or left). These lines describe an angle Turnover Right (TOR) or Turnover Left (TOL). They are expressed as:

$$\measuredangle\ TOR = \tan^{-1}\left(\frac{TODR}{WL_W - WL_G}\right) \quad (12)$$

$$\measuredangle\ TOL = \tan^{-1}\left(\frac{TODL}{WL_W - WL_G}\right) \quad (13)$$

They describe the angle between a vector from the CG normal to the right turnover and the left turnover and the vertical.

Similar boundaries are computed for the pitchback condition. The hinge line about which the aircraft is likely to pitch back is the line which joins the right to left main gear. The distance from the CG to the hinge line is defined as pitch back distance (PBD) and expressed as:

$$PBD = (CGx - MGx) \quad (14)$$

where,

PBD= pitchback distance
CGx= aircraft CG station
MGx= aircraft main gear station

The associated pitchback angle (PBA) is given:

$$\measuredangle\ PBA = \tan^{-1}\left(\frac{MGx\text{-}CGx}{WL_W\text{-}WL_G}\right) \quad (15)$$

where,

WLw = Waterline to the aircraft CG
WLg = Waterline to the ship deck

Turnover incidents are static or dynamic in character. Static turnover is the same as on shore when the weight vector migrates beyond either the friction forces or the reaction forces causing the aircraft to turnover. Figure 5 shows dynamic turnover caused by the uneven loading of the rotor disk or by ship motion.

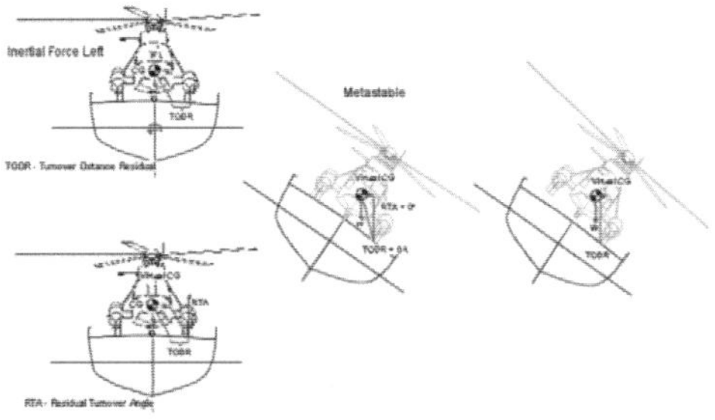

Figure 5- On-deck Dynamic Turnover

The aircraft CG is in motion. In the sum of forces, the weight vector is continually modified in response to inertial forces applied by either the rotor disk, ship motion or both. The distances TODR, TODL, and PBD reflect system stability. At the point where a distance becomes negative, the system is unstable and will seek to find a more stable, but usually undesirable geometric solution. In similar fashion, when the landing gear friction values are exceeded by the combination of aircraft apparent weight and induced inertial forces, slippage will occur. Aircraft slide will continue until the aircraft frictional forces are greater than the disturbing inertial forces. Finally, when the vertical inertial force equals and opposes the aircraft weight, the deck friction goes to zero and an unintentional liftoff is indicated. The sum of these incidents becomes aircraft-ship operational envelopes. Using the computational flow, it is possible to compute the encountered force an AV receives from the ship as a result of the ship's motions (figure 6).

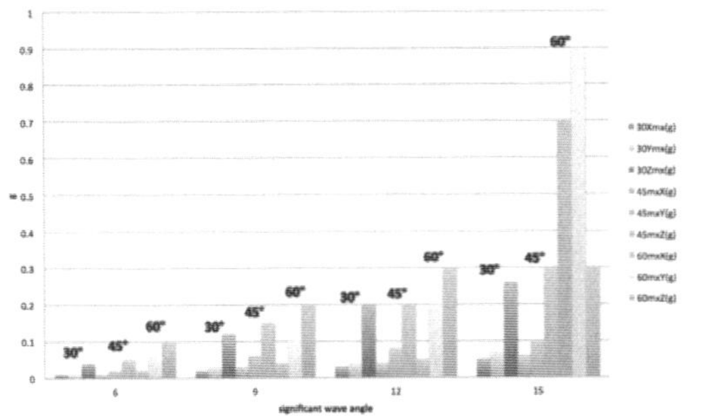

Figure 6 – Deck Loads computed by Probabilistic Synthetic time histories

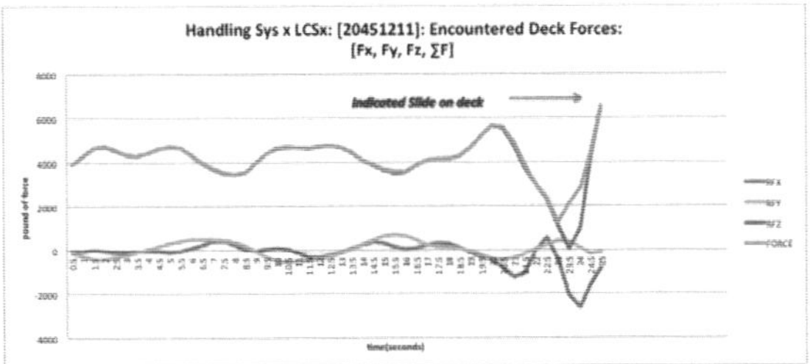

Figure 7 – Encountered deck forces

Figures 8 and 9 display the corresponding simulated ship's motions showing angular and rates along with rise time computations.

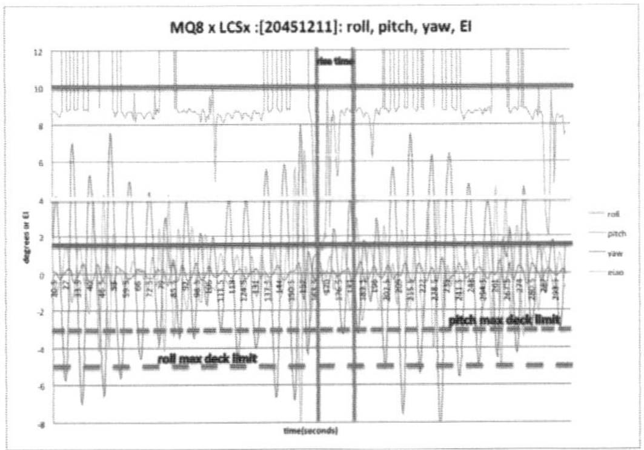

Figure 8 – Angular time history computed

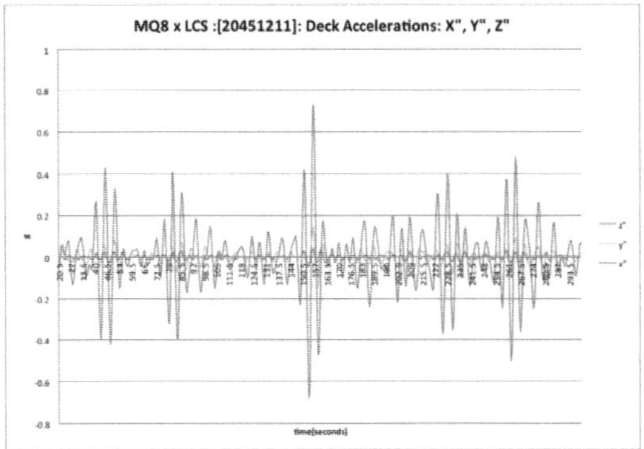

Figure 9 – translational measures

The overall deck envelope is presented in figure 10. This is a summary of the simulated motions and deck limits computed for a given device.

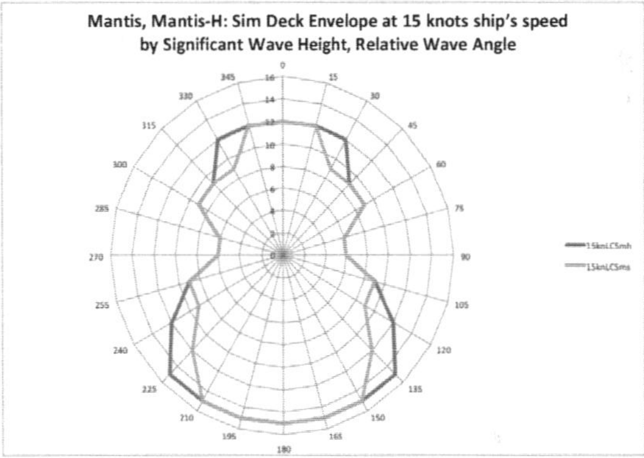

Figure 10 – Handling Sys deck limits

Until relatively recently, the capability for the Navy to measure deck rates and accelerations was limited. For this reason guidance criteria in the form of Naval Air Training and Operating Procedures Standardizations definitions and Military Specifications tended to set the limits conservatively. Recently, at-sea trial data recordings were compared to the simulated versions. More than 248 hours of ship motions were recorded. The trial program experienced up to sea state 5 but flew in more benign conditions. There were 88 L&R events with a significant number of deck rise times recorded.

REFERENCES

1. Raymer, Michael (2009). A Comparative Analysis of the Army MQ-8B Fire Scout
2. Applebee, T., Baitis, E., Meyers, W. (1981). "User's Manual for the Standard Ship Motion Program, SMP". David W. Taylor Naval Ship R&D Center Report. **DTNSRDC/SPD-0936-01**. Washington.
3. O'Reilly, Dr. PJF. (1987). Aircraft/Deck Interface Dynamics for Destroyers. **Marine Technology**. **Volume 24**, *Number 1*. Society of Naval Architects and Marine Engineers. New York.
4. Ferrier, Dr B., Duncan, Dr. J., Manning, T. (2003). "Minimum Air Wake and Quiescent Motion Indicator Simulator Test Program". Proceedings of the American Helicopter Society. Ft. Worth.
5. Blackwell, Dr. J and Feik, R. A (1988). "A Mathematical Model of the On-Deck Helicopter/Ship Dynamic Interface (U)". **Aerodynamics Technical Memorandum 405**. Aeronautical Research Laboratory. Melbourne.

Constructing a system for utilizing unmanned logistical VTOL aircrafts

Yu Ito
yuito@kuronekoyamato.co.jp
IT Planning Division, Yamato Holdings Co., Ltd.
Tokyo, Japan

Abstract

This study is aiming to give several original practical ideas into the argument on industrializing unmanned vertical taking-off and landing (VTOL) aircrafts in the field of logistics. This study will examine the use of drones from three interactive viewpoints: transport efficiency, maneuverability, and safety. These issues include the argument on how to construct an accommodating traffic control system as a social infrastructure. The ideas introduced in this study is aimed to help aircraft manufactures, logistics providers, and air traffic policy makers consider the design on aircrafts, facilities and control system that includes development management, operation, profitability, or sustainability. This study will also give insights among researchers, aircraft designers, project managers and logistics practitioners when designing the logistical use of VDs in the future.[1]

Introduction

Recent development over unmanned VTOL aircrafts, or so to say, *verti-drones* (VDs), has been proceeding in each ways of technologies, additionally introducing various new kinds of aircrafts into its field. This study is aimed to present how the evolving unmanned VTOL aviation technology could be adopted to make its maximum utility for logistical use in the society.

The main aim of this study is to design terminal facilities and traffic control (ATC) system of VDs for cargo carrying. These will be the significant components of the whole acting system for VDs to be adopted and utilized in the real society; if VDs were compared to rail carriages, terminal facilities would be stations, and air traffic control would be the track.

The system to be consolidated in this paper should be not understood solely as one of the miscellaneous plans or ideas of utilizing VDs in civilian industry, but as the initial proposal to start a discussion towards constructing a unique policy on unmanned VTOL aircraft traffics. Hence, this argument should be held in a social, political, and global manner, and aimed to make a de jure standard in this field. It should of course maintain safety, security, and discipline though, utility and the potential social value of VDs as a mean of transport should also be considered at the same time. This paper will argue air traffic policy for future VD transport regarding business affordability and sustainability.

Presented at the 7th AHS Technical Meeting on VTOL Unmanned Aircraft Systems at Mesa, Arizona, held from 24th to 26th January, 2017.

The former section of this paper consists of arguments on three elements for the usage of unmanned VTOL technology: transport efficiency, maneuverability, and safety. Transacting among each other, the three elements coordinately decides the utility of the system of drones to be realized. Making discussion on each of the three and the multiple interference among each other, this study will conduct a structural argument on constructing the use of VDs in the real society. In this part of the study, ideas form systems thinking [1] will be referred for argument.

In the latter section, design of an ATC system fulfilling all the requirements derived in the former part will be discussed. Since the system will be constructed on a zero-base, highly theoretical studies on ATC related to the idea of multi agent thinking such as [2] [3] [4]are referred. In addition, idea of traffic control of trains was also introduced for enriching the argument [5].

At the last of the introduction, it should be stressed that this study has been held privately and has no direct nor indirect connection with the author's affiliation. The purpose of this study is purely to give academic contribution to the science of VTOL aviation, and has no intention to benefit any specific enterprises.

Method

This research will examine the method to foresee and design the whole system as a social infrastructure to utilize VDs from social and macroscopic prospects in a deductive manner. The significance of study like this in advance to the realization of industrial VDs is derived from the fact that their social value is intimately related to the economic activity on the ground below.

[1] The marketing and business strategy perspective of this study partially originates in [33], a study previously presented by the same author. This study has been held privately and has no direct or indirect connection with the company the author works for.

Regarding fixed wing aircrafts that are the great majority in today's industrial aviation and have shaped air traffic policy for their own sake, they have provided benefit in the air with less relation to the ground; significant speed hardly caught up by any ground transports has been their major profit, and airports were at many cases isolated from economical areas forcing customers an inconvenient ground journey for it. Flight regulation for such fixed aircraft was allowed to construct their own superordinate ecosystem. In comparison, efficient VDs should fly over crowded urban areas, and take-off from or land on wherever necessary to support economic and social activities on ground that will be changed organically. For such future transportation, theory to support adopt the operation to the business bellow will be needed.

Before the achievement of this goal, a contradictory requirement obstructs: the more accuracy on foresight pursued, the less decisive the policy for construction of system will become. If it were to scientifically illustrate down all the available events and technical advancements that can happen on the universe in the undecided future, it will be infinite sets of possibilities. The commutation will be mathematically stated as an ever-continuing surface laying in what Wittgenstein calls the *logical space* [6], a multidimensional space that contains every aspect of visible and invisible factors.

On the other hand, a blue print for a system in the context of business should be uniquely decided for development process afterwards. Only providing highly philosophical statements won't help enterprises make plans and decision towards an actual construction. How to derive at least limited solutions from the explanative formula for strategy planning. Accuracy of modeling should improve if the business area on consideration limited, although, such measurement won't help enterprises and the society fundamentally since such narrow use of VDs will be hard to collect the massive investment. To make the formulated system benefitable and sustainable, it should be opened to a broad area of business and social logistics activities.

To pursue both of this contradictory requirement, this study will first theoretically derive out sine qua nons for a logistical VD use, regardless of their inclusion relation between each other. Secondly from a business prospective, the elements drawn out above will be examined to uniquely compose a design for the VD traffic infrastructure.

To avoid ambiguity of having multiple conclusions standing abreast in the second phase, establishment of a principle for choosing out the best policy from among the candidates is required. That will be to evaluate how the ideas systematically fulfil all the multiple requirements for a logistical VD use derived in the first phase. This idea will be a reverse engineering of systems thinking; as systems theory initially introduced by von Bertalanffy [1] orients itself from the rational phenomenon to the transacting system structure and their object, this study will compose the unique configuration from component purposes regarding them as a total interacting systems while noticing how they act among each other.

The common utility of VDs in logistical use

To design a basic system to be applicated in the undecided future, the system must contain somewhat tolerance on the aircrafts which are to use it. Additionally, the virtue of the whole system should be based on the usage of all the kinds of VDs to fly on. Hence, various common utility of VDs should be argued and re-understood in a unified manner.

The utility of VDs in freight transportation should be divided into virtue as a robot and the strength of VTOL aviation.

Primary, unmanned transport will cut down human labor cost tautologically. If a vehicle could be automated and requires no human labor for its accident-less navigation and operation, it shall benefit a lot its user making the business much maneuverable. Enterprises newly entering the field of logistics should achieve even much advantage once when this self-driving transport developed, because if so, they won't need experience nor tacit knowledge which occupies a great part of the current logistics business resource.

Flying without pilot(s) can also enhance payload efficiency. Thanks to the rapid improvement on computer engineering, the weight of onboard computers required for automatic or ground controlled unmanned navigation has become much smaller than that of living humans (See Table 1). The lightened weight required for navigation will save much of the craft's lifting capacity.

Table 1 List of major light onboard computers capable of auto-piloting

Manufacturer	Name of system	Weight of system (g)
Cloud Cap Technology	Piccolo SL	110
UAV Navigation	VECTOR	180
EuroLink Systems	Veronte	200
Ascending Technologies	Mastermind	325
Extreme Engineering Solutions	XPand6000	1,814
UAVOS	AP9.0	5,460

Secondly, vertical aviation can carry cargo linearly in high speed regardless of terrain or traffic jam on the ground requiring only a little space for taking-off and landing. DHL has already utilized this character of VDs into delivering parcels to isolated islands [7]. Another scene that this property of VTOL seems to have advantage the most is logistics in crowded urban areas. Table 2 shows the traffic congestion level in cities that are rated within the world's worst 10 in 2015. For example, in Bangkok in the evening rush hour, automobile trip took more than double the time it takes if the road were free from traffic jam. In such circumstance, the actual speed of VTOL aircraft inside the metro area would have attraction to some services that the length of lead time makes important or critical value, e.g. express delivery, on-demand delivery, medical instrument transport. If vertical vehicles can make rooftop of buildings – one of the least used space in urban areas – terminal points as helipads, they can supply an overwhelming logistics service that cannot be alternated by any other transports.

Therefore, VDs which are to contain both strengths should benefit enterprises to make use of them in proper field of business that requires rapid, urgent, or on-demand transportation. It should be noted that the usage of such transportation is not limited to freight but also to passenger travel that requires rapidity, such as air taxi or ambulance. Enterprises such as Airbus [8] and Uber [9] have already this vision of services using human-carrying drones. Concept of drone ambulance capable of carrying a patient and a paramedic has been also designed [10]. To developing a socially valuable platform for VDs, both human and non-human carrying VDs should be considered together to keep the initial introducing number of crafts large and make initial cost per operating craft small. This means that the size of aircraft considered to be operated on this system is not limited to an inhabitable small one but up to about that of middle-sized helicopters. Enterprises to join and share the VD system will be to choose the size and specs of crafts that benefit each of their own business. The ongoing military's plan for developing heavy cargo VDs would give insight that such craft will have some efficiency on logistical use [11].

Table 2 Traffic congestion level of the world's worst 10 cities

	City	Congestion level	Morning peak	Evening peak
1	Mexico City	59%	97%	94%
2	Bangkok	57%	85%	114%
3	Istanbul	50%	62%	94%
4	Rio de Janeiro	47%	66%	79%
5	Moscow	44%	71%	91%
6	Bucharest	43%	83%	87%
7	Salvador	43%	67%	74%
8	Recife	43%	72%	75%
9	Chengdu	41%	73%	81%
10	Los Angeles	41%	60%	81%

[a] Ranking and figures based on [12].
[b] Each percentage shows the average additional rate of time required for traveling compared to an uncongested situation.

Although the possibility of commonality of airframe with the passenger carrying business as pointed out above, it should also be noted that construction of the use for cargo carrying is the main argument of this total study. Even if it is to share or develop the craft both for passenger and freight, the method of utilizing them should be considered separately to be rational enough for each other. This is mainly due to the difference of the movement of humans and cargos alighted: while men and women can change transports by their own will towards their final destination at each transport node, parcels have to be sorted by proper machines or human labor in a logistical hub for proper distribution. Hence, this study afterwards will still focus strictly onto how to realize the use of VDs for physical distribution.

Development scenario considering alternative transports

The benefit of utilizing VDs has been argued in the section above. However, the process towards the goal of a completely autonomous aviation would rather be the real key of robotic transport business. In other words, how to automate the job currently for humans will decide whether the whole development could financially succeed or not. Unless the development could earn a tremendous initial capital enough to go on a series of non-profitable experience, it should somehow share the field with the current human-involved service operation. Even if the system were to be developed without financial restriction, it still contains the risk of drifting out of the true demands from the market. Therefore, development involving the current service, such like the so-called agile development or DevOps in computer system development, will be required for a commercial success in the future. Also, it can be said that field where the current logistics service providers can take part in, make their experience and foresight into value, to secure and maintain their status in the coming age of drones.

In fact, some of the actual development of autonomous transport has made use on optionally piloted vehicles (OPVs). For instance, a driverless ground delivery system on going by Yamato Transport in Japan puts a device for self-driving onto a conventional van, planning to use both human drivers in the first phase for testing of the system, and auto navigation in the later phase for confirmation of the test [13]. Moreover, also in the field of vertical aviation, Kaman aerospace has modified its K-1200 K-MAX "aerial truck" helicopter into a remote controlled OPV and had it tested by the US Army deployed in Afghanistan [14].

What this means is that the system to be constructed must take a broad range of navigation method (e.g. OPV, remote controlled, full autonomous) into consideration; a system that can contain various way of flight including human involved, will be the one to have the scope for future expansion. Contained into consideration, currently human-involving navigation can gradually advance towards a full automatic guiding system (See Figure 1). Therefore, it should have room to allow human guided vehicles share the system partially. However, this does not necessarily mean that VDs and human-guided VTOL aircrafts will be allowed to join the system totally, without any exception. Since the segregation effects the efficiency and safety of the VDs to be used on the system, the standard on where to divide human-involved and autonomous crafts will be argued in the latter section of this paper.

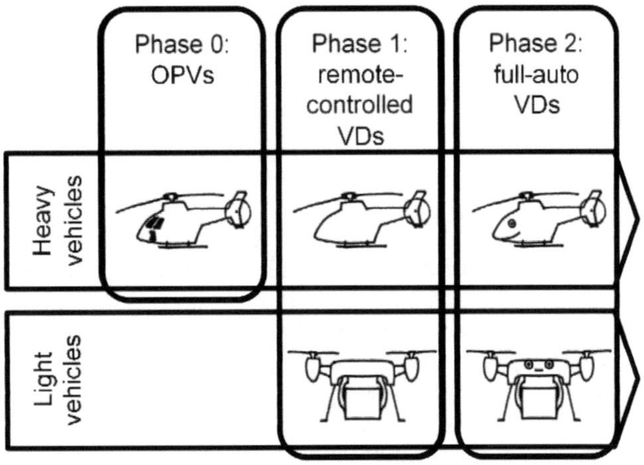

Figure 1 Development scenario towards a full-auto VD (illustrated by the author)

The logistics of VD transport

What 'logistical' originally means

Today, the proposition that logistics thought is needed in business widely has become a common understanding, however what logistics is and how can it be provided is not [15]. The adjective 'logistical' as the objective of this study is not to indicate the state of the current business logistics practice that is to be, but to represent the philosophical idea of what logistics ought to mean; this study focuses on *how* VD can provide logistics, rather than *what* they can achieve.

Since its introduction by Jomini [16], the military terminology 'logistics' have been regarded commonly throughout history in the basic context of consideration on non-combat movements of personnel and goods in military science. However, as technology on weapons and transports improved, the practice and necessity of logistics has fluctuated a lot. Reference [15] points out that once disregarded as an army idea because of its abbreviation due to the introduction of railroad, the concept of logistics has achieved its modern sense as "total management on physical distribution" when the Allies successfully executed the landing on Omaha beach on June 6th 1944, making its practical significance large enough to be referred by business science within the later several decades. In a more metaphysical description, from this point, the thought of logistics has gained its positive status in military science as what makes operation of a force available on any location on the globe regardless of distance, terrain, or existing structure, which was then geographically restrained by water, road, rail, and location of supply.

The importance of maneuverability and efficiency in business logistics

Though the marketing method of business logistics contains additionally the consideration on economical profit, it shares the basic idea mentioned above with modern military logistics thought. Practically, as well as transport efficiency, the importance of maneuverability for logistics is common among military and business.

Because business logistics is to make enterprises operate wherever they want to, the lines of supply should be connected or diverted as soon as possible whenever necessary from some business reason. In this context in a longer term, maneuverability means the ability to easily deploy and retreat local facilities, or in other words, to have less unmovable fixed capitals. This understanding can be seen in the increasing use of third-party logistics services and lease services as one way for enterprises to reduce its logistical fixed capital [17].

When designing transports, maneuverability should be considered in the equilibrium with transport efficiency. For instance, the special underground light rail formerly used by the Royal Mail connecting post offices and sorting hubs in London have had high efficiency on transport at its opening, however gradually lost its efficiency as the sorting hubs moved out from its wayside; the Royal Mail has announced that running it had costed five times more than simple truck transportation at its closure in 2003 [18]. Such transport design that lacks maneuverability should be evaluated as least logistical, whatever operation efficiency it has in the initial.

Considering the three major elements

In this section, three major elements that decides the utility and sustainability of logistical VD use: transport efficiency, maneuverability, and safety design, will be argued. Measurements considered to secure each of the elements will contribute to the consolidation of the VD logistics operating system to be argued in the later section of this study.

Transport efficiency

When constructing VD transports in the near future, it will become in comparison between truck transports. In other words, it will be to state what virtues of possible VD will rationally urge the modal shift from current ground transport.

In order to suggest at what circumstance the modal shift from today's ground transportation to VD will make sense, mathematical idea of boundary is used. The mathematical concept of boundary is defined as: "the set of points, known as boundary points, which are members of the set closure of a given *S* and the set closure of its complement set" [19]. This surface concept can represent the future state to be determined only as an infinite combination of possibility into a single formula. If one should predict a unique boundary point of modal shift, he/she has to rely on his/her subjective probability of each changes to occur in the future. The formula expressing the boundary will be independent from such ambiguity, with a tolerance to a width of performance of future technologies.

When a providing method of a service changes, market – including customer and the good itself – also changes; remind how the introduction of wide body passenger jets has changed the flight into a cheap and popular means of travel. However, on the other hand, at the precise point of that change, both quality and quantity of the demand hardly changes; Boeing 747 was at its debut designed for the then existing relatively few luxury passengers rather than the potential customers, which as some fact like the installation of a vast piano lounge space instead of extra seats by some carriers shows [20]. Because of this slow-acting nature of innovation in service provision, which resembles to what is called the introduction stage in product life cycle phenomenon [21], the boundary will be stated simply as if the newly adopted VD could handle the quantity and economy provided by conventional transports.

Before conducting the formula that represents the boundary, the author must explain the effect that VD will have on transfer hubs (THs). Many present cargo carriers bring each parcel into THs where they are sorted and aggregated for rationalizing its service. The combined cargo is then transported THs in each region, where it will be sorted again for their final destination. The current operation puts emphasis on night hours for this aggregation and transportation, which most of their customers are asleep or off business, and the traffic (including road, railway, and air) is free from crowded passenger traveling. On the contrary, by using intermittent helicopter flights that can avoid traffic jams and land just near the THs, logistics providers will be able to run THs and transfer cargo all through the day. This will not only increase the schedule speed of the service, but also minimize the cost needed for THs. How efficient will the VD help the running cost for THs is hard to calculate in a generalized matter, however, it can be re-described as 'how effectively they can eliminate the deviation in quantity of parcel sorting throughout the day'. Suppose a TH equipped with a fully automated sorting system, which at the current, handles 75% of the daily number of parcels in the midnight 6 hours, while the residual 25% in the remaining 18 hours. If the enterprises using this TH could flatten this fluctuation by using VDs, the needed amount of sorting facilities, which were then required to meet the needs for the peak hours, could be one third of before the modal shift.

Thus, at the precise time point of the transition, the position of the modal shift boundary will be stated as a true proposition as:

$$\exists W_V \exists F_V \exists Cv_V P(W_V, F_V, Cv_V) \tag{1}$$

$$\text{s.t.} \quad W_V \times F_V \times n_V \times r - Cv_V \times n_V - Ch_V \geq W_T \times F_T \times n_T \times r - Cv_T \times n_T - Ch_T \tag{2}$$

$$W_V \times F_V \times n_V = W_T \times F_T \times n_T = q \tag{3}$$

$$Ch_V = x \times Ch_T \qquad 0 < x \leq 1 \tag{4}$$

$$W_T \geq 0, F_T \geq 0, Ch_T \geq 0, n_T \geq 0, q \geq 0 \tag{5}$$

The nomenclature of the formula above is given with proviso as follows:

W	weight of cargo per service/flight (tons)
F	transportation frequency per day, each vehicle
r	usual benefit rate earned by moving cargo, per ton
n	size of the running fleet
Cv	running cost per a vehicle, a day

Ch cost caused by transfer hub using each transport, a day

q average daily quantity of parcels needed to be transported

(Subscript)

V VDs

T trucks (or ground vehicles)

Each of the equation expresses the following statement: The boundary is explained by three variables representing capacity of the newly introduced helicopter that is, payload, transportation frequency, and running cost (1). Benefits earned by VDs; pure benefit earned by transferring the amount of cargo, minus total running cost of the entire fleet and the cost required for TH while using VDs, is larger than that by conventional transports (2). As mentioned above, the quantity of demand met by using helicopters can be assumed as the same as that of conventional transports (3). The cost required for THs when using helicopters is smaller than using conventional transports (4). Finally, all the figures are positive real numbers (5).

The proposition could be simplified as:

$$\exists W_V \exists F_V \exists Cv_V P(W_V, F_V, Cv_V) \quad (1)$$

$$\text{s.t.} \quad Cv_V \times n_V \times q + x \times Ch_T \leq Cv_T \times n_T \times q + Ch_T \quad (2\text{-bis})$$

$$0 < x \leq 1$$

$$n_V = \frac{q}{W_V \times F_V}, n_T = \frac{q}{W_T \times F_T} \quad (3\text{-bis})$$

$$W_T \geq 0, F_T \geq 0, Ch_T \geq 0$$
$$n_T \geq 0, q \geq 0 \quad (5)$$

$$W_V \geq 0, F_V \geq 0, Cv_V \geq 0 \quad (6)$$

This proposition can be visualized in a three-dimensional matter by using n_V instead of W_V and F_V, such as Fig. (1). Derived from equation (2-bis), the vertical z-axis represents the total cost (C) for using each transport; the boundary is represented as the VD's smaller cost than the present transports. When the precise cost needed for conventional transports decided uniquely as c_T by each fixed n_T and Cr_T, the boundary will be the intersecting line and below of the horizon containing the point c_T and the set surface of C_V (see Figure 2 for geometrical understandings).

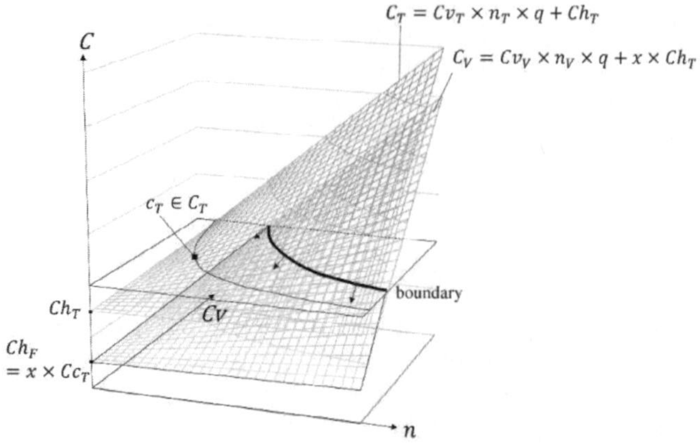

Figure 2 A geometrical explanation of the boundary

Of course, the model is simplified that it wouldn't represent the reality as it would be. Even if VDs were to play a leading act in freight service, it won't completely expel other existing means of transportation; conventional transports do have some advantages, e.g. trucks and freight trains more economical during night time, and it should be auxiliary means for helicopters. However, the aim of this concept is to show both the VD manufacturers and cargo carriers in what condition could helicopter enter the main role of civilian regular cargo service. Hence, the simplified equation will contribute more to an implication on what is needed on VD or when to make the decision to replace the current transports.

Additionally, it should be considered that in reality, Cv is not completely independent but feebly subject to n because the fleet size should affect somewhat the running cost of each vehicles. Simply theorizing, the unit cost for repair and maintenance should reduce when the size of fleet gets bigger because it would enable to share precious facility, personnel, and knowledge by existing and additional vehicles. However, on the other hand, it is also well assumable that the large number of assets causes confusion on identifying and managing them and affect negatively on rationalization. After all, it depends on the design of management towards the entire fleet. Since there haven't been any logistics provider that have possessed and managed such big fleet of VD so far, it is impossible to state the equation between Cv and n that has never recorded or thought to differ in occasions.

Moreover, the speed of the helicopter will absolutely create a new value that will earn extra benefits for logistics service providers, which seems to be ignored in the formula above. Since logistics is to make its customer available to have goods whenever they want, it means that shorter lead time makes better logistics.

Effort on business logistics to make this pursuit towards the shorter time of order-to-delivery has largely been made on designing the informational platform so that it urged a reverse insight towards the military field and provoked a new idea of *net-centric operation* [22]. However, it is assumable that also the design on transports could shorten total lead time.

However, as mentioned before, this proposition represents the exact moment of the modal shift that the existing customers hasn't changed their preference towards the service. Besides, however there should be some hidden needs for such service, estimating them rationally is impossible. By excluding such ambiguous term that will cause danger of losing existing customer or causing massive loss, the formula could provide a safe standard on a decision making to step into an innovation waiting to be occurred.

A certain fact visible from the model is that smaller n_V or Cv_V is required in order to make the efficiency of VD superiority and achieve the modal shift. Smaller n_V could be achieved by introducing heavy VDs that handle more cargo at once. One of the solution to make Cv_V smaller might be the development on propulsion. For instance, since conventional helicopters required high maintenance cost on their rotor head, alternative technologies such as hinge-less rotor to reduce the cost has been developed [23]. Logistics provider aiming to enter the field of vertical flight should pay attention to the recent rapid development of VD that are to equipped with new type of propulsion in the point of their running cost.

The conclusion derived from the argument in this section is, firstly, that the constancy of transportation could be one of the biggest advantage of VDs to conventional ground transport if realized. Secondly, in order to gain maximum efficiency of VD system, the infrastructure should be open to various type of aircrafts that should be evolve in the latest sophistication.

Maneuverability

As mentioned before, long termed maneuverability is represented in lesser number of fixed assets that is to be wasted when the receiver changes its location. This means that if a transport doesn't require any terminal facility for its service, the maneuverability would be maximum. VD will have this advantage against, at least, fixed-wing aircrafts since they require less area of ground for operation. Moreover, it apparently seems that an ultra-light VD capable of landing itself or dropping cargo directly to inside the customers' estate without requiring any modification on there will fulfill this requirement. Major ongoing conception of VDs, such as Amazon's Prime Air, are based on this understanding. However, two problems should be concerned.

The first is that such transport would apt to become one-way service. Current delivery service is constructed on the use of delivery vehicles with enough payload for multiple delivery and pickups. The amount of delivery and pickup parcels at a time is of course not always the same, and therefore, if the VD is to carry single parcel, the service flight should become one-way. In sum, such flight should contain a large amount of empty flight time that won't earn any benefits. This inefficiency becomes more when the length of route extends. Therefore, the author predict that logistical VD will get bigger and more complexed as technology grows.

Secondly, the structure of the aircraft will become big or complicated not enough to be handled without terminal facilities when craft performance pursued after all. For instance, the series of Parcelcopter developed by DHL has increased its size and payload, and the latest Parcelcopter 3.0 (though payload limited to 2kg) is designed to land on a helipad called "SkyPort" which contains parcel locker function [7].

Regarding these problems, a solution will be to make maximum use of human labor and turn operational human resource into service resource. Ground support should be inevitable when VD is designed to handle multiple cargo for rationalization, and making personnel to handle loading and unloading parcels from the craft, the same personnel could provide customer service. Additionally, this will also prevent from losing extra marketing benefits on collecting customer's unexpressed demand and handling unexpected trouble which human deliverers currently has.

Specifically, the service conceived above will be like as, arranging proper land for vertical landing and taking-off near the customer(s), such as an open parking space, and carry parcels from and to that extempore verti-field by ground personnel. It would be better to share one verti-field for several customers. To assure maneuverability, the verti-field has to be very simple, and so, security of the flight will be partially (e.g. clearance, craft observation) could be ensured by ground staffs. In short, the ground personnel would more look like a delivery driver without trucks.

Although this so-to-call 'human terminal' system would make VD service internal and not to make service design completely free from human labor, it should benefit logistics provider in reducing wasteful labor of drivers to just drive vehicles with his/herself, and turn them into a more service-oriented business resource sticking to the precise area of business. Also, as the size of the craft gets bigger and the airframe gets more sophisticated, it would be more difficult and dangerous to handle such VDs without proper training. Therefore, it is predictable that the future use of VDs in logistics service will be internal to service providers, and the style of delivery won't change dramatically even VD technologies evolve.

Another solution will be as a consequence just the opposite to the above: to replace exactly what human porters are doing currently with small payload VDs. A truck manufacturer Workhorse Group Inc. is developing a system combining truck and a light drone [24]. This design will be to compensate for the current mobile VD's absolute lack of payload. It can also solve the problem of the immobility of the helipads, since the platform – truck – itself has enough mobility. When even the trucks would be automated, this autonomous delivery system should cause some impact to the market, because the labor cost will be reduced, while size and amount of the parcel won't be reduced. However, homes and estates that are able to be directly delivered by drones should be limited than that by human due to the problem of clearance for approach and landing. Moreover, the problem of the collection of customer's hidden needs mentioned before would be left as it is. Therefore, the service to be achieved should be relatively limited to a narrow area of the market.

Safety design

In addition to transport efficiency and maneuverability, safety should be maintained on a proper level to make operation in the reality. In this section, lessons for VD safety measurements will be learned from accident record on conventional human-guided vehicles. Though human and machine should differ their cause of error, the occasion of the aviation accidents would possibly resemble, because more accidents on an occasion tend to be related to more complexed decision required that takes fatal time [25]. Complicated decision process should also cost computers at the current level critical time, and cause potential danger, unless a revolutionarily high-process computer would be developed in the far future.

Table 3 shows each number of in what occasion civilian helicopter accidents occurred throughout the year 2014 inside and outside the USA[2].

Table 3 Occasion of helicopter accidents (NTSB, 2014)

Contents of occasion	Frequency	Percentage
fundamentally avoidable with regular logistics operation		
training or experimental flight	35	19.9%
T-O or landing outside public airports	25	14.2%
extremely low altitude flight	25	14.2%
hovering: observation	1	0.6%
Subtotal	87	48.8%
potential danger in regular transport operations		
cruise flight	19	10.8%
T-O or landing inside a public airport	14	8.0%
taxing or parking	10	5.7%
approach or descent for landing	10	5.7%
ascent	4	2.3%
hovering: external cargo loading	1	0.6%
other/unknown	32	18.2%
subtotal	89	51.2%
Total	176	100.0%

In total, nearly 30% of the accidents occurred due to the hard use unique to VTOL aircrafts; during takeoff and landing outside an airport, or extremely low altitude flight that are usually served for temporal use, i.e. pesticide spraying, geographical survey, hunting, cherry drying. These flexible yet dangerous actions could be removed from the operation of civilian regular freight service if users could manage to build landing facilities with proper assistance equipment. Or perhaps even without such expensive facilities, if ground personnel mentioned above could attend the occasion of landing, he/she can check the clearance. Also, almost 20% of the events took place during instruction or test flight, which can be at least kept away from taking place near the operational area.

In sum, VD used by logistics supplier with proper ground assistance can be kept out of nearly half of the potential helicopter accident, at least inside or near their facilities. On the other hand, the remaining half of potential danger should be removed by making device on operation.

More than 10 % of the accident took place during cruise flight. However, this occasion should be less dangerous in VD flight that necessarily rely on instrument navigation. Though it is impossible to figure

2 The data is collected by the author from accidents which took place in the year 2014 and have been reported by NTSB including those non-fatal [32]. All the 176 reports are among which occurred inside the USA, American-built helicopters involved, or both.

the exact rate of the visual flight during the accidents, since some of the record lacks the information of whether the flight was under instrument guidance or not, several statements indicating that the flight of accident took place during visual flight under low visibility. The danger of not using instrument can be seen in the new regulation of FAA [26] that requires all manned helicopters to install on-board instruments for reducing accidents.

8% of the accidents occurred during take-off and landing though using public airport. The major reason seems due to the blind spot of pilots; they tend to have crashed their rotors and tail booms in such occasion. The design of a safe VD on-board system should be one to decide the maneuver by processing information coming from multiple sensors simultaneously, which actual humans are not good at [27]. Additionally, the introduction of multicopters should reduce potential danger because of their smaller configuration compared to conventional single rotor helicopters due to absence of their long tail boom.

5.7% of the accidents occurred during taxing or parking inside an airport. Most of this type of accidents was caused by losing control during hover taxing, which could be solved by increasing stability of the craft while taxing. The primary requirement is to install wheeled landing gear. Craft arresting devices such as the "Beartrap" developed by the Royal Canadian Navy for flight deck landing assistance would possibly help improve safety. The accident on top of the Pan Am Building in 1977 shows that these kinds of safety must be kept tight if logistics providers want to keep make use of the rooftop of their facilities as helipads and operate heavy VDs[3].

Also, it should be noted that descent is as dangerous as taxing/parking and as twice as ascent. In the operation constructed, number of descent (and ascent, that necessarily should cause the same number of descent) should be decreased as possible.

It is quite discovering that accidents occurred while hovering figures only 1.1% of the total accident, and moreover, only one accident out of 176 took place as for external loading. External loading of cargo is one of the unique benefits of VTOL aircrafts, which will shorten the time of service, reduce the number of airports needed, and increase payload. Thanks to this low rate of accident, it would be possible to compose a service relying on this unique maneuver of VTOL aircrafts.

In addition to the lessons learned from the actual accident record caused by the current crafts and way of use, special device on ATC should be produced in urban areas. If the VD infrastructure grows and density of aircrafts increases, ATC system should act to avoid mid-air collision. Since the danger and fear will become the major barrier against the social implementation of new technology, this requirement presently hidden to VTOL aircrafts because of their contemporary minor use will be vital for the future VD system.

Constructing the needed ATC system for VDs

Based on the consideration on the three elements for profitable and continuous VD operation in the former section, a united theory that supports the composition of logistical VD operation system – the main fruit of this study – will be conducted in this section. Looking back into the former argument, six crude principles needed when deploying and operating VDs were derived from each of the three elements consisting the logistical value as follows.

- *a* Transport efficiency – superior constancy of the transport service that reduces the running cost for each logistics hub.
- *b* Transport efficiency – openness to various type of VTOL aircrafts evolving themselves daily.
- *c* Maneuverability – high versatility/less investment on each helipad and ground facility, or, mobility of the helipad itself.
- *d* Safety design – requirement for a regular route setting and proper ground support systems.
- *e* Safety design – the number of ascent/descent should be less as possible.
- *f* Safety design – requirement for avoiding collision in a high-density of VDs.

Based on the philosophy of systems theory that an ordered active configuration consists of interacting systems with each of their own purpose [1], the sustainable and affordable VD system argued here will be composed on the transaction of invisible functional systems to achieve the six principles above. Hence, as to construct the use of VD from the five principles regarding their systematical transaction, the relation which contradiction or reciprocal enhancement could occur have to be noted. By locating where trade-off relevance exists, the point of development to achieve all the six requirements could be specified.

3 On May 16th, 1977, a broken landing gear of the New York Airways Sikorsky S-61L on a passenger ferry flight caused a roll over on the heliport on the roof of the Pan Am Building (the Pan Am Metro Heliport), killing four passengers waiting to board by its rotor, and one pedestrian on the street below its by broken parts. This accident had terminated the use of the heliport for good. The airline went bankrupt 2 years later due to this accident [31].

How to express and recognize the relation among the principles

In order to fulfill the requirement above, transaction among elements have to be introduced. In other words, roughly finding the line in a six-dimensional space consisting from axis a to f is needed. In this study, interaction among set of requirements will be stated as the following protocol.

1. If requirement x were to be enhanced by enhancing y and vice versa, the relation of this set will be stated using the letter R for relation as:

$$R(x,y) \sim R(+,+) \tag{7}$$

2. If requirement x were to be enhanced only by diminishing y and vice versa, the relation of this set will be stated as:

$$R(x,y) \sim R(+,-) \tag{8}$$

3. A set resembles to its reversed composition, such as:

$$R(+,+) \sim -R(+,+) = R(-,-) \tag{9}$$
$$R(+,-) \sim -R(+,-) = R(-,+) \tag{10}$$

This law is adopted even when the number of elements is bigger than two, such as:

$$R(+,+,-) \sim R(-,-,+) \tag{11}$$

As for this expansion to three or more dimension, simply assume a line in a three-dimensional graph, e.g.

$$x = y = -z$$

, and track both ways increasing and decreasing the volume of x, and observe how y and z changes. Regarding this definition, formula (7) can be also stated as:

$$R(x,y) \sim R(+,+) \sim R(-,-) \tag{12}$$

, or also:

$$R(x,y) \sim R(+,+), R(-,-) \tag{13}$$

4. The state of an element y will be expressed when all the other element(s) told using character $R^y_{x,y}$. For instance, if formula (7) is given, it can be stated as:

$$R^y_{x,y}(+,y) = + \tag{14}$$
$$R^y_{x,y}(-,y) = - \tag{15}$$

5. Relation between three elements x, y, z can be stated if both the relation $R(x,y)$ and $R(x,z)$ are both known as follows:

$$R(x,y,z) \sim R\left(+, R^y_{x,y}(+,y), R^z_{x,z}(+,z)\right), \quad R\left(-, R^y_{x,y}(-,y), R^z_{x,z}(-,z)\right) \tag{16}$$

However, this composition is only true when the following relation exists:

$$R(y,z) \sim R\left(R^y_{x,y}(+,y), R^z_{x,z}(+,z)\right) \tag{17}$$

6. In reverse, this means that any relation regardless of the number of elements could be expressed by examining all the relations between the combination of available two elements.

This expression will help recognize the whole interaction among the six principles. Without this notation, it would be hard to express the whole system because each of them have several possible alternatives to realize and are necessarily not stated in a clear linear algebra manner nor by dimensional analysis. Hence however, all available combinations of two from a to f is required, because the total relation is not mathematically guaranteed, and aren't certain enough to be expanded to other relations without verification. If any confliction found, the elements at stake have to be re-examined and split into more simple ones.

Relation among the six principles and basic thought for mutual realization

The relations between combination of two from six principles from a to f will be listed as in Table 4. Though all the six are aimed for the long span prospect, they come controversial in the short run.

Table 4 Relations between two principles from *a* to *f*

Seq.	Relation	Note
1	$R(a,b)\sim R(+,-),R(-,+)$	Mixing aircrafts of various specs, i.e. velocity, has the risk of confusing traffic, cause inefficiency, and lose transport constancy.
2	$R(a,c)\sim R(+,-),R(-,+)$	Cheap and/or mobile helipads will limit transport constancy. However, this negative correlation is not so much.
3	$R(a,d)\sim R(+,+),R(-,-)$	Regular routing will help gaining transport constancy, and vice versa.
4	$R(a,e)\sim R(+,+),R(-,-)$	Aviation with least number of ascent/decent will make the operation simple and help gaining transport constancy.
5	$R(a,f)\sim R(+,+),R(-,-)$	Two requirements are related to each other, or moreover, almost the same.
6	$R(b,c)\sim R(+,+),R(-,-)$	Smaller investment on ground facilities will help introducing new type of aircrafts, avoiding path dependency.
7	$R(b,d)\sim R(+,-),R(-,+)$	Mixing aircrafts of various specs, i.e. velocity, could cost more for route setting and ground support systems.
8	$R(b,e)\sim R(+,-),R(-,+)$	In order to reduce the number of ascent/decent, the climb ability of the crafts have to be similar to each other, which relates to limitation of aircrafts to be used.
9	$R(b,f)\sim R(+,-),R(-,+)$	Excessively mixed types of requirement will affect against anti-collision measurement.
10	$R(c,d)\sim R(+,-),R(-,+)$	Fixed regular routing and expensive ground support facilities will contradict to maneuverability/mobility of helipads and landing facilities.
11	$R(c,e)\sim R(+,-),R(-,+)$	Disordered and complexed placement of helipads can, without proper device, force flight change altitude.
12	$R(c,f)\sim R(+,-),R(-,+)$	Disordered and complexed placement of helipads will lead to dangerous air traffic jam.
13	$R(d,e)\sim R(+,+),R(-,-)$	Regular routing and ground support system will help constructing simple flight path with minimum ascent/decent.
14	$R(d,f)\sim R(+,+),R(-,-)$	Regular routing will enhance safety against mid-air collision.
15	$R(e,f)\sim R(+,+),R(-,-)$	Less ascent/decent will help avoid mid-air collision in vertical direction.

From this result, the six fulfill the requirement to be completely transactional among each other, and will be stated as:

$$R(a,b,c,d,e,f)\sim R(+,-,-,+,+,+),\quad R(-,+,+,-,-,-) \tag{13}$$

This means two groups of different orientation of enhancement exists within this set of requirements, whose nature has already been reflected rightfully in previous plans for development. The *skyways* concept proposed by Airbus Group [8] for cargo drone operation to be tested in cooperation with the National University of Singapore will help involve principles a,d,e and f much, that matches the major group of formula (18), but will act against, or at least very weakly to b and c. This concept is to construct static landing points called *parcel stations* and aerial corridors linking them together for drone flight. In such system that resembles to visions of aerial cars from space opera films such as Star Wars [28] or The Fifth Element [29] has the advantage of achieving proper safety standard inexpensively, because flying static aerial corridor will keep them out from colliding with the terrain, and mid-air collision can be avoided just by securing proper distance with the craft ahead and aft one-dimensionally, just as automobiles running the highway. However, because such traffic system will have difficulty on adding or changing nodes and links of the network, and limit the total speed of traffic to that of the vehicle with the lowest top speed in the flow, which act against respectively principles b and c, this measurement won't be the most logistical.

Moreover, ambulance and other emergency use of VTOL aircrafts that need airspace and landing points nearest to the location of disaster will contradict this kind of restrictive operation, because the emergency vehicle cannot cross or merge into the aerial corridor, or otherwise all the vehicles has to be equipped with high velocity anti-mid-air-collision devices only for emergency that will waste precious onboard electricity and payload in vain for all the regular operation. Also in crowded traffic, open space off the corridor becomes inefficient. Rather than tracing the road between buildings, flying freely above the urban skyline should enhance the utility of drones.

Then, what means for a less restrictive ATC can be taken? The worst will be to rely on human controllers like as conventional aviation: as already has been pointed, even the complicated task of current air traffic controllers to make decision and direct human pilots has come near to the limit of human ability [3]. The ability of VDs capable of hovering and flying at low speed without changing altitude could increase the density of crafts to an incomparably high rate, that should go easily beyond the process capability of human brain. If it is to increase the number of air traffic controllers to contemporary crowded areas, it will act negatively to principle *c*. Above all else, the design of human giving instructions to computer is not only inefficient, but moreover, becomes needlessly inaccurate intricated that covers and hides danger it contains. Hence, all the decision on the newly constructed high-density airborne traffic configuration of drone aviation has to be made purely by computers.

Because of this incompatibility of human and machine guidance, this high-density VD traffic (HDVDT) should be divided from each other. Practically, the altitude of flight should be strictly separated. Also, air zones around airport where the human controlled layer comes down to the ground should be free from machine controlled traffic to avoid collision with arriving and departing airliners (see Figure 3).

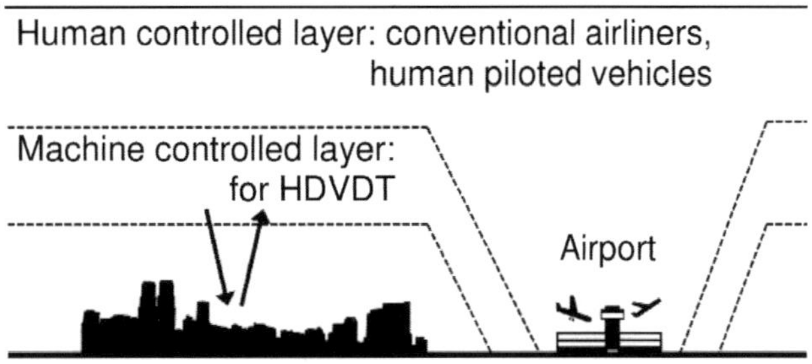

Figure 3 Distribution of each human and machine controlled layers

OPVs or remote controlled drones also have to be also guided automatically when joining HDVDT while at cruise. This does not necessarily deny the use of human guidance, because piloting the vehicle could at some occasion have advantage, such as landing and take-off, or flight outside the highly-crowded area. It is assumable that especially at the early phase of the development where autonomous guidance is not matured yet, the complicated and dangerous maneuver near the ground should be entrusted to pilots' hands and machine-learned. Moreover, because of the divided distribution between human and machine controlled as mentioned above, VDs entering air zones near the airport should be guided somehow by human pilots' intervention[4].

Regarding ATC free from human intervention, computerized air control on conventional aviation previously studied have mainly focused on the intercommunication among aircrafts based on the idea of multi agents [2] [3] [4]. Since they consider mainly conventional fast fixed wing aircrafts with relatively low rate of density, these concepts are effective only in traffic coming cross with limited number (usually only one in conventional aviation) of other aircrafts at once. However, in the HDVDT, number of craft would be unlimited because of the ability of VDs to stop or change speed and direction instantly in crowded traffic; their path will be like that of pedestrians walking through the Shibuya Crossing. Setting excessive number of course confirmation towards other crafts to onboard computers and transponders will not only become expensive but also place load on onboard computers and increase determinate electric consumption on the craft that restrict the potential of each VDs. Even if it were to centrically control by ground computers, data transmission between the control server and multiple aircrafts will crunch the bandwidth of the datalink, declining safety level or decelerating the traffic.

Cellular ATC system (CATCS)

In order to fulfill the requirement for new guiding method for HDVDT, a new method of flight control has to be established. To make things simple however efficient, the author has focused on a different field of transport, trains.

Trains which only move linearly (or, in a one-dimensional manner) are controlled by signals given to separated track sections with electrical existence detection called blocks in relation with the other train traveling just ahead of one: red light shows the next block is occupied by another train suggesting to stop; yellow indicates train on near and orders caution and slow speed; green means the track is free of other trains for a distance and allow maximum speed to enter the next section [5]. When the *Shinkansen* was developed in Japan as a high-speed track standard, technicians developed a new way of exclusive control to alternate conventional signal in fear of drivers missing the signal in high velocity and colliding into the train ahead. The newly introduced system was to transfer the signaling information to onboard devices and apply brakes automatically and designedly without

[4] In this context, the meaning of the term "pilot" will be related to that of maritime which indicates ship handler who navigates ships specially inside harbors, rather than that of aviation.

overrun [5] [5]. The point of this system for the *Shinkansen* was that location of trains are recognized by pre-divided blocks free from observation error, and trains can be controlled safely however simply in relation with just the occupation status of the blocks directly given to the onboard device.

Adapting and adjusting this idea into a two-dimensional (not three in order to match principle e) manner, the needed concept of ATC for HDVDT matching all the six requirements from a to f will be constructed. In this concept, the linear blocks on railroad track will become cells on maps, that gives itself provisional name in this study as, *cellular ATC system* (CATCS). The configuration of CATCS will be stated as follows:

1. Air space will be divided into cells of several-hundred-meters to a kilometer square in the virtual chart[6].
2. Each VD will be equipped with GPS locator, broad-band communication, onboard navigation computer, and an anti-mid-air-collision system to be used in low velocity.
3. The status of each cell reflects its occupation and speed limitation subjectively to a VD as:
    a. Green: The cell is reserved exclusively for the vehicle. Entering the cell in maximum velocity is allowed. The status is shown as red to other VDs.
    b. Yellow: Several crafts are allowed to enter the cell. Speed inside this cell is restricted to a proper level that VDs can stop suddenly for collision avoidance. When the cell is occupied with maximum number of crafts allowed, status turns red for other VDs.
    c. Red: The cell is occupied or reserved exclusively by other craft(s). Entering the cell is prohibited.
    d. White: No status registered. The cell can be reserved into green or yellow for entering.
    e. Grey: HDVDT prohibited zone, e.g. high mountains, airports, military and government facilities, or nuclear power stations. The cell cannot be entered nor reserved.
4. The existence area of the craft on the chart will be ranged in about a hundred-meters diameter from the center of the airframe. Vehicles with this existence area to enter whether entirely or partially will be regarded as "inside" the cell (Figure 4).

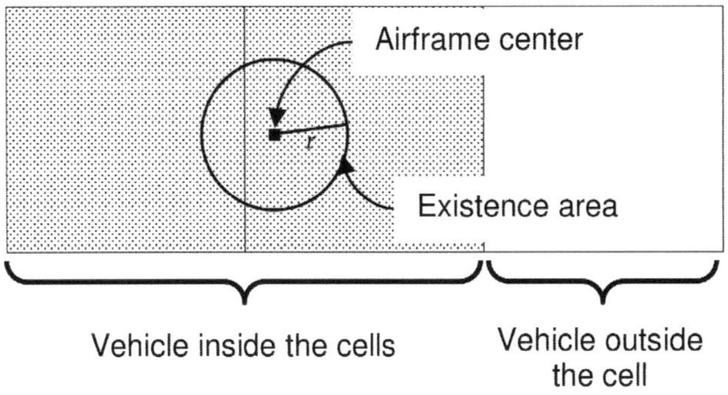

Figure 4 Definition of the location of the vehicle and occupation of cells

5. If the cell is white, a VD can reserve maximum 9 cells, up to next 3 sections in one direction, based on its flight plan (Figure 5). As the vehicle moves to its next cell(s), the next cell(s) becoming the third section ahead can be newly reserved if they are white.

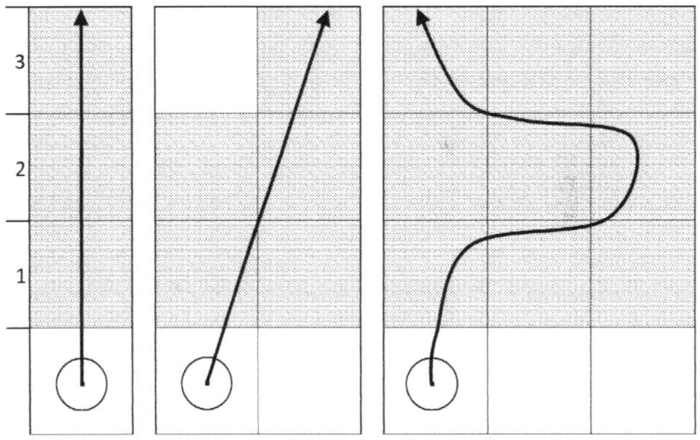

Figure 5 Position of cells allowed to reserve as green when the current status is white (filled with dots)

6. White cells neighboring the cell to be passed by one VD exclusively in maximum speed (green cell for the vehicle) will turn yellow for other VDs. The definition of "neighborhood" is adopted from

[5] This system for *Shinkansen* is called ATC (automatic train control), however, it is not to be introduced in the text in order to avoid confusion with the abbreviation of air traffic control.

[6] Though the shape of each cells could also be another homotypic tessellation, such as isosceles right triangle, equilateral triangle, or equilateral hexagon instead, the shape of square has the advantage of being able to be simulated easily by using existing spreadsheet program such as Microsoft Excel.

Moore's in his study on cellular automata [30] (Figure 6, 7).

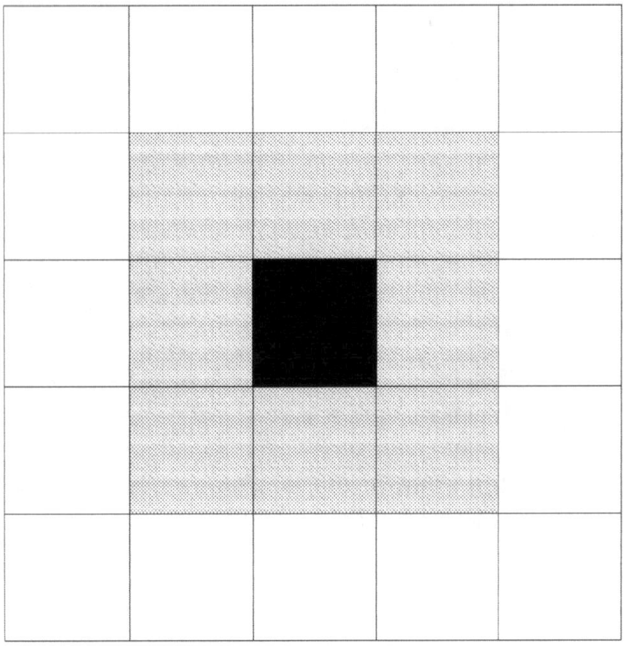

Figure 6 Cells filled with dots defined as neighboring the black cell

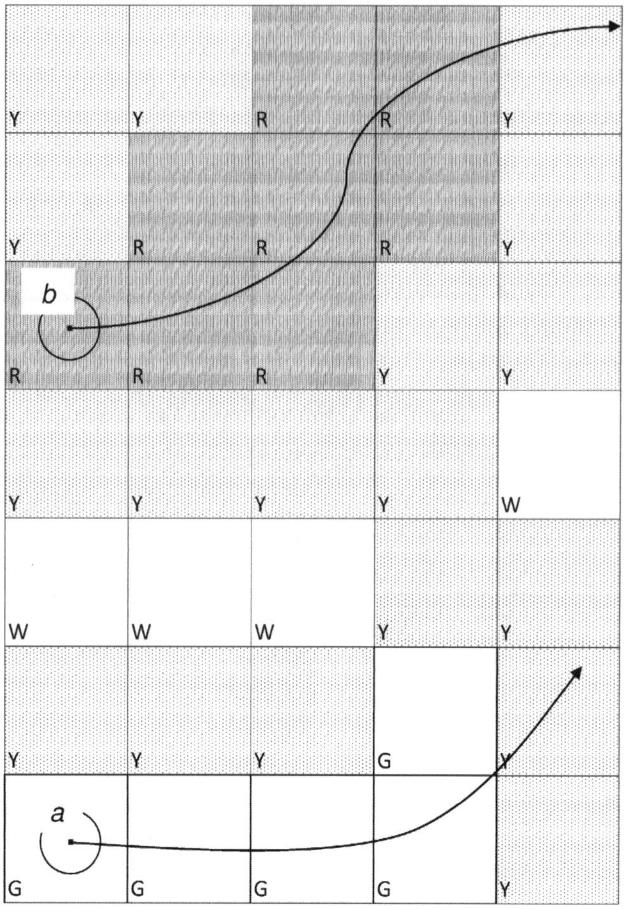

Abbreviation on each cell
G: green, Y: yellow, R: red, W: white

Figure 7 Cell status shown to vehicle *a* when course of another vehicle *b* is plotted simultaneously

In a two-dimensional manner, relation among multiple vehicles have to be considered, while regarding only the single relation between the train ahead is required on signaling block. This causes the difference that CATCS have to show the status of the cells to each craft individually (for example, the cell status in which vehicle *b* is currently located is shown as green to *b*, but red to *a*), while railway signals show objective status, which becomes a little more complicated in the system. However, by following this regulation, each VDs can simply adjust speed and course obeying the indication given as the status of each cell while following essentially the initial flight plan without multi-level crossing. This will firstly match principles d, e and f, and also b because it can contain various type of aircraft without confusing the total system.

In this CATCS, the only function needed for ground navigation computers is to receive GPS location from each vehicle, reflect them on the virtual map, and send the information of the map to each vehicle. Developing and operating such computer to handle location information won't be difficult or expensive because they have already been realized in presenting smart phone games, such as *Ingress* or *Pokémon-go*. The ability of enhancing speed planning and course plot on each onboard computer will help the craft travel more efficiently, the improvement will be up to each carriers, keeping the basic cost for running the infrastructure.

Also, creating new vertiports on CATCS will be very easy, because all needed is just to register the status of the cells on and near the location of the new vertiport to yellow parmanently to limit the airspeed to avoid collision of arriving and departing flights, at the same time allowing multiple aircrafts to enter the zone. The phisical structure of vertiports will be independent from ATC, making able to construct the most suitable ground facility in each location. Not only to match principle c, this maneuverability of landing area will enhance the social importance, because it can provide safe air space for observation, search, rescue, ambulance or emergency logistics both for CATCS aircrafts and human piloted vehicles instantly by changing the cell status to yellow or red in occasion of disaster, such as fire, earthquake, tsunami, or terorism.

Even a cell on a basic route becomes unavailable for disasters such as referred above or because of bad weather, VDs can find alternative bypass automatically simply by referring status of the cells. This will help maintain the constancy and business continuity even in vast possible situations, securing principle a.

Verified as above, CATCS can realize all six principles from a to f, which will make the services to join truly logistical. Though the argument was constructed from the perspective of freight carrying, the method of ATC seems to be suitable also for passenger crafts, possessing the potential to become the new standard for all on-demand flights.

Conclusion

This study has firstly deduced three elements needed for constructing a logistical VD service regarding their general nature and current alternative means. These elements were derived into six more tangible principles needed to be realized at the same time. However, it was shown in a mathematical manner that these principles are hard to conciliate especially by conventional ATC system or existing future concepts. Alternatively, a new configuration called CATCS was conceived referring to the control system for railways, and have been verified to have the ability of realizing all the six principles simultaneously.

Limitation and further research

In this study, how operational speed of VDs under current ATC system, existing service plans, and the newly introduced CATCS haven't been simulated systematically. Also, CATCS resembling to cellular automata, how the status of each cell changes in relation with other cells would have and have to be studied in a mathematically in further research. Such research will also help finding where to construct vertiports for overall optimization of the whole system.

Acknowledgments

My deepest appreciation goes to Mr. Tanaka, Mr. Tanokuchi, and all the other colleagues in IT planning division, Yamato Holdings who have personally provided helpful comments and suggestions. I would also like to thank Mr. Makiura, Yamato Holdings who have privately encouraged me to write down all this paper. Without their kind support, it would have been impossible to finish this study. Finally, I would like to give all my gratitude to my wife, Nanami, who have given me both mental and academic support throughout this study. The responsibility of any errors is of course mine.

Reference

[1] L. v. Bertalanffy, General System Theory:, New York: George Braziller, 1968.

[2] G. J. Pappas, C. Tomlin and S. Sastry, "Conflict resolution for multi-agent hybrid system," in *35th Conference on Decision and Control*, Kobe, 1996.

[3] C. Tomlin, G. J. Pappas and S. Sastry, "Conflict resolution for air traffic management: a study in multiagent hybrid systems," *IEEE Transactions on Automatic Control,* vol. 43, no. 4, pp. 509-521, 1998.

[4] Z. Wei and J. Li, "Design of air traffic control operation system using multi agent technology and simulation," *The Open Fuels & Energy Science Journal,* vol. 8, pp. 368-374, 2015.

[5] T. Takashige, "Signalling systems for safe railway transport," *Japan Railway & Transport Review,* vol. 21, pp. 44-50, 1999.

[6] L. Wittgenstein, Tractatus Logico-philosophicus, London: Kegan Paul, Trench, Trubner & Co., Ltd., 1922.

[7] Deutsche Post DHL Group, "DHL Parcelcopter 3.0," 9 May 2016. [Online]. [Accessed 21 August 2016].

[8] Airbus Group, "A new perspective on urban mobility," *Forum,* vol. 88, pp. 6-13, 2016.

[9] Uber, "Fast-forwarding to a future of on-demand urban air transport," 27 October 2016. [Online]. Available: https://www.uber.com/elevate.pdf. [Accessed 6 December 2016].

[10] S. Pozzebon, "This concept of a drone ambulance promises to revolutionise emergency operations," Business Insider, 16 February 2015. [Online]. Available: http://uk.businessinsider.com/drone-ambulance-concept-with-amazing-images-2015-2. [Accessed 8 January 2017].

[11] Defense Avdanced Research Projects Agency, "ARES Aims to Provide More Front-line Units with Mission-tailored VTOL Capabilities," Defense Avdanced Research Projects Agency, 11 Feburary 2014. [Online]. Available: http://www.darpa.mil/news-events/2014-02-11. [Accessed 13 November 2016].

[12] TomTom International BV., "TomTom Traffic Index - Measuring congestion worldwide," TomTom International BV., 2016. [Online]. Available: https://www.tomtom.com/en_gb/trafficindex/list. [Accessed 18 November 2016].

[13] R. Oliner, "The RoboNeko Yamato service will use self-driving cars to bring purchases to customers' homes starting in 2017," 28 July 2016. [Online]. Available: http://www.psfk.com/2016/07/driverless-delivery-service-will-launch-in-japan.html. [Accessed 14 August 2016].

[14] D. Lamothe, " Kaman aerospace has modified its K-1200 K-MAX "aerial truck" into a remote control OPV for military and civilian hazardous use.," The Washington Post, 25 May 2014. [Online]. Available: https://www.washingtonpost.com/news/checkpoint/wp/2014/07/25/robotic-helicopter-completes-afghanistan-mission-back-in-u-s/. [Accessed 1 November 2016].

[15] Y. Ito, "D-Day: When Logistics Came into Modern Sense," in *Japan Marketing Academy Conference 2014*, Tokyo, 2014.

[16] A. H. Jomini, Précis de l'art de la guerre ou nouveau tableau analytique, Paris: Anselin, 1838.

[17] M. Christopher, Logistics and Supply Chain Manegement (fourth edition), London: Pearson, 2013.

[18] A. Dangerfield, "Mail Rail: What is it like on the 'secret' Tube?," BBC News, 28 January 2014. [Online]. Available: http://www.bbc.com/news/uk-england-london-25145632. [Accessed 30 1 2016].

[19] E. W. Weisstein, "Boundary," MathWorld--A Wolfram Web Resource, [Online]. Available: http://mathworld.wolfram.com/Boundary.html. [Accessed 19 January 2016].

[20] M. Bowman, Boeing 747: A History: Delivering the Dream, Barnsley: Pen and Sword, 2014.

[21] C. R. Anderson and C. P. Zeithaml, "Stage of the Product Life Cycle, Business Strategy, and Business Performance," *Academy of Management Journal,* vol. 27, no. 1, pp. 5-24, 1984.

[22] J. G. Grimes, "(DoD) Net-centric services strategy," Department of Defense, Washington D.C., 2007.

[23] J. Watkinson, Art of the Helicopter, Oxford: Butterworth-Heinemann, 2013.

[24] S. Sparks, "Workhorse CEO Burns explains how drones will change delivery trucks," Tucks.com, 5 May 2016. [Online]. Available: https://www.trucks.com/2016/05/05/workhorse-ceo-explains-drones-will-changedelivery-trucks/. [Accessed 8 November 2016].

[25] J. Orasanu and L. Martin, "Errors in aviation decision making: a factor in accidents and incidents," *Proceedings of 2nd Workshop on Human Error, Safety, and System Development,* pp. 100-107, 1998.

[26] Federal Aviation Administration, Department of Transportation, "RIN 2120–AJ53 Helicopter Air Ambulance, Commercial Helicopter, and Part 91 Helicopter Operations," Federal Aviation Administration, DOT, Washington D.C., 2014.

[27] T. Dubois, "Pilot eye tracking reveals poor visual scanning," Vertical Magazine, 7 December 2016. [Online]. Available: https://verticalmag.com/news/pilot-eye-tracking-reveals-poor-visual-scanning/. [Accessed 11 December 2016].

[28] G. Lucas, Director, *Star Wars Episode I: Phantom Menace.* [Film]. USA: 20th Century Fox, 1999.

[29] L. Besson, Director, *Le Cinquième élément.* [Film]. France: Gaumont, 1997.

[30] E. F. Moore, "Machine models of self-reproduction," *Proceedings of Symposia in Applied Mathematics,* vol. 14, pp. 17-33, 1962.

[31] J. R. Chiles, The God Machine - From boomerangs to Black Hawks: The story of the helicopter, New York: Bantam, 2007.

[32] National Transportation Safety Board, "Aviation Accident Database & Synopses," National Transportation Safety Board, 2014-2015. [Online]. Available: http://www.ntsb.gov/_layouts/ntsb.aviation/Index.aspx. [Accessed 18 December 2015].

[33] Y. Ito, "Insights for logistical realization of unmanned vertical take-off and landing aircrafts (UVTOLAs)," in *Japan Marketing Conference 2016*, Tokyo, 2016.

Automatic Landing on a Moving Platform Using Deep Neural Network Estimation

Toshinobu Watanabe
twatanabe8@gatech.edu
Ph.D. Student
Georgia Institute of Technology
Atlanta, Georgia, USA

Eric N. Johnson
eric.johnson@ae.gatech.edu
Associate Professor
Georgia Institute of Technology
Atlanta, Georgia, USA

This paper describes a control architecture designed to enable autonomous landing on a moving vehicle. This architecture can be separated into two parts: prediction and trajectory generation. For the prediction method, we use a Deep Neural Network (DNN), which can estimate multiple future states of the landing target by using past measurement data. For trajectory generation, we employ differential dynamic programming (DDP), which involves the interior penalty function with time-varying weight in the cost function. As a result, we can predict the state for a future time by using the multi-point prediction technique using DNN and generate the trajectory with time-varying weight using DPP. We provide the architecture of DNN and DDP with implementation techniques and sample results.

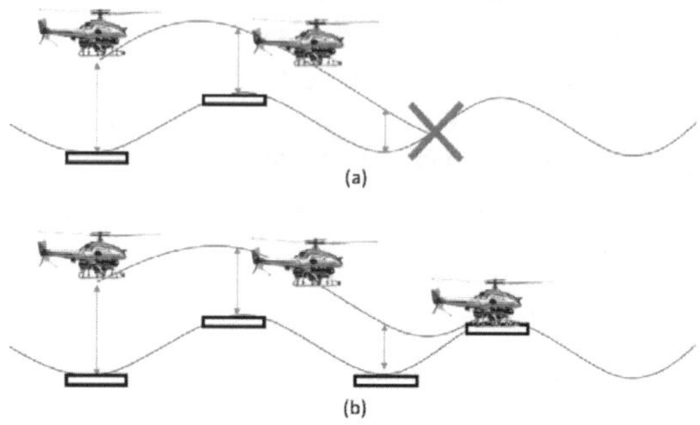

Figure 1. The flight path to control the relative distance on the moving target. (top) (a) is a case of collision from the reason why the target moves faster than the UAVs. (bottom) (b) is a case of the inefficient energy consumption from the wavy trajectory.

I. Introduction

Autonomous landing is often a challenging task for unmanned aerial vehicles (UAVs) operation. A normal landing of the UAVs assumes a fixed target. An even more challenging goal is landing on a moving target.[1] The movement of the target landing location is sometimes assumed to be constrained to a two-dimensional plane and does not consider the heave movement and attitude change. Including altitude changes for the target is required when landing on a ship on a rough sea. In this situation, the target moves in six degrees of freedom including an attitude change. To achieve this landing task, however, the conventional approach for controlling the relative distance has two problems. One is: if the speed of the target on a rough sea is faster than the UAV, then the UAV may collide with the target prior to the intended landing (Figure 1-a). The other is, since the path of the vehicle is not smooth, the energy efficiency is reduced (Figure 1-b). To solve these problems, this paper proposes an architecture that involves the estimation of the future states and trajectory generation based upon this prediction.

There are many research studies to deal with state prediction. If we already know the dynamics of the target for state prediction, the Kalman filter is an effective method. We, however, usually do not know the dynamics of the target in detail. We have two approaches to obtain the future state. One uses the free dynamics model for the Kalman filter, and the other uses a neural network. A conventional neural network cannot estimate future states about the complicated model. We, however, can consider the Deep Neural Network (DNN) to solve this problem. The DNN shows impressive outcomes in the several competitions in terms of the Image recognition, Speech recognition, and Natural language processing.[18] Currently, the DNN has the potential to spread to apply finance, control and so on. In particular, to solve the control problem, the model of DNN for the future state estimation is generally the recurrent neural network (RNN).[5] However, for the longer-term prediction, the usage of this method tend to increase errors from the accumulation of errors.

A trajectory generation is one of the most important topics in Robotics. When the operator of UAVs generates a trajectory, he or she sets several waypoints and an endpoint. The algorithm generates the trajectory based on the environment including obstacles. To generate the trajectory, a traditional approach includes a curve interpolation method using a combination of linear or arcuate line, Bezier Curves, and Spline Curves.[6] These curve interpolation methods generate a path using a start point, several waypoints, and an endpoint. These techniques, however, have some problems. For instance, since the spline curves draw the N-order interpolation smooth curve, the curve has a discontinuity of curvature. When the vehicle moves on this curve, the path leads to a wasteful energy consumption. Furthermore, since these methodologies only output a set of positions, they do not consider a dynamics of the vehicle. This sometimes leads to create an infeasible maneuver.

This paper proposes a novel method of future state prediction and trajectory generation to achieve landing on a moving target. With regard to the state prediction method, we employ the deep hierarchical model, which is one of the DNN models. In this model, a set of past states is an input and a set of future states is an output. For the trajectory generation method, we adopt differential dynamic programming (DDP) [7], which can draw the path to optimize the cost and consider the vehicle dynamics.

Before elaborating on the method, we describe our strategy in section II. Section III and IV will describe the detailed contents of the DNN and DDP. Section V presents simulation results to demonstrate and validate the approach. Finally, Section VI will conclude the paper.

II. Strategy

Figure 2 shows a comparison of the conventional procedure, which controls relative distance, and the predictive landing procedure. This paper assumes the vehicle can estimate the target pose with noise. We replace the "Control relative position" into two parts: "Predict Future states" and "Generate Trajectory". The reason is that we need to predict future positions of the target to generate the trajectory to make a smooth flight path as well as avoid collisions. In addition, in order to avoid the collision with the target, the landing target must be treated as the obstacle any time before landing time. Therefore, we need to predict several points during the landing phase. In our neural network for prediction, an input is current and set of past states and an output is future states. Figure 3 shows this structure for a simultaneous prediction of the multiple points, which is known Multi-point prediction. Additionally, to solve the problem we mentioned before, this paper proposes the implementation of interior penalty function with the time-varying weight. By the effect of the time-varying weight, the DDP including the new weight generates the trajectory to avoid the target treated as an obstacle before the landing time and to land on the target at the landing time.

III. Prediction using DNN

This section describes some of the techniques to use the DNN for prediction of the state. We introduce an activation function, which we employ. Next, when the DNN estimates the future state, it is efficient technique to implement a normalization. Although normalization is a general technique in the neural network, since units like distance and angle, which is used in control, are different, normalization is necessary in order to maintain the uniform accuracy. After the introduction of a structure of the network, we explain the training method.

A. Leaky ReLU

The ReLU (Rectified Linear Unit), which is one of the kinds of activation function in a neural network, output same value with input when x>=0 and zero when x<0. From this simplicity, the calculation of output and gradient is fast. In addition, from that characteristic, since much of values in weight matrix are zero, the network become the sparse neural network. Furthermore, if the network has many layers, an error is propagated without a vanish of the gradient. From them, this activation function is effective to the DNN. As the improvement of ReLU, leaky rectified linear unit (Leaky ReLU) is first introduced in the acoustic model.[9] Leaky ReLUs are one attempt to fix the "dying ReLU" problem. Instead of the function being zero when x < 0, a leaky ReLU will instead have a small negative slope (of 0.01, or so). The function is shown below.

$$f(x) = \begin{cases} x & if\ x > 0 \\ \alpha x & otherwise \end{cases}, \tag{1}$$

where α is a small constant value. In the solution of the state space form of control systems, the exponential function is mostly involved. Similarly, the sine function is used for the computation of change of the frame. In our experience using the three-layer model shown in part III-C, the leaky ReLU shows the better result to estimate their function.

B. Normalization

In neural network systems, normalization is known as an effective method to enhance a generalization capacity and reduce an error. Furthermore, since control systems involve various kinds of units, a variation in the relative maximum value occurs. This variation leads a reduction of precision to influence values of each other. Therefore, we should

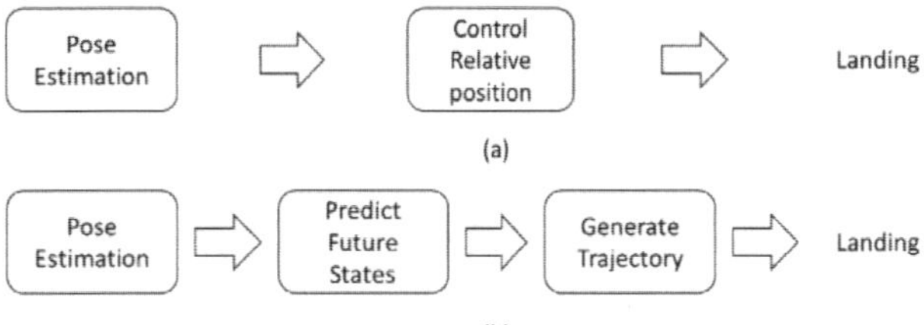

Figure 2. Comparison between the conventional approach and the proposed approach. Figure (a) is the conventional landing procedure and Figure (b) is predictive landing procedure.

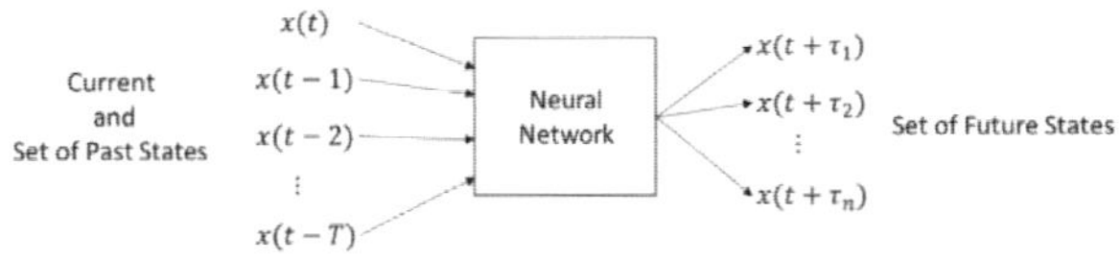

Figure 3. Input and output of Multi-point prediction. The left-hand side values are input values. The right-hand side values are output values from the neural network. τ_n indicates the landing time.

normalize each state before values are inputted to the neural network. The equation for normalization is as follows:

$$\hat{x}_{i,j} = \frac{x_{i,j} - \bar{x}_i}{\left|\max_j x_{i,j}\right|} \quad (2)$$

where $x_{i,j}$ is state value, $\bar{x}_i$ is mean value and $\hat{x}_{i,j}$ is normalized state value. i indicates i-th state and j indicates state at j-th time step.

C. Model Structure

Figure 4 shows the neural network model, which we use. The network model consists of three layers of neurons. The numbers of a neuron of each layer are 64, 32, and 16 respectively. Leaky ReLU is adopted as the activation function in section III-A. The numbers of neurons in the input and output layers depend on the source dataset and the number of prediction points. To realize multi-point prediction described in Section II, the consecutive historical dataset is treated as input and the states of all prediction time points are treated as the output. We show this structure in Figure 5.

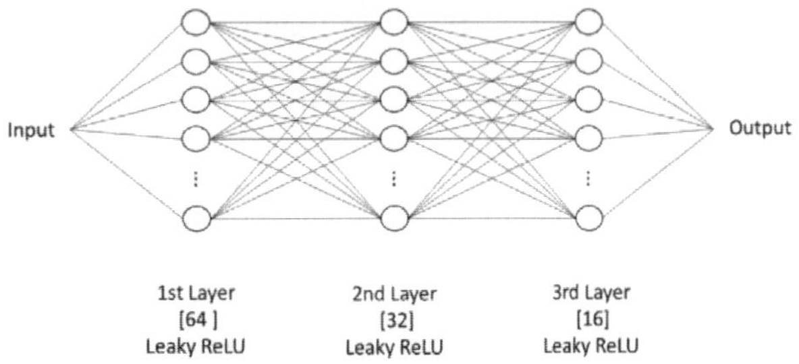

Figure 4. Neural network model. The circles indicate neurons and the lines indicate the connection between neurons. The numbers inside brackets indicate the number of neurons and the last row indicates the activation function.

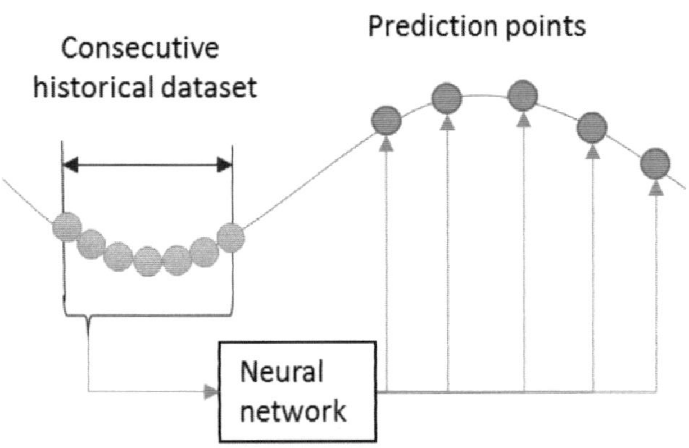

Figure 5. The image of the multipoint prediction. The neural network predicts the set of future states simultaneously from the consecutive historical dataset through.

D. Training

The neural network primarily computes a gradient from all training cases. However, since this is quite difficult in particular for control problems, we use the stochastic gradient descent method (SGD).[10] In SGD, we employ the Adam optimization algorithm[11] with a learning rate of 0.001 to train the neural network model, which is shown in Part C. Additionally, if the training cases are limited, since the estimation of the gradient may become unstable, the time to converge may tend to be longer. To solve this problem, we employ the mini-batch method, which computes the gradient from several dozen or hundred cases.

IV. Trajectory Generation by Using DDP

In this section, we provide an implementation of the DDP[3], which has an interior penalty function.

A. DDP equations

In this section, we provide an implementation of the DDP, which has an interior penalty function in a cost function, to solve a constrained optimization problem.[1]

We consider the cost under minimization. The cost function is

$$\begin{aligned} &V(x(t_0), t_0) \\ &= \min_u \left[\phi(x(t_f), t_f) + \int_{t_0}^{t_f} \mathcal{L}(x, u, t)dt\right], \end{aligned} \quad (3)$$

subject to the dynamics

$$\frac{dx}{dt} = f(x, u, t), x_0 = x(t_0). \quad (4)$$

We can rewrite this equation in discrete time.

$$V(x(t_k), t_k) = \min_u [l(x(i), u(i), i) + V(x(i), i)]. \quad (5)$$

We introduce the term $Q(x, u)$, which is known as the state-action value function.

$$\begin{aligned} &Q(x(t_k), u(t_k), t_k) \\ &\quad = l(x(t_k), u(t_k), t_k) + V(x(t_k), t_k). \end{aligned} \quad (6)$$

We can expand a Q-function to the second order in terms of δx.

$$\begin{aligned} &Q(x(t_k) + \delta x_{t_k}, u(t_k) + \delta u_{t_k}, t_k) \\ &= Q(t_k) + Q_x(t_k)\delta x_{t_k} + Q_u(t_k)\delta u_{t_k} \\ &+ \frac{1}{2}\begin{bmatrix}\delta x_{t_k}{}^T & \delta u_{t_k}{}^T\end{bmatrix}\begin{bmatrix}Q_{xx}(t_k) & Q_{xu}(t_k) \\ Q_{ux}(t_k) & Q_{uu}(t_k)\end{bmatrix}\begin{bmatrix}\delta x_{t_k} \\ \delta u_{t_k}\end{bmatrix}. \end{aligned} \quad (7)$$

Here, the second order local model of the Q-function can be propagated backward in time using these functions.

$$
\begin{aligned}
Q_x(t_k) &= V_x(t_{k+1})F_x + l_x(t_k) \\
Q_u(t_k) &= V_x(t_{k+1})F_u + l_u(i) \\
Q_{xx}(t_k) &= F_u V_x(t_{k+1})F_x + V_x(t_{k+1})F_{xx} \\
&\quad + l_{xx}(t_k) \\
Q_{xu}(t_k) &= F_x V_x(t_{k+1})F_u + V_x(t_{k+1})F_{xu} \\
&\quad + l_{xu}(t_k) \\
Q_{uu}(t_k) &= F_u V_x(t_{k+1})F_u + V_x(t_{k+1})F_{uu} \\
&\quad + l_{uu}(t_k),
\end{aligned} \tag{8}
$$

where $x(t_{k+1}) = F(x(t_k), u(t_k))$ is a model of the dynamics.

The optimal δu_{t_k} is derived from equations above.

$$\delta u_{t_k} = -Q_{uu}^{-1}(t_k)\left[Q_{ux}(t_k)\delta x_{t_k} + Q_u(t_k)\right]. \tag{9}$$

After updating the control output $u(t_k)$ from this equation, the second order local model of the value function is given as

$$
\begin{aligned}
V(t_k) &= V(t_{k+1}) - Q_u(t_k)Q_{uu}^{-1}(t_k)Q_u(t_k) \\
V_x(t_k) &= Q_x(t_k) - Q_u(t_k)Q_{uu}^{-1}(t_k)Q_{ux}(t_k) \\
V_{xx}(t_k) &= Q_{xx}(t_k) - Q_{xu}(t_k)Q_{uu}^{-1}(t_k)Q_{ux}(t_k).
\end{aligned} \tag{10}
$$

B. Cost Function

One of the approaches for an avoidance of obstacles is to add a penalty function to the cost function and enforce appropriate nonlinear constraints. We apply this concept to the cost function of DDP. The additional function is selected from an interior penalty function and an exterior penalty function. If a vehicle is in the feasible region, we can adopt an interior penalty function called "barrier function." If the vehicle is inside the infeasible region, we use an exterior penalty function. Since the vehicle state must be in the feasible region when landing on the ship, we employ the interior penalty function, here.

A barrier function $B(x)$ is a function that is continuous and nonnegative over the interior of $\{x \mid g(x) \leq 0\}$, i.e., over the set $\{x \mid g(x) < 0\}$ and approaches infinity as the boundary is approached from the interior.

$$B\big(x(t_k)\big) = -\sum_{j=1}^{m} \frac{1}{g_j(x(t_k))} \tag{11}$$

$$g_j\big(x(t_k)\big) = \hat{x}(t_k) - x(t_k), \tag{12}$$

where m is a number of constrained state and $\hat{x}$ is an estimated state.

We add this interior penalty function into equation (6).

$$
\begin{aligned}
&l'(x(t_k),u(t_k),t_k) \\
&= l(x(t_k), u(t_k), t_k) + \mu B\big(x(t_k)\big),
\end{aligned} \tag{13}
$$

where μ is a weight of the Barrier function.

In the landing mission, it is necessary to match the vehicle states and the target states involving position and attitude at the landing time. In addition, the target itself is treated as an obstacle before landing time. To satisfy them, the value of penalty function is required to be zero at landing time and a certain value before the landing time. However, since this original penalty function has a certain value at all time steps, the vehicle cannot reach the target at landing time. Therefore, this paper proposes to transform the time-invariant weight (μ) to linear time-varying weight ($\mu(t_k)$). The modified value of weight is shown below.

$$\mu(t_k) = k\big(t_f - t_k\big), \tag{14}$$

where k is a constant value.

Figure 6 shows the graphical image of the cost using the equation (14). The figure shows that the cost is increased with the reduction of relative distance and is decreased linearly with the reduction of remaining time.

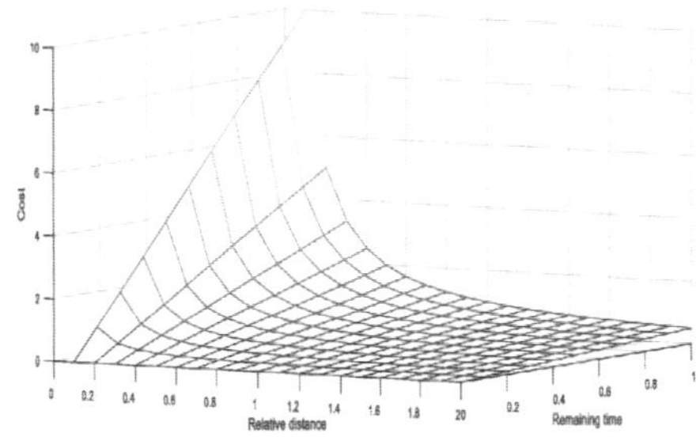

Figure 6. Cost change with respect to the relative distance and time-varying weight.

V. EXPERIMENT

Since this proposed architecture can run as the open loop, this procedure can be divided into the state prediction and the trajectory generation parts. Therefore, we evaluate them separately.

A. State Prediction

The DNN predicts the states. To implement the DNN with our algorithm, we adopt the Chainer[4], which is a software framework for deep neural networks. We show the target model and simulation result.

1) Target Model

The real ship movement has six degrees of freedom (DOF) and the ship has a complex movement by external forces, which mainly arise from sea waves. In addition, although the number of frequency of sea wave is infinity, to simplify the problem, we assume that the ship movement consists of three sine waves with different frequencies and phases, and predict only the three dominant axes: x, z, and pitch axes. Here, the x-axis is a direction of travel of the ship with velocity v. The Z-axis is the vertical direction. The pitch axis is an axis of nose up or down. The results of the formulation are shown below.

$$x(t_k) = \sum_{j=1}^{3} A_{1,j} \sin\big(\varpi_{1,j} t_k + \varphi_{1,j}\big) + vt + w \tag{15-1}$$

$$\dot{x}(t_k) = \sum_{j=1}^{3} A_{1,j}\varpi_{1,j}\cos(\varpi_{1,j}t_k + \varphi_{1,j}) + v + w$$

$$z(t_k) = \sum_{j=1}^{3} A_{2,j}\sin(\varpi_{2,j}t_k + \varphi_{2,j}) + w$$

$$\dot{z}(t_k) = \sum_{j=1}^{3} A_{2,j}\varpi_{i,j}\cos(\varpi_{2,j}t_k + \varphi_{2,j}) + w, \quad (15\text{-}2)$$

$$\theta(t_k) = \sum_{j=1}^{3} A_{3,j}\sin(\varpi_{3,j}t_k + \varphi_{3,j}) + w$$

$$\dot{\theta}(t_k) = \sum_{j=1}^{3} A_{3,j}\varpi_{3,j}\sin(\varpi_{3,j}t_k + \varphi_{3,j}) + w,$$

where index i is axis number and index j is wave index. A is a magnitude of wave and φ is phase angle of the wave. w is a noise.

2) Simulation Test

The period of the learning data set is 100 seconds. A neural network predicts the future states at 1.5, 3, 4.5, and 6 seconds ahead from a historical dataset of the past 10 seconds. The result of the prediction of the target model is shown in Figure 7. This figure compares the target movement with noise with the result of the prediction 6 seconds ahead. This result indicates that, although there are slight errors in the peak, it can be seen a good prediction at all times. About the prediction of the position in X-axis, the speed is predicted well, but ripples have been difficult to predict. The reason is since the amplitude of the ripples is reduced by the effect of normalization, this is considered to have become difficult to predict the amplitude of the ripples. In addition, Table 1 shows the RMS error of each state and each time. This result indicates that there is no significant difference in the accuracy of prediction in each time.

Table 1. RMS Error

	RMS Error			
	1.5 sec ahead	3 sec ahead	4.5 sec ahead	6 sec ahead
$x[m]$	0.294	0.281	0.255	0.337
$\dot{x}[m/s]$	0.555	0.411	0.675	0.536
$z[m]$	0.264	0.275	0.222	0.236
$\dot{z}[m/s]$	0.023	0.019	0.018	0.023
$\theta[rad]$	0.073	0.056	0.056	0.052
$\dot{\theta}[rad/s]$	0.040	0.035	0.037	0.037

*B. **Trajectory Generation***

To verify the trajectory generation by the DDP with interior penalty function, we assume the vehicle is a quadrotor. This simulation includes six degrees of freedom.

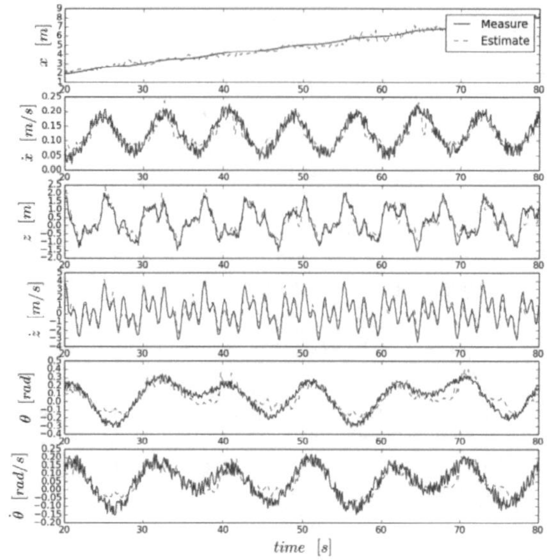

Figure 7. The result of prediction (six seconds ahead). A solid line write measurements with noise, the result of the estimation by using a dotted line writes the DNN.

*1) **Quadrotor dynamics model***

The following equations are equations of motion of Quadrotor.[8]

$$\begin{aligned}
\ddot{X} &= (\sin(\psi)\sin(\phi) + \cos(\psi)\sin(\theta)\cos(\phi))\frac{U_1}{m}\\
\ddot{Y} &= (-\cos(\psi)\sin(\phi) + \sin(\psi)\sin(\theta)\cos(\phi))\frac{U_1}{m}\\
\ddot{Z} &= -g + (\cos(\theta)\cos(\phi))\frac{U_1}{m}\\
\dot{p} &= \frac{I_{YY} - I_{zz}}{I_{XX}}qr - \frac{J_{TP}}{I_{XX}}q\Omega_{all} + \frac{U_2}{I_{XX}}\\
\dot{q} &= \frac{I_{ZZ} - I_{XX}}{I_{YY}}rp + \frac{J_{TP}}{I_{XX}}p\Omega_{all} + \frac{U_3}{I_{YY}}\\
\dot{r} &= \frac{I_{XX} - I_{YY}}{I_{ZZ}}qr + \frac{U_4}{I_{ZZ}},
\end{aligned} \quad (16)$$

where

$$\begin{aligned}
U_1 &= T_1 + T_2 + T_3 + T_4\\
U_2 &= l(-T_2 + T_4)\\
U_3 &= l(-T_1 + T_3)\\
U_4 &= -M_1 + M_2 - M_3 + M_4\\
\Omega_{all} &= -\varpi_1 + \varpi_2 - \varpi_3 + \varpi_4,
\end{aligned} \quad (17)$$

where $\ddot{X}, \ddot{Y}$, and $\ddot{Z}$ are accelerations with respect to the body frame. $\dot{p}, \dot{q}$, and $\dot{r}$ are angular accelerations at each axis with respect to the body frame. l is the distance between the center of the vehicle and the center of the propeller, and we assume that the centers of the propeller are on the plane of the center of propellers. T_1, T_2, T_3, and T_4 are the thrusts of each rotor.

M_1, M_2, M_3, and M_4 are moments of each rotor, respectively. $\varpi_1, \varpi_2, \varpi_3$, and ϖ_4 are angular velocities of each rotor. I_{XX}, I_{YY}, and I_{ZZ} indicate a mass moment of inertia at each axis. J_{TP} is the total rotational moment of inertia around the propeller axis.

As the aerodynamics model, we use the following equation.

$$\begin{aligned} T_i &= b\varpi_i^2 \\ M_i &= d\varpi_i^2, \end{aligned} \tag{18}$$

where b and d are the propeller coefficient that translates from the angular velocity. ϖ is an angular velocity of the propeller. Index i indicates the number of the motor.

States are set as:

$$X = [x, \dot{x}, y, \dot{y}, z, \dot{z}, \phi, p, \theta, q, \psi, r]^T. \tag{19}$$

The relationship between Euler-Angle Rates and Body-Axis Rates is as following:

$$\begin{bmatrix} p \\ q \\ r \end{bmatrix} = \begin{bmatrix} 1 & 0 & -\sin(\theta) \\ 0 & \cos(\phi) & \sin(\phi)\cos(\theta) \\ 0 & -\sin(\phi) & \cos(\theta)\cos(\theta) \end{bmatrix} \begin{bmatrix} \dot{\phi} \\ \dot{\theta} \\ \dot{\psi} \end{bmatrix}. \tag{20}$$

2) *Simulation Test*

The movement of the Target is the same as that used in state prediction of section V-A. We set an objective as landing on the target at six seconds ahead. Therefore, the simulation objective is that all states (position, attitude, and velocity) of a vehicle and target coincide at that time. In addition, since the position of the Z-axis is treated as the obstacle before the landing time, the vehicle avoids it by the interior penalty function with time-varying weight about the position of Z-axis.

As shown in Figure 8, the vehicle trajectory of the Z-axis avoids the target before the landing time and the vehicle states match with the target states at the landing time. For the X-axis, the vehicle overtakes the target in the first half. After this, the vehicle slows down after passing the target, and the position and velocity of the vehicle and target at X-axis correspond. For the pitch angle, because there is a large influence on the position and velocity for Quadrotor if varying the attitude, the vehicle changes the attitude in accordance with the translation until the landing time. Finally, the vehicle changes the attitude to coincide with the target attitude at the landing time.

VI. Conclusion

This paper introduces a multi-point prediction technique using DNN and trajectory generation technique with time-varying weight. Simulation results show the validity of the methods. By using these methods, we simulate the autonomous landing on a moving target. As a result, the multi-point prediction using DNN can estimate the future states of the periodic ship movement. Additionally, DDP with the interior penalty function with time-varying weight can generate a trajectory to avoid a target before the landing time.

In conclusion, with the proposed procedures, the proposed method can land on a moving target.

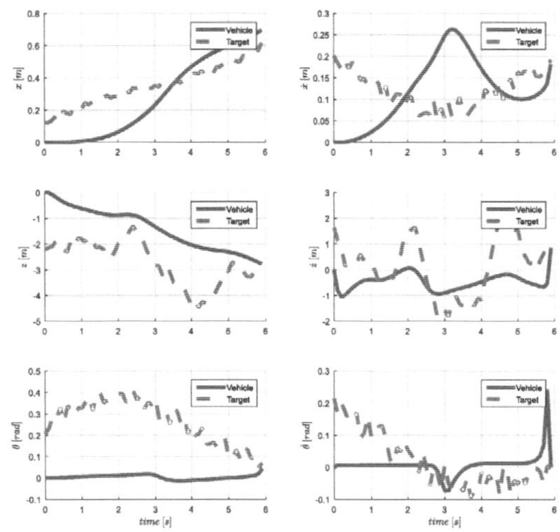

Figure 8. Trajectory generation result. A solid line writes the generated trajectories of the vehicle, and a dotted line writes the trajectories of the target.

References

[1] Takuma Nakamura, Stephen Haviland, Dmitry Bershadsky, and Eric N. Johnson, " Vision-Based Optimal Landing On a Moving Platform.", *AHS 72nd Annual Forum*, West Palm Beach, Florida, May 17-19, 2016

[2] Maas, Andrew L, Hannun, Awni Y, and Ng, Andrew Y, "Rectifier nonlinearities improve neural network acoustic models. ", *ICML*, Vol 30, 2013.

[3] David H. Jacobson and David Q. Mayne. "Differential Dynamic Programming," Elsevier, New York, NY, 1970.

[4] Ren Wu, Shengen Yan, Yi Shanm Qingqing Dang, and Gang Sun, "Deep image: Scaling up image recognition," *CoRP*, Vol. abs/1501.02876, 2015.

[5] Rumelhart, David E., Geoffrey E. Hinton, and Ronald J. Williams. "Learning internal representations by error propagation." No. ICS-8506. CALIFORNIA UNIV SAN DIEGO LA JOLLA INST FOR COGNITIVE SCIENCE, 1985.

[6] Jolly, K. G., R. Sreerama Kumar, and R. Vijayakumar. "A Bezier curve based path planning in a multi-agent robot soccer system without violating the acceleration limits.", *Robotics and Autonomous Systems* 57(1) 23-33, 2009

[7] David H. Jacobson and David Q. Mayne. "Differential Dynamic Programming," Elsevier, New York, NY, 1970.

[8] Bresciani, Tammaso. "Modelling, identification and control of a quadrotor helicopter," *MSc Theses*, 2008.

[9] Maas, Andrew L, Hannun, Awni Y, and Ng, Andrew Y, "Rectifier nonlinearities improve neural network acoustic models. ", *ICML*, Vol 30, 2013.

[10] Ren Wu, Shengen Yan, Yi Shanm Qingqing Dang, and Gang Sun, "Deep image: Scaling up image recognition," *CoRP*, Vol. abs/1501.02876, 2015

[11] Kingma, Diederik, and Jimmy Ba. "Adam: A method for stochastic optimization." arXiv preprint arXiv:1412.6980, 2014

Runway Independent High Speed Unmanned: CarterCopter Jet VTOL UAS

Jay Carter, Jr.
Chief Engineer and CEO
jay.carter@carteraero.com

Jeffrey R. Lewis
Engineering Manager
jeff.lewis@carteraero.com

Carter Aviation Technologies, LLC
Wichita Falls, Texas, USA

ABSTRACT

Carter Aviation Technologies, LLC has been developing aircraft concepts utilizing its Slowed Rotor/Compound (SR/C™) technology for over 20 years – a technology that offers VTOL capability with fixed wing cruise performance. Key to this technology is a rotor and related control system that can be slowed dramatically in flight while remaining stable. This technology has been demonstrated in flight with multiple manned demonstrators. The U.S. Armed Forces have expressed a desire for VTOL capability in future UASs . Carter's SR/C technology enables a UAS with this VTOL capability, but with the speed to reach targets much more quickly than conventional VTOL technologies. This paper briefly describes Carter's new Jet VTOL UAS and its potential applications.

INTRODUCTION

Carter has developed a high speed UAS concept, the CarterCopter Jet VTOL UAS (CC-JVU). This aircraft utilizes the benefits of the company's demonstrated slowed rotor / compound (SR/C™) technology, with turbofans for propulsion, and an innovative use of an electric motor for prerotation and yaw control in flight. The 19,000 lb (4870 kg) gross weight aircraft has a 55 ft (16.8 m) diameter rotor and wingspan, and is powered by twin 3000 lb (13.3 kN) thrust turbofan engines, enabling a top speed in excess of 380 kts (704 km/hr).

SR/C DEVELOPMENT BACKGROUND

Slowing the rotor provides two key benefits. As per Foster (Ref. 1), rotor profile drag is proportional to RPM^3. So, slowing the rotor down to 1/3 of its normal operational rpm will reduce rotational profile drag to 1/27 of what it would have been, and overall rotor profile drag to 1/5.5 of what it would have been. Secondly, slowing the rotor keeps the advancing blade tip speed from becoming too high, especially at airspeeds where it would exceed Mach 1 with a conventional rotorcraft. This is all possible to do in an SR/C aircraft because retreating blade stall can be avoided by unloading the rotor by transferring lift to the wings, as well as other innovative design features Carter developed to maintain rotor stability. The end result is a vehicle with efficiency on par with a general aviation fixed wing aircraft, or 2.5 – 3 times better than conventional helicopters.

Presented at the AHS International Specialists' Meeting on Unmanned Rotorcraft,

Almost as soon as rotorcraft were successfully demonstrated, designers understood the potential performance benefits of slowing the rotor. Two of the most successful and earliest attempts were the McDonnell XV-1 (Ref. 2) and the Fairey Rotodyne, both of which first flew in the late 1950s. Neither exceeded μ-1 (where μ, the advance ratio, is the ratio of the aircraft speed to the rotor tip speed), but both achieved higher advance ratios than previous rotorcraft, demonstrating the potential for speed and efficiency improvements that come with slowing the rotor. However, vibration and other technical issues led to the cancellations of those programs. No aircraft was able to exceed μ-1 until Carter solved these technical issues and demonstrated SR/C technology on the CarterCopter (see Figure 1).

Figure 1. Carter CCTD Prototype –First Aircraft to Reach Mu-1

Carter began work on a slowed rotor concept in 1994. The first demonstration of the viability of Carter's approach was a 6 ft (1.8 m) diameter rotor mounted on the front of a vehicle driven down the highway at up to 80 mph (129 km/hr) and stably slowed the rotor to an advance ratio of 8. Following that successful demonstration, Carter built its first flight demonstrator, the CarterCopter Technology Demonstrator, or CCTD (see Figure 1). The CCTD was the first aircraft ever to slow the rotor in flight to μ-1. Based on lessons learned from the CCTD, Carter developed its next demonstrator, the Personal Air Vehicle, or PAV (Ref 3, see Figure 2). The PAV exceeded μ-1 routinely, and demonstrated an L/D greater than 10, with further improvements expected in continued flight tests.

Figure 2. Carter PAV Prototype – L/D > 10

CARTER CC-JVU CONFIGURATION

Carter has developed a turbofan powered UAV concept utilizing SR/C technology, the CC-JVU, with properties as listed in Table 1. With the capability of runway independence, this aircraft has the flexibility to be operated from forward operating bases without airfields, or even to push forward with troops, while still offering the speed, range, and efficiency of a fixed wing aircraft. When airfields are available, the aircraft can perform very short rolling takeoffs for even more capability. The aircraft can carry up to 8 Hellfire missiles in an internal weapons bay for strike missions, reconfigure the modular mission payload bay for ISR missions, or carry sensors or ordnance on wing hard points.

Table 1. Carter CC-JVU Aircraft Characteristics

Characteristic	*English*	*Metric*
Engines	2 x Williams FJ44-3A	
SLS Takeoff Thrust	2 x 3000 lbs	2 x 13,345 N
Wingspan	55 ft	16.7 m
Rotor Diameter	55 ft	16.7 m
Max Useful Load	10,737 lbs	4,870 kg
Max Gross Weight	19,000 lb	8,618 kg
Max Range	3769 nm	6980 km
Service Ceiling @ MTOW	37.2k ft	11.3 km
Max GW TO Altitude	7500 ft	2286 m
Vmax @ MTOW	387 ktas	716 km/hr

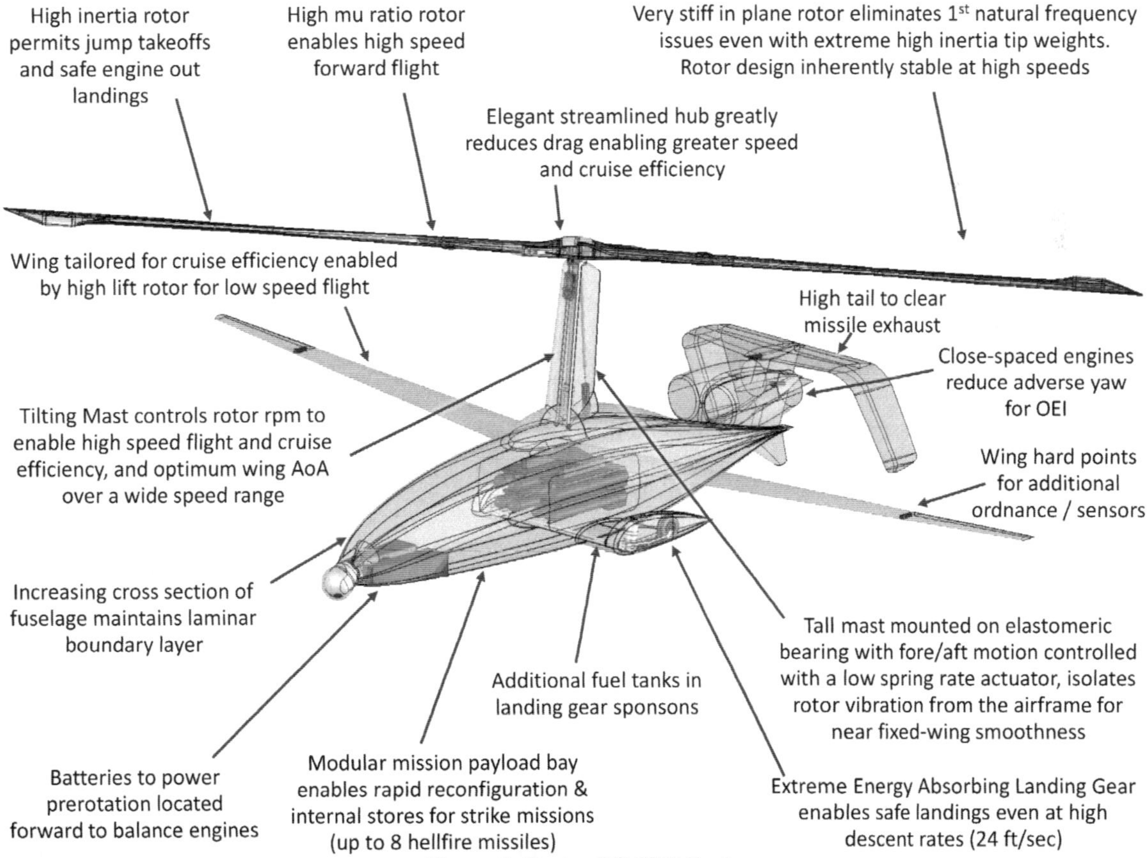

Figure 3. Carter CC-JVU Features

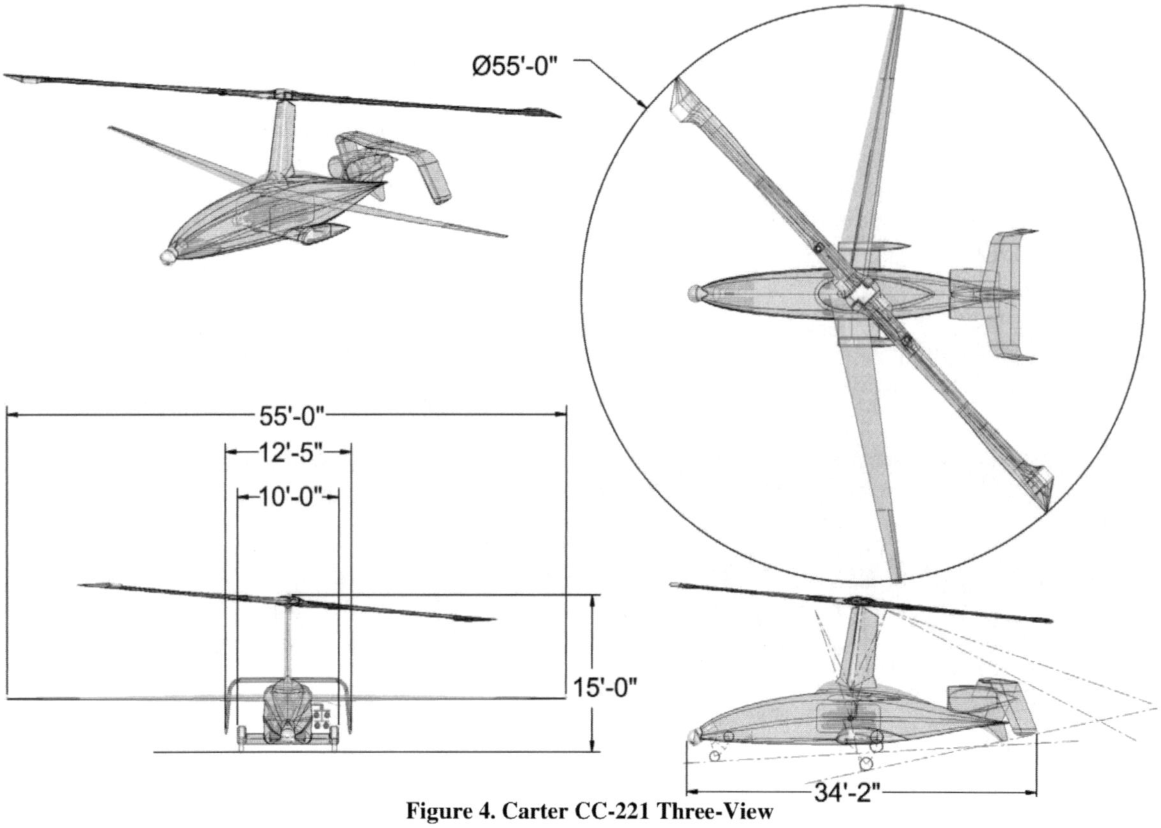

Figure 4. Carter CC-221 Three-View

Electric Prerotator and Yaw Control

The aircraft uses an electric motor to prerotate the rotor on the ground prior to flight. By accelerating the rotor to an overspeed rpm, the aircraft can perform zero roll jump takeoffs using the stored energy in the rotor, as has previously been demonstrated with Carter's manned prototypes. Once in the air, the rotor operates in autorotation, with the rotor rpm controlled primarily aerodynamically, but with the electric motor providing an additional means under certain conditions (e.g. when slowing the airspeed rapidly). In low speed flight when the low dynamic pressure reduces rudder authority, the electric motor provides yaw control to the aircraft, applying positive or negative torque to the rotor (operating as a generator to apply negative torque) to create a corresponding yawing moment on the aircraft. Because of the high inertia of the Carter rotor, this can be accomplished with very little effect on the rotor rpm. To help reduce adverse yaw in case of an engine failure, the engines are located very close to the aircraft centerline. This is especially important during low speed flight when rudder authority is reduced.

Empty Weight Estimate

Carter used multiple sources for the weight estimate. The primary source for conventional components was Raymer (Ref 4), and the empirical trends he developed using historical data from multiple aircraft. For components related specifically to SR/C technology and other innovative Carter components (e.g. rotor, landing gear), Carter used its own historical data from previous design efforts, including both actual weights from fabricated parts and estimated weights from detailed design of components. Carter also used previous design experience for UAV specific components, since Raymer's trends were for manned systems.

The structures group summary is shown in Table 2. The actual airframe components (wing, horizontal tail, vertical tail, and fuselage) were all calculated using the methods from Raymer (Ref 4), applying a 50% scale factor to reflect the weight savings possible with carbon composite construction as opposed to aluminum construction which was the basis for Raymer's historical trends. The landing gear weight was based on Carter's landing gear design experience.

Table 2. Structures Group Weight Estimate

Component	*Weight, lbs*	*Mass, kg*
W_wing	600.8	272.5
W_horizontal tail	44.2	20.0
W_vertical tail	13.7	6.2
W_fuselage	458.5	208.0
W_main landing gear	481.3	218.3
W_nose landing gear	120.3	54.6
Total Structural	**1,718.7**	**779.6**

The propulsion group summary is shown in Table 3. All components are based on Raymer (Ref 4).

Table 3. Propulsion Group Weight Estimate

Component	*Weight, lbs*	*Mass, kg*
W_installed engine	1,554.7	705.2
W_engine controls	21.9	9.9
W_starter	55.8	25.3
W_fuel system	103.9	47.1
Total Propulsion	**1,736.4**	**787.6**

Table 4 gives the weight estimate for the equipment group. All of these components used the methods from Raymer (Ref 4).

Table 4. Equipment Group Weight Estimate

Component	*Weight, lbs*	*Mass, kg*
W_flight controls	347.3	157.6
W_hydraulics	19.0	8.6
W_electrical	225.0	102.1
W_avionics	1,207.9	547.9
W_EO/IR	193.0	87.5
W_anti ice	42.0	19.1
Total Equipment	**2,034.2**	**922.7**

The weight summary for the rotorcraft elements is given in Table 5. The rotor is based on previous Carter designs. Jump takeoff analysis ensured proper inertia for jump takeoff performance.

Table 5. Rotorcraft Elements Weight Estimate

Component	*Weight, lbs*	*Mass, kg*
Rotor	1,200.0	544.3
Complete Prerotator System	800.0	362.9
Total SR/C Elements	**2,000.0**	**907.2**

The complete empty weight build-up is given in Table 6. The aircraft empty weight is 8263 lbs (3748 kg), for an empty weight fraction of 0.43.

Table 6. Empty Weight Build-Up

Component	*Weight, lbs*	*Mass, kg*
Total Empty Weight Before Margin	7,489.3	3397.1
Margin % of Empty Weight	0.100	0.1
Total Empty Weight Including Margin	8,238.3	3736.8
Empty Weight Fraction Including Margin	0.434	0.434
Basic Weight (Including fluids)	8,263.3	3748.2

Takeoff Performance

The CC-JVU can takeoff using various strategies depending on circumstances. Three in particular were examined for this study.

1. Jump takeoff, immediately transitioning to a climb to clear any obstacles – This offers the best takeoff performance, but limits the aircraft gross weight. For this study, the takeoff criterion was clearing a 50 ft obstacle within 400 ft of takeoff (15 m obstacle within 122 m).
2. Jump takeoff, but staying in ground effect to build airspeed – This offers higher takeoff weights than the first strategy while still allowing the aircraft to operate independently of a runway, but does require suitable terrain.
3. Very short rolling takeoff – This strategy offers the highest takeoff weight, but requires the aircraft to takeoff from a prepared runway. For this study, a 500 ft (152 m) roll was used. Additionally, the aircraft stayed in ground effect after liftoff to build airspeed to the climb airspeed.

To perform zero roll jump takeoffs, the rotor is first accelerated to 330 rpm at flat collective pitch. During this prerotation, the engines are brought up to max thrust. Once the rotor reaches 330 rpm, the torque to the rotor is reduced to a fraction of the prerotate torque, then the rotor collective pitch is increased to jump into the air. The aircraft will use the stored rotor energy to accelerate and climb, transitioning into autorotation once the rotor rpm decays.

Rolling takeoffs are similar. The rotor is first accelerated to 330 rpm while the engines are brought up to max thrust. Once the rotor reaches 330 rpm, the torque to the rotor is reduced to that required to hold 330 rpm at flat pitch. Next, the brakes are released, allowing the aircraft to accelerate down the runway. At 475 ft (145 m), the rotor collective begins increasing, such that the aircraft will actually lift off at 500 ft (152 m). As the rotor rpm decays, the aircraft transitions into autorotation.

To model takeoff performance, Carter built a model of the aircraft in the X-Plane flight simulator. X-Plane is a physics based simulator, allowing users to define their own aircraft models. The models are built by defining critical geometry (wings, stabilizers, etc.), specifying the lift, drag, and moment curves for each of the airfoils to be used in the model, and specifying other key aircraft parameters (engine thrust, bypass ratio, etc.). The simulator divides the different components of the aircraft into multiple elements, and performs real time blade element analysis during flight. In addition, Carter has developed a proprietary program that integrates with X-Plane to handle unique systems & automation.

Results for the three different takeoff strategies are shown in Figure 5. Using a rolling takeoff, the aircraft can take off at max gross weight up to 7500 ft (2286 m) density altitude. For a jump takeoff staying in ground effect, the aircraft can lift off at 16,000 lb at 7500 ft (7257 kg at 2286 m). With a jump takeoff to clear a 50 ft obstacle, the aircraft can lift off at 16,000 lb at 5000 ft (7257 kg at 1524 m).

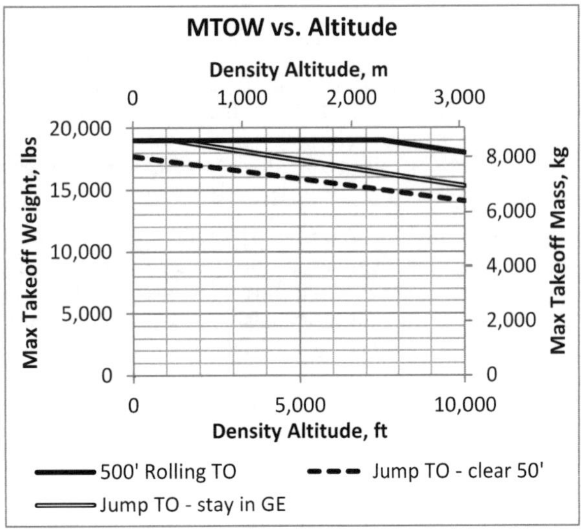

Figure 5. Takeoff Performance

Cruise Performance Estimate

Carter has developed a proprietary cruise performance model. Drag prediction is based on the methods from Foster (Ref. 1), but modified to improve rotor drag prediction at high advance ratios. This model has been calibrated to measured flight data from the PAV prototype. The data reduction methods used on the measured flight data were reviewed and validated by the U.S. Army Aviation Applied Technology Directorate (Ref 5).

Predicted lift to drag is shown in Figure 6. All curves are at max gross weight.

Figure 7 shows the level flight envelope when operating in autorotation. The aircraft could fly faster or slower in autorotation in a descent. Additionally, for transient periods when the rotor is powered by inertia (i.e. takeoff and landing), the airspeed can be reduced to zero.

Figure 8 shows the rate of climb vs. airspeed at various altitudes at max gross weight.

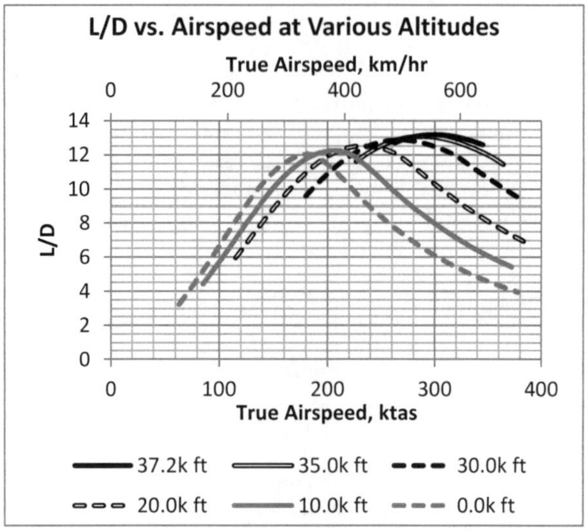

Figure 6. Carter CC-JVU L/D at Max Gross Weight

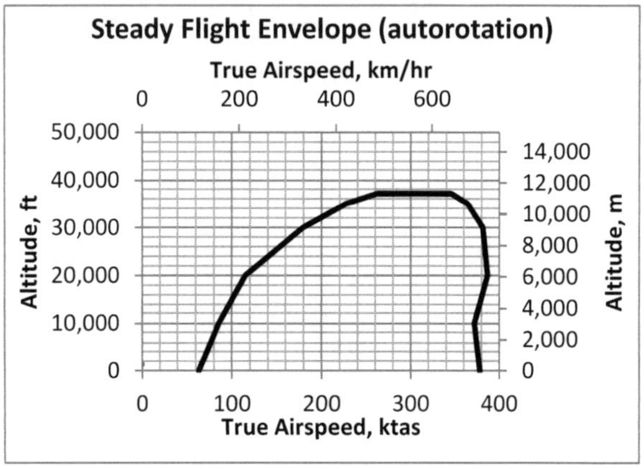

Figure 7. Autorotation Flight Envelope at Max GW

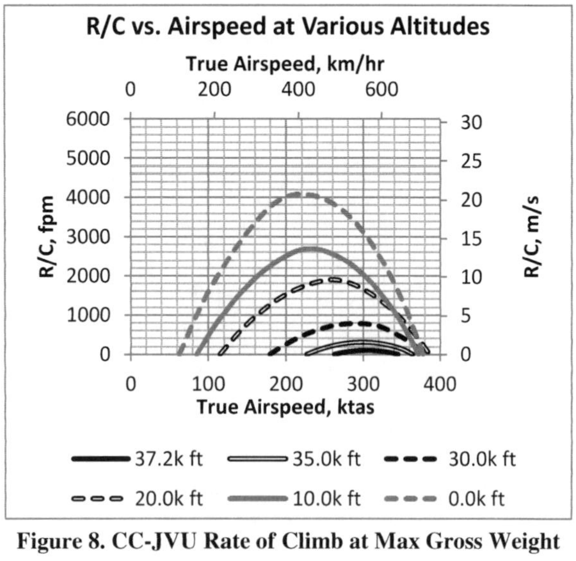

Figure 8. CC-JVU Rate of Climb at Max Gross Weight

CC-JVU Mission Performance Estimate

Three different missions were considered for cruise performance, as summarized below, and detailed in the rest of this section.

- Mission 1 – One way, 30 min resurve
- Mission 2 – Simulated strike, round trip, 30 min reserve
- Mission 3 – Simulated ISR, round trip, 30 min reserve

The first mission was a one way flight, as detailed in Table 7. This included a warm-up segment, climb, cruise at 30 – 45k ft, descent, and a 30 minute reserve. The results of payload vs. range for Mission 1 are shown in Figure 9.

Table 7. Mission 1 Profile - One way, 30 min reserve

Segment	*Avg Speed, ktas*	*Avg Speed, km/hr*
1. Warm up for 1 minute at Max Thrust at S.L.	0	0
2. Climb at Max ROC to Cruise Alt (30k – 45k ft)	258	478
3. Cruise at best range speed	324	600
4. Descend to Sea Level	246	455
5. 30 minute reserve at best endurance	147	272

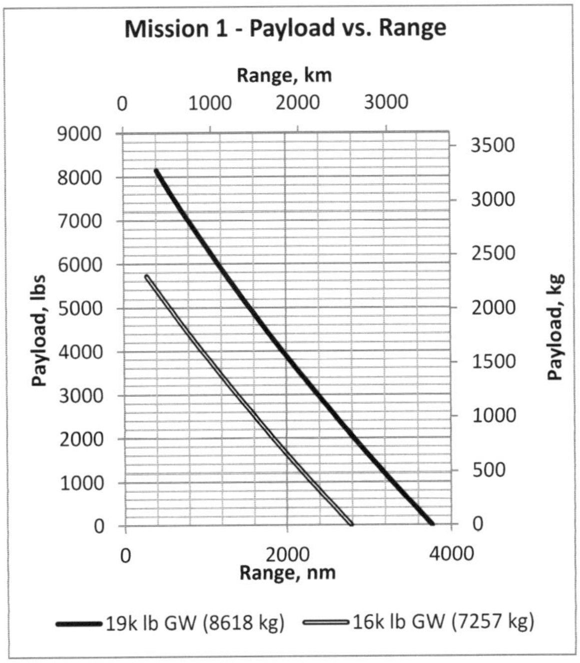

Figure 9. Payload vs. Range, Mission 1 – One way, 30 min reserve

Mission 2 simulated a strike mission, as detailed in Table 8. It was a round trip mission, with an outbound cruise at 30k – 40k ft, 5 minutes at 15k ft at the destination, then a return cruise at 40k – 45k ft, including a 30 minute reserve, as well as the warm-up, climb, and descent portions. The results of payload vs. mission radius for Mission 2 are shown in Figure 10. The payload was carried the entire mission.

Table 8. Mission 2 Profile – Simulated strike, round trip, 30 min reserve

Segment	*Avg Speed, ktas*	*Avg Speed, km/hr*
1. Warm up for 1 minute at Max Thrust at S.L.	0	0
2. Climb at Max ROC to Cruise Alt (30k – 40k ft)	255	473
3. Cruise at max speed	361	669
4. Descend to 15k ft Loiter	262	486
5. Loiter for 5 minutes	175	323
6. Climb at Max ROC to Cruise Alt (40k – 45k ft)	264	489
7. Cruise at best range speed	316	585
8. Descend to Sea Level	244	451
9. 30 minute reserve at best endurance	143	165

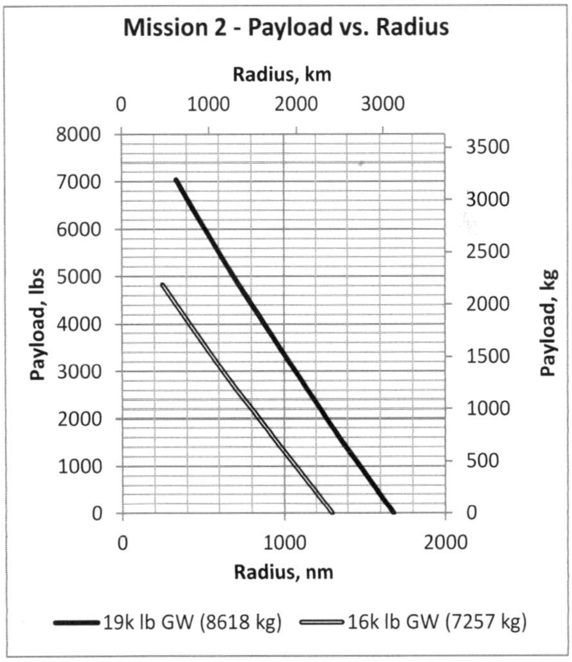

Figure 10. Payload vs. Mission Radius, Mission 2 – Simulated strike, round trip, 30 min reserve

Mission 3 simulated a reconnaissance type mission, as detailed in Table 9. It was a round trip mission, with an outbound cruise at 30 – 40k ft, loiter at 15k ft at the destination, then a return cruise at 40k – 45k ft, including a 30 minute reserve, as well as the warm-up, climb, and descent portions. During loiter, one engine was shut down so that the other could operate more efficiently (drag increase was accounted for). Only the built-in sensors were included in the mission – no additional payload was added. The results of time on station vs. mission radius for Mission 3 are shown in Figure 11.

Table 9. Mission 3 Profile – ISR, round trip, 30 min reserve

Segment	*Avg Speed, ktas*	*Avg Speed, km/hr*
1. Warm up for 1 minute at Max Thrust at S.L.	0	0
2. Climb at Max ROC to Cruise Alt (30k – 40k ft)	255	473
3. Cruise at max speed	361	668
4. Descend to 15k ft Loiter	265	491
5. Loiter (fuel limited)	168	311
6. Climb at Max ROC to Cruise Alt (40k – 45k ft)	250	463
7. Cruise at best range speed	279	517
8. Descend to Sea Level	237	438
9. 30 minute reserve at best endurance	133	247

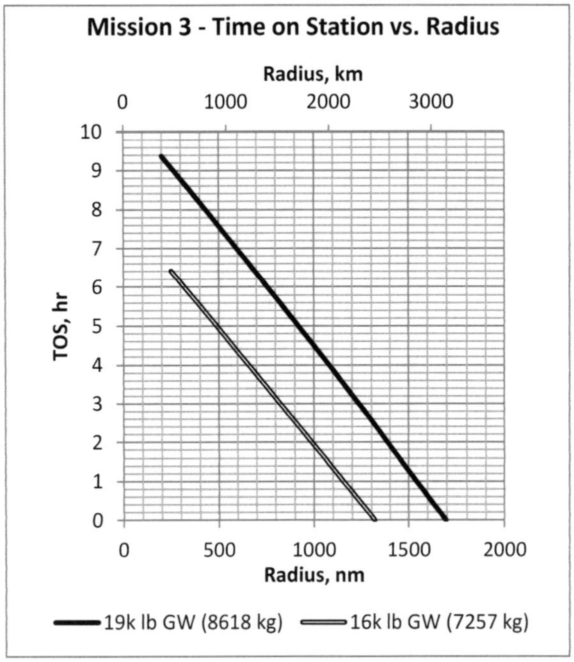

Figure 11. Time on Station vs. Mission Radius, Mission 3 – ISR, round trip, 30 min reserve (built-in sensors only)

CONCLUSIONS

1. CarterCopters with SR/C technology offer runway independence combined with the speed, range, and efficiency of fixed wing aircraft.

2. Carter has successfully demonstrated stably slowing the rotor, both in flight in excess of μ-1, and in ground testing up to μ-8.

3. The CC-JVU utilizes SR/C technology for a UAV with exceptional flexibility, able to be operated in close support of forward units, flying at jet speeds.

4. By utilizing different takeoff strategies, when conditions permit, takeoff at max gross weight can be performed at up to 7500 ft (2286 m) density altitude.

5. Applying Carter's SR/C technology enables a VTOL UAV with top speeds in excess of 380 kts (704 km/hr), a service ceiling of 37.2k ft (11.3 km), a one way range of 3769 nm (6980 km), or a round trip mission radius of 1695 nm (3139 m).

6. An internal weapons bay, modular mission payload bay, and external wing hard points, offer exceptional mission flexibility.

REFERENCES

[1]Foster, R. D., "A Rapid Performance Prediction Method for Compound Type Rotorcraft," WADC TR 55-410, 1957.

[2]Harris, F. D., "The XV-1's Rotor System," AHS 4th Decennial Specialist's Conference on Aeromechanics, San Francisco, CA, January 2004.

[3]Lewis, J. R., Carter, J., Peryea, M., Pichon, E., "Carter Personal Air Vehicle," American Helicopter Society Specialists' Meeting, Fort Worth, TX, October 2008.

[4]Raymer, D. P., *Aircraft Design: A Conceptual Approach*, Third ed, American Institute of Aeronautics and Astronautics, Reston, VA, 1999, Chapter 6.

[5]Cox, J. A., "Assessment of Current State of Technical Development of the Personal Air Vehicle Developed by Carter Aviation Technologies, LLC," Aviation Development Directorate, Aviation Applied Technology Directorate, Flight Test Branch, 2015.

A Multi-Modality Mobility Concept for a Small Package Delivery UAV

L.A. Young
Aeromechanics Office
NASA Ames Research Center
Moffett Field, CA 94035

Abstract

This paper will discuss an alternate approach to the typical notional small package delivery drone concept. Most delivery drone concepts employ a point-to-point aerial delivery CONOPS from a warehouse directly to the front or back yards of a customer's residence or a commercial office space. Instead, the proposed approach is somewhat analogous to current postal deliveries: a small aerial vehicle flies from a warehouse to designated neighborhood VTOL landing spots where the aerial vehicle then converts to a "roadable" (ground-mobility) vehicle that transits on sidewalks and/or bicycle paths till it arrives at the residence/office drop-off points. This concept and its associated platform or vehicle will be referred in this paper as the MICHAEL (Multimodal Intra-City Hauling and Aerial-Effected Logistics) concept. It is suggested that the MICHAEL concept potentially results in a more community friendly "delivery drone" approach.

Nomenclature

b	(Primary) wing span, m
C	Circuituity, $C=(d_A - d_M)/d_M$, nondim.
c	(Primary) wing mean chord, m
C_D	Vehicle in-flight (cruise) drag coefficient, D/qS
C_L	Vehicle in-flight lift (cruise) coefficient, L/qS
C_P	Vehicle total rotor-shaft-output hover power coefficient
C_T	Vehicle hover rotor (aggregate) thrust coefficient
d_A	Actual distance traveled, miles or km
d_{AG}	Actual distance traveled on the ground (versus in the air), miles or km
d_M	Minimum point-to-point ("straight line") distance, miles or km
ESC	Electronic speed controllers
G	Ground travel ratio, $G=d_{AG}/d_A$
L/D	Vehicle in-flight (cruise) lift-over-drag ratio

Presented at the Seventh AHS Technical Meeting on VTOL Unmanned Aircraft Systems, Mesa, Arizona, USA, January 24-26, 2017.

q	Freestream dynamic pressure, lb_f/ft^2 or N/m^2
S	Wing planform area, S=bc, ft^2 or m^2
α	Vehicle angle-of-attack, AOA, Deg.

Introduction

There is considerable socio-economic and public-service potential to small autonomous aerial vehicles, Refs. 1-4. In particular, the economic potential of small "delivery drones," e.g. Ref. 1, has captured considerable interest. But, despite the potential, there are many challenges to be overcome to see the realization of such mission applications. Among those challenges is that these vehicles and their associated CONOPS must be seen as being community friendly in terms of minimizing their noise, emissions, and invasiveness. If these challenges can be overcome, then society will benefit in terms of economic growth while minimizing environmental impact.

For any small package-carrying autonomous aerial vehicle concept to ultimately prove viable it must achieve one or more of the following goals: 1. reduce delivery time as compared to ground transportation alternates; 2. improve economics of package delivery service thorough reduced labor and increased customer satisfaction; 3. improve energy efficiency of delivery service; 4. reduce environmental emissions; 5. reduce roadway and tranportation system infrasture development pressure; 6. improve reliability and quality of potentially critical delivery services, particularly those impacting health-care services to seniors and underserved populations. Additionally, in general, any viable small package delivery drone must also meet the following design and operational constraints: 1. be as safe as or safer than the baseline delivery ground transportation system; 2. generate less than current community annoyance levels for emissions and noise from ground transportation and other community noise sources; 3. be seen as minimally invasive as to community/personal privacy; 4. must be all-weather reliable as to yield timely package deliveries; 5. be seen as providing secondary public services and community enhancements in addition to the primary mission of commercial small package deliveries.

It is still generally unproven whether some of the many delivery goods and services distribution concepts being proposed over the past few years can successfully meet the above noted goals and design and operational constraints. This paper seeks to closely consider some of these issues and propose an alternate goods and services aerial vehicle distribution system. Accordingly, a novel approach to the delivery drone paradigm is proposed. Throughout this paper this approach will be referred to as the MICHAEL (Multimodal Intra-City Hauling and Aerial-Effected Logistics) concept and its associated platform or vehicle.

General Problem Area:

How to improve the community friendliness of uninhabited aerial vehicles (UAVs) performing on-demand small package delivery or courier services?

1. Are there alternative concepts of operation (CONOPs) of delivery drones that could satisfy the potential market for fast, on-demand, delivery of small packages/cargo without requiring their close flyover (<30m or <100 feet AGL) over residences and takeoff and landing onto personal property?
2. Are there safe and efficient VTOL aerial vehicle configurations that could support – in fact be ideally uniquely tailored to – such a CONOPs?

Proposed Solution:

The MICHAEL concept will focus on the examination of a small "roadable" (or, more correctly, a sidewalk or bike lane compatible ground mobility) hybrid aerial vehicle that can not only vertically takeoff and land but also ideally cruise with airplane-like efficiency.

Instead of landing in someone's yard to deliver a package, the vehicle would land at a neighborhood landing site and then have the vehicle move with wheeled locomotion, on sidewalks and bike lanes, to the individual residences. This alternate approach to delivery drones might be seen as more community friendly than concepts of operation that require (very close) proximity to residences, personal property, and people/animals.

The solution space to be explored, then, is the use "roadable aircraft" in the context of small VTOL UAVs – versus past studies in the literature examining larger passenger-carrying aerial vehicles – to make such vehicles multimodal (air and ground) mobility platforms to act as community robotic "postal carriers" for intra-city deliveries.

Fig. 1 – Package Delivery Using a MICHAEL System (package being shown extracted from payload fairing)

Mission Requirements

Figure 2 shows a notional mission profile for a MICHAEL platform. This mission profile is roughly consistent with other delivery drone CONOPS being proposed for the in-flight portion of the MICHAEL mission. It is also consistent with early work by the author in Ref. 1. However, midway through the mission – instead of landing or hovering over a residence or commercial delivery point – instead the vehicle lands at a designated neighborhood landing spot.

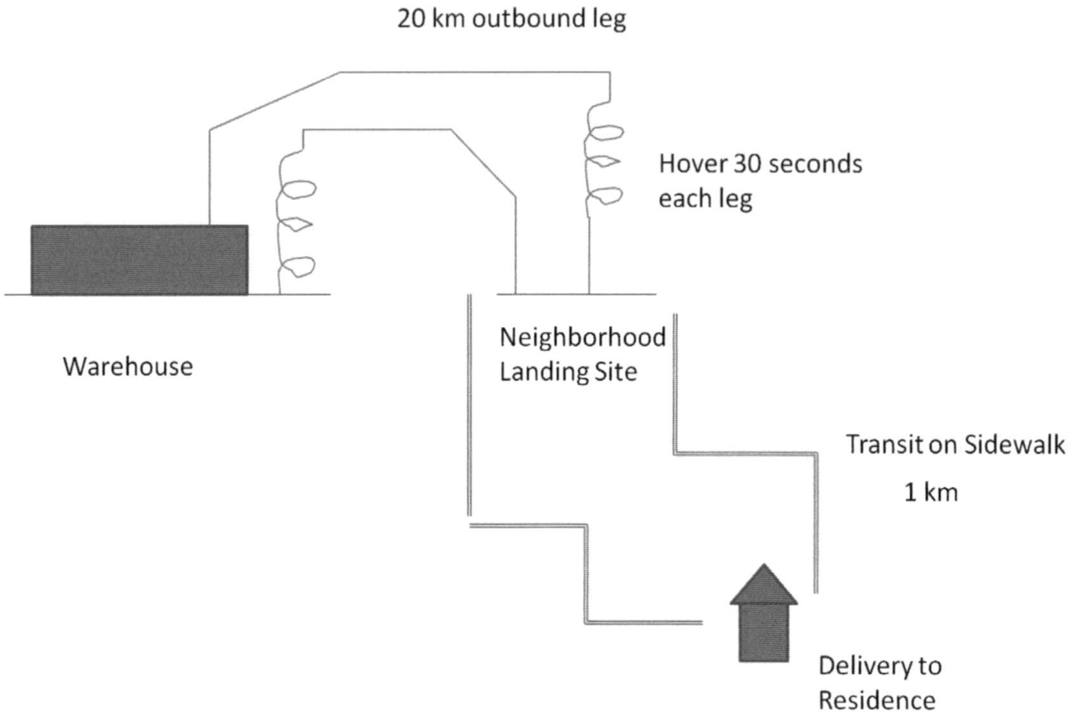

Fig. 2 – MICHAEL Mission Profile

Among the potential neighborhood VTOL landing sites for the MICHAEL mission are municipal parks, nearby office or retail building rooftops, etc. The above notional MICHAEL mission profile is not optimized on the basis of a detailed package-delivery network analysis/simulation. Instead it is a generic profile provided so as to define some reasonable or plausible mission requirements for vehicle conceptual design discussion later in the paper.

The key element of the MICHAEL concept – versus the point-to-point (warehouse-to-backyard) delivery drone concepts – is the ground mobility phase of the MICHAEL mission. The required ground distance to be traveled and the average ground speed attained will both substantially impact the relative productivity and overall success of the MICHAEL concept. The more time spent on the ground, versus the air, will reduce the greatest advantage of delivery drones versus automotive (truck or van) delivery: time-to-delivery to the consumer.

Notional Baseline Vehicle:

"Roadable" aerial vehicles have been proposed for decades, continuing to this very day (Ref. 9). And, yet, except for a small number of proof-of-concept vehicles, roadable vehicles have yet to be successfully developed. Despite this mixed development history, however, it may be quite possible to develop a small (<90kg or <200 lb_f) vehicle that travels at a relatively low speed (<16kph or <10 mph) on bicycle paths or sidewalks versus the far more challenging design problem of larger passenger-carrying vehicles that operate on roadways.

The MICHAEL robotic multimodal platforms are also a good mission application test case for the implementation of all-electric or hybrid-electric propulsion for small VTOL aerial vehicles. Hybrid-electric propulsion has recently gained a considerable amount of interest within NASA, including its application to rotorcraft (Ref. 9). A small electric propulsion UAV is a far more tractable problem than a larger passenger-carrying vehicle.

Figure 3 illustrates one notional MICHAEL configuration; Fig. 3 is a set of "time sequence" images showing the vehicle in its ground-mobile (with folded/stowed wings) form, followed by it unfolding/unstowing its wings prior to and during takeoff. The baseline MICHAEL configuration shown in Fig. 3 is a ducted-fan tailsitter vehicle. Upon vertical takeoff, the vehicle would pitch forward with increasing forward speed and transition to level-flight cruise mode. A variety of means of successfully trimming the vehicle pitching moment throughout transition can be devised. Among those methods are the use of vanes with flaps within the duct – and within the rotor downwash – so as to provide the required control moment authority. The baseline MICHAEL vehicle employs a set of fixed-pitch coaxial rotors within the duct.

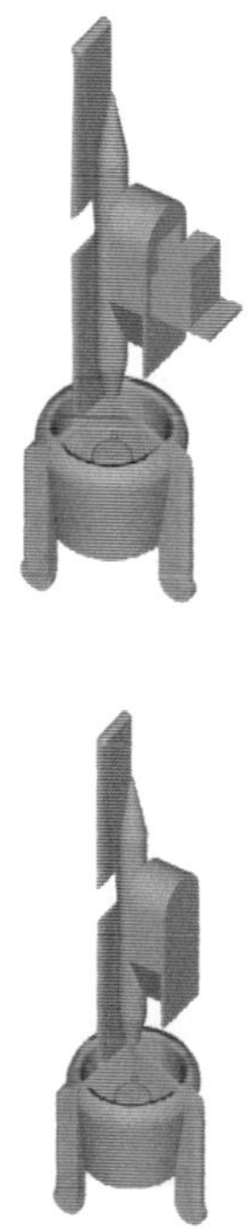

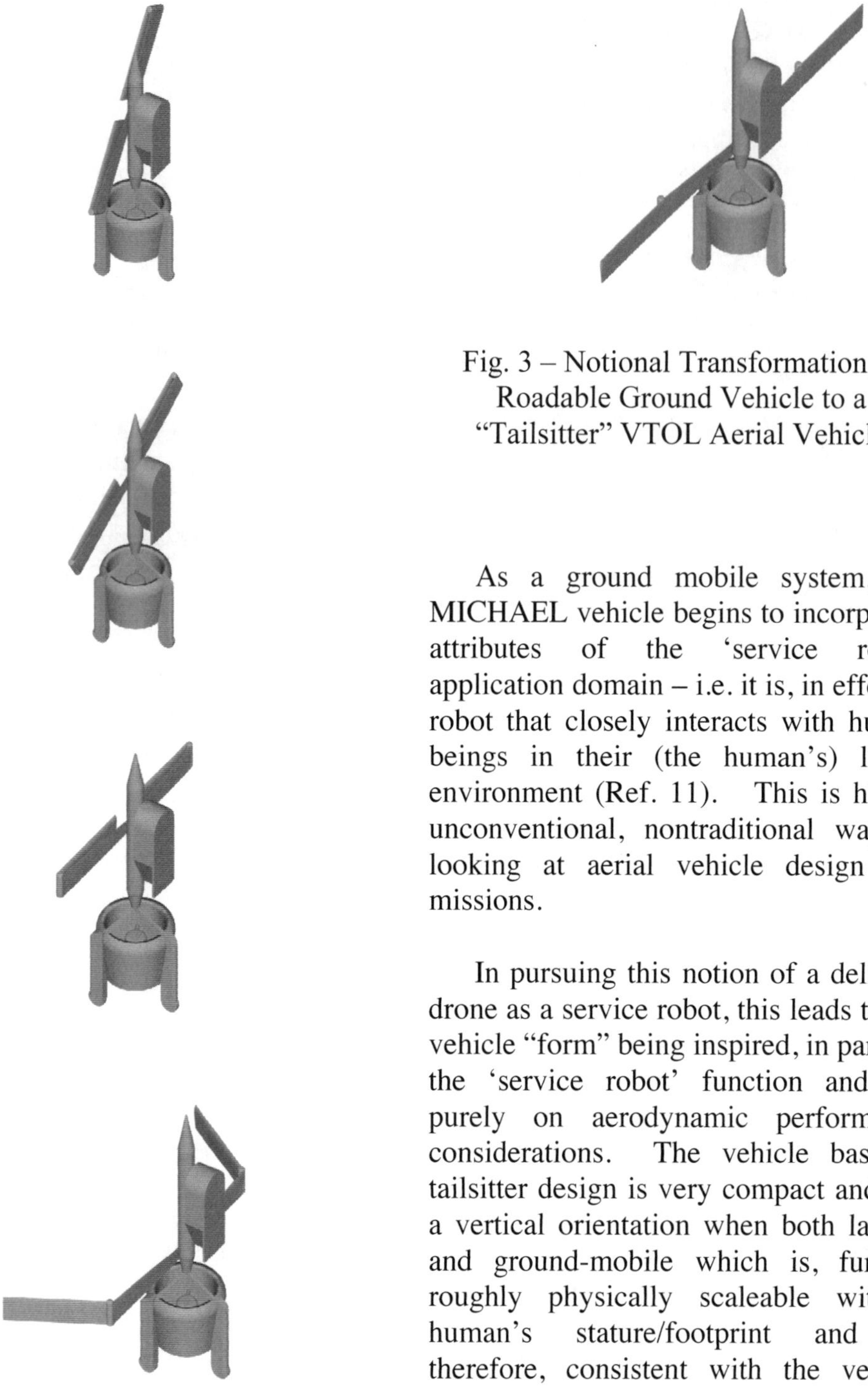

Fig. 3 – Notional Transformation of Roadable Ground Vehicle to a "Tailsitter" VTOL Aerial Vehicle

As a ground mobile system, the MICHAEL vehicle begins to incorporate attributes of the 'service robot' application domain – i.e. it is, in effect, a robot that closely interacts with human beings in their (the human's) living environment (Ref. 11). This is highly unconventional, nontraditional way of looking at aerial vehicle design and missions.

In pursuing this notion of a delivery drone as a service robot, this leads to the vehicle "form" being inspired, in part, by the 'service robot' function and not purely on aerodynamic performance considerations. The vehicle baseline tailsitter design is very compact and has a vertical orientation when both landed and ground-mobile which is, further, roughly physically scaleable with a human's stature/footprint and is, therefore, consistent with the vehicle acting as a service robot in a social interaction.

An alternate MICHAEL-like concept that has been previously studied at NASA Ames Research Center has been the use of a VTOL aerial vehicle to transport and deploy (once on the ground at the neighborhood landing site) an independent ground-mobile robotic system for final package delivery. The ground-mobile robot would in effect be "cargo" for a utility-type rotary-wing aerial platform.

There are also many analogous aspects of the MICHAEL delivery drone concept with respect to urban metro/regional aerial transportation concepts (Refs. 6-8): i.e. low-altitude flight over urban/suburban areas, the use of all-electric or hybrid-electric propulsion for community friendliness as to emissions and noise, the employment of high-levels of autonomous system technology, and the extremely complex nature of the air traffic management problem with respect to the coordination of hundreds to thousands of autonomous aerial vehicles safely interacting with manned aircraft.

System Analysis of Mission Tradeoffs between Ground- and Aerial-Mobility

MICHAEL delivery times will fall somewhere in between delivery times for ground transportation deliveries by means of automotive platforms (with or without drivers) and the warehouse-to-doorstep aerial transportation model that has to-date been the default paradigm for most small package delivery drone concepts.

Early work on this problem was performed in Ref. 1. In that early work, the potential for small autonomous aerial vehicles was identified. The follow-on question for this current work is what magnitude of compromise in time-to-delivery is accepted if a point-to-point air delivery was substituted by the MICHAEL CONOPS model wherein multimodal air and ground mobility were used for package delivery?

Figures 4-5 introduce the concept of ciruituity of distance traveled by air- and ground mobile systems – with emphasis of delivery drones versus automobile/truck deliveries. Circuituity will be a key parameter for the system analysis to follow. Note that because of the low altitude and extremely short ranges of vehicle travel, that a straight line approximation can be made in the system analysis and discussion that follows. The straight line approximation can be used instead of having to consider in more detail ascent and descent profiles and great circle, etc., flight paths typically required for the analysis of conventional aircraft in air traffic management simulations.

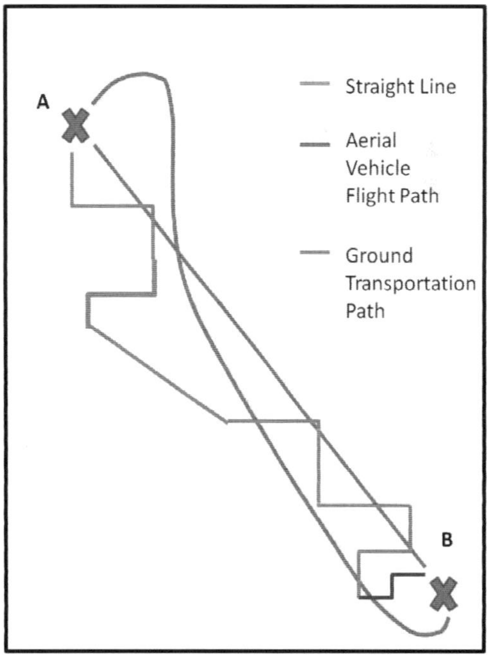

Fig. 4 – Ground versus Aerial Vehicle Point-to-Point Path "Circuituity"

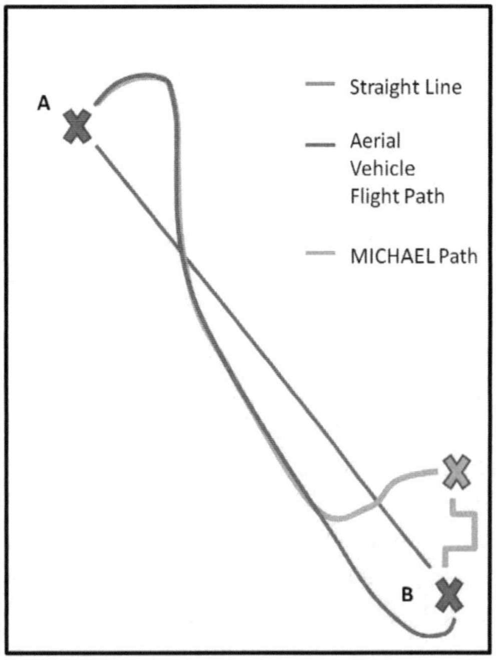

Fig. 5 – Notional Comparison between Point-to-Point Aerial Flight Path and the MICHAEL Flight/Ground Path

The initial system analysis performed in this paper looks at two aspects of the problem, for the various delivery options (ground/truck, pure aerial vehicle point-to-point, and the MICHAEL multimodal mobility): energy expenditure and time.

Relative energy expenditure is assessed between automobile/truck versus pure point-to-point aerial vehicle package delivery for a prescribed nominal travel distance (42 km) in Fig. 6. As expected, the energy expended by an automobile or truck is significantly larger than an equivalent distance flown by a small aerial vehicle.

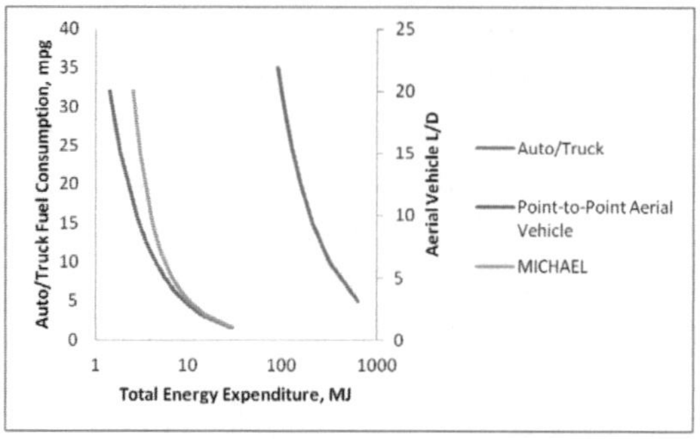

Fig. 6 – Relative energy expenditure

A "breakeven" analysis is now provided whereby an assessment is made of the maximum number of flights required (one package at a time) that would have to be flown to breakeven with the overall energy expenditure of delivery by automobile or truck. The truck performance is assumed to be only secondarily affected by the number of packages onboard the truck. A delivery truck would typically carry 100-300 packages. As can be seen in Fig. 7, depending on the fuel milage of the delivery trucks and the L/D of the competing delivery drones, this breakeven number of packages can range from just over ten packages to close to 200 packages. Vehicle L/D has, not surprisingly, a significant effect on this breakeven package count estimate. Providing a high L/D for a VTOL platform has historically been a challenge; this challenge is compounded when considering including "roadable" capability. Nonetheless, such a challenge is potentially addressable with new technologies and innovative vehicle designs. But even with a high L/D, the results of Fig. 7 (and earlier work in Ref. 1) would suggest that delivery drones will primarily be focused on small, time critical, and high-value packages instead of bulk shipments of low-value items.

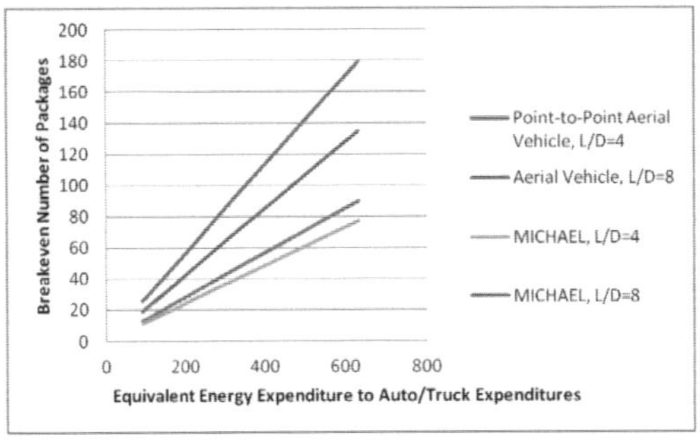

Fig. 7 – "Breakeven" number of flights to expend the same amount of energy as an automobile/truck

Next the relative energy expenditure is assessed of a multimodal mobility delivery approach (i.e. MICHAEL) as a function of the ground travel ratio, G, in Fig. 8. If G=0, then a pure point-to-point aerial vehicle delivery is assumed. Nonzero values of G imply some level of multimodal mobility. G=1 is a fully ground-mobile system (though at reduced energy expenditure as compared to an automobile or truck given the vehicle's lightweight and all-electric nature). Additionally, with G=1 there is a step change in energy expenditure in that hover and an inflight reserve are no longer required. The energy expenditures are predicated on a 42 km total mission range.

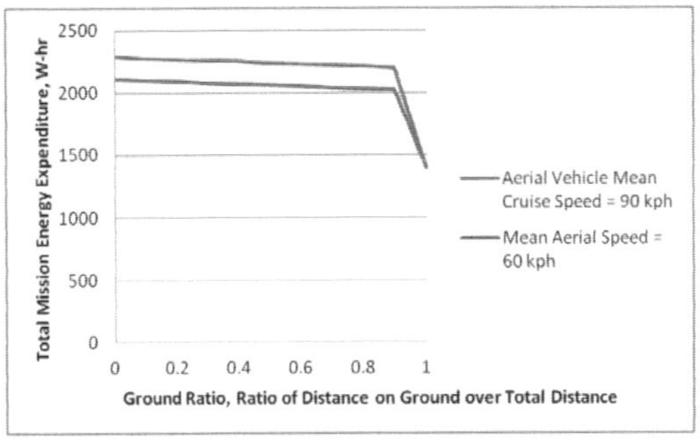

Fig. 8 – Energy Expenditure as a function of Ground Travel Ratio, G

Delivery time is now considered instead of energy expenditure. First, relative time saved between an automobile/truck package delivery (for one package at some prescribed nominally distance) is compared to the delivery times for a pure point-to-point aerial vehicle for various assumed vehicle speeds and circuituity levels. The auto/truck estimates do not account for delays due to intersection lights, traffic, or multiple package deliveries.

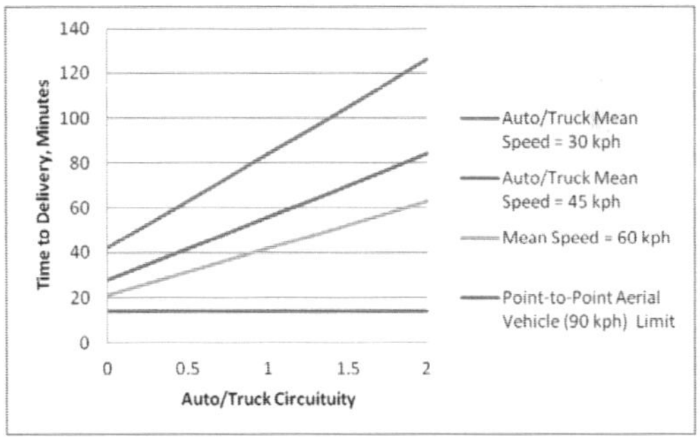

Fig. 9 – Time-to-delivery Tradeoffs between Auto/Truck and Point-to-Point Aerial Vehicle

Finally, the relative time saved between a pure point-to-point aerial vehicle versus a multimodal mobility vehicle (i.e. MICHAEL) is assessed as a function of G, the ground travel ratio. The below time to delivery estimates assume that the vehicle flies at a cruise speed of 90kph and moves on the ground at speed of 15kph. The warehouse package preparation and launch time is not factored in the delivery estimate. Additionally, delays on the ground due to roadway intersection crossing and traffic congestion are also not factored in to the delivery time estimates.

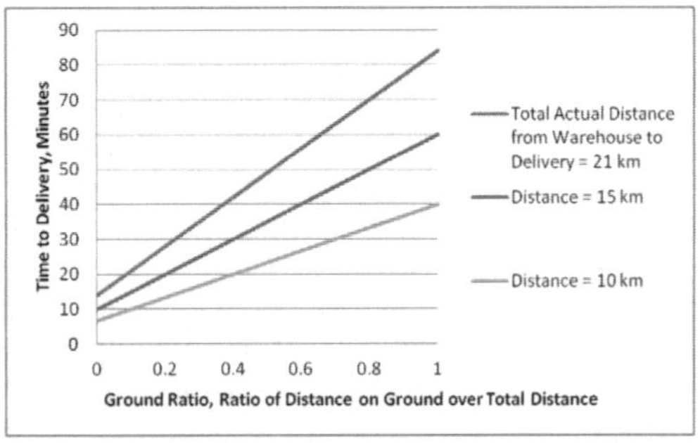

Fig. 10 – Time-to-delivery as a function of G, ground travel ratio

Ultimately, this simplified type of system analysis needs to be superceded by local airspace simulation modeling. This level of analysis, though, can be used to guide both MICHAEL vehicle design as well as develop network designs for neighborhood landing sites.

Cargo Delivery "Drone" Trade Space Examination

There are many different rotary-wing and VTOL aerial vehicle configurations that could be relevant to the MICHAEL mission design requirements and CONOPS. The ducted-fan tailsitter configuration introduced earlier is just one possible configuration – other potential configurations include a whole gamut of multi-rotor configurations, including distributed, modular, and heterogeneous rotor systems (Ref. 4). Other VTOL concepts of merit include versions of autonomous aerial vehicles studied in Refs. 2 and 3 and, perhaps, smaller versions of the electric VTOL vehicles ("Hoppers") explored in Refs. 6-8. Nonetheless, the ducted fan tailsitter is adopted as a baseline configuration for the remainder of the discussion in this paper primarily of its compactness and maximum cruise speed capability.

A well-known graphical means of showing the global design trade space of VTOL vehicles is the "wheel of V/STOL aircraft and propulsion concepts" (Ref. 12). An analogous "wheel of delivery drones" is proposed and presented below in Fig. 11 to help foster a global understanding of the design trade space of this emerging application domain.

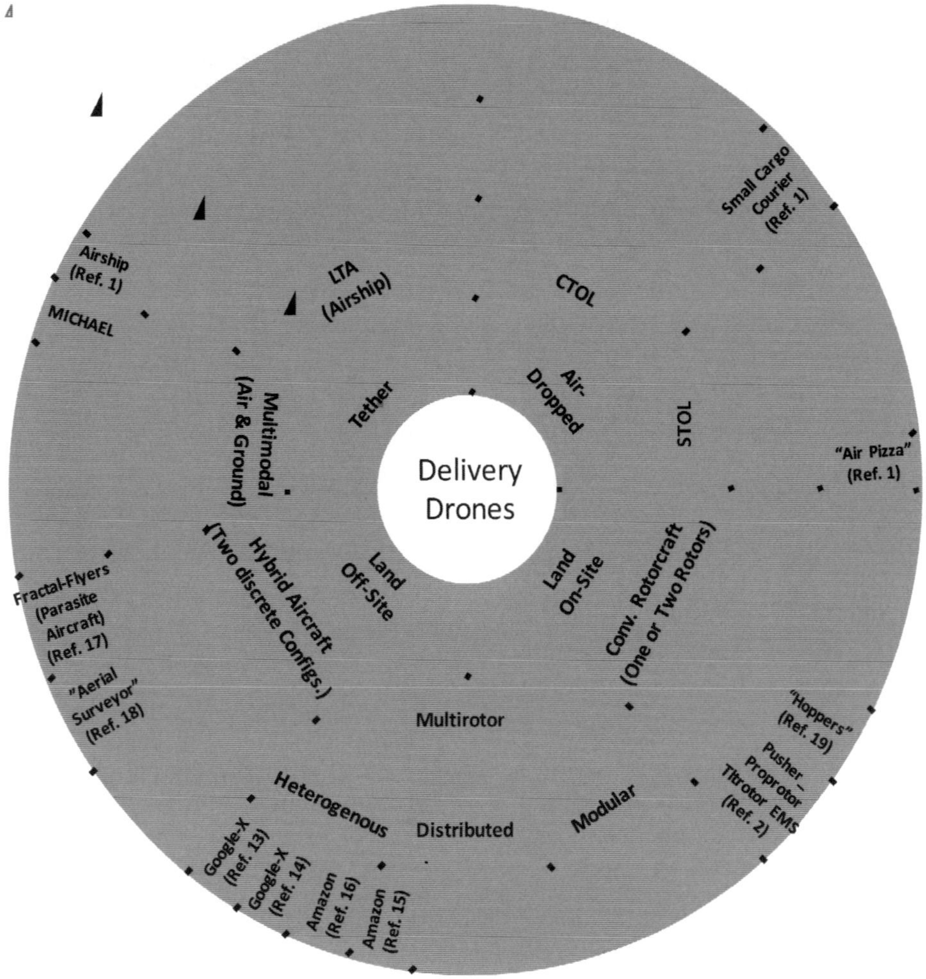

Fig. 11 – Wheel of Delivery Drones

Using a conceptualization device such as the Fig. 11 "wheel of delivery drones" provides not only a convenient way to summarize past work in this technical area but, through gaps or empty sectors in the wheel, it also provides insight into possible future design approaches to explore. The MICHAEL concept has been included on the Fig. 11 "wheel of delivery drones."

One key secondary design attribute for delivery drones is, as noted earlier, hybrid electric propulsion. Internal combustion engines might possibly be employed for in-flight cruise propulsion whereas electric propulsion might be used for takeoff and landing and, as well, ground mobility (especially near residences or, even more so, if used for mobility inside commercial properties or buildings). There is a spectrum of hybrid-electric propulsion options for small cargo delivery drones such as MICHAEL. This hybrid-electric design trade space will undoubtedly foster novel technical insights.

Another secondary design attribute being debated between delivery drone proponents is the exact method of final delivery of packages. In particular, some proponents believe the package should lowered to the ground via a reelable tether that can be deployed from the vehicle hovering some distance above the ground (Refs. 13-14). Alternatively, some proponents believe the package should be deployed from the vehicle while it is stationary on the ground (Refs. 15-16). And, finally, some proponents believe an air-drop would be the best method of delivery. In some cases, the type of aerial vehicle being considered automatically dictates which final delivery method is employed: e.g. a small lighter than air (LTA) airship – because of its comparatively large volume/size – cannot closely approach the ground and would have to deploy packages via a tether or air-drop; alternatively, a conventional takeoff and landing aerial vehicle (because of inadequate available landing area) would likely have to precision air-drop packages.

Baseline MICHAEL Vehicle Concept Definition

Table 1 summarizes some high-level notional mission/design requirements for the baseline MICHAEL conceptual design.

Table 1 – Conceptual Design Requirements

Requirements	
Max. Payload/package Mass (kg)	2.5
Max. Payload/package Dimensions (cm)	30x30x30
Total Flight Range (km)	40
Total Hover Duration (min.)	2
Reserve (in cruise, min.)	10
Max Cruise Speed (kph)	90
Cruise Altitude, AGL (m)	120
Total Ground Distance (km)	2
Max Ground Speed (kph)	15
Max. Terrain Grade (Deg.)	10
Max. Surface Unevenness (cm)	1
Braking distance at max. ground speed (m)	3
All-Electric propulsion	Ground & Air
Max. Dimension of Vehicle Footprint (m)	1
Max. Vehicle Height (m)	2.2

The vehicle height and max footprint dimensions are defined to be consistent with the ability of the vehicle to operate within the entrances and interiors of residences and office buildings.

Aerodynamic Design and Analysis

The rotorcraft computational fluid dynamics software tool RotCFD is being used to perform initial studies of the aerodynamic characteristics of the baseline MICHAEL vehicle.

(a)

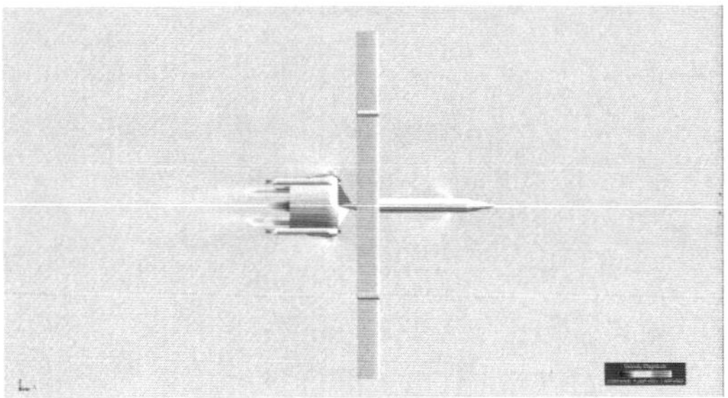

(b)

Fig. 12 – MICHAEL: (a) Hover and (b) Cruise flow field predictions

(a)

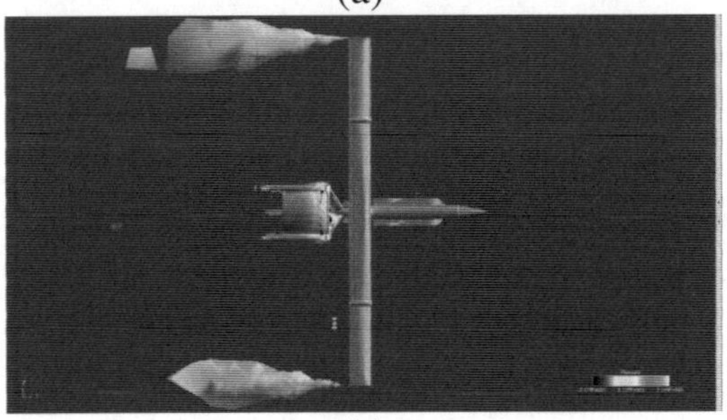

(b)

(c)

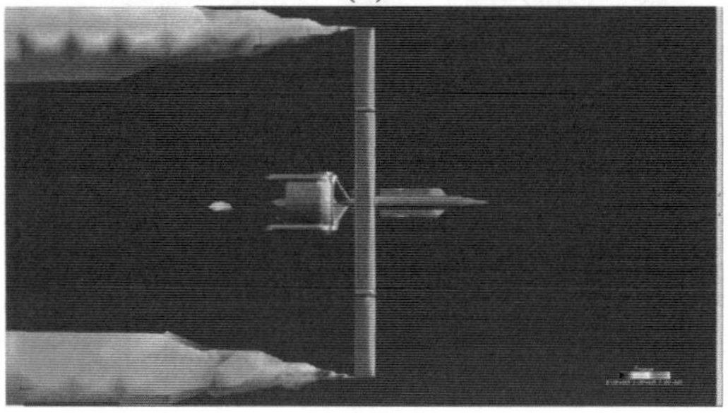

(d)

Fig. 13 – Body Surface Pressure Distributions (top view) as a function of Angle-of-Attack: (a) and (b) AOA=3 and (c) and (d) AOA=7

Initial predictions were made at a forward-flight speed of 50 ft/s (or ~34 mph or ~55 kph). This speed should be considered the lower bound of the tailsitter conversion/transition from lift being provided primarily from the propellers to fixed-wing-borne flight. As noted in Table 1, the target design maximum cruise speed is 90kph. The reference area used for the vehicle coefficients is the wing area which is 5 ft^2 or 0.47 m^2 (note that the reference area does not include any of the duct "planform" area, even though the duct is carrying net lift in forward flight). The wing span is 8.9 ft or 2.7 m and the wing chord length is 0.56 ft or 0.17 m. The predicted vehicle lift coefficients are indeed as high as they are because a significant portion of the vehicle lift (or perhaps more properly net vertical force) in cruise comes from duct and the payload fairing contributions (as can be seen by the body surface pressures shown in Fig. 13) – but, of course, at the cost of very high drag levels with the current design.

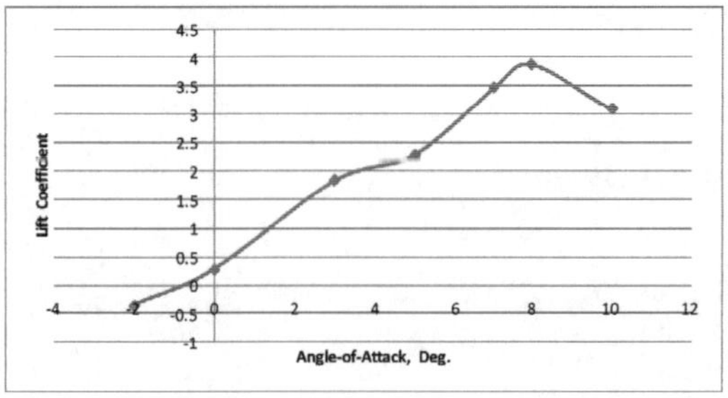

Fig. 14 – Lift Coefficent versus Angle-of-Attack of Baseline Vehicle

The predicted drag is quite large because of a number of factors. First, the baseline design's duct employs a rather thick, cambered airfoil. Second, the external payload fairing (in which the small package to be delivered is contained) in the current baseline design is approximately 37% thick, a very low finite-span aspect ratio of 0.24, and only roughly airfoil-like in geometry; as can be seen in Fig. 13(a) and (c), there is a significant amount of pressure drag due to surface pressures at the leading and trailing-edges of the payload/package fairing. The splitter plane below the payload fairing helps moderate some of that pressure drag but it appears that this is an area for design improvement. Third, support vanes attaching the main fuselage with the duct show elevated levels of positive pressure on their leading-edges. Fourth, and finally, a rather bluff cylindrical body is used in the CFD model to represent the housing notionally containing the (electric) motors driving the vehicle propellers; this bluff body also has a significant pressure drag contribution to the overall vehicle's drag predictions. The above issues will be addressed in subsequent design iterations and modeling refinements.

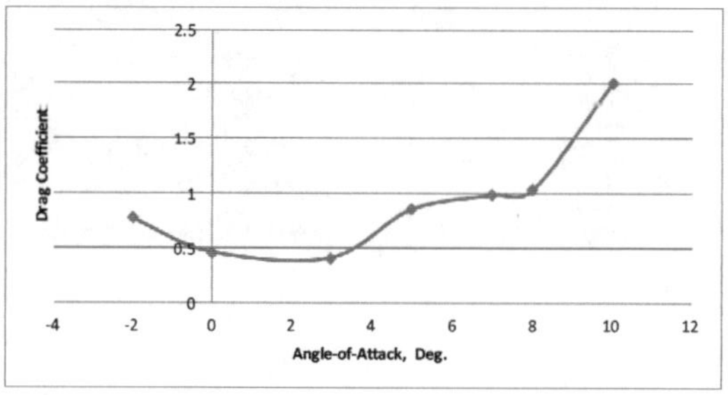

Fig. 15 – Drag versus Angle-of-Attack of Baseline Vehicle

The lift-to-drag ratios currently predicted for the current baseline design are very modest.

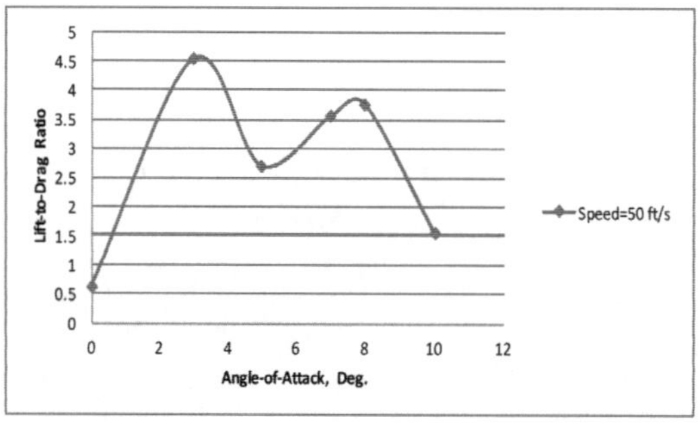

Fig. 16 – Lift-to-Drag Ratio as a function of Angle-of-attack

These aerodynamic analysis results were factored into the vehicle sizing analysis discussed in the next section of the paper.

Vehicle Sizing

Vehicle sizing of a roadable vehicle is unlike that of a conventional aircraft. Vehicle sizing of a VTOL vehicle acting, in part, as a socially interactive service robot is even more unconventional. The initial focus should first be on defining and sizing the ground-mobile subsystems of the vehicles (Fig. 17). Secondly, after the ground mobile elements have been initially sized then the second set of subsystems to be defined and sized is the wing(s) – and/or (as appropriate) rotor(s) – folding/stowing mechanisms. Only then, after that these initial critical subsystem sizing efforts, can sizing be performed on the aerial vehicle itself.

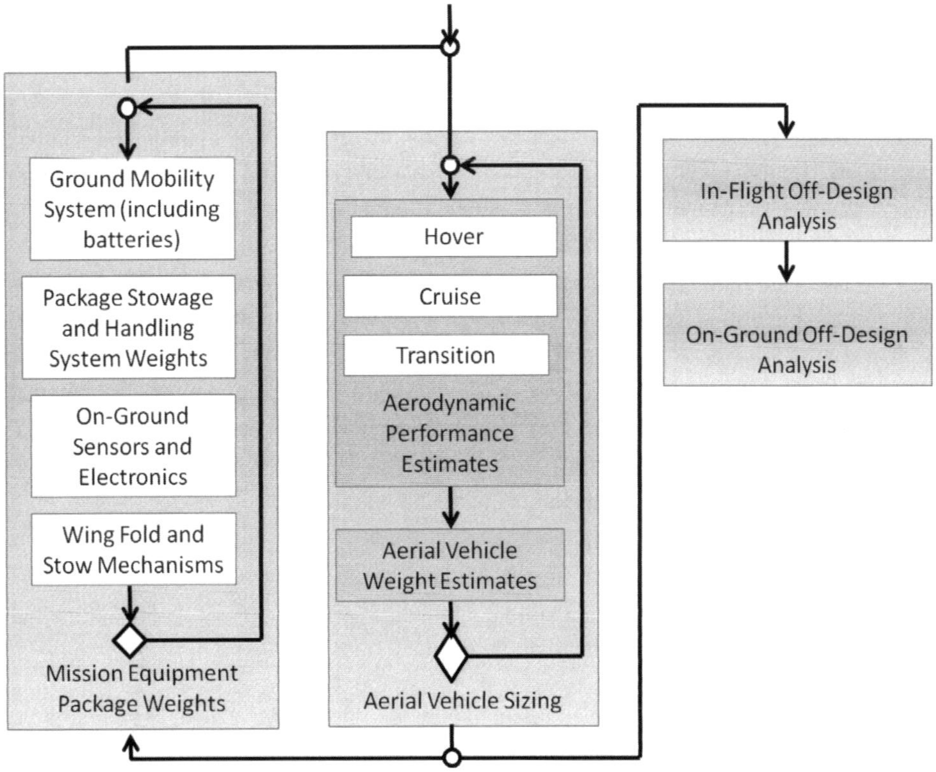

Fig. 17 – MICHAEL Vehicle Sizing: Treating "Roadable" and Package Handling Subsystems as a "Mission Equipment Package"

The first-order vehicle sizing analysis performed herein is not based on a clean-sheet paper airplane exercise. This initial analysis effort has been very much informed by available COTS mechnical and electric components readily available for the proof-of-concept and prototyping work discussed later in the paper. Accordingly, non-optimal components result in a heavier, less efficient, and nominally less capable vehicle than what is theoretically achievable with a clean-sheet approach. The advantage of the current approach, though, is that it is less dependent on statistical or historical data and regression-analysis weight equation approach for vehicle components and subsystems. This is particularly an important consideration in that regression-analysis-based weight equations don't exist for subsystems such as the ground mobility and wing fold/stow elements.

Overall duct size is a key consideration in the MICHAEL design. Sizing of the duct is primarily driven by three factors: first, propeller size (driven, in turn, by the VTOL requirement) drives the duct diameter, second, the necessity for the duct and it's nominal airfoil thickness is driven by safety considerations (to protect the propeller blades and, in turn, people and property from damage if the rotating or nonrotating blades collide with anything) and, third, the incorporation of the ground-mobility propulsion subsystems requires a structurally robust and stiff duct to be mounted to or otherwise support. Additionally, there are secondary considerations such as duct size and geometry influencing propeller static thrust aumentation. As already seen from discussion of initial baseline design CFD results, though, the duct is a significant contributor to both vehicle lift and drag. Future design iterations will have to look closely at whether the duct is oversized with respect to meeting its ground-mobility and safety requirements and whether it might be possible, accordingly, to be reduced in size and otherwise slimmed down (such as using less cambered and thinner airfoils).

The baseline design looks tail-heavy because of the large duct. Mass balancing would be achieved by locating the large mass of batteries as far forward along the vehicle longitudinal axis as possible. Additionally, the primary wing will have a mechanism to translate/traverse the wing along the longitudinal axis so that the aerodynamic center is made coincident with the center of gravity. Figure 18 schematically illustrates a notional layout of key subsystems for the vehicle. This layout assumes an all-electric propulsion system for the vehicle. Later design iterations may well consider a hybrid-electric system (where a small piston-based internal combustion engine (ICE) either direct-drives the propellers in cruise or, alternatively, this ICE drives a generator that then feeds current to the propulsion electric motors, i.e. a "range extender" type system, e.g. Ref. 19).

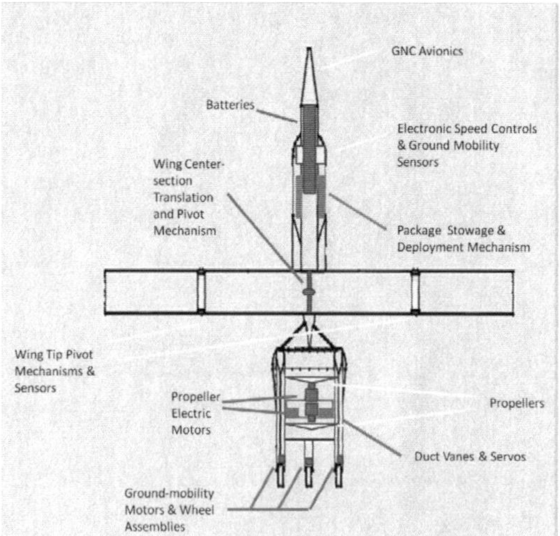

Fig. 18 – MICHAEL Baseline Design Subsystem Overview Schematic

Tables 2 and 3 summarize the current first-order sizing of a MICHAEL vehicle. Table 2 summarizes the vehicle geometry and aerodynamic characteristics and Table 3 summarizes the weight breakdown estimates or measurements (in the case of actual COTS hardware being employed). The below tables do not represent a optimized point design but rather a work-in-progress assessment of a proof-of-concept vehicle.

Table 2 -- MICHAEL Vehicle Geometry and Aerodynamics Characteristics

Parameter	**Value**
Wing/duct airfoil	NACA 4412
Primary wing span	2.71 m
Primary wing constant chord	0.17 m
Vehicle Cruise L/D	4
Coaxial propeller cruise effective efficiency	0.9
Coaxial propeller static thrust figure-of-merit	0.6
Duct fairing diameter	0.73 m
Duct fairing length	0.39 m
Number of (2-bladed) propellers	2
Propellers diameter	0.69 m

Table 3 -- MICHAEL Vehicle Weight Breakdown

Component/Subsytem	**Weight (kg)**
Payload Capacity	2.5
Avionics and Sensors	0.5
Batteries (Li-Ion; 150 W-hr/kg)	15.2
ESC for propeller and wheel motors (QTY 5)	0.31
Fuselage structure	3.2
Primary wing structure	1.6
Duct Fairing, Vanes, Legs	7.0
Vane servos	0.25
Wing fold, pivot, and traverse servos	0.5
Propellers (QTY 2)	0.54
Propeller electric motors (QTY 2)	2.6
Wheel gearmotors (QTY 3)	2.39
Wheels (QTY 6)	1.12
Total =	37.7

Note that it would have been perhaps desirable to have the vehicle weigh at or less than the current FAA commercial UAV weight limit of 25 kg but this is not perceived as a hard design requirement. The vehicle sans batteries does weigh less than 25 kg. To complete the mission profile summarized in Table 1 the vehicle has been estimated by means of first-order analysis to weigh approximately 37.7 kg, with 15.2 kg of batteries.

System Sensors & Controls

A rudimentary control perspective is offered below as to distinct mission phasess for the MICHAEL platform: ground mobility; conversion from ground to aerial vehicle configuration (wing folding/stowing); hover; low-speed transition from hover to cruise forward flight; cruise. During each of these distinct mission phases there will be a different subset (though sometimes overlapping) of sensors, controls, and control laws required for those mission phases. This is summarized in Tables 4-6 immediately below.

Eliminating unneccesary redundancy of subsystems will be a key challenge to reduce overall vehicle weight and, consequently, the viability of a small roadable aerial vehicle. Table 4 summarizes the anticipated sensors and controls required for the ground mobility phase of the MICHAEL mission. Table 5 summarizes the sensors and controls required for conversion from ground mobility configuration of MICHAEL to its aerial configuration, including the critical elements of the wing unfolding, pivoting, and translating. Table 6 summarizes the sensors and controls for VTOL takeoff and landing, tailsitter transition from vertical to horizontal orientation, and cruise. Tables 4-6 identifies the sensors and controls, indicates their purpose, notes potential dependencies with other sensors and controls and notes their potential priority to safely and effectively accomplish the overall mission. Table 4 is unique to roadable aerial vehicles. Several sensors and controls are unique to roadable aerial vehicles.

Table 4 – Sensors/Controls Required for Ground Mobility

#	**Descrip.**	**Purpose**	**Depend-ancies**	**Priority (1,2,3)**
S1	Kinectix	Indoor Visual Nav.		2
S2	wifi	Package Exchange		2
S3	Ultrasound	Indoor Forward Prox.Sensor		2
S4	Thermistor	Motor Temp.	C1	3
S5	Thermistor	Motor Temp.	C2	3
S6	Thermistor	Motor Temp.	C3	3
S7	Current	Wheel 1 Motor	C1	3
S8	Current	Wheel 2 Motor	C2	3
S9	Current	Wheel 3 Motor	C3	3
S10	Contact Sensor	Package hatch in Payload Fairing		2
C1	E-motor	Wheel 1 Drive	C1, C3	1
C2	E-motor	Wheel 2 Drive	C1, C3	1
C3	E-motor	Wheel 3 Drive	C1, C2	1
C4	E-brake	Wheel 1		2
C5	E-brake	Wheel 2		2
C6	E-brake	Wheel 3		2
C7	Linear Actuator	Package Hatch Open/Closed		2

Table 5 – Sensors/Controls Required for Conversion from Ground to Aerial Vehicle

#	**Descrip.**	**Purpose**	**Depend-ancies**	**Priority (1,2,3)**
S11	IMU	GNC		2
S12	GPS	Outdoor Nav.		1
S13	Rotor 1 RPM	Failsafe: C1, C2, C3 Disabled		1

		when nonzero		
S14	Rotor 2 RPM	Failsafe: C1, C2, C3 Disabled when nonzero		1
S15	Contact Sensor	Failsafe: Wing Tip 1 Folded	C1, C2, C3	3
S16	Contact Sensor	Failsafe Wing Tip2 Folded	C1, C2, C3	3
S17	Contact Sensor	Failsafe: Wing Center-section Pivoted Closed	C1, C2, C3	3
S18	Current	Wing Tip 1 Servo		
S19	Current	Wing Tip 2 Servo		
S20	Current	Wing Center-section Pivot Servo		
S21	Current	Wing Center-section Translation Motor		
C8	Servo	Wing Tip 1 Pivot		
C9	Servo	Wing Tip 2 Pivot		
C10	Servo	Wing Center-section Pivot		
C11	E-motor	Wing Center-section Translation		
C12	Solenoid	Wing Tip 1 Lockout Pin		
C13	Solenoid	Wing Tip 2 Lockout Pin		
C14	E-brake	Wing Center-section Pivot Lockout		
C15	Solenoid	Wing Center Translation Lockout Pin		

Table 6 – Sensors/Controls Required for Aerial Vehicle Flight

#	Descrip.	Purpose	Depend-ancies	Priority (1,2,3)
S22	LIDAR	Nadir Visual Nav. for Landing		1
S23	Potentio-meter	Duct Vane 1 Position		1
S24	Potentio-meter	Duct Vane 2 Position		1
S25	Potentio-meter	Duct Vane 3 Position		1
S26	Inclinometer	Vehicle Orientation (100 Deg. Range)		1
C16	Servo	Duct Vane 1 for Trim		1
C17	Servo	Duct Vane 2 for Trim		1
C18	Servo	Duct Vane 3 for Trim		1
C19	E-motor current; ESC	Drive propeller 1	S13 and S14	1
C20	E-motor current; ESC	Drive propeller 2	S13 and S14	1

As can be seen in Tables 4-6, in many cases similar sensors are used during different multimodality phases of the vehicle's mission. For example, a LIDAR system or an imaging camera could have great utility during both the in-flight and the gorund mobility phases of the mission. However, bcause of the different requirements for both mission phases (LIDAR and camera aligned with the vehicle's longitudinal axis for in-flight and along a lateral axis when ground mobile), either separate and indepedent sensors might have to be used or some sort of repositioning and focusing mechanisms might need to be implremented to use the same set of sensors for the two different mission phases.

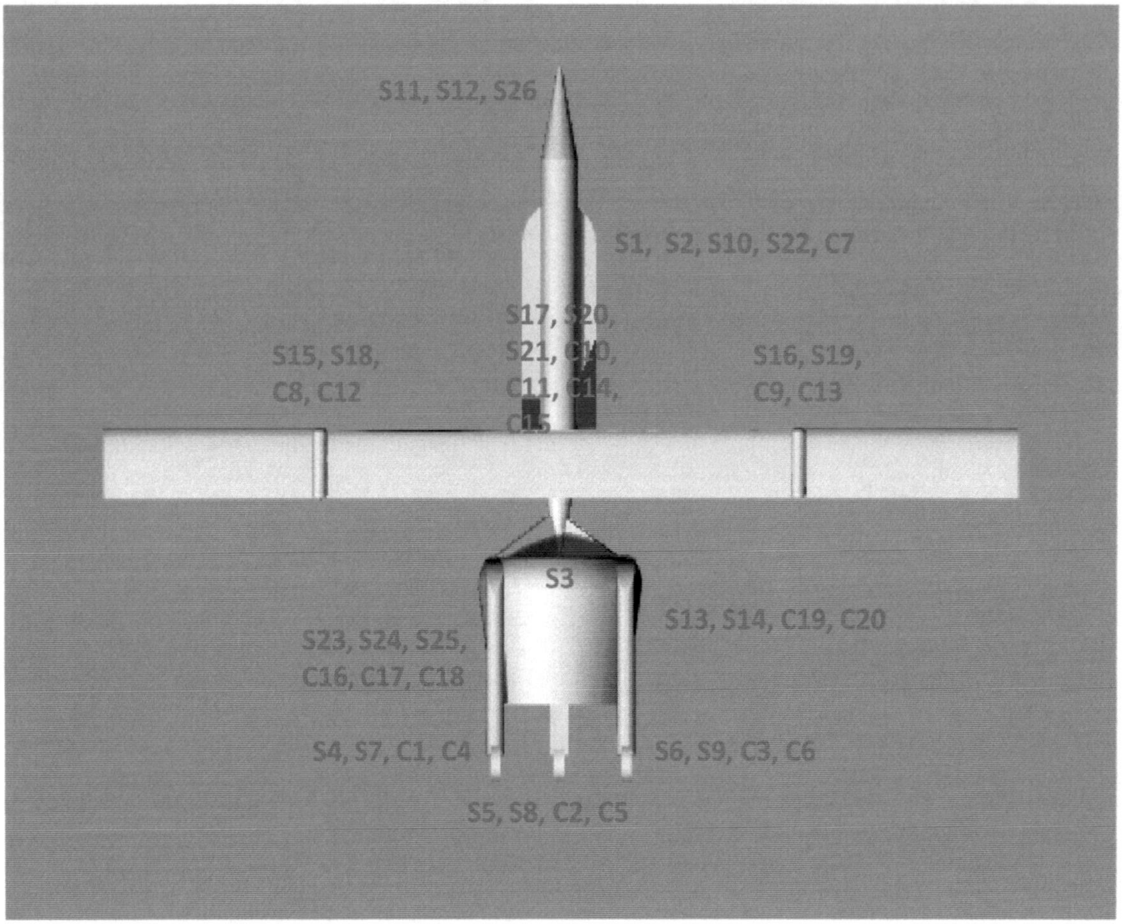

Fig. 19 – Overall Vehicle I/O Sensor/Controls Schematic

Initial Proof-of-Concept and Prototyping Activities

The body of proof-of-concept work is focused on vehicle prototyping. Appropriately so, proof-of-concept work to date has focused on using 3D printing additive manufacturing as a first-choice integral part of the design/development process (Fig. 20).

One of the advantages of using 3D printing for the development of the MICHAEL proof-of-concept test articles is that it potentially allows for a future open source approach for other research teams that might want to build upon the MICHAEL concept and the design work performed to date. MICHAEL was originally conceptualized as a student intern engineering project. Use of 3D designs and 3D print files potentially allows for an easy "handoff" to successive student teams for future work on the overall MICHAEL concept.

Large COTS clockwise- and counterclockwise-rotating propellers (27 inch, or 0.69m, diameter) are currently being used for the proof-of-concept test article development, Ref. 26. Correspondingly, COTS multirotor compatible electric motors are being used as the primary propulsion system for the coaxial propellers.

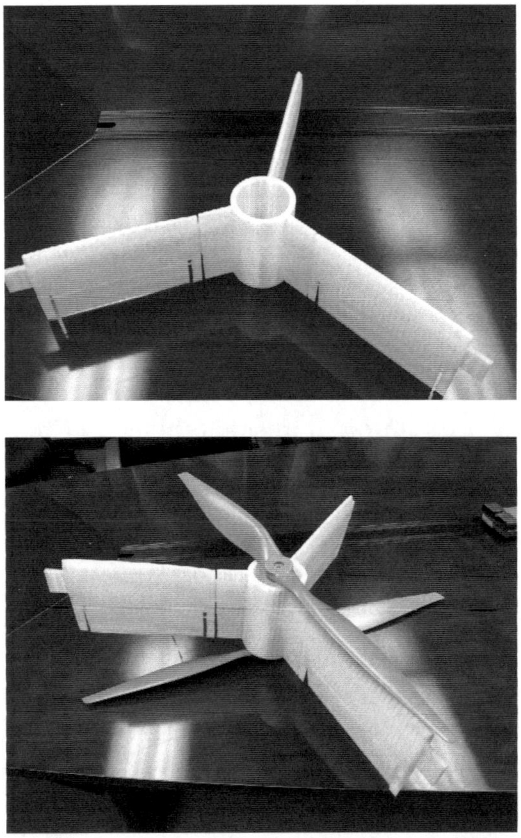

Fig. 20 – Proof-of-Concept Test Articles (in-development)

Two of the major downsides of attempting to use 3D printing for small aerial vehicle prototyping is that, first, it does not necessarily result in the most weight-effcient solution to aircraft structures and, second, the strength of the resulting components is not necessarily on par with other fabrication approaches and materials. In this regards, use of 3D printing was embraced more because of its design/development flexibility and fabrication convenience than weight and strength considerations.

Ground Mobility

It has been asserted in this paper that the design of a MICHAEL vehicle should first start from the premise that the system is a service robot first and an aerial vehicle second.

Wheel count and the number of electric motors required for ground mobility is a very important set of design questions. Commercially demonstrated technologies already exist that allow wheel counts to range from one to greater than four wheels. One and two wheel/motor cofigurations require gyroscopically balanced control systems for the ground mobility subsystem. The minimal wheel count for a passively stable wheeled system is three wheels. Additionally, creative wheel designs stemming from the consumer robotics market are increasing the inherent flexibility for precision positioning of wheeled vehicles/robots. The baseline MICHAEL design would use a tripod support with three wheels, each driven by an electric gearmotor with a right-angle output, each terminating with a pair of Vex robotics-kit wheels (in the particular case of the proof-of-concept work reported in this paper this would be sets of 4-inch Mecanum wheels; Ref. 20). Refer to Fig. 21. A simpler alternate to the baseline design would be to use a non-motor-driven "tail dragger" type wheel assembly for one of the three wheel assemblies; however, for precision positioning capability in confined spaces such as might exist for MICHAEL deliveries in office spaces this alternate was not incorporated into the baseline design. The electric motors for the MICHAEL design are of the same class of motors used for comparable applications such as automotive, large robots, electric bicycles, electric golf carts, and electric wheel chairs: i.e., DC electric motors with reasonably low mass, low rpm output, and high-torque capability.

Fig. 21 – Initial Wheel Assembly Layout

Wing Folding/Stowing

A reliable and robust actuated wing folding/stowage subsystem has long been a not completely resolved technical issue for "roadable" aerial vehicle concepts. Advances in materials and actuators are making this problem more tractable than it has been in the past but it is still nonetheless challenging. What makes this somewhat more viable than the personal air vehicle "flying car" design efforts of the past is that the MICHAEL vehicle is intentionally defined as being a small (<200lbf or <90kg mass) vehicle that travels along the ground at slow speeds (<10 mph or <16 kph). Consequently, it is anticipated that the loads and, therefore, overall vehicle structure and propulsion requirements are more manageable design-wise.

The baseline MICHAEL vehicle conceptual design in Fig. 3 illustrates one specific approach to the wing fold/stow problem. The proposed unfolding/unstowing process is as follows: 1. the folded/stowed wing assembly swings from a vertical orientation (along the fuselage longitudinal axis) to a horizontal orientation; 2. the wing assembly pivot point traverses downward from aproximately midway along the fuselage axis to just above the ducted fan lip (coincident with the assumed c.g. of the vehicle); 3. two outboard wing fold sections unfold at discrete hinge lines to slowly align themselves with the center-span wing section; 4. once aligned, lockout pins are engaged and the unfolded wing assembly forms one continuous large aspect-ratio wing structure; 5. the above unfolding and unstowing process is reversed (i.e. folding and stowing) just prior to the ground-mobility phase of the mission. Note, though, that the above is just one possible approach to the general wing fold/stow problem.

The are two polar extremes of mechanically actuating the wing fold/stow process. At one extreme, each wing degree-of-freedon would require a minimum of one actuator, servo, or electric motorr and one lock-out mechanism (a solenoid driven pin or electric brake, etc.). Add in, as needed, redundancy and the result is a fairly high component count. The degree of freedom count for the baseline MICHAEL design is four degrees of freedom: two wing folds (of the outer tip span sections of the wing) and wing center section pivot and its longitudinal axis translation. The component count for a non-redundant acutation sytem for the baseline MICHAEL design would be: four actuators/servos/motors and four solenoid-driven-lockout-pins/electirc-brakes. The sizing (mass and power) of these actuators and lockout mechanisms can be moderated somewhat by creative mechanical design and tailoring of the fold/stow process so as to minimize actuation forces and moments. On the other extreme, it is possible to rely mostly on passive deplyment for the wing folding and stowing. This could be accomplished by the use of springs, wing section center of gravity tailoring, gravity loads, and use of inerital loads stemming from aceleration/deceleration of the vehicle/wing due to spinning or rocking backwards and forwards the vehicle by its electric-motor-drive wheels. Additionally, the necessity for lockout pins could be somewhat alleviated by tailoring the direction of wing unfolding/folding. The key determining factors for the wing fold/stow mechanical subsystem design approach are reliability, robustness, mechanical system complexity, and mass, power, and cost. A mostly passive design that is unreliable or prone to failure is not viable irrespecitve of its conceptual simplicity; on the other hand, a very reliable and robust system that weughs too much or draws too much power is also unacceptable.

(Service) Robotics

MICHAEL needs to closely interact with people during the final stages of the package delivery, to navigate amongst people within their neighborhoods, to operate in close proximity to and maybe within residences and businesses. It is this social interaction, among other things, that makes the MICHAEL vehicle/system quite different from a conventional autonomous aerial vehicle.

Designing a socially interacting service robot is still an ongoing area of research for robotists (Ref. 11). To illustate the robotics challenge underlying the MICHAEL concept a relatively simple social interaction task is now described: the final delivery of the package itself. Assume for the moment that the package is to be directly delivered to a specific individual within a office work space. Among the many subtasks that might need to be performed to deliver the package are some of the following: 1. the robot must able to remotely ring/acess door bells, transit through open doorways and enter, exit and operate elevators; 2. the robot must be capable of indoor navigation to a predesignated dropoff point; 3. the robot must engage in hazard (furniture and people) avoidance when navigating indoors; 4. the robot would have to enter into communication exchanges or otherwise negotiate an authorization or signature for the package handoff; 5. an automated package release or deployment from the robot fuselage or chassis; 6. automated resealing or closure of the fuselage or chassis upon package delivery; 7. indoor navigation to back-track to and through the building exit. This list of subtasks is quite challenging even by state-of-the-art robotics standards. Further, this is just one task of several required to address the social interactivity required of MICHAEL as a service robot.

Finally, from a vehicle design perspective, there are interesting questions as to hazardous materials and fire hazards posed by a MICHAEL vehicle acting also as a "service robot" in close proximity to people and even operating inside dwellings and office-spaces. High-energy batteries and fuel-storage subsystems (for aerial vehicles having hybrid-electric propulsion) will have to be carefully considered in the context of public safety. Operation of such systems may require refinement of municipal building codes.

Beyond Cargo Delivery

As noted in the introduction of this paper, one of the high-level constraints of any proposed delivery drone transportation system concept is that not only does there need to be a compelling commercial business case made for such a system but it is asserted that it is also important that such systems are seen as providing secondary public services and/or community enhancements in addition to primary mission of commercial small package deliveries. There are many public service missions (Ref. 5) that have analogous mission profiles to the cargo delivery mission discussed in this paper.

Transportation provided by small autonomous aerial vehicles goes beyond just cargo or small packages. There is a whole spectrum of transporation that could nominally be provided by autonomous aerial vehicles (Fig. 22).

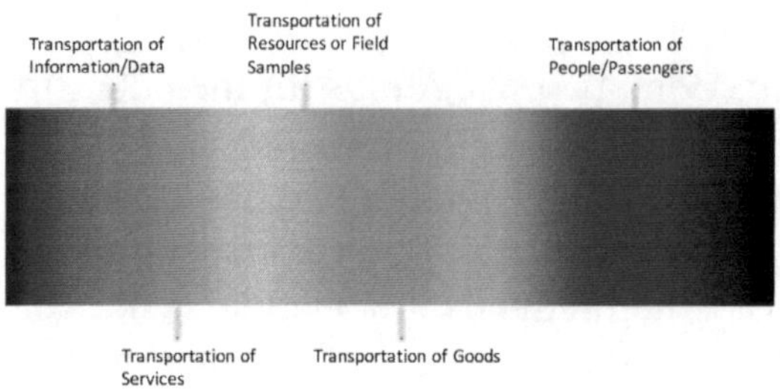

Fig. 22 – Autonomous Aerial Vehicle Transportation "Spectrum" Perspective

Potential Regulatory and Local/State Government Challenges

Regulations related to municipal parks, sidewalks, and bike lanes are all primarily under the purview of local governments. To enable MICHAEL systems using these municipal resources will require a concerted municipality-by-municipality corrdination effort.

There are many examples of open issues that need to be addressed for delivery drones including the requirements for all-weather operations, vehicle inflight power/fuel reserves, beyond-line-of-sight sensors for autonomous aerial vehicles flying at low altitudes (<120m AGL). Commercial operations of small VTOL UAVs are also currently limited by the FAA to under 25 kg; it is likely that a VTOL delivery drone with a total range of approximately 40 km will need to be larger than that weight limit. All of these considerations will likely need future FAA rulemaking.

Ancillary Benefits (and Challenges) of Implementing MICHAEL

Because of the ground mobility attribute of a MICHAEL vehicle, the result may be a cross-cutting expansion of wheel-chair and disability access to residences as well as expansion of bike-paths and lanes along or on roadways.

Because of the necessity for neighborhood VTOL landing sites for MICHAEL vehicles there could also be an expansion and/or improvement of public parks and facilities to accommodate not only their traditional usage but their potential dual-use for MICHAEL takeoff and landing.

Correspondingly, though, there will still be significant challenges to the introduction/adoption of MICHAEL cargo delivery.

Future implementation of a MICHAEL-based small package cargo distribution system would no doubt lead to a reexamination of community urban-planning concepts.

Future Work

The concept will be examined both computationally and also through system proof-of-concept prototyping. The mission CONOPS will also continue to be refined and assessed in detail.

Concluding Remarks

Delivery drones, roadable aircraft, and service robot assistants have each in their own way been considered to be "visionary" technologies that futurists have been anticipating for years if not decades. It is perhaps most appropriate that the promise of each of the above technologies might one day find their realization through a natural synergism resulting from their combined application. The MICHAEL concept – embodying all three general sets of technologies -- has considerable merit in addressing a number of socioeconomic pressures facing us in the future.

This work continues ongoing research investigations into "rotorcraft as robots." There is much promise but also much concern as to this ongoing fusion of intelligent systems, robotics, and aerial vehicle design: we have to be wise as much as we are smart. If we succeed, the outcome will be new knowledge, new services, and new capabilities that we could only previously imagine.

Acknowledgments

The author would like to acknowledge the work of Mary Anderson, New Mexico Institute of Mining and Technology, Cara Moulton, Sitting Bull College, and Rachal Skyving, University of Idaho, for their contributions to this work.

References

1. Young, L.A., "Aerobots as a Ubiquitous Part of Society," AHS Vertical Lift Aircraft Design Conference, San Francisco, California, January 18-20, 2006.

2. Aiken, E.W., Ormiston, R.A., and Young, L.A., "Future Directions in Rotorcraft Technology at Ames Research Center," 56th Annual Forum of the American Helicopter Society, International, Virginia Beach, VA, May 2-4, 2000.

3. Young, L.A., Aiken, E.W., Johnson, J.L., Demblewski, R., Andrews, J., and Klem, J., "New Concepts and Perspectives on Micro-Rotorcraft and Small Autonomous Rotary-Wing Vehicles," AIAA 20th Applied Aerodynamics Conference, St Louis, MO, June 24-27, 2002.

4. Young, L.A., "Conceptual Design Aspects of Three General Sub-Classes of Multi-Rotor Configurations: Distributed, Modular, and Heterogeneous," Sixth AHS International Specialists Meeting on Unmanned Rotorcraft Systems, Scottsdale, AZ, January 20-22, 2015.

5. Young, L.A., "Future Roles for Autonomous Vertical Lift in Disaster Relief and Emergency Response," Heli-Japan 2006: AHS International Meeting on Advanced Rotorcraft Technology and Life Saving Activities, Nagoya, Japan, November 15-17, 2006.

6. Sinsay, J.D., Alonso, J.J., Kontinos, D.A., Melton, J.E., and Grabbe, S., "Air Vehicle Design and Technology Considerations for an Electric VTOL Metro-Regional Public Transportation System," 12th AIAA Aviation Technology, Integration, and Operations (ATIO) Conference and 14th AIAA/ISSMO Multidisciplinary Analysis and Optimization Conference, Indianapolis, IN, September 2012.

7. Melton, J., Kontinos, D., Grabbe, S., Alonso, J., Sinsay, J., and Tracey, B., "Combined Electric Aircraft and Airspace Management Design for Metro-Regional Public Transportation," NASA TM-2014-216626, October 2014.

8. Alonso, J.J., Arneson, H.M., Melton, J.E., Vegh, J.M., Walker, C., and Young, L.A., "System-of-Systems Considerations in the Notional Development of a Metropolitan Aerial Transportation System: Implications as to the Identification of Enabling Technologies and Reference Designs for Extreme Short Haul VTOL Vehicles with Electric Propulsion," NASA TM (To be Published); draft report deliverable: https://www.nari.arc.nasa.gov/sites/default/files/Young_HopperII_NASA_TM15K.pdf

9. Russell, C.R.; Young, L.A.; Yamauchi, G.K.; Johnson, W.; Boyd, D.D.; Gorton, S.A.; Snyder, C.A.; and Kohlman, L.W., "Greener Helicopters," Encyclopedia of Aerospace Engineering, Eds. R. Blockley and W. Shyy, John Wiley, Chichester, DOI: 10.1002/9780470686652.eae1009, 12/29/2015.

10. Anders, S.G., Asbury. S.C, Brentner, K.S., Bushnell, D.M., Glass, C.E., Hodges, W.T., Morris, Jr., S.J., and Scott, M.A., "The Personal Aircraft – Status and Issues," NASA TM 109174, December 1994.

11. Fong, T., Nourbakhsh, I., and Dautenhahn, K., "A Survey of Socially Interactive Robots," Robotics and Autonomous Systems, 42 (2003), Elsevier Science B.V., pp. 143-166.

12. Anom., V/STOL Aircraft and Propulsion Concepts; http://vertipedia.vtol.org/vstol/wheel.htm.

13. Google-X Project Wing – First Generation Aircraft; https://x.company/projects/wing/.

14. Google-X Project Wing – Second Generation Aircraft; https://x.company/projects/wing/.

15. Amazon Prime Air Drone – First Generation Aircraft; https://m.youtube.com/watch?v=cRTNvWcx9Oo.

16. Amazon Prime Air Drone – Second Generation Aircraft; https://m.youtube.com/watch?v=MXo_d6tNWuY.

17. Young, L.A., "Small Autonomous Air/Sea System Concepts for Coast Guard Missions," USCG Maritime Domain Awareness Requirements, Capabilities, and Technology (MDA RCT) Forum, Santa Clara, CA, May 2, 2005.

18. Young, L.A., Aiken, E.W., and Briggs, G.A., "Smart Rotorcraft Field Assistants for Terrestrial and Planetary Science," 2004 IEEE Aerospace Conference, Big Sky, MT, March 2004.

19. Example of "Range Extender" Technology: http://www.launchpnt.com/news/news/bid/105014/LaunchPoint-Develops-High-Specific-Power-Genset-for-UAVs (website last accessed January 9, 2017).

20. Vex Robotics educational/competition website: https://www.vexrobotics.com/wiki/Games_%26_Competitions

21. Pfaender, H., DeLaurentis, D., and Mavris, D.N., "An Object Oriented Approach for Conceptual Design Exploration of UAV-Based System-of-Systems," 2nd AIAA "Unmanned Unlimited" Systems, Technologies, and Operations, AIAA-2003-6521, September 15-18, 2003, San Diego, CA.

22. Preisner, L.J., DeLaurentis, D., Mavris, D., and Schrage, D., "Application of a Technology Identification, Selection, and Evaluation Method for the Conceptual Design of a VTOL UAV," AIAA-2002-3465, AIAA 1st Technical Conference and Workshop on Unmanned Aerospace Vehicles, May 20-23, 2002, Portsmouth, VA.

23. Bellotto, N. and Hu, H., "Multisensor-Based Human Detection and Tracking for Mobile Service Robots," IEEE TRANSACTIONS ON SYSTEMS, MAN, AND CYBERNETICS—PART B: CYBERNETICS, VOL. 39, NO. 1, FEBRUARY 2009.

24. Baeg, S-H, et al, "Building a Smart Home Environment for Service Robots based on RFID and Sensor Networks," International Conference on Control, Automation and Systems (ICCAS '07), 2007.

25. Ground robot package delivery: https://www.starship.xyz/; (website last accessed January 2, 2017).

26. COTS 27x13 Propeller: http://apcserve.w20.wh-2.com/v/PERFILES_WEB/PER3_27x13E.dat; (website last accessed January 3, 2017).

27. Multirotor Electric Motor: http://www.rctigermotor.com/html/2016/Power-Type_0414/309.html (website last accessed 1-11-2017)

28. Multirotor Electronic Speed Controller: http://www.rctigermotor.com/html/2013/esc_0916/86.html (website last accessed 1-11-2017)

AUTHOR INDEX